IFAC

International Federation of Automatic Control

MODELLING AND CONTROL OF BIOTECHNOLOGICAL PROCESSES

IFAC Proceedings Series, 1986. Number 10

IFAC PROCEEDINGS SERIES

Editor-in-Chief
JANOS GERTLER, Department of Computer and Electrical Engineering,
George Mason University, Fairfax, Virginia, USA

GERTLER & KEVICZKY (*General Editors*): A Bridge Between Control Science & Technology (*Ninth Triennial World Congress, in 6 volumes*)

- Analysis and Synthesis of Control Systems (*1985, No. 1*)
- Identification, Adaptive and Stochastic Control (*1985, No. 2*)
- Large-scale Systems, Decision-making, Mathematics of Control (*1985, No. 3*)
- Process Industries, Power Systems (*1985, No. 4*)
- Manufacturing, Man–Machine Systems, Computers, Components, Traffic Control, Space Applications (*1985, No. 5*)
- Biomedical Applications, Water Resources, Environment, Energy Systems, Development, Social Effects, SWIIS, Education (*1985, No. 6*)

BARKER & YOUNG: Identification and System Parameter Estimation (1985) (*1985, No. 7*)

NORRIE & TURNER: Automation for Mineral Resource Development (*1986, No. 1*)

CHRETIEN: Automatic Control in Space (*1986, No. 2*)

DA CUNHA: Planning and Operation of Electric Energy Systems (*1986, No. 3*)

VALADARES TAVARES & EVARISTO DA SILVA: Systems Analysis Applied to Water and Related Land Resources (*1986, No. 4*)

LARSEN & HANSEN: Computer Aided Design in Control and Engineering Systems (*1986, No. 5*)

PAUL: Digital Computer Applications to Process Control (*1986, No. 6*)

YANG JIACHI: Control Science & Technology for Development (*1986, No. 7*)

MANCINI, JOHANNSEN & MARTENSSON: Analysis, Design and Evaluation of Man–Machine Systems (*1986, No. 8*)

GELLIE, FERRATE & BASANEZ: Robot Control "Syroco '85" (*1986, No. 9*)

JOHNSON: Modelling and Control of Biotechnological Processes (*1986, No. 10*)

IFAC Related Titles

BROADBENT & MASUBUCHI: Multilingual Glossary of Automatic Control Technology
EYKHOFF: Trends and Progress in System Identification
ISERMANN: System Identification Tutorials (*Automatica Special Issue*)

MODELLING AND CONTROL OF BIOTECHNOLOGICAL PROCESSES

Proceedings of the 1st IFAC Symposium
Noordwijkerhout, The Netherlands
11 – 13 December 1985

Edited by

A. JOHNSON

Technische Hogeschool Delft, The Netherlands

Published for the

INTERNATIONAL FEDERATION OF AUTOMATIC CONTROL

by

PERGAMON PRESS

OXFORD · NEW YORK · BEIJING · FRANKFURT
SÃO PAULO · SYDNEY · TOKYO · TORONTO

U.K.	Pergamon Press, Headington Hill Hall, Oxford OX3 0BW, England
U.S.A.	Pergamon Press, Maxwell House, Fairview Park, Elmsford, New York 10523, U.S.A.
PEOPLE'S REPUBLIC OF CHINA	Pergamon Press, Qianmen Hotel, Beijing, People's Republic of China
FEDERAL REPUBLIC OF GERMANY	Pergamon Press, Hammerweg 6, D-6242 Kronberg, Federal Republic of Germany
BRAZIL	Pergamon Editora, Rua Eça de Queiros, 346, CEP 04011, São Paulo, Brazil
AUSTRALIA	Pergamon Press Australia, P.O. Box 544, Potts Point, N.S.W. 2011, Australia
JAPAN	Pergamon Press, 8th Floor, Matsuoka Central Building, 1-7-1 Nishishinjuku, Shinjuku-ku, Tokyo 160, Japan
CANADA	Pergamon Press Canada, Suite 104, 150 Consumers Road, Willowdale, Ontario M2J 1P9, Canada

---

First edition 1986

Library of Congress Cataloging in Publication Data

Modelling and control of biotechnological processes.
(IFAC proceedings series)
Papers presented at the IFAC Symposium on
Modelling and Control of Biotechnological Processes,
sponsors, International Federation of Automatic
Control (IFAC) ... [et al.]
1. Biotechnology—Mathematical models—Congresses.
2. Fermentation—Mathematical models—Congresses.
3. Process control—Congresses. I. Johnson, A.
II. IFAC Symposium on Modelling and Control of
Biotechnological Processes. III. International
Federation of Automatic Control. IV. Series.
TP248.2.M64 1985 550'.6 85–19185

British Library Cataloguing in Publication Data

IFAC Symposium (1st: 1985: *Noordwijkerhout)*
Modelling and control of biotechnological processes
1. Biotechnology
I. Johnson, A. II. International Federation of
Automatic Control
660'.6 TP248.2
ISBN 0-08-032565-3

These proceedings were reproduced by means of the photo-offset process using the manuscripts supplied by the authors of the different papers. The manuscripts have been typed using different typewriters and typefaces. The lay-out, figures and tables of some papers did not agree completely with the standard requirements: consequently the reproduction does not display complete uniformity. To ensure rapid publication this discrepancy could not be changed: nor could the English be checked completely. Therefore, the readers are asked to excuse any deficiencies of this publication which may be due to the above mentioned reasons.

The Editor

Printed in Great Britain by A. Wheaton & Co. Ltd., Exeter

1ST IFAC SYMPOSIUM ON MODELLING AND CONTROL OF BIOTECHNOLOGICAL PROCESSES

Organized by
Division of Automatic Control of the Royal Institution of Engineers in the Netherlands (KIvI) in co-operation with the Netherlands Biotechnological Society (NBV) and the Departments of Chemical Technology of KIvI and the Royal Netherlands Chemical Society (KNCV)

Further support is acknowledged from
AKZO, Arnhem
Delft University of Technology, Delft
Gist-Brocades, Delft
Heineken, Zoeterwoude
Honeywell, Amsterdam
Nederlandse Unilever Bedrijven, Rotterdam
Shell Nederland, Rotterdam

Sponsored by
International Federation of Automatic Control (IFAC)
IFAC Committee on Applications (APCOM)
IFAC Committee on Theory (THEORY)

Co-sponsors
European Federation of Biotechnology (EFB)
Royal Netherlands Academy of Sciences (KNAW)

International Programme Committee
A. Johnson, Netherlands (Chairman)
V. V. Biryukov, USSR
A. Cheruy, France
I. J. Dunn, Switzerland
A. Halme, Finland
M. Kümmel, Denmark
L. Ljung, Sweden
G. Olsson, Sweden
J. A. Roels, Netherlands
K. Schügerl, Germany
T. Takamatsu, Japan
G. C. Vansteenkiste, Belgium
D. I. C. Wang, USA
D. W. Zabriskie, USA

National Organizing Committee
A. Johnson (Chairman and Symposium Editor)
R. M. Dekkers (Secretary)
P. M. van den Broecke
F. J. Kylstra
E. A. G. van der Meijden
K. Vellenga

CONTENTS

MODELLING OF MICROKINETICS

OPTIMIZATION THEORY

THE CONTROL OF FERMENTATION PROCESSES

A. Johnson

Department of Applied Physics, Delft University of Technology, Prins Bernhardlaan 6, 2628 BW Delft, The Netherlands

Abstract. A survey of the control of fermentation processes operated in fed-batch mode is presented. Truely optimal conditions can only be found by minimizing (or maximizing) a cost criterion subject to the constraints imposed by a mathematical model of the process. This approach leads to the design of a tracking controller, or servomechnism as it is also called. A different approach uses a physiological model of the process to determine a control strategy, which usually results in keeping a measured or estimated variable at its setpoint by means of a regulator.

Keywords. Biocontrol; fermentation; fed-batch processes; modelling; estimation.

1. INTRODUCTION

Most biotechnological processes consist of the fermentation, oxidation and/or reduction of feedstuff by microorganisms such as yeasts or bacteria. These processes may be conveniently classified according to the mode chosen for process operation: either batch, fed-batch or continuous. During batch operation of a process no feedstuff is added to the initial change nor is product removed until the end of the process. Some pharmaceutical preparations are made in this way but generally batch operation is not commercially attractive. More economic is continuous operation where feedstuff is continually added and product continually removed. Examples are the continuous fermentation of milk in the production of margarine and the biological purification of waste-water.

From the control engineer's viewpoint it is the fed-batch processes, however, which present the greatest challenge. In fed-batches, the feed rate may be changed during the process but no product is removed until the end. Baker's yeast and antibiotics such as penicillin are made in fed-batches commercially, and there is an enormous economic incentive to optimize such processes. Such activity has been confined in the past to improving equipment design and, of course, to microbiological investigations. Thus the efficiency of penicillin production at Gist Brocades, The Netherlands, has been increased three to four times during the last 15 years (v.d. Meer and Valkema, 1985). Recently, attention has turned towards the optimization of the feed rate and better control during the fed-batch.

It must be said from the outset that the quest to maximize productivity is not the only reason for the control engineer's interest in the fed-batch process. Not only are these processes commercially important but they are technically challenging: the process variables are difficult to measure, the "quality" of the product plays an important role, the process model usually contains strongly time-varying parameters etc. etc. But above all the technical challenge arises from the nature of the process optimization since it is a *dynamical problem*.

Two approaches have emerged to solve this dynamic optimization problem and the principal contribution of this survey is to compare and contrast both approaches. The approach termed here physiological control attempts to emulate the routine process control of, say, the chemical industry where process variables are held at set-points by means of controllers called regulators. Of paramount interest here is not the design of the regulator (about which so much has been written in the control engineering literature) but rather how the set-points are chosen.

The other approach is to minimize (or maximize) a cost criterion while not violating the dynamic behaviour of a mathematical model describing the process. This text book approach, which is henceforth referred to simply as dynamic optimization, results in a number of variables being required to follow non-steady optimal trajectories. This tracking of the variables is accomplished using a servomechanism, or tracking controller as it is sometimes called. Here, it is not the design of the servomechanism which is of particular interest since again a substantial body of literature exists on the subject, rather the choice of mathematical model, cost criterion and optimization technique.

In constrast to previous surveys (e.g. Halme and Holmberg, 1978; Zabriskie, 1979) only fed-batch biotechnological process control has been reviewed. By this means a complete in-depth study has been possible. Moreover, if control engineers are confronted by problems concerning continuous processes the technique of physiological control described in this paper may offer an attractive solution, without the need for mathematical modelling.

2. MODELLING FOR CONTROL

We may distinguish three different types of model used for control. The first is the physiological model where knowledge of the physiology of the biological growth process is expressed in consequent, but usually non-mathematical, statements. The second is the structured model where a system of partial differential equations (sometimes simplified to ordinary differential equation) is used to characterize the age distribution, for example, of the growing cells in the culture. Thirdly comes the unstructured model where the fermentation is assumed to be dominated by a single, homogeneously growing organism. Of the three types of model, the first and the last are most frequently encountered in controller synthesis. Since it is usual to control the environmental variables (temperature, pH, aeration and agitation) to be constant the models given will be understood to be valid only in these circumstances.

2.1. Control without a mathematical model

The earliest attempts at control of the fed-batch/batch fermentation process used no model at all. Successful state trajectories from previous batches which had been stored in the process computer were tracked using open-loop control (Yamashita, Hoshi and Inagaki, 1969). Many industrial fermentations are still operated using this method.

Constructing an accurate mathematical model of the fermentation process is a formidable task; moreover suitable values must be found for the model's parameters. It is not surprising, therefore, that considerable attention has been paid to developing controllers without the need for a mathematical model. Physiological models based upon an understanding of the Pasteur, Crabtree etc. (see e.g. Roels, 1983) effects are important here. Consider the aerobic cultivation of baker's yeast using molasses, which is a commercially important fed-batch fermentation. From physiological models, (at least) four control strategies can be envisaged viz:

A) Cell growth rate is maximal without byproduct (ethanol) formation if substrate concentration is held at its critical value.
B) Cell growth is optimal when the specific growth rate follows a predetermined trajectory.
C) Cell growth is optimal when the ethanol concentration is constant.
D) Maximum cell growth occurs when there is pure respiratory growth without ethanol production and simultaneously maximum molasses feed.

Obviously, refinements and hybrids of these strategies are possible. All need direct measurements or estimates to produce the required variables for feedback purposes. Estimators and the implementation of control strategies derived from physiological models are discussed in Sections 3 and 4, respectively.

Another promising approach, subject to much current research activity, is that of fuzzy control theory (Tong, Beck and Latten, 1980; Nakamura, Kuratani and Morita, 1985) where no formal model is required but where the process operator's experience can be exploited in the controller design.

2.2. Control with a structured model

The most mathematically complex fermentation models are those describing the effects of the internal state of the organisms on process operation. The model variables are classified or structured according to the age, volume, mass, species of the organism or the different organic carbon compounds in the growth medium (Alberghina, Martegani and Mariani, 1982; Holmberg and Ranta, 1982). The most important special case of this type of model is the cell age distribution model because the growth, decay, production of secondary products and consumption of oxygen of the organisms is known to be age dependent.

By introducing various assumptions regarding the cell metabolism, it is possible to reduce the distributed parameter cell age model to a lumped parameter (or finite dimensional) model of tractable size for controller design (Dairaku, Izumoko Morikawa, Shioya and Takamatsu, 1982b; Ranta 1982; Nestaas and Wang, 1983). However, apart from this approach there does not yet appear to be any application of distributed parameter models for control purposes.

2.3. Control with a homogeneous, single organism model

The last approach to controller design is to use an unstructured fermentation model where the culture is assumed to consist of a single, homogeneously growing organism. Moreover, it is often assumed that there is a single growth limiting substrate. An expression for the growth rate of the micro-organism and the consumption rate of the substrate is given by the equations

$$r_x(t) = \mu(t)x(t) - k_d(t)x(t) \tag{2.1}$$

$$r_s(t) = -\frac{\mu(t)}{Y(t)}x(t) \tag{2.2}$$

where x(t) is the biomass concentration (kg/m^3), $\mu(t)$ is the specific growth rate (hr^{-1}), $k_d(t)$ is the biomass decay rate (hr^{-1}) and Y(t) is the yield factor. For a perfectly mixed fermentor to which substrate of concentration s_i (kg/m^3) is fed, diluting the culture at a rate D(hr^{-1}), then a mass balance using the above equations gives

$$\dot{x}(t) = \mu(t)x(t) - D(t)x(t) - k_d(t)x(t) \tag{2.3}$$

$$\dot{s}(t) = -\frac{\mu(t)}{Y(t)}x(t) + D(t)[s_i(t) - s(t)] \tag{2.4}$$

$$x(t_0) = x_0 \tag{2.5}$$

$$s(t_0) = s_0 \tag{2.6}$$

Here, s(t) is the substrate concentration (kg/m^3) and the dilution rate D(t) = F(t)/V(t), where F(t) is the volumetric feed flowrate (m^3/hr) and V(t) the volume of the culture (m^3). In some circumstances such as antibiotic production it is necessary to augment the above equations with an equation describing the product (e.g. antibiotic) concentration. An equation for the concentration of dissolved oxygen may also be required.

There are many analytical expressions for $\mu(t)$. Spriet (1982) lists nine possibilities, although more exist. The most popular is the Monod equation (Monod, 1942)

$$\mu(t) = \frac{\mu_m \cdot s(t)}{k_s + s(t)} \tag{2.7}$$

with μ_m the maximum specific growth rate and k_s the Monod constant for growth on the substrate. An alternative approach is to assume that no analytical expression is available for $\mu(t)$, which leads to the design of an adaptive controller (Dochain and Bastin, 1984).

Much the same can be said for the yield factor, Y(t). In controller design it is often assumed to be constant

$$Y(t) = Y \tag{2.8}$$

However, this neglects the influence of other factors, e.g. the synthesis of (by-)products. Dekkers (1983) has shown that when changes are occurring slowly the yield factor for baker's yeast should be expressed as a relatively simple function of the specific growth rate and the critical specific growth rate, $\mu_c(t)$.

The model is not complete, of course, without the output equations relating the measured variables to the two (or more) state variables. Sensors meauring the biomass and substrate concentrations would be most advantageous but as yet there are no such commercially available sensors suitable for industrial applications. Discussion of the output equations is therefore deferred to Section 3 on estimation.

We consider next some properties of the unstructured fermentation model presented. For control purposes it is important to distinguish between two situations. Firstly (because it is easier), we investigate what happens when small deviations occur in the desired, or reference, trajectories of x(t) and (t). For instance, how long does it take before such perturbations are damped out? This sort of question is vitally important to setpoint and tracking control. Secondly, the dynamical behaviour of the model must be understood "in the large", when deviations can no longer be considered to be small. This understanding is essential for dynamic optimisation. Note that wherever numerical values of parameters have been used they relate to the aerobic cultivation of baker's yeast using molasses.

The perturbation analysis may proceed by the linearization of eqns. (2.3) and (2.4). However, we choose to transform these equations first into equations in terms of M(t), the weight of biomass (kg), and W(t) the weight of substrate (kg):

$$\overset{o}{M}(t) = \mu(t)M(t) \tag{2.9}$$

$$\overset{o}{W}(t) = \frac{-\mu(t)}{Y(t)} M(t) + s_i(t)F(t) \tag{2.10}$$

$$\text{and} \quad \overset{o}{V}(t) = F(t) \tag{2.11}$$

$$\text{with} \quad M(t_o) = M_o \tag{2.12}$$

$$W(t_o) = W_o \tag{2.13}$$

$$V(t_o) = V_o \tag{2.14}$$

It has been assumed that $k_d(t)$ is so small compared with $\mu(t)$ that it may be neglected. Note that if $\mu(t)$ and Y(t) were known a priori the above two equations would be linear. Moreover, if $\mu(t)$ was known to be a constant, any perturbation in the biomass from a desired trajectory would never decay away, nor could it be controlled away (i.e. the system is unstabilizable). More realistically, let both $\mu(t)$ and Y(t) be known linear functions of the substrate concentration:

$$\mu(t) = \alpha s(t) + \beta \tag{2.15}$$

$$Y(t) = \gamma s(t) + \delta \tag{2.16}$$

Linearization of eqns. (2.9) - (2.16) about some not necessarily steady state gives

$$\begin{bmatrix} \Delta\overset{o}{M}(t) \\ \Delta\overset{o}{W}(t) \\ \Delta\overset{o}{V}(t) \end{bmatrix} = \begin{bmatrix} a_{11} & a_{12} & a_{13} \\ a_{21} & a_{22} & a_{23} \\ 0 & 0 & 0 \end{bmatrix} \begin{bmatrix} \Delta M(t) \\ \Delta W(t) \\ \Delta V(t) \end{bmatrix} + \begin{bmatrix} 0 \\ s_i \\ 1 \end{bmatrix} \Delta F(t) \tag{2.17}$$

where $\Delta M \overset{\Delta}{=} M(t) - M_r(t)$ etc. Consideration of the numerical values of the parameters of the above equation shows that the eigenvalue corresponding to the volume is zero while the eigenvalues of the biomass and substrate are widely separated. The biomass responds very slowly when perturbated from its reference value (and never returns to it) while the substrate feed is more or less immediately consumed. There are two important conclusions to be drawn from this analysis. Firstly that regulatory control is necessary for maintaining the biomass trajectory at its desired value in the face of disturbances, and secondly (Dekkers, 1983) that because of the separation of the eigenvalues the two major problems of control, that of optimization and that of regulation, may be separated.

It is worthwhile noting here that it would not be prudent to use the continuous-time, linearized model, equation (2.17) to design a controller. Rather, we would prefer to use a discrete-time model so that we could easily implement the controller in a digital computer and also so that stochastic (noise) processes corrupting the system could easily be modelled. The type of discrete-time model required is one whose behaviour at the sampling instants exactly matches that of eqn. (2.17); such a model has been termed the *equivalent discrete system*. Once the sampling period, T, is known the equivalent discrete system is easily derived (Johnson, 1985):

$$\begin{bmatrix} \Delta M(k+1) \\ \Delta W(k+1) \\ \Delta V(k+1) \end{bmatrix} = \begin{bmatrix} 1-a_{11}T & a_{12}T & a_{13}T \\ a_{21}T & 1+a_{22}T & a_{23}T \\ 0 & 0 & 1 \end{bmatrix} \begin{bmatrix} \Delta M(k) \\ \Delta W(k) \\ \Delta V(k) \end{bmatrix} + \begin{bmatrix} 0 \\ s_i T \\ T \end{bmatrix} \Delta F(k) + n(K) \tag{2.18}$$

Here, k = 0,1,2,..., $\Delta M(k)$ is the value of $\Delta M(t)$ after kT and n(k) is a vector of zero-mean discrete white noise.

For large changes the above perturbation analysis no longer holds. Some insights into the global dynamic behaviour of the fermentation model now follow. Dochain and Bastin (1984) have proved under certain dynamic conditions that the variables x(t) and s(t) of the model (2.3) - (2.6) are bounded for any dilution rate. To be precise, if the following assumptions are made:

(A1) $Y(t) = Y$, constant

(A2) $k_d(t) \equiv 0$

(A3) $\mu(t) = 0$ if $s(t) = 0$

(A4) $\mu(t) > 0$

(A5) $0 \leq s_i(t) \leq s_i^{max}$

(A6) $\frac{x_o}{Y} + s_o \leq s_i^{max}$

then for any nonnegative, but not necessarily bounded, dilution rate,

$$0 \leq x(t) \leq Y s_i^{max} \tag{2.19}$$

$$0 \leq s(t) \leq s_i^{max} \tag{2.20}$$

The upper bounds on the state variables are, of course, very conservative and the assumption of a constant yield factor is not very realistic, but this result is nevertheless important for the convergence analysis of adaptive control and estimation schemes.

If dynamic optimization (section 5) is implemented by a computer then the controlled variable F(t) will be piecewise-constant. It is possible to show that the dynamic reponse of the model (2.9) - (2.16) between sampling instants is in that case governed by a Bernoulli-type equation. For fast sampling rates the linear model, equation (2.17), would be an adequate approximation to reality and its response to a piecewise-constant control signal is readily calculated.

3. ESTIMATION

Many different types of sensors and monitoring systems for fermentors have been described or postulated in the literature (see e.g. Proceedings 1st IFAC Workshop on Modelling and Control of Biotechnical Processes, Helsinki, 1982; Proceedings Symposium on Analytical Methods and Problems in Biotechnology, Noordwijkerhout, 1984). Some of the more common types of sensor are listed in Table 1.

TABLE 1

VARIABLE	SENSOR	REFERENCE
biomass growth	filtration probe	Nestaas et al (1983)
	optical density fluorescence	Shioya et al (1982)
dissolved oxygen	polarographic PTFE tubing galvanic cell	Flynn (1982)
gaseous oxygen	paramagnetic zirconia mass spectro-meter	
gaseous carbon dioxide	infra-red mass spectro-meter	
substrate concentration	affinity microbial enzyme electrode	Shioya et al (1982)
	enzyme thermis-tor	Axelson et al (1982)
ethanol concentration	teflon tubing	Shioya et al (1982)
air flow rate	flow meters	
pH	glass electrode	
temperature	resistance thermometer	
agitated speed	counter	
volume	differential pressure cell	
foam	contact electrode	

If we exclude variables that are usually encountered in process control (e.g. temperature, pressure etc.) then many of the remaining sensors which register directly biomass, byproduct or substrate concentrations are not robust enough or are too expensive for routine industrial control applications. Sometimes measurements are available but they must be analysed in a laboratory before a value of the variable of interest can be obtained. Therefore it is common practice to use measurements of related variables, such as gaseous oxygen and gaseous carbon dioxide, to estimate the unobtainable variables.

The simplest, but least accurate, way to obtain the estimate is to neglect all measurement errors, instrument noise etc. and to correlate experimentally obtained measurements of the desired and related variables. A more sophisticated approach is to model the neglected errors and noise and to employ an estimator. If linearity and additivity assumptions on the model are justified, the optimal estimator is the Kalman Filter; if not, recourse is normally made to a suboptimal estimator such as the Extended Kalman Filter. For a description of these estimators and their properties the reader is referred to one of the many texts on the subject (e.g. Anderson and Moore, 1979; Johnson, 1985) while applications of both optimal and suboptimal estimators to fermentation processes have been well documented (Svrcek, Elliott and Zajic, 1974; Dekkers, 1982; Nihtila, Harmo and Perttula, 1984). In the remainder of this section attention is focussed on just one aspect of estimation for control, namely the theoretical and experimentally derived relationships between measured and unobtainable variables that we wish to estimate. The relationships serve as the output equations for the models discussed in the previous section.

We begin by defining three intermediate variables OUR, CER and RQ. It is assumed that we measure the partial fraction, λ_o(mol/mol), of gaseous oxygen and the partial fraction, λ_c(mol/mol), of gaseous carbon dioxide in the dried effluent gas, together with the total dry air inlet flowrate, ϕ(mol/hr). From these three measurements the *oxygen uptake rate*, OUR(mol/hr), and the *carbon dioxide evolution rate*, CER(mol/hr) can be calculated (Cooney, Wang and Wang, 1977):

$$OUR(t) = \phi(t)\left[\frac{0.20945\lambda_c(t)+0.99967\lambda_o(t)-0.20945}{\lambda_c(t)+\lambda_o(t)-1.0}\right] \quad (3.1)$$

$$CER(t) = \phi(t)\left[\frac{0.79055\lambda_c(t)+0.00033\lambda_o(t)-0.00033}{1.0-\lambda_c(t)-\lambda_o(t)}\right] \quad (3.2)$$

Dekkers (1982) has calculated additional terms for the above expressions to account for the influence of various dynamical accumulation effects. The order of magnitude of the correction terms depends on the process circumstances. For control purposes it may be desirable to use as the controlled variable the *respiratory quotient*, RQ(t), which is the ratio of the amount of carbon dioxide produced to the amount of oxygen consumed, i.e.

$$RQ(t) = \frac{CER(t)}{OUR(t)} \quad (3.3)$$

$$= \frac{0.79055\lambda_c(t)+0.00033\lambda_o(t)-0.00033}{0.20945-0.20945\lambda_c(t)-0.99967\lambda_o(t)} \quad (3.4)$$

3.1. Estimating the biomass

It has been found experimentally that the behaviour of OUR(t), CER(t) and RQ(t) is very dependent upon the type of fermentation, and upon the operating conditions. For fed-batch baker's yeast fermentation where the critical specific growth rate is exceeded, OUR(t) varies only slowly and is correlated to the biomass growth; CER(t) or RQ(t), by contrast, change more quickly and are correlated to the uptake of the substrate feed (Dekkers, 1985). This is not true with many other organisms, where the RQ(t) is more or less constant during the fermentation; in such cases the biomass growth may be correlated with either the OUR(t) or CER(t).

Using the stochiometric equation for the growth of biomass, Dekkers (1982) has shown that for baker's yeast fermentation

$$BPR(t) = 2.17\ OUR(t) - 3.26\ CER(t) + 6.52\ GFR(t) \qquad (3.5)$$

is a good approximation to reality, with BPR(t) the biomass production rate (mol/hr) and GFR(t) the glucose feedrate (mol/hr). Equation (3.5) could also be written as

$$\dot{M}(t) = 0.06\ OUR(t) - 0.09\ CER(t) + 297.75\ F(t) \qquad (3.6)$$

Subject to knowledge of the reaction mechanism being available, the above approach may be applied to any fed-batch fermentation. In certain circumstances, when the fermentation proceeds with a near constant respiratory quotient for example, the correlation may be considerably simplified. For cell growth in fed-batch penicillin production, Mou and Cooney (1983) have found that the amount of biomass is accurately estimated from the relation

$$M(t) = K \int_0^t CER(\lambda)d\lambda \qquad (3.7)$$

where K is an experimentally determined constant.

Because of measurement noise and the uncertainty of the initial estimate of the biomass in the inoculum it is not usually acceptable (Stephanopoulos and San, 1981) to obtain the estimate of the biomass from the estimate of the biomass production rate by straightforward integration, such as is done in equation (3.7). A much better approach, see Fig. 1, is to augment output equation (3.6) or similar with a model having two states, M(t) and $\dot{M}(t)$, and to use the Extended Kalman Filter to produce estimates of both the amount of biomass and its production rate (Dekkers, 1982). For baker's yeast the state model is simply

$$\dot{M}(t) = BPR'(t) \qquad (3.8)$$

$$\dot{BPR}'(t) = [BPR'(t)]^2/M(t) \qquad (3.9)$$

where BPR'(t) is BPR(t) expressed in kg/hr.

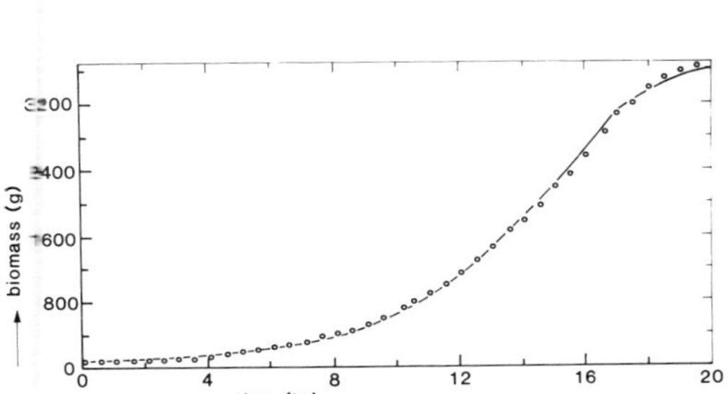

fig.1. Biomass estimation (Dekkers, 1982).
—— on-line estimates
∘ ∘ ∘ off-line measurements.

3.2. Estimating the substrate concentration

This topic has received little attention in the literature. The only noteworthy approach is based on equation (2.10) which models the substrate accumulation during the fed-batch. It is assumed that the composition of the exit gas is in some way correlated to the amount of substrate consumed by the growing biomass; for example in the case of fed-batch fermentation of glutamic acid that OUR(t) is directly proportional (Hong et al, 1984) to the substrate consumption $\mu(t)x(t)/Y(t)$. If s_i is constant and since $W(t) = V(t)s(t)$, where $d/dt(V(t)) = F(t)$, integration of eqn. (2.10) gives

$$s(t) = \frac{V(o)s(o)}{V(t)} + [1 - \frac{V(o)}{V(t)}]s_i - \frac{K}{V(t)} \int_0^t OUR(\lambda)d\lambda \qquad (3.10)$$

It would appear that the estimate of the substrate concentration obtained in this way, even with exactly known initial concentrations and volumes, would be subject to an accumulation of errors as mentioned in the previous section for biomass estimation.

3.3. Estimating the specific growth rate

If estimates of both the biomass production rate, $\dot{M}(t)$, and the amount of biomass, M(t), are available then an estimate of the specific growth rate can be made from equation (2.9).
When both estimates of M(t) and $\dot{M}(t)$ have been produced by an Extended Kalman Filter using a two state model, this approach gives quite satisfactory results (Dekkers, 1982) for a fed-batch baker's yeast fermentation. See Fig. 2.

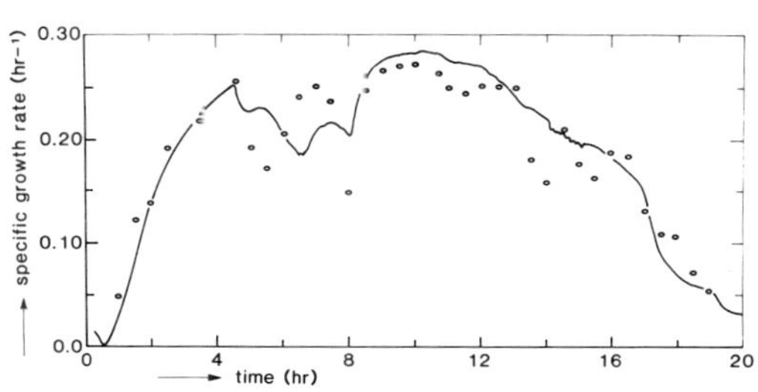

fig.2. Specific growth rate estimation (Dekkers, 1982).
—— on-line estimate
• • • off-line measurements.

Sometimes it is possible to regard the specific growth rate as a state variable in a model of the type

$$\dot{\mu}(t) = 0 \qquad (3.11)$$

$$\dot{M}(t) = \mu(t)M(t) \qquad (3.12)$$

and to estimate $\mu(t)$ directly by means of the Extended Kalman Filter, using measurements of CER(t) and OUR(t). The suboptimal estimates can once again be quite satisfactory (Nihtila, Harmo and Pettula, 1984).

If only one estimate of the amount of biomass is available, the derivative in equation (2.9) could be approximated by a backward-difference term to give (the biomass decay rate is zero)

$$\mu(t) \simeq \frac{1}{\Delta t} \ln[M(t)/M(t-\Delta t)] \qquad (3.13)$$

Mou and Cooney (1983) have applied this formula to estimate the specific growth rate in a penicillin fermentation. The approach is, however, very sensitive to small errors in the estimates of the biomass.

4. PHYSIOLOGICAL CONTROL

In section 2.1 we indicated that a control strategy could be formulated without a mathematical model, and we gave four examples of strategies derived from physiological models. The examples illustrate what has recently been termed *physiological control* (Mou and Cooney, 1983) although such control strategies have been in use since the fifties (see e.g. Dietrich, 1959). The two stages of the synthesis of a physiological controller are (i) the identification, by means of a physiological model, of one or more variables that must either be held constant or required to follow specified time-trajectories (ii) the development of a controller to ensure that those variables are, indeed, maintained at their constant (setpoint) values, or track the specified time trajectories without error. This second stage is the usual setpoint or servomechanism (tracking) control problem usually involving the estimation of one or more unmeasurable variables; the many different approaches to the controller design will not be discussed further here. We only remark that to thisend, the controller parameters may either be adjusted on-line using a technique such as that known as the Ziegler-Nichols method or be calculated offline with the aid of a mathematical model. A third possibility, exploited in adaptive controllers, is to automate the on-line adjustment of the controller parameters.

Physiological control has the advantage that the mathematical relationships between the process variables need not be known; at most a linearised model expressing the dynamics of perturbed variables is required. As a result it is possible in some circumstances to control the product quality. Moreover, physiological control is usually computationally less demanding than other approaches.
Its chief disadvantage is that it may not produce truly optimal control. It may also require special sensors or costly offline laboratory analyses, which can introduce a variable time delay into the system and thus complicate the control.

Some aspects of the physiological control of a fed-batch baker's yeast fermentation are now briefly explained. Other fermentations are usually more complex and less well understood. Starting point is the assumption that temperature and pH are controlled to be constant; the two manipulated variables are then the air supply and the feedrate of sugar. In general, the growth rate of the biomass (dM(t)/dt) can be assumed to be limited by the available amount of sugar. Recalling equation (2.9) we see that the rate of growth of yeast is proportional to the value of the specific growth rate, $\mu(t)$: the larger $\mu(t)$ is, the more yeast is produced. But $\mu(t)$ is controlled by the substrate concentration in the broth, which is in turn determined by the sugar feedrate, so that increasing the sugar feedrate increases the specific growth rate, producing more biomass. Two complications upset this straightforward view of the physiology, however. Firstly, if the substrate concentration in the broth is too high, ethanol will be produced (the glucose or negative Pasteur effect). Secondly, if not enough oxygen is supplied to combust the substrate, ethanol will again be produced. Ethanol is unwanted since it represents wasted sugar (although the quality of the yeast may be improved if some alcohol is present in the broth at the end of the fermentation). From stochiometric considerations it can be shown that when ethanol is produced CER(t) increases while OUR(t) remains constant. This has the effect of increasing RQ(t), see equation (3.3), so that the respiratory quotient is a good indicator of ethanol production.

The above outline of the physiology of yeast growth is by no means complete. However, it is comprehensive enough to explain nearly all the physiological control strategies which have been proposed to date in the literature, some of which are now described. Unless otherwise stated, all refer to baker's yeast fermentation.

4.1. Substrate control

Assume that there is sufficient oxygen dissolved in the broth to ensure the combustion of the sugar present. Then control, in some way, the substrate concentration to be constant during the whole fermentation. This will ensure that the specific growth rate is constant and, if it is less than some critical value, $\mu_c(t)$, no ethanol will be produced, leading to maximum productivity (Peringe and Blachere, 1978). According to equation (2.9) there will be an exponential growth of yeast resulting from this substrate control:

$$M(t) = M_o e^{\mu t} \tag{4.1}$$

How can both the dissolved oxygen concentration and the substrate concentration be held constant? The dissolved oxygen concentration may be measured and its value used by a feedback setpoint controller to adjust the air flow rate and/or the stirring speed. However, one of the major constraints in a yeast fermentation is the oxygen transfer capacity of the fermentor (Wang, Cooney and Wang, 1979) so constant dissolved oxygen concentration control may not be feasible throughout the fermentation.

Various techniques have been proposed to keep the substrate concentration constant. The most obvious is to use a feedback controller where the sugar in the broth is measured and the sugar feedrate is varied accordingly (Axelson et al, 1982). If both the substrate concentration and the biomass production rate are measured, the adaptive minimum variance control algorithm proposed by Dochain and Bastin (1984) could be used to keep the substrate concentration constant.

If a reliable sensor is not available then an estimate of the substrate concentration could be used in its place (see section 3.2). Failing this, open-loop control could be used: from equation (2.4) the feedrate to achieve a constant substrate concentration is given by

$$F(t) = \frac{\mu M_o e^{\mu t}}{Y(s_i - s)} \simeq \frac{\mu M_o e^{\mu t}}{Y s_i} \tag{4.2}$$

However, open-loop control cannot compensate for any disturbances occurring in the process conditions during the fermentation. To counteract this, Cooney and Swartz (1982) proposed incorporating equation (4.1) in (4.2) to give

$$F(t) = \frac{\mu M(t)}{Y s_i} \tag{4.3}$$

Based on stochiometric balances and measurements of CER(t) and OUR(t) the biomass, μ and Y can be estimated from an expression similar to equation (3.6) so that feedback control can be used. With the specific growth rate kept just below 0.22 hr^{-1} and the substrate concentration constant and near zero (< 0.15 g/l) the authors claim high productivities and yields approaching the maximum theoretically achievable of 0.5 g cell/g sugar.

Hong et al (1984) exploited the correlation between substrate concentration and OUR(t) to maintain a constant substrate level during glutamic acid fermentation. Their predictions of increased productivity were verified by experiments using a 30 l fermentor.

4.2. Specific growth rate control

As we have seen, the philosophy of substrate control implies that the specific growth rate is held at some constant value, μ_C, during the whole fermentation. Recently an adaptive controller has been proposed (Wu, Chen and Chiou, 1985) to accomplish this for a fed-batch baker's yeast fermentation. The actual specific growth rate is estimated from measurements of OUR(t). If the heuristic Certainty Equivalence approach is valid, which the experimental results of the authors seem to imply, then good control around the chosen specific growth rate setpoint could be expected; no proof of convergence is given, however.

The optimization of comprehensive mathematical models of fed-batch fermentations subject to realistic cost criteria indicate that maintaining the specific growth rate at a constant value seldom or never will lead to optimal productivities or yields (Dekkers, 1983). Instead, the specific growth rate must follow a time-dependent trajectory since the critical specific growth rate (above which ethanol is produced) is not constant as had hitherto been assumed but time-varying. In fact, it appears that for baker's yeast fermentation, the value of the critical specific growth rate depends upon whether the fermentation is at the stage of (i) start-up or exponential growth, or (ii) limited oxygen supply (Dekkers, 1984). Anyway, since the desired specific growth rate is time dependent, the controller for specific growth rate control is not of the set-point type, but of the tracking variety.

For the fed-batch culture of baker's yeast Dairaku and co-workers (1982a) have calculated the glucose feedrate necessary to achieve time optimal transfer from one specific growth rate to another. Using a measurement of the ethanol concentration by means of a porous teflon tube, and a feedback controller also regulating the glucose feedrate, it could be assumed that there was no ethanol consumption or production during the fermentation. The resulting optimal control strategy consisted of firstly applying the maximum feedrate until the critical substrate concentration was achieved and then maintaining that critical concentration until the desired specific growth rate is reached.

The principle of specific growth rate control has found primarily application in the fed-batch fermentation of penicillin. The objectives there are to maximize the specific growth rate during the fast-growth period and then to keep the production-phase specific growth rate at a lower, fixed value (Mou and Cooney, 1983). The specific growth rate is calculated from an on-line estimate of the cell mass, found by either stochiometric balances or from the correlation with CER(t). Refinements to the control, such as provision to ensure no substrate accumulation in the broth or to ensure that the dissolved oxygen never falls below a certain value (Nelligan and Calam, 1983) have been proposed and verified experimentally.

4.3. Ethanol control

According to the physiological model, any ethanol which is produced during a baker's yeast fermentation represents a wasted conversion of sugar. Using a porous teflon tubing ethanol sensor, a control system to regulate the ethanol concentration about a setpoint has been proposed (Dairaku et al, 1983). The objective of the controller is to maintain a maximum feedrate of sugar without ethanol production, and it is basically a three-term (PID) controller. The controller parameters have been determined by a series of experiments, and are tuned on-line according to the biomass concentration, estimated from a simple mathematical model.

It may be desirable to allow some alcohol to be produced during the last stage of the yeast fermentation. This is because it has been noted (Reed and Peppler, 1973) that the baking power of yeast is diminished if significant alcohol consumption takes place at the end of the batch.

4.4. RQ control

The respiratory quotient can be easily measured. If the RQ is regulated at 1.03 - 1.06 during the yeast fermentation while at the same time the sugar feed rate is maximized then there will only be pure respiratory growth without ethanol production, and hence maximum productivity. These two factors have made RQ control very popular over the years.

Nearly all the examples of RQ control in the literature employ the setpoint, feedback type of controller (Aiba et al, 1976; Wang, Cooney and Wang, 1979; Pons, Bordet and Engasser, 1982). There has been little attention paid to choosing the optimal parameter values of the controller (i.e. the PID action coefficients). This is a pity, since the RQ is very often hardly controlled at all during the start or the end of the batch, due to the different stages during a fermentation requiring different controllers. See Fig. 3.

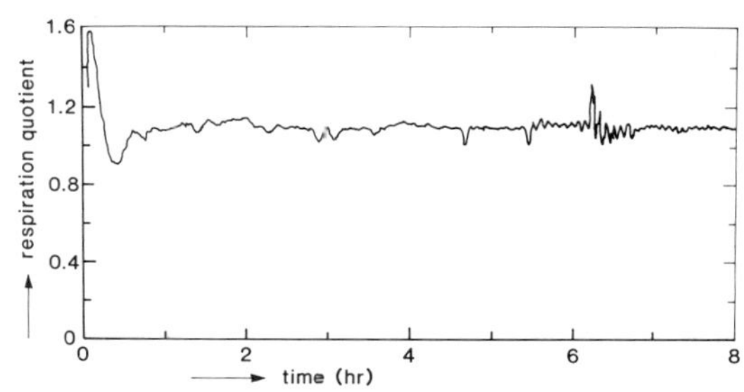

fig.3. RQ control (Dekkers, 1984).

Recently, studies of the optimal control of a baker's yeast fermentation have shown that the optimal RQ(t) trajectory will not, in general, be a constant straight line (Dekkers, 1983). That is if an economically realistic cost criterion is used for the optimization (see section 5). Two adaptive RQ controllers which can, if necessary, follow a predetermined RQ trajectory have been proposed and tested experimentally (Dekkers and Voetter, 1985). Both appear to offer much better RQ regulation than was previously considered possible.

4.5. Environment and quality control

Often the environmental variables such as temperature, pH, agitation and aeration are controlled to be constant. If this is not so, or indeed if process conditions are deliberately changed during the fermentation the consequences can be far-reaching and may be beneficial (Biryukov, 1982). For example, it is commonplace during some fermentations to control the pH of the broth to increase from about 3 initially (so reducing the chance of infection of the microorganism during the period of greatest vulnerability) to a final value of 7. Another beneficial change could be to lower the temperature of the broth in order to prevent oxygen starvation. However, although oxygen dissolves easier at a lower temperature this benefit may be offset by the fall in the mass transfer coefficient, K_La (see Fig. 4).

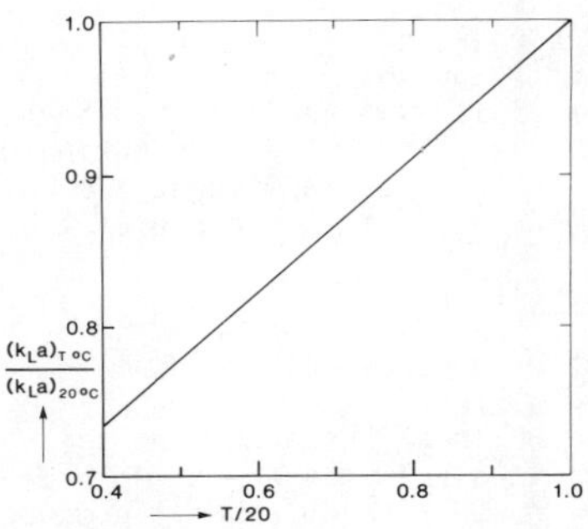

fig.4. k_La temperature dependence (Smith et al, 1977).

Quality control also lends itself to the physiological model approach. Baker's yeast quality depends, for example, on its baking power and its "shelf-life" - the time the product can be stored before its properties seriously degenerate. These two factors are mutually competitive, i.e. higher baking power generally implies a shorter shelf-life, since baking power requires a high nitrogen content of the yeast while shelf-life is determined by how much oxygen is present. A physiological controller could be designed to ensure good product quality by regulating the production of new cell buds at the end phase of the fermentation.

5. DYNAMIC OPTIMIZATION

The physiological model was used in the preceeding section to obtain one or more process variable which must be kept constant or allowed to follow some specified trajectory. From an economic viewpoint it is unlikely that even if the process variables could be controlled at precisely the required values, the fermentation productivity or yield would be truely optimal. This is because of the lack of sophistication and precision in the physiological model. It is therefore important to explore other means to obtain optimal set-point values, or trajectories, that must be achieved during the fermentation. The approach described in this section replaces the physiological model with a mathematical model and a cost (or performance) criterion: it is the approach to controller design that has been applied in other fields (e.g. electronics, space and process industries) since the early sixties, but with one important difference which will become apparent shortly.

First, let us state that this alternative approach to control consists of two stages which are (i) solving the dynamic optimization problem to produce one or more variables that must be held constant or required to follow specified time-trajectories and (ii) the development of a controller to ensure that those variables are maintained at their setpoint values or tracked with the minimum of error. Comparing this with the stages of physiological control we see that only the first stage is different. The second stage is the usual (setpoint or servomechanism) control problem, often involving the estimation of one or more unmeasurable variable and is not discussed further in this section. The dynamic optimization problem referred to is the task of minimizing, or maximizing, the cost criterion subject to the constraints imposed by the mathematical model of the fermentation process.

Because of the fed-batch mode of operation the optimization problem is essentially dynamic, i.e. a dynamic mathematical model is required together with a finite-horizon cost criterion. This is in sharp contrast to the vast majority of optimization problems in other fields where, since the processes are continuous by nature, steady state optimization may be applied to produce optimal setpoints. For fed-batch fermentations we usually expect to find non-constant optimal trajectories between the given startup and shutdown times. Actually, since real fermentation processes can consist of a train of fermentors, each growing the organism to some particular stage of maturity, the duration of each fed-batch fermentation is quite critical. Several authors (Blachere et al, 1976; Nagamune et al, 1982; Dourado and Calvet, 1982) have considered this overall (static) optimization problem of determining the optimal size, duration and initial concentration of the inoculum of a cascade of fermentors.

The best known dynamic optimization problem consists of minimizing a quadratic cost criterion subject to a linear model but no other constraints: a closed-form of solution is known in this case (Kalman, 1960). This solution can be utilized thanks to the observation (Dekkers, 1983) that at least for baker's yeast fermentation, on the time-scale of optimization the consumption of the substrate feed is instantaneous. If the substrate feed concentration is constant, equations (2.8), (2.9) and (2.10) give the linear model

$$\dot{M}(t) = bF(t) \tag{5.1}$$

We would like to maximize the final biomass yield while minimizing the substrate feed rate. Consider, therefore, minimizing the quadratic cost criterion

$$J = (M(t_f)-M*)^2 + \rho \int_0^{t_f} F^2(t)dt \tag{5.2}$$

where $M* > M(t_f)$ and $\rho > 0$ is a weighting factor. The optimal feedrate which achieves this is

$$F(t) = F = \frac{b}{\rho+b^2 t_f}(M*-M_0) \tag{5.3}$$

i.e. a constant feed. The corresponding biomass and specific growth rate profiles are given by

$$M(t) = \frac{b^2(M*-M_0)}{\rho+b^2 t_f}.t + M_0 \tag{5.4}$$

$$\mu(t) = \frac{b^2(M*-M_0)}{(M*-M_0)b^2t+M_0(\rho+b^2t_f)} \tag{5.5}$$

See Fig. 5 (broken lines). The maximum specific growth rate, μ_{max}, occurs at t = 0 and from the above equations the maximum attainable biomass is fixed by μ_{max}:

$$M(t_f) = M_0(1+\mu_{max}t_f) \tag{5.6}$$

To remove this unrealistic restriction we can include a quadratic biomass weighting term in the cost criterion:

$$J = c(M(t_f)-M*)^2 + \int_0^{t_f}\{q^2(M(t)-M*)^2 + \rho^2F^2(t)\}dt \tag{5.7}$$

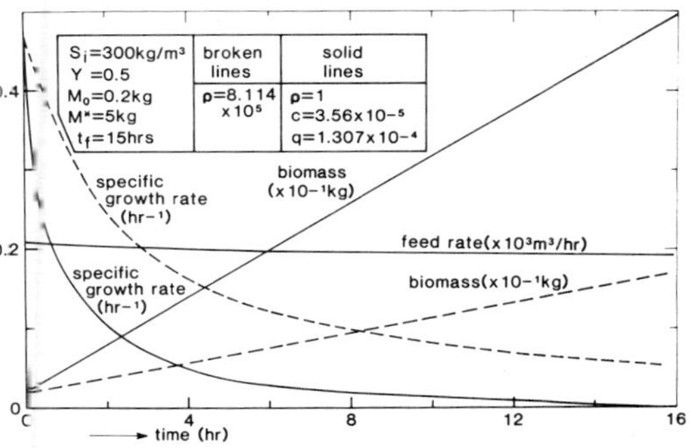

fig.5.

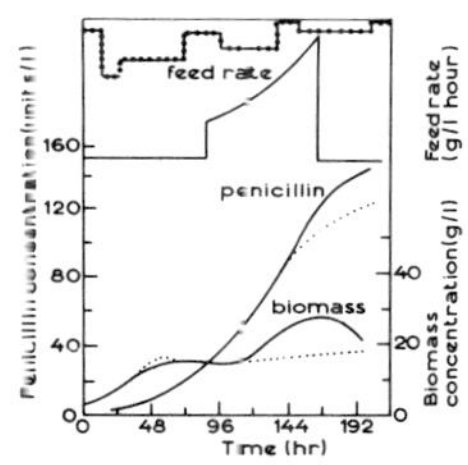

fig.6. Optimally controlled penicillin fermentation.
—— (Fishman and Biryukov, 1974).
--- (Kishimoto et al, 1982).

The optimal feedrate is no longer constant but decreases (slightly):

$$F(t) = \frac{q}{\rho}(M^*-M_o)\left[\frac{1+\sigma e^{\gamma(t-t_f)}}{1-\sigma e^{-\gamma t_f}}\right] e^{-\gamma\frac{t}{2}} \tag{5.8}$$

where $\sigma = (cb-q\rho)/(cb+q\rho)$ and $\gamma = 2\,bq/\rho$. The optimal biomass and specific growth rate trajectories become:

$$M(t) = M^*-(M^*-M_o)\frac{(1-\sigma e^{\gamma(t-t_f)})}{(1-\sigma e^{-\gamma t_f})}.e^{-\gamma\frac{t}{2}} \tag{5.9}$$

$$\mu(t) = \frac{\gamma}{2}\cdot\frac{1}{\dfrac{M^*(1-\sigma e^{-\gamma t_f})e^{\gamma\frac{t}{2}}}{(M^*-M_o)(1-\sigma e^{\gamma(t-t_f)})}-1} \tag{5.10}$$

By suitable choice of the parameters c, q, ρ and M*, a more realistic value of $M(t_f)$ can be ensured (see Fig. 5 solid lines). From Fig. 5 we conclude that the extra weighting term in the cost criterion (5.7) does not drastically change the shape of the feed rate, biomass or specific growth rate optimal trajectories. Note that this approach does not result in an exponentially increasing feed rate (see section 4.1) which was derived by considering μ(t) to be constant throughout the fermentation. We will return to the question of just how realistic the above linear-quadratic problem is later in this section.

Because the cost criterion is not quadratic, or the unknown parameters such as μ(t) and Y(t) exclude the use of a linear model, or because we wish to introduce constraints in the process variables corresponding to physical limitations, realistic dynamic optimization of a fed-batch fermentation process usually involves iteration towards the optimum, often by one of four techniques based on Green's Theorem, Pontryagin's Maximum Principle, Variational Calculus or Dynamic Programming. We will now survey the published results.

Fishman and Biryukov (1974) used Pontryagin's Maximum Principle on a homogeneous, single organism type model of a fed-batch penicillin fermentation to maximize the penicillin yield. The cost criterion consisted of just one term to account for the terminal state. Two constraints were imposed on the control variable (= $F(t)s_i(t)/V(t)$), namely that it was bounded from above and positive, and that the total amount of glucose fed must not exceed some fixed value. The technique of quasilinearization (see e.g. Seinfeld and Lapidus, 1974) was used to estimate the unknown model parameters, which included the cell age. The result, a rather discontinuous glucose feedrate, is shown in Fig. 6. Simulations predicted a 12% improvement in penicillin yield compared to that obtained with a constant glucose feed.

Some years later the same optimization problem was resolved using dynamic programming (Kishimoto et al, 1982). It is interesting to note that the optimal control was not the same, due either to the errors introduced by discretization or to the fact that all the process variables were constrained to lie within a certain distance from the experimental data already collected. See Fig. 6. It was shown that the optimization results were significantly affected when the unknown parameters could not be estimated by linear multiple regression to better than 10% of their true values. The adaptive optimal control approach was also applied to a fed-batch culture for glutamic acid production with ethanol feeding, and verified experimentally.

Various researchers have investigated very similar problems - the same terminal-state-constrained cost criterion, type of model and physical constraints on the process variables - as Fishman and Biryukov. Maximization of the yield of amino acids from an auxotrophic mutant was studied by applying Green's Theorem, which transforms a line integral into a surface integral (Ohno, Nakanishi and Takamatsu, 1976). Recently, Staniskis and Levisauskas (1984) again used Pontryagin's Maximum Principle to optimize the yield of α-amylase enzyme. The unknown parameters (the specific rates of product formation and substrate consumption) were modelled as polynomial functions of the substrate concentration; the coefficients of these functions were estimated on-line, although the estimator appears to be very sensitive to the initial values chosen to start the algorithm.

So far, dynamic optimization has consisted of maximizing the product yield. The result is usually a quite discontinuous control trajectory. The same can be said of time-optimal control (Dairaku et al, 1982a). By considering only a part of the fermentation (this time the fed-batch culture of baker's yeast) during which the biomass concentration changes very little, a model consisting of just equation (2.4) could be used. Minimization of the cost criterion

$$J = \int_{t_o}^{t_f} dt \tag{5.11}$$

was performed using Green's Theorem, resulting in a bang-bang type control. In order to optimize the product quality it has been suggested (Dairaku et al, 1982b) that the ratio of budding to total cells be minimized at the termination of the fermentation of baker's yeast. The ratio is obtained from a structured model, where the specific growth rate $\mu(t)$ is the control variable. One again, a bang-bang trajectory for $\mu(t)$ results from the optimization, leading to undesirable ethanol formation.

In an attempt to solve a more realistic optimization problem Dekkers (1983) considered minimizing a cost criterion:

$$J = -M(t_f) + \rho \int_{t_0}^{t_f} s_i F(t)dt \tag{5.12}$$

where the cost of glucose fed to the fermentor can be traded off against the ultimate yield of product (baker's yeast in this case). This problem is quite similar to that posed previously in this section (cf. equations 5.1 and 5.2). Because there is instantaneous consumption of the glucose feed (as far as the time-scale of optimization is concerned) the model used (equation 5.1) is particularly simple; however, here the yield factor, $Y(t)$, is not assumed to be constant. With the restriction that $\mu(t) > \mu_c(t)$ holding, the optimization problem was solved using variational calculus to give an optimal $RQ(t)$ trajectory, see fig. 7. Comparing Fig. 5 and Fig. 7 we see that the increased realism of a time-varying yield factor and the constraint on the specific growth rate have a considerable effect on the optimal trajectories. It was noted that these trajectories were quite sensitive to disturbances in the initial conditions.

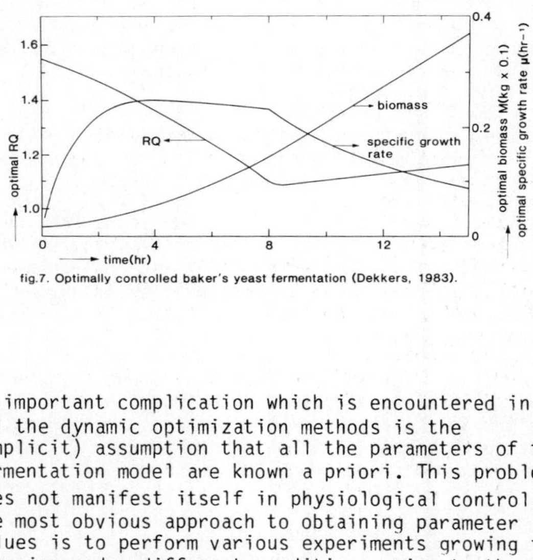

fig.7. Optimally controlled baker's yeast fermentation (Dekkers, 1983).

An important complication which is encountered in all the dynamic optimization methods is the (implicit) assumption that all the parameters of the fermentation model are known a priori. This problem does not manifest itself in physiological control. The most obvious approach to obtaining parameter values is to perform various experiments growing the organism under different conditions prior to the commencement of the fermentation. More realistically, we require an *adaptive dynamic optimization* technique, where a parameter estimation algorithm and an optimization algorithm are used sequentially during the actual fermentation. However, such an approach will only succeed if both algorithms, *when used in combination*, converge to their correct (i.e. optimal) solutions, and special precautions must be taken to ensure this (Kumar, 1983; Roberts and Williams, 1981). It would be sensible to use a discrete-time formulation of the dynamic optimization problem (such as is usual with dynamic programming for example) to facilitate its use with a parameter estimation algorithm.

6. CONCLUSIONS

A survey has been presented of the control of fed-batch fermentors. The fed-batch mode of operation is used in industry to produce amongst other things vaccines, enzymes, antibiotics and baker's yeast and is commercially very important.

The synthesis of a fed-batch fermentor controller consists of two stages. Firstly, the time trajectories of one or more process variables are found which, when followed, produce a suitable end product at a required price. Secondly a setpoint or tracking controller is designed to hold the process variables at the specified value. Environmental variables such as temperature, pH, agitation and aeration are maintained at constant values using conventional setpoint controllers.

There are many interrelated and competing reactions occurring during fermentation. These may be expressed by a linear mathematical model with time-varying parameters or by a physiological model. The same mathematical model, given in Section 2, has been used with only minor modifications to describe many different fermentations. An example of a useful physiological model has been presented in Section 4.

It is only the first stage of the controller design which is unfamiliar to the control engineer. There are two possible approaches, based on whether a mathematical or a physiological model is chosen to describe the fermentation. The physiological controller, of which six variants have been described in Section 4 has found widespread appeal in biotechnology. Such controllers do away with the need for a mathematical analysis of the process but are seldom optimal. Although they allow a degree of quality control which is difficult to achieve by other means, the physiological controller may require special sensors or costly laboratory analysis.

The alternative is to use a dynamic mathematical model for which the minimum (or maximum) of a finite-horizon cost criterion is sought. The solution of this dynamic optimization problem provides the required time-trajectories of the variables. There are indications that more use could be made of existing results pertaining to linear models and quadratic cost criteria to prevent, for example, too many abrupt changes in the control variable. It may even be possible to develop a closed form solution to the nonlinear optimization problem based on the nonlinear mathematical model widely used to describe a variety of fermentation processes. There is also a need for an adaptive dynamic optimization technique with guaranteed convergence properties.

At the time of this survey physiological control looks more promising than dynamic optimization. However, physiological models must be improved so that nearly optimal results can be obtained. The controllers to achieve this will be more complicated than at present since more states of the process will have to be incorporated to account for the changing process parameters.

Aiba, S., S. Nagai and Y. Nishizawa (1976). Fed-batch culture of saccharomyces cerevisiae: a perspective of computer control to enhance the productivity in baker's yeast cultivation. Biotech. and Bioeng., 28, 1001-1016.

Alberghina, L., E. Martegani and L. Mariani (1982). Analysis of protein distribution in populations of budding yeast based on a structured model of cell growth. Proc. 1st IFAC Workshop on Modelling and Control of Biotechnical Processes, Helsinki. Pergamon Press, 83-91.

Anderson, B.D.O. and J.B. Moore (1979). Optimal Filtering. Prentice Hall, New Jersey.

J.P. Axelson, P. Hagander, C.F. Mandenius and B. Mattiasson (1982). Computer control of sucrose concentration in a fermentor with continuous flow. Proc. Ist IFAC Workshop on Modelling and Control of Biotechnical Processes, Helsinki. Pergamon Press, 273-280.

Biryukov, V.V. (1982). Computer control and optimisation of microbial metabolite production. Proc. Ist IFAC Workshop on Modelling and Control of Biotechnical Processes, Helsinki. Pergamon Press, 135-144.

Blachere, H.T., P. Peringer and G.V. Corrieu (1976). Optimization of fermentation plants. Workshop on Comp. Appl. in Ferm. Technology. Verlag Chemie, Weinheim, 1-10.

Cooney, C.L., Y.W. Wang and D.I.C. Wang (1977). Computer aided material balancing for prediction of fermentation parameters. Biotech. and Bioeng., 29, 55-67.

Cooney, C.L. and J.R. Swartz (1982). Application of computer control to yeast fermentation. Proc. Ist IFAC Workshop on Modelling and Control of Biotechnical Processes, Helsinki. Pergamon Press, 243-252.

Dairaku, K., Y. Yamasaki, H. Morikawa, S. Shioya and T. Takamatsu (1982a). Experimental study of time-optimal control in fed-batch culture of baker's yeast. J. Ferm. Tech., 60, 1, 67-75.

Dairaku, K., E. Izumoto, H. Morikawa, S. Shioya and T. Takamatsu (1982b). Optimal quality control of baker's yeast fed-batch culture using population dynamics. Biotech. and Bioeng., 24, 2661-2674.

Dairaku, K., E. Izumoto, H. Morikawa, S. Shioya and T. Takamatsu (1983). An advanced micro-computer coupled control system in a baker's yeast fed-batch culture using a tubing method. J. Ferm. Tech., 61, 2, 189-196.

Dekkers, R.M. (1982). State estimation of a fed-batch baker's yeast fermentation. Proc. Ist IFAC Workshop on Modelling and Control of BiotechnicalProcesses, Helsinki. Pergamon Press, 201-211.

Dekkers, R.M. (1983). Dynamic optimization of a fed-batch fermentation process. Preprints 1st IASTED Int. Symp. on Appl. Control and Identification. Copenhagen, 334-339.

Dekkers, R.M. (1984). Optimal control of a fed-batch fermentation. In "Innovations in Biotechnology". Eds. E.H. Houwink and R.R. v.d. Meer. Elsevier. 313-330.

Dekkers, R.M. and M. Voetter (1985). Adaptive control of a fed-batch baker's yeast fermentation. Preprints Ist IFAC Symposium on Modelling and Control of Biotechnological Processes, Noordwijkerhout. Pergamon Press.

Dietrich, K.R. (1959). Albaufuerwertung und Abwasserreinigung. Springer, 1959.

Dochain, D. and G. Bastin (1984). Adaptive identification and control algorithms for nonlinear bacterial growth systems. Automatica, 20, 5, 621-634.

Durado, A. and J.L. Calvet (1982). Static optimization of the ethanol production in a cascade reactor. Proc. 1st IFAC Workshop on Modelling and Control of Biotechnical Processes, Helsinki, Pergamon Press, 177-185.

Fishman, V.M. and V.V. Biryukov (1974). Kinetic model of secondary metabolite production and its use in computation of optimal conditions. Biotech. and Bioeng., 4, 647-662.

Flynn, D.S. (1982). Instrumentation for fermentation processes. Proc. Ist IFAC Workshop on Modelling and Control of Biotechnical Processes, Helsinki. Pergamon Press, 5-12.

Halme, A. and A. Holmberg (1978). Application of control engineering to fermentation processes. Proc. 7th IFAC World Congress, Helsinki. Pergamon Press, 291-299.

Holmberg, A. and J. Ranta (1982). Procedures for parameter and state estimation of microbial growth process models. Automatica, 18, 2, 181-193.

Hong, K.T., S.H. Park, J.H. Lee, C.Y. Choi and J.C. Bae (1984). Control of sugar feeding for glutamic acid fermentation. J. Ferment. Tech., 62, 1, 49-54.

Johnson, A. (1985). Process Dynamics, Estimation and Control. Peter Peregrirus, London.

Kalman, R.E. (1960). Contributions to the theory of optimal control. Bol. Soc. Mat. Mexicana, 5, 102-119.

Kishimoto, M., T. Sawano, T. Yoshida and H. Taguchi (1982). Optimization of a fed-batch culture by statistical data analysis. Proc. 1st IFAC Workshop on Modelling and Control of Biotechnical Processes, Helsinki. Pergamon Press, 161-168.

Kumar, P.R. (1983). Optimal adaptive control of linear-quadratic-gaussian systems. SIAM J. Control and Opt., 21, 2, 163-178.

Meer, R.R. v.d. and F. Valkema (1985). Nederlandse bedrijvigheid in de biotechnologie. Biotechnologie in Nederland, 2,3, 100-102.

Monod, J. (1942). Recherches sur la Croissance des Cultures Bactériennes. 2nd ed. Hermann, Paris.

Mou, D.G. and C.L. Cooney (1983). Growth monitoring and control through computer-aided on-line mass balancing in a fed-batch penicillin fermentation. Biotech. and Bioeng., 25, 225-255.

Nagamune, T., I. Endo and I. Inoue (1982). Optimal operation for the seed culture process. J. Chem. Eng. of Japan, 15, 6, 481-486.

Nakamura, T., T. Kuratani, Y. Morita (1985). Fuzzy control application to glutamic acid fermentation. Preprints Ist IFAC Symp. on Modelling and Control of Biotechnical Processes, Noordwijkerhout. Pergamon Press.

Nelligan, I. and C.T. Calam (1983). Optimal control of penicillin production using a mini-computer. Biotech. Letters, 5, 8, 561-566.

Nestaas, E. and D.I.C. Wang (1983). Computer control of the penicillin fermentation using the filtration probe in conjunction with a structured process model. Biotech. and Bioeng., 25, 781-796.

Nihtila, M., P. Harmo and M. Perttula (1984). Real-time growth estimation in batch fermentation. Proc. 9th World Congress IFAC, Budapest, 225-230.

Ohno, H., E. Nakanishi and T. Takamatsu (1976). Optimal control of a semibatch fermentation. Biotech. and Bioeng., 18, 847-864.

Peringer, P. and H.T. Blachere (1978). Modelling and optimal control of baker's yeast production in repeated fed-batch culture. Proc. 2nd Int. Conf. on Computer Appl. in Ferm. Tech., Univ. of Pennsylvania, Philadelphia.

Pons, M.N., J. Bordet and J.M. Engasser (1982). A micro-mini computer hierarchical control system for a laboratory scale fermentor. Proc. Ist IFAC Workshop on Modelling and Control of BiotechnicalProcesses, Helsinki. Pergamon Press, 259-267.

Ranka, J. (1982). Some properties of microbial age dynamics models. Proc. 1st IFAC Workshop on Modelling and Control of Biotechnical Processes, Helsinki. Pergamon Press, 93-100

Reed, G. and H.J. Peppler (1973). Yeast Technology. The Avi Publ. Co., Westpoint, Connecticut.

Roberts, P.D. and Williams, T.W.C. (1981). On an algorithm for combined system optimization and parameter estimation. Automatica, 17, 1, 199-209.

Roels, J.A. (1983). Energetics and Kinetics in Biotechnology. Elsevier, Amsterdam.

Seinfeld, J.H. and L. Lapidus (1974). Mathematical methods in chemical engineering. Volume 3. Process modelling, estimation and identification. Prentice Hall.

Shioya, S., T. Takamatsu and K. Dairaku (1982). Measurement of state variables and controlling biochemical reaction processes. Proc. Ist IFAC Workshop on Modelling and Control of Biotechnical Processes, Helsinki. Pergamon Press, 13-25.

Smith, J.M., K. van 't Riet and J.C. Middleton (1977). Scale-up of agitated gas-liquid reactors for mass transfer. Preprints 2nd European Conf. on Mixing, Cambridge, U.K. F4, 51-66.

Spriet, J.A. (1982). Modelling of the growth of micro-organisms: a critical appraisal. In "Environmental Systems Analysis and Management". Rinaldi (Ed.) N.Holland, 451-456.

Staniskis, J. and D. Levisauskas (1984). An adaptive control algorithm for fed-batch culture. Biotech. and Bioeng., 26, 419-425.

Stephanopoulos, G. and K.Y. San (1981). State estimation for computer control of biochemical reactors. Advances in Biotechnology, 1, 399.

Svrcek, W.Y., R.F. Elliott and J.E. Zajic (1974). The Extended Kalman Filter applied to a continuous culture model. Biotech. and Bioeng., 16, 827-846.

Tong, R.M., M.B. Beck, A. Latten (1980). Fuzzy control of the activated sludge waste-water treatment process. Automatica, 16, 695-701.

Wang, H.Y., C.L. Cooney and D.I.C. Wang (1979). Computer control of baker's yeast production. Biotech. and Bioeng., 21, 975-995.

Wu, W-T, K-C Chen and H-W Chiou (1985). On-line optimal control for fed-batch culture of baker's yeast production. Biotech. and Bioeng., 27, 75[illegible]-760.

Yamashita, S., H. Hoshi and T. Inagaki (1959). Automatic control and optimization of fermentation processes. In Fermentation Advances, Ed. D. Perlman, Academic Press, 441-463.

Zabriskie, D.W. (1979). Real-time applications of computers in fermentation processes. Annals New York Acad. Sci., 233-239.

MODELLING OF BIOTECHNOLOGICAL PROCESSES

K. Schügerl

Institut für Technische Chemie der Universität Hannover, Callinstrasse 3, D-3000 Hannover 1, FRG

INTRODUCTION

In biotechnology mathematical models can be developed on different levels:

- molecular or enzyme level (enzyme synthesis model)
- intracellular component level (structured cell models)
- cellular level (kinetic models)
- cellular environmental level (unstructured reactor models)
- dynamic cellular environmental level (structured reactor models).

The prerequisite for the development of these models is a system analysis based on experimental data of sufficient amount and quality.
For the development of dynamic models (e.g. for process control), on-line analysis of the system is necessary. The bottle-neck of model development is the insufficient measuring technique.

ENZYME SYNTHESIS MODELS

Since constitutional enzymes are formed at constant rates independent of the cell's surroundings - due to their large number of copies in the cell - only enzymes are considered here which are only present in a few copies in the cell, if they are not needed. As soon as substrates are present which can be utilized by the cell, the proper enzymes are formed. The enzyme synthesis is subject to genetic control /1/. In the case of <u>inducible enzymes</u>, the inductor inactivates the repressor. The operator gene is free and transcription occurs (Fig. 1a). E.g. as long as enough amino acid is present in the medium, the enzymes for the amino acid synthesis are <u>repressed</u>. In presence of a (co-)repressor, a <u>repressor</u> is formed by the co-repressor and aporepressor. The repressor can block the expression of the structural gene by binding to the repressor gene (Fig. 1b). This lac-operon theory of Monod et al. /2-4/ was further developed by different authors /5/.
Van Dedem and Moo-Young developed an induction and repression model for cell extracellular enzyme synthesis by assuming that the interactions between the corepressor or the inducer and the cytoplasmic repressor can be treated as chemical equilibria /6/:

$$R + nP \underset{}{\overset{k_1}{\rightleftharpoons}} RP_n \tag{1}$$

$$O + R \overset{k_3}{\rightleftharpoons} OR \tag{2}$$

$$O + RP_n \overset{k_4}{\rightleftharpoons} ORP_n, \tag{3}$$

where R = cytoplasmatic repressor
P = effector
RP_n = effector-cytoplasmatic repressor complex
O = operator
OR = operator-repressor complex (induction)
ORP_n = operator-repressor complex (repression)
n = number of binding sites for effector on R (n = 2 is assumed).

The fraction of free operator for the induction

$$Q_i = \frac{(O)}{(O)_t} = \frac{1+k_1(P)^n}{1+k_1(P)^n - k_3(R_t)} \tag{4}$$

and repression

$$Q_r = \frac{(O)}{(O)_t} = \frac{1+k_1(P)^n}{1+k_1(P)^n + k_1k_4(P)^n(R_t)} \tag{5}$$

is given as a function of the concentration of the effector P and total repressor R_t per unit cell mass.
If one assumes that the rate of enzyme synthesis is limited by the amount of m-RNA and there is a linear relationship between the cell mass concentration (X) and m-RNA concentration (mRNA), a relationship (6) can be developed /6/:

$$\frac{d(E)}{d(X)} = k_g Q \tag{6}$$

In Fig. 2, the dimensionless enzyme concentration

$e = \frac{(E)}{k_g(S_o)Y}$ is shown as a function of

dimensionless cell mass concentration:

$x = \frac{(X)}{(S_o)Y}$ for <u>induction</u> and

for constant $k_3' = k_3(R_t)$ and different

$k_1' = k_1(S_o)^n$: 10^4, 10^3, 10^2, 10 (from top to bottom) and in Fig. 3 for <u>repression</u> for constant $k_4' = k_4(R_t)$ and different k_1': 0.1, 1.0, 10, 100 (from top to bottom). In the case of induction, all curves level off at the end of the growth cycle, when the inducer can no longer saturate the repressor.
In the case of repression, the enzyme production increases at the end of the growth due to reduction of the repressor concentration.
By assuming that the permease E_1 for the favored substrate is constitutional, whereas the permease

E_2 for the second substrate is subject to induction and catabolite repression, Van Dedem and Moo-Young developed a model for diauxic growth /7/ (Fig. 4). The product of the two permeases is a common intermediate (F), which is the substrate of the branched pathway leading to either cell mass or ATP.
The two key enzymes E_4 and E_3 regulate the magnitudes of the energy and cell mass generating fluxes, respectively. The two enzymes controlling the branched pathway are allosteric and under heterotropic control by ATP (A in Fig. 4). Induction of the inducible permease E_2 is dependent on the rate of uptake of S_2.
Accumulation of A causes the inhibition of enzyme E_4 through a negative heterotropic interactionand stimulation of E_3 through a positive heterotropic reaction. In Fig. 5, the dimensionless cell mass concentration $X = (X)V_c$, and substrate concentrations $\sigma_1 = \frac{(S_1)}{(S_1)+(S_2)}$, $\sigma_2 = \frac{(S_2)}{(S_1)+(S_2)}$ are shown as functions of the dimensionless time $\Theta = k_{15}'t$, where V_c is the specific cell volume and k_{15}' is defined by (7)

$$\frac{d(X)}{dt} = k_{15}'r_3(F,A)(X) \qquad (7)$$

$r_3(F,A)$ flux of material into cell mass.
This model was also applied for the description of the diauxic growth of *Candida lipolytica* growing on olecic acid and glucose, and *Trichoderma viride* growing on mixed substrates of glucose and lactose /11/.
As a function of the dilution rate, the specific enzyme activity can increase, decrease, pass a maximum or minimum /8/. This phenomenon was modelled by Toda by assuming simultaneous effects of induction and repression on two sites of the operator loci for enzyme formation (Fig. 6). He assumed that the synthesis of the enzyme is able to begin only then, when both operator genes O_1 and O_2 work concurrently /8/. It is possible to describe the variation of ratio of the specific activity of the invertase to its theoretical maximum (Q_E) in *Saccharomyces carlsbergiensis* as a function of the dilution rate by means of this model (Fig. 7). The same was applied to describe the α-Amylase production by *Bacillus subtilis*, the β-Galactosidase production by *Saccharomyces fragilis* and the β-Fructosidase production by *Saccharomyces carlsbergiensis* in a phosphate-poor glucose medium /12/. In the first case, the enzyme production was linearly associated with the growth rate in a glucose-limited chemostat culture, in the second case, the enzyme production was rather invariable with the growth rate in a galactose medium, and in the third case, the enzyme production was independent of the growth rate, except with low dilution rates, where it increased with the growth rate.
The production of acid phosphatase by *S. Carlsbergiensis* showed a maximum at a particular intermediate dilution rate. Toda et al. succeeded to describe all these enzyme syntheses by means of their models (Fig. 8) /12/.
Imanaka and Aiba /9/ modified the dual-control model of Toda /8/ by reinterpreting the parameters of his model (Table 1).

The modified Toda model is shown in Fig. 9. According to this model there is only one binding site between repressor and operator. The second operator site (O_2) is a promotor gene P according to the authors.

Table 1

Corresponding symbols in the two models /9/

Toda	Imanaka/Aiba
S_2	cAMP
R_2	CAP
O_2	P
$S_2{}^nR_2$	$(cAMP)^n \cdot CAP$
$S_2{}^nR_2O_2$	$(cAMP)^n \cdot CAP.P$
$o_2 = \frac{(O_2)}{(O_2)_t}$	$1 - \frac{(P)}{(P)_t}$

The symbols in Table 1 have the following meaning:

cCAMP = cyclic 3', 5'-adenosin monophosphate
CAP = catabolite gene activator protein
P = promotor gene

A model was developed for catabolite repression of inducible enzyme biosynthesis by Gondo et al. /10/. The inducible enzyme is also subjected to repression by a specific (regulatory) repressor. Fig. 10 depicts the model for catabolite repression and the induction-repression mechanism involving the operon-coding for the inducible enzyme. The glucose controls the intracellular level of cAMP and in doing so regulates the transcription rate of mRNA specific to the inducible enzyme.
cAMP combines with the cAMP receptor protein (CRP) to form a complex. The RNA polymerase (RNAP) generally binds its recognition site on the promotor (P) and forms an inert complex which does not permit transmission. Transcription of mRNA occurs, when the cAMP-CRP complex attaches to the promotor site of the gene. In this case, the RNAP shifts the inert (close) complex to the open (active) complex (Fig. 11).
It is assumed that the production rate of mRNA, specific to translating the inducible enzyme, is proportional to the fraction of repressor-free operator Q ("induction index") coding for the enzyme synthesis and the "catabolite repression index" = (fraction of open complex to the total complex of DNA and RNAP).
They used this model to calculate the glucose isomerase production in *Streptomyces venezuelae* induced by xylose. Fig. 12 shows the variation of the concentration of glucose (G_B), xylose (X_B), cell mass (C) and enzyme (E) as a function of the dilution rate. This model was extended by Vieth et al. /13/ by assuming the following relationship between bulk and intracellular substrate concentrations:

$$\frac{d}{dt}(L_{in}) = \frac{A(t)(L_B)}{B(t)+(L_B)}, \qquad (8)$$

where
A(t) is the transport modulus (mol/L h)
B(t) transport constant (mol/L)
(L_{in}) intracellular substrate (lactose)
(L_B) bulk substrate (lactose) concentration (g/L),

between the fraction of repressor-free operators Q and the intracellular substrate concentration

$$Q = \frac{1+b_1(L_{in})}{1+b_2(L_{in})+b_3} \qquad (9)$$

and between the dimensionless cAMP concentration and intracellular ATP and bulk glucose concentrations

$$\alpha = -\frac{k'(ATP)/(cAMP_m)}{1+k_1'(ATP)+(G_B)[k_e'+k_3'(ATP)]} \qquad (10)$$

where k', k_1', k_2', k_3' are lumped constants, (G_B) bulk glucose concentration (g/L).
This modified model was verified by means of the experimental data of Kaczorowski et al. /14,15/ and Kanshik /16/.
The published models are suitable to describe the enzyme synthesis in different microorganisms under different cultivation conditions.

MODELLING OF PLASMID STABILITY OF CELLS IN CONTINUOUS CULTURE

Ollis and Hung-Teh Chang developed a simple kinetic model for the batch-cultivation of unstable recombinant cells /17/ based on the investigations of Imanaka and Aiba /18/. According to the latter, the mixed batch-culture of plasmid-bearing (+) and plasmid-free (-) cells can be described by the following simple network:

$$X^+ \xrightarrow{\mu^+} (2-p)X^+ \qquad (11a)$$
$$X^+ \xrightarrow{\mu^+} pX^-$$

$$X^- \xrightarrow{\mu^-} 2X^- \qquad (11b)$$

where p is the (small) probability per generation that a loss of the desired phenotype would occur, and μ^- and μ^+ are the specific growth rates of plasmid-free (-) and plasmid-bearing (+) cells.
By assuming exponential cell mass growth, the following relationship was found for the fraction of plasmic-bearing cells F_n after n-generations:

$$F_n = \frac{(X^+)}{(X^+)+(X^-)} = \frac{\alpha+p-1}{(\alpha-1)+p2^{n(\alpha+p-1)}} \qquad (12)$$

where $\alpha = \mu^-/\mu^+ > 1$.

From eq. (12) follows for $\alpha > 1$ and $p > 0$ that $F_\infty = 0$. However, the rate of approach to this asymptote may be very slow. Ollis and Hung-Teh Chang /17/ and Ollis /19/ extended this model by the substrate and product balances by assuming that the plasmid-free (-) cells do not form products.
The ratio of the cell mass balances

$$\frac{d(X^-)}{d(X^+)} = \frac{\mu_m^-(X^-)+\mu_m^+(X^+)p}{(1-p)\mu_m^+(X^+)} \qquad (13)$$

yields by integration the ratio X:

$$X = \frac{(X^-)}{(X^+)} = \frac{[(\alpha+p-1)(X_o)+p][(X^+)/(X_o^+)]^{(\alpha/1-p)-1}-p}{(\alpha+p-1)} \qquad (14)$$

When assuming that the yield coefficients are constant: $Y^+ = Y^- \approx Y$, the substrate utilization rate is given by (15):

$$\frac{d(S)}{dt} = -\frac{1}{Y}\frac{(S)}{K+(S)}\{\mu_m^+ + \mu_m^-(X)\}(X^+) \qquad (15)$$

For the exponential as well as for the Monod model, the steady-state solution is given by

$$(X^-)(t) = 0$$

$$(X^-)(t) \neq 0$$

However, when values of p are small, e.g. $p < 0.01$, the time constant of the plasmid loss $1/p\mu$ is much longer than the characteristic growth time. Hence, the culture may be operated continuously for estimable lengths of time, but a "steady state" is never achieved /19/. The influence of the growth rate on the E. Coli plasmid content and expression of a cloned-gene product was described by means of a mathematical model by Sun Bok Lee and Bailey /20/ using the λdv plasmid replication model and known relationships between growth rate and transcription and translation activities of the host cell.
Figure 13 shows the model and Table 2 the model equations. In Fig. 13 and in Table 2, the symbols have the following meaning:

G	number of plasmid molecules per cell
n	transcription efficiency
$mRNA_R$	repressor gene message
$mRNA_I$	initiator gene message
ORI^+	activated origin
R	repressor protein
I	initiator protein
REP	replication complex
k_m^o	overall transcription rate constant
k_{-m}	decay rate constant for mRNA
k_p^o	overall translation rate constant
k_{-p}^R	decay rate constant for repressor protein
k_{-p}^I	decay rate constant for initiator protein
Θ	dimensionless degree of origin activation
N_A	Avogadro number
V	cell volume
K_c	constant

In Fig. 13, the thick lines present the flow of information encoded in the plasmid, thin lines represent activities provided by the host function. The dotted line shows the regulatory action of repressor protein and initiator protein. Traces (a), (b) and (c) represent the replication, transcription and translation processes and d) the involvement of the replication protein in the formation of replication-initiation complexes. The replication of λdv plasmid is regulated by cro-repressor protein (R) and initiator protein (I). Transcription from the promotor-operator region (P_RO_R) results in the synthesis of repressor and initiator messages and also transcriptional activation of the origin region. It was assumed that the repressor and initiator synthesis follow linear kinetics (eq. 1-4 in Table 2), as well as the transcriptional activation of the origin region (eq. 5 in Table 2), which is the prerequisite for the formation of replication initiation complex REP (eq. 6 in Table 2). The number of plasmid molecules at time t is given by

$$G(t) = G(0) + m, \qquad (16)$$

where G(o) represents the number of plasmid molecules in a newborn cell (t = o);
m is the number of initiation events between t = 0 and t.

The plasmid segregation at cell division causes a difference between the plasmid content of dividing (τ) and newborn (o) cells:

$$G_{n+1}(0) = INT\ G_n(\tau)/2 \qquad (17)$$

where n is the number of generations
τ the double time
INT (...) the integer function.

The gene concentration of plasmid DNA in eq. 1 and 3 of Table 2 is given by

$$(G) = \frac{G}{VN_A} \tag{18}$$

For the gene expression, the following eq. was developed:

$$\frac{d(mRNA)_p}{dt} = k_m^o(G) - k_{-m}(mRNA)_p - \mu(mRNA)_p \tag{19}$$

$$\frac{d(P)}{dt} = k_p^o(mRNA)_p - k_{-p}^P(P) - \mu(P), \tag{20}$$

where $mRNA_p$ is the cloned gene message
P the cloned gene product
k_{-p}^P the decay rate constant for the cloned gene product

This model was used to simulate the experimental data (o) from Engberg and Nordström /28/. The dotted line shows the plasmid copy number (Fig. 14). The model simulates correctly the decrease in the plasmid content with increasing growth rate, as observed experimentally for different plasmids. The population balance model /29,30/ was used by Hjorstso and Bailey /21/ to describe the plasmid stability in budding yeast populations following a shift to a nonrestrictive (e.g. antibiotic-free) medium, where plasmid-free cells are able to grow also. The difference between nonsegregated and segregated models are most pronounced in asymmetric cell division at low growth rates. Sun Bok Lee and Bailey developed genetically structured models for lac Promoter-Operator Function in the chromosome and in multicopy plasmids /22,23/. The average number of a particular gene in the cell depends on the growth rate and map position of a gene. When C and D represent the time for replication of the chromosome and the time between arrival of the replication fork at the terminus and chromosome and cell division, the average number of genes encoded at the chromosomal map position f is given by

$$\tilde{G}_f(t) = \exp \int_{t-C-D}^{t-fC} \mu(\tau)d\tau \tag{21}$$

For the modelling of the transcription (the production of a single strand of RNA complementary to a specific DNA sequence), the rate of production of mRNA is needed:

$$\frac{d(mRNA)_i}{dt} = k_{p,i}^o \eta_i(G_i) - k_{d,i}(mRNA)_i - \mu(mRNA)_i \tag{22}$$

$k_{p,i}^o$ is the overall transcription elongation rate of gene i
η_i the overall transcription efficiency
$k_{d,i}$ the RNA decay rate constant

For the description of translation, the product P (protein) balance is used:

$$\frac{d(P)_i}{dt} = k_{q,i}^o \xi_i(mNRA)_i - k_{e,i}(P) - \mu(P), \tag{23}$$

where $k_{q,i}^o$ is the overall translation rate constant of protein i
ξ_i the overall translation efficiency
$k_{e,i}$ the protein decay rate constant

The regulation of the transcription of the lac-operon has already been considered (Fig. 11). The initiation of transcription was described by the following reaction:

$$[RNP\text{-}P_{II}]_{closed} \underset{}{\overset{K\ open}{\rightleftharpoons}} [RNP\text{-}P_{II}]_{open} \xrightarrow{k_p} mRNA, \tag{24}$$

where K_{open} is the equilibrium constant for the closed ⟶ isomerization step
RNP RNA-polymerase core enzyme
P_{II} RNP-binding site

Once initiated, the transcription proceeds at a uniform rate. The rate of the mRNA synthesis (r) is given by

$$r = k_p K_{open}[RNP\text{-}P_{II}]_{closed} \tag{25}$$

or

$$r = k_p^o \eta[G] \tag{26}$$

The model was applied to investigate the influence of mutations in the lac I gene, which change the level of activity of the lac-repressor in the cell, and in the lac O sequence, which influences the repressor-operator interaction and is extended for diploid cells and multicopy plasmids /22/. The model was also extended and applied to investigate the influence of certain host cell mutations and also mutations in the lac P promotor sequence /23/. According to this extended model, the copy number of the cloning vector has a major effect on the efficiency of modulator binding to the promotor-operator. With the increasing number of plasmid copies in the cell, the number of specific and nonspecific binding sites for regulatory proteins also increases. The fraction of binding sites occupied by their corresponding modulators decreases as the copy number of the vector increases. The net effect is a maximum in the cloned gene transcription rate with respect to the plasmid copy number (Fig. 15). This agrees with the experimental results: The gene expression level is proportional to the copy number for certain relatively low copy number systems, and the cells die, if the copy number is too large. Thus, a maximum must occur. The describe the plasmid content distribution and population product synthesis activity in unstable recombinant organisms with binary fission, a segregated population-balance model was used by Seo and Bailey /24/.
In steady-state exponential growth, the population balance based on cell age is given by

$$\frac{dW(a)}{da} = -\nu W(a) \tag{27}$$

where W(a) is the cell age distribution
ν the overall specific growth rate of plasmid-containing cells.

The relation between dividing (a = P) and newborn (a = 0) cells is

$$W(0) = (2-\Theta)W(P) \tag{28}$$

where Θ is the probability of birth of a plasmid-free cell upon division of a plasmid-containing cell. Since $\int_0^P W(a)da = 1$, the solution of eq. (27) is subject to the given condition (28):

$$W(a) = \frac{2-\Theta}{1-\Theta}\nu e^{-\nu a} \quad (29)$$

Eqs. (28) and (29) yield the relationship between ν, generation time P of the cells and the parameter Θ, which characterizes plasmid instability:

$$\nu = \frac{\ln(2-\Theta)}{F} \quad (30)$$

It is assumed that the age distribution of cells which have x plasmids at birth is given

$$W_x(a) = C_x e^{-\nu a} \quad (31)$$

The sum of eq. (31) overall x-values yields the age distribution W(a). Depending on the plasmid number distribution at birth, different models can be developed:

- When assuming random distribution of plasmids to daughter cells at cell division and that all cells contain N plasmids at division, the coefficient C_x in eq. (31) is given by:

$$C_x = \frac{\nu}{(1-\Theta)}[f(x;N) + f(N-x;N)] \quad (32)$$

$\Theta = f(0;N) + f(N;N)$ and eq. (32) yields a binomial distribution (Model I)

$$f(x;N) = \frac{N!}{x!(N-x)!}\gamma^x(1-\gamma)^{N-x} \quad (33)$$

where f(x; N) is the probability that from a cell f(·) a daughter forms with f(x). γ is the probability that a plasmid in the dividing cells remains in one of the particular daughters upon division (one usually assumes: $\gamma = 0.5$).

- When assuming that some plasmids, 2 M in number - partitioned evenly between the daughter cells - the remaining plasmids are segregated at random, C_x is given (Model II):

$$C_x = \nu[f(x-M;N-2M) + f(N-x-M;N-2M)]$$
$$\text{for } x \geq M,\ 2M < N \quad (34)$$

- When assuming that all plasmids are randomly distributed to the daughter cells and K plasmids are formed during each cell cycle - unless the plasmid number reaches some maximum value (N_m) - two C_x-values are possible, depending on the x-value: $1 \leq x \leq K+1$ or $K+2 \leq x \leq N_m$ (Model III).

In Fig. 16, the plasmid number distribution at birth are shown for Models I and III. The difference between them is slight. The age distribution and plasmid content distribution at birth were used to obtain the distribution of the single-cell plasmid content in the population by assuming a zero order kinetics for the plasmid synthesis:

$$n = x + k_n a$$

where k_n is the plasmid synthesis rate constant.

A comparison of the different single-cell plasmid content distributions indicate that according to Model III the cellular plasmid content is distributed over a much broader range than according to Model I (Model II yields results which are between those of Models I and III).

The specific rate of product synthesis q_p is given by:

$$q_p = \int_0^\infty W(n) r_e(n) dn, \quad (35)$$

where W(n) is the plasmic content distribution

$r_e = k_p(n(1-cn)$ the single cell gene expression rate

k_p the product synthesis rate constant

C the constant

In Fig. 17, the dimensionless specific product synthesis rate $q_p c/k_p$ is plotted as a function of the dimensionless plasmid number cN (Model I) and/or 2ck (Model III). There is an optimum plasmid number to maximize the specific product formation rate. A simplified population model for baker's yeast was developed by Takamatsu et al. /39/ by assuming symmetric cell division to determine the ratio of budding cells to total cells and were used for the control of fed-batch culture.
Different operational strategies can be chosen for enhancing plasmid stability in chemostat systems. Ryder and Di Biasio /26/ carried out steady-state stability analyses for cell cultures, in which the faster-growing plasmid-free cells are more sensitive to substrate (methanol) inhibition of the cell growth than the slower-growing plasmid-harboring cells. At higher substrate concentration, it is possible to maintain a higher specific growth rate μ^+ (for plasmid-harboring cells) then μ^- (for plasmid-free cells), thus a coexistence of the two cells is possible. These results are based on model calculations using simple mass balance equations:

$$(\dot{X}^-) = -D(X^-) + \mu^-(S)(X^-) + \alpha\mu(S)(X^+) \quad (36a)$$

$$(\dot{X}^+) = -D(X^+) + \mu^+(S)(X^+)(1-\alpha) \quad (36b)$$

$$(\dot{S}) = D(S_F - S) - \sigma^-(S)(X^-) - \sigma^+(S)(X^+) \quad (36c)$$

The tendency of the X^+ strain to loose plasmid and turn into X^- phenotype is accounted for by the $\alpha\mu^+(X^+)$ function. Here it is assumed that the plasmid loss rate is proportional to the growth rate of X^+ cells. The specific substrate consumption rates are equal:

$$\sigma(S) = \sigma^+(S) = \sigma^-(S) \quad (37)$$

The results of the steady-state analysis indicate that the isolation of plasmid-bearing cells by means of their lower specific growth rate is unfeasible and the coexistent state is constrained within an operating range. The steady-state plasmid cell to total-cell concentration ratio is given by:

$$\frac{(X^+)}{(X^+)+(X^-)} = 1 - \frac{\alpha_o\mu^+(S)}{\mu^+(S)-\mu^-(S)} \quad (38)$$

Figure 17b shows the specific rates as a function of the substrate concentration for the growth model.

$$\mu^- = k_1 \frac{(S)}{k^1 + (S) + \frac{(S)^2}{K_I}} \quad (39a)$$

$$\mu^+ = k_2 \frac{(S)}{k_1 + (S)} \quad (39b)$$

$$\sigma = k_3 \frac{(S)}{k_2 + (S)} \quad (39c)$$

However, the desired coexistence state is unstable. Feed-back control utilizing the PI-control algorithm and measuring the substrate concentration (instead of the cell mass concentration) to control the dilution rate was simulated, and the stabilization at the desired state was attained.

Other possibilities were modelled by Ryu et al. /25/: selective pressure as well as the separation of the growth and production stages. The cells are grown in the repressed state with very little expression of the cloned gene product and thereby meant to minimize the expression-related stability problem. In the second stage, the appropriate inducer is added, and the gene expression for production is turned on. The calculations were carried out with a temperature-sensitive cI repressor (temperature-sensitive phage λcI 857), which blocks transcription of the cloned gene. Raising the culture temperature above 38°C denaturates the repressor and causes expression. The experimental control of these models is very important. Besides the usual process parameters, like substrate and cell mass concentration /27/, the product concentration (enzyme activity), the copy number and the cell size (by laser flow cytometer) can be determined.

STRUCTURED CELL MODELS

Harder and Roels gave an excelent review on simple structured models in bioengineering /5/. They divide these models into two- and three-compartment models. In two-compartment models, the first compartment often consists of RNA and precursor molecules and the second compartment of protein and DNA. Other models distinguish between the assimilating compartment, which takes up nutrients and transforms them to energy and biomass precursors, and the synthetic compartment which produces cell mass /5/.
In the model of Jansen et al. /31/ for the growth and product kinetics of Klebsiella pneumoniae, the key parameter is ATP. In the first phase of the batch fermentation, ATP is produced by the oxydation of substrate through the TCA cycle. However, in a later phase of the cultivation, the culture is oxygen-transfer-limited. This limits the amount of ATP that can be produced by the electron-transfer system. When the cells require more ATP than can be produced oxidatively, fermentative production of 2, 3 butanediol occurs.

Jöbses et al. /32/ modelled the growth and substrate conversion of Zymomonas mobilis by means of a two-compartment model. The maintenance energy requirements are included into a non-zero rate of the G-compartment turnover.
The substrate consumption rate is given by

$$r_S = \frac{1}{Y_{SK}} \frac{\mu_m(S)\ (X)}{K_S+(S)}$$

and the conversion rate in the k-compartment:

$$r_K = \frac{\mu_m(S)\ (X)}{K_S+(S)} - k_K KG(X) + k_G G(X),$$

where k_K is the reaction rate constant of G-biomass formation,
k_G the reaction rate constant of G-biomass degradation.

The conversion of K- into the G-compartment (biomass formation) is proportional to the biomass K-content and the G-content and the conversion of the G- into the K-compartment (biomass depolymerisation) is proportional to the G-content of the biomass. The net formation of the G-compartment is

$$r_G = Y_{KG} k_K KG(X) - k_G G(X) \qquad (40)$$

and the net formation of total biomass

$$r_X = r_K + r_G, \qquad (41)$$

where K + G = 1.

For steady-state continuous culture, the specific substrate consumption is /32/:

$$q_S = 1/Y_{SK}[(k_K+k_G-k_G/Y_{KG})/k_K Y_{KG}) - k_G + (2k_G+k_K-2k_G/Y_{KG})/ /(k_K Y_{KG})D - (1/Y_{KG}-1)/(k_K Y_{KG})D^2] \qquad (42)$$

In eq. (42), the first term is growth-rate-independent, the second is linearly dependent on μ and/or D, and the third is a nonlinear second-order function of μ.
The model predicts a linear increase in the K-compartment on increasing growth rates according to eq. (43):

$$K = (D+k_G)/k_K Y_{KG} \qquad (43)$$

However, this increase of the K-compartment is quantitatively different from the estimated RNA contents. This may be due to the linear relationships between the K-turnover rate and K and G assumed in eq. (40). It is quite possible that the RNA turnover rate is not proportional to the RNA content, but might be controlled by growth-rate-dependent factors /32/.
The two-compartment model yields a better description of the growth behaviour of Z.mobilis than the unstructured model. From eq. (42) one can see that maintenance energy demand is partly growth-rate-independent, partly linearly growth-rate-dependent, and partly nonlinearly growth-rate-dependent, which agrees with the experiments.
A compartment model was used by Scheiding et al. /35,37/ to describe the growth of Chaetemium cellulolyticum on glucose substrate, since the unstructured model was not able to describe the cell growth satisfactorily. This model has the following components: the constant basic level of protein PROT, which together with the total inner permease concentration (PER_t) yields the overall cell protein concentration ZEPROT, the metabolite M1 which induces the synthesis of the carrier enzyme, PER, from metabolite M2. An increasing substrate supply causes, first of all, the increase of the metabolite concentration M1. The formation of permease is induced by M1. The substrate uptake rate and the specific cell growth rate are increased due to the adaption of the transport system to the new substrate concentration /35-37/. Other models use simplified metabolic pathways of the microorganisms /33,34,38/.
These models are very flexible. E.g. the yeast model of Bellgardt et al. /34/ can describe oxidative and aerobic fermentative growth on glucose, oxidative growth on ethanol, oxygen-limited growth on glucose and/or ethanol and anaerobic growth. To simplify the model, the Blackman's master reaction concept and a quasi steady-state approach for cellular metabolic reactions was applied.

KINETIC MODELS

Simple kinetic models (e.g of Monod) are often used to evaluate the experimental data on cell growth, substrate consumption or product formation. Takamatsu et al. (46) compared these unstructured growth models of microorganisms. However, these models should not be reviewed here. The situation becomes more complex, if the growth of cells in mixed cultures are taken into consideration /40-47/. It is well known that two microbial populations competing for a single resource in a homogeneous environment with time-invariant inputs cannot coexist in a steady-state /45/ unless the growth of the populations is limited by at least as many environmental variables as the existing populations /48/. The influence of spatial inhomogenities on the coexistence of competing microbial populations was investigated by Stephanopoulos and Frederickson /47/, and the influence of wall attachment of one of the populations was considered by Baltziz and Frederickson /44/. Because od the cell's attachment to the walls, the environment is no longer homogeneous, thus, in most cases the two competitors are able to coexist in a stable steady state in a wide range of the operating parameters.

A special case of mixed cultures is the commensalism, in which one population - the commensal - receives some benefit from the presence of the other population - the host - whereas the host population is not affected substantially by the presence of the commensal population /40-42/.

The mixed culture of Lactobacillus plantarum and Propionibacterium shermanii grown anaerobically in glucose, a minimal medium exhibits features typical of a commensal interaction. L. plantarum, the host, grows on glucose and produces lactic acid as a by-product. P. shermanii, the commensal, grows on lactic acid and produces propionic acid and CO_2 as by-products. P. shermanii grown anaerobically is able to utilize either glucose or lactate as a carbon source, however, it preferentially uses acetate to glucose. After inoculation into a medium-containing glucose, they compete for the glucose. However, as soon as L. plantarum has produced a small quantity of lactate, P. shermanii will use lactate. This system was modelled by Lee et al. /41/:

$$\frac{d(X_1)}{dt} = \frac{\mu_1(X_1)(S)}{K_1+(S_1)} - \frac{(X_1)}{\theta} \tag{44a}$$

$$\frac{d(X_2)}{dt} = \frac{\mu_2(X_2)(S_2)}{K_2+(S_2)} - \frac{(X_2)}{\theta} \tag{44b}$$

$$\frac{d(S_1)}{dt} = -a\frac{\mu_1(X_1)(S_1)}{K_1+(S_1)} - (X_1) + \frac{S_{10}-S_1}{\theta} \tag{44c}$$

$$\frac{d(S_2)}{dt} = b\frac{\mu_2(X_1)(S_2)}{K_2+(S_2)} + \alpha\frac{\mu_1(X_1)(S_1)}{K_1+(S_1)} + \beta(X_1) - \frac{S_2}{\theta} \tag{44d}$$

where

X_1	is the cell mass of host (L. plantarum)
X_2	the cell mass of commensal (P. shermanii)
S_1	glucose
(S_{10})	glucose concentration in feed
S_2	lactose
K_1	Monod constant of L. plantarum
K_2	Monod constant of P. shermanii
μ_1	max. specific growth rate of L. plantarum
μ_2	max. specific growth rate of P. shermanii
a	stoichometric coefficient for glucose consumption by L. plantarum
b	stoichometric coefficient for lactate consumption by P. shermanii
α	stochiometric coefficient for growth-associated production of lactate by L. plantarum
ν	specific rate of glucose consumption for maintenance of L. plantarum
β	specific rate of nongrowth-associated production of lactate by L. plantarum
θ	holding time

In chemostat cultures eq. (44) allow three steady states /41/:

(a) total washout steady state:

$$(\bar{X}_2) = (\bar{X}_1) = (\bar{S}_2) = 0;\ (S_1) = (S_{10}) \tag{45a}$$

(b) the steady state of washout of X_2:

$$(\bar{X}_2) = 0;\ (\bar{X}_1),\ (\bar{S}_2) > 0;\ 0 < (S_1) < (S_{10}) \tag{45b}$$

(c) commensal steady state:

$$(\bar{X}_1),\ (\bar{X}_2),\ (\bar{S}_1) > 0;\ 0 < (\bar{S}_1) < (S_{10}) \tag{45c}$$

In this special case, the commensal stable steady state is a node, so that the approach to this steady state should be without oscillations. The experimental results shown in Fig. 18 exhibit no oscillations during the transient approach to the new steady state, in agreement with the model.

UNSTRUCTURED REACTOR MODELS

Unstructured reactor models are the well-stirred batch reactor, well-stirred contrinuous tank reactor, cascade of well-stirred tanks and tubular reactor with axial dispersion and their combinations. Well-mixed reactors can be described with lumped parameter models. Tubular and tubular loop reactors have usually been described by models with distributed parameters. A review on modelling of tubular reactors is given by Luttmann et al. /49/ and Schügerl /50/. These models are not considered here.

STRUCTURED REACTOR MODELS

Unstructured reactor models are often unsuitable for the description of bioreactor performance, especially on a commercial scale.

Several review articles considered different aspects of bioreactor modelling /52-55/. From black-box models to grey-box models with stuctures on different levels in details were recommended. It is well-known that in stirred tank reactors - besides the primary liquid rotation - a secondary liquid circulation pattern prevails (Fig. 19). The consequences of this pattern on the performance of the reactor was considered by several authors /57,63,64,65,66,68,69/.

Also, in tower reactors, a circulation pattern prevails /56-62,67/ which can influence the reactor performance.

Different models were recommended to describe this flow pattern. Joshi developed an axially symmetric multiple circulation cell model by energy balance method (Fig. 20), which can explain the longitudinal dispersion in tower reactors fairly well /57/. According to this model, the circulation rate V_c is given by

$$V_c = 1.31\ [gD_t(W_{SG} - \varepsilon_G W_{SG\infty})]^{1/3} \tag{46}$$

The relationship between the longitudinal dispersion coefficient D_L and V_c is very simple:

$$D_L = 0.33(V_c+W_{SL})D_t \tag{47a}$$

Since usually $V_c \gg W_{SL}$, eq. (47a) can be reduced to (47b)

$$D_L = 0.33\ V_c D_t \tag{47b}$$

In eqs. (46) and (47), D_t is the tower diameter

W_{SL} the superficial liquid velocity

W_{SG} the superficial gas velocity

ε_G the relative gas holdup

Joshi developed a similar model for stirred tank reactors in the turbulent range /57/:

$$D_L = 0.25\ D_t V_c \tag{48}$$

$$V_c = \frac{1.86\ Nd_N^2}{D_t} \text{ for } D_t = H_L, \tag{49}$$

where D_t is the tank diameter
H_L the liquid height in the tank
N the impeller speed
d_N the impeller diameter

Zehner developed a two-dimensional rolling vortex cell model (Fig. 21) which yields the following relationship for the longitudinal dispersion coefficient /60-63/:

$$D_L = \frac{D_t}{2} \sqrt[3]{\frac{1}{\xi} \frac{\Delta\rho}{\rho L} g D_t W_{SG}} \tag{50a}$$

$$\text{or } D_L \alpha D_t^{4/3} W_{SG}^{1/3} \tag{50b}$$

where ξ is the residance factor
ρ_L the liquid density
$\Delta\rho$ the difference between densities of liquid and gas phases
g the acceleration of gravity

This model assumes that the concentrations in the adjacent vortices equalize.
Also, this model can describe the experimental results satisfactorily.
Lübbert /67/ showed by means of a synergetic two-phase model that depending on the Rayleigh numbers of the two-phase system, different flow patterns are possible (Fig. 22), the flow patterns of Joshi and Zehner models also. At high energy dissipation rates the regular pattern becomes chaotic. The experimental investigations mainly considered the variation of dissolved oxygen concentration along the vortex stream lines. A system consisting of two well-stirred tanks - one of them aerated, the other not - and the medium is recirculated through them, was used to physically simulate the variation of oxygen concentration along the vortex lines in the reactor /63,66/.
Sinclair and Brown /65/ developed a two-region mixing model. Both regions are perfectly mixed, but the transfer of material between regions is limited. It is assumed that nutrient solutions and cells, both enter and leave only from the upper region of the culture (Fig. 23). They succeeded to describe the X-D diagram measured by Herbst et al. (Fig. 24). Another model assumes a well-mixed region at the impeller (Fig. 25) or two regions with forced interchange (Fig. 26). Fig. 27 shows the difference between X-D diagrams with perfect mixing and with forced interchange between two regions. Bajpai and Reuss /68/ and Reuss /69/ developed a circulation model to describe the partial segregation of medium components in fermentation broths. The model consists of a micromixing region with a high mixing intensity, which prevails at and around the impeller, and a macromixing region, far away from the impeller, with low mixing intensity. The fluid elements follow a circulation pattern and move from the micromixer to the macromixer region and back again to the one of the micromixer (Fig. 28). If the fluid elements of the broth are saturated by oxygen in the micromixer, the oxygen concentration decreases during their passage through the macromixer in the discrete elements of the model (Fig. 29). The variation of the local concentration depends on the circulation time distribution of the fluid elements in the macromixer region (Fig. 30). This model was used to describe the oxygen supply of the cells in highly viscous mycelial broth of *Aspergillus niger*. A similar model was used to explain the oxidative and fermentative growth of *S. cerevisiae* in continuous culture as a function of glucose concentration variation in the broth. The agreement between model calculations and measurements is satisfactory (Fig. 31).
The yield coefficient of cell mass with regard to the substrate strongly varies with the circulation rate in tower loop reactors /71/. The variation of the respiratory quotient of *Methylophilus methylotrophus* was found by Field and Slater in an airlift loop reactor with the air flow rate and liquid circulation rate /72/. They explained this with the local segregation of methanol, which was used as a substrate.
Páca reported on the effect of multistream ethanol feeding on growth and physiological characteristics of *Candida utilis* in a multistage tower fermentor /73/. However, no models have been developed yet to simulate these phenomena.

CONCLUSIONS

Several mathematical models have been developed at different levels of the cells and their environement. Their improvement depends on more and better experimental data.
Very few papers have been published on the interconnection of these models. In this field, new measurements are needed to determine the typical time constants of the particular processes and the dynamic behaviour of the interconnected systems.

LITERATURE

1. J.E. Bailey, D.F. Ollis. Biochemical Engineering Fundamentals, McGraw Hill Co. 1977.
2. J. Monod, M. Cohn. Adv. Enzymol. 13 (1952) 67.
3. G.N. Cohen, J. Monod. J. Bacteriol. Rev. 2 (1957) 169.
4. F. Jacob, J. Monod. J. Mol. Biol. 3 (1961) 318.
5. A. Harder, J.A. Roels. Adv. in Biochem. Eng. 21 (1982) 55-107.
6. G. van Dedem, M. Moo-Young. Biotechnol. Bioeng. 15 (1973) 419-439.
7. G. van Dedem, M. Moo-Young. Biotechnol. Bioeng. 17 (1975) 1301-1312.
8. K. Toda. Biotechnol. Bioeng. 18 (1976) 1117-1124.
9. T. Imanaka, S. Aiba. Biotechnol. Bioeng. 19 (1977) 757-764.
10. S. Gondo, K. Venkatasubramarian, W.R. Vieth, A. Constantonides. Biotechnol. Bioeng. 20 (1978) 1797-1815.
11. A.R. Moreira, G. van Dedem, M. Moo-Young. Biotechnol. Bioeng. Symp. No. 9 (1979) 179-203.
12. K. Toda, T. Takeuchi, H. Sano. J. Chem. Tech. Biotechnol. 29 (1979) 747-755.
13. W.R. Vieth, K. Kaushik, K. Venkatasubramarian. Biotechnol. Bioeng. 24 (1982) 1455-1460.
14. G.Z. Kaczorowski, D.E. Robertson, M.L. Garcia, E. Padau, L. Patel, G. Le Blanc and H.R. Kaback. Ann. N.Y. Acad. Sci. 358 (1980) 307.
15. K.J. Kaczorowski, D.E. Robertson, H.R. Kaback. Biochemistry 18 (1979) 3697.
16. K.R. Kaushik. Ph. D. Thesis, Rutgers University, New Brunswick, N.J. 1981.

17. D.F. Ollis, Hung-Teh Chang. Biotechnol. Bioeng. 24 (1982) 2583-2586.

18. T. Imanaka, S. Aiba. Biomedical Eng. II in: A. Constantinides, W.R. Vieth and K. Venkatasubramarian, Eds. Ann. N.Y. Acad. Sci. 369 (1981) 1.

19. D.F. Ollis. Phil. Trans. R. Soc. Lond. B 297 (1982) 617-629.

20. Sun Bok Lee, J.E. Bailey. Biotechnol. Bioeng. 26 (1984) 66-73.

21. M.A. Hjortso, J. E. Bailey. Biotechnol. Bioeng. 26 (1984) 814-819.

22. Sun Bok Lee, J.E. Bailey. Biotechnol. Bioeng. 26 (1984) 1372-1382.

23. Sun Bok Lee, J.E. Bailey. Biotechnol. Bioeng. 26 (1984) 1383-1389.

24. Jin-Ho Seo, J.E. Bailey. Biotechnol. Bioeng. 27 (1985) 156-165.

25. D.D.Y. Ryu, R. Siegel, Sun Bok Lee. Bio '85 Japan, Osaka, 5-8 Nov. 1985.

26. D.F. Ryder, D. Di Biasio, Biotechnol. Bioeng. 26 (1984) 942-947.

27. K. Dennis, F. Srienc, J.E. Bailey. Biotechnol. Bioeng. (1985) 1490-1494.

28. B. Engberg, K. Nordström. J. Bacteriol. 123 (1975) 179.

29. J.M. Eakman, A.G. Fredrickson, H.M. Tsuchiya. Chem. Eng. Symp. Ser. 62 (1966) 37.

30. A.G. Frederickson, D. Ramkrishna, H.M. Tsuchiya. Math. Biosci. 1 (1967) 327.

31. N.B. Jansen, M.C. Flickinger, G.T. Tsao. Adv. of Biotechnol. Vol. I (1980) Ed. M. Moo-Young, Pergamon Press 309-313.

32. I.M.L. Jöbses, G.T.C. Egberts, A. van Baaken, J.A. Roels. Biotechnol. Bioeng. 27 (1985) 984-995.

33. M.M. Ataai, M.L. Shuler. Biotechnol. Bioeng. 27 (1985) 1027-1035.

34. K.-H. Bellgardt, W. Kuhlmann, H.D. Meyer. In "Modelling and Control of Biotechnological Processes", IFAC-Workshop Helsinki 1982, Pergamon Press, Ed. A. Halme 67-74.

35. W. Scheiding, V. Hecht. In "Modelling and Control of Biotechnological Processes", IFAC-Workshop Helsinki 1982, Pergamon Press, Ed. A. Halme 75-82.

36. W. Scheiding, M. Thoma, V. Hecht, W. Rosen, K. Schügerl. J. of Biotechnology 2 (1985) 207-224.

37. W. Scheiding, M. Thoma, V. Hecht, W. Rosen, K. Schügerl. Appl. Microbiol. Biotechnol. 21 (1985) 299-308.

38. G. Auling, K.-H. Bellgardt, H. Diekmann, M. Thoma. Appl. Microbiol. Biotechnol. 19 (1984) 353-357.

39. T. Takamatsu, S. Shioya, H. Chikatana, K. Dairaku. Chem. Eng. Sci. 40 (1985) 499-507.

40. Chih-Cheng Chao, R.J. Reilly. Biotechnol. Bioeng. 14 (1972) 75-92.

41. Inn Hee Lee, A.G. Frederickson, H.M. Tsuchiya. Biotechnol. Bioeng. 18 (1976) 513-526.

42. G. Stephanopoulos. Biotechnol. Bioeng. 23 (1981) 2243-2255.

43. T. Takamatsu, S. Shioya, H. Kurome. IFAC-Workshop on Modelling and Control of Biotechnol. Processes, Helsinki 1982, Ed. A. Halme, Pergamon Press, 103-109.

44. B.C. Baltziz, A.G. Frederickson. Biotechnol. Bioeng. 25 (1983) 2419-2439.

45. R. Aris, A.E. Humphrey. Biotechnol. Bioeng. 19 (1977) 1375.

46. T. Takamatsu, S. Shioya, K. Okuda. J. of Ferment. Technol. 59 (1981) 131-136.

47. G. Stephanopoulos, A.G. Frederickson. Biotechnol. Bioeng. 21 (1979) 1499-1505.

48. P.A. Taylor, P.J.L.B. Williams. Can. J. Microbiol. 21 (1975) 90.

49. R. Luttmann, A. Munack, M. Thoma. Adv. in Biochem. Eng. Vol. 32 (1985) Ed. A. Fiechter, Springer Verlag 95-205.

50. K. Schügerl. Adv. in Biochem. Eng. Vol. 22 (1982) Ed. A. Fiechter, Springer Verlag 93-224.

51. R.K. Bajpai, M. Reuss, W. Held. Biotechnol. Bioeng. 23 (1981) 499-521.

52. N.W.F. Kossen. In "Microbial Technology", Society for General Microbiology Symposium 29, Bull, Ellwood, Ratledge, Eds. (1979) 327-357.

53. J.A. Roels. In "Computer Applications in Ferment. Technology", Manchester 1981; London: Society of Chemical Industry 37-45.

54. N.W.F. Kossen. In "Computer Applications in Ferment. Technology", Manchester 1981; London: Society of Chem. Industry 23-30.

55. J.A. Roels. J. Chem. Tech. Biotechnol. 32 (1982) 59-72.

56. J.B. Joshi, M.M. Sharma. Trans. I. Chem. E. 57 (1979) 244-251.

57. J.B. Joshi. Trans. I. Chem. E. 58 (1980) 155-165.

58. M.H.I. Baird, R.G. Rice. Trans. I. Chem. E. 59 (1981) 138-139.

59. J.B. Joshi. Trans. I. Chem. E. 59 (1981) 139-143.

60. P. Zehner. Verfahrenstechnik 16 (1982) 347-351.

61. P. Zehner. Verfahrenstechnik 16 (1982) 514.

62. P. Zehner. Chem. Ing. Tech. 54 (1982) 248-251.

63. N.M.G. Oosterhuis. Scale-up of Bioreactors. A scale down approach.

64. N.M.G. Oosterhuis, N.W.F. Kossen. Biotechnol. Bioeng. 26 (1984) 546-550.

65. G.C. Sinclair, D.E. Brown. Biotechnol. Bioeng. 12 (1970) 1001-1017.

66. R.D. Tanner, I.J. Dunn, J.R. Bourne, M.K. Klu. Chem. Eng. Sci. 40 (1985) 1213-1219.

67. A. Lübbert. Paper at Achema 1985.

68. R.K. Bajpai, M. Reuss. Can. J. of Chem. Eng. 60 (1982) 384-392.

69. M. Reuss. IFAC-Workshop on Modelling and Control of Biotechnical Processes. Ed. A. Halme, Pergamon Press 1982, 43-55.

70. H. Brauer. Adv. in Biochem. Eng. Vol. 13 (1979). Ed. T.K. Ghose, A. Fiechter, N. Blakebrough, Springer-Verlag, 87-119.

71. A. Gebauer. Dissertation, Hannover 1985.

72. P.R. Fields, K.H. Slater. Biotechnol. Bioeng. 26 (1984) 719-726.

73. J. Páca. Enzyme Microbiol. Technol. 3 (1981) 123-128.

Fig. 1 (Bailey, Ollis 1977)

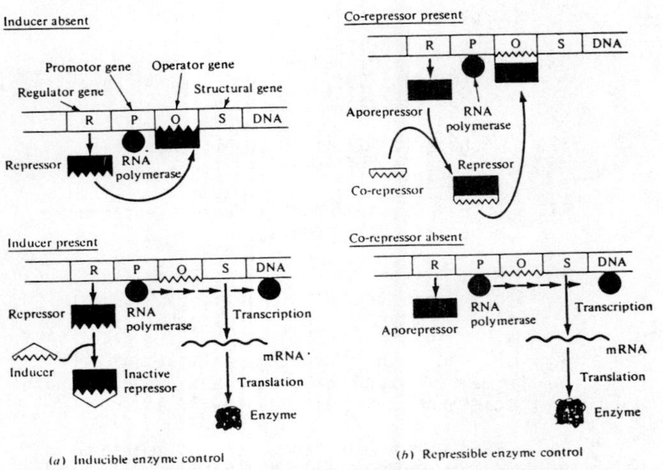

Figure 1. (*a*) An inducer acts by inactivating repressor so that the repressor does not bind to the operator gene and block transcription of the structural gene. (*b*) A (co-) repressor combines with an aporepressor to form an active repressor, which can block expression of the structural gene by binding at the repressor gene. In this manner synthesis of repressible enzymes is controlled. [*Reprinted from A. L. Demain, Theoretical and Applied Aspects of Enzyme Regulation and Biosynthesis in Microbial Cells, in L. B. Wingard, Jr. (ed.), "Enzyme Engineering," Interscience, New York,* 1972.]

Fig. 2 (Van Dedem, Moo-Young 1973)

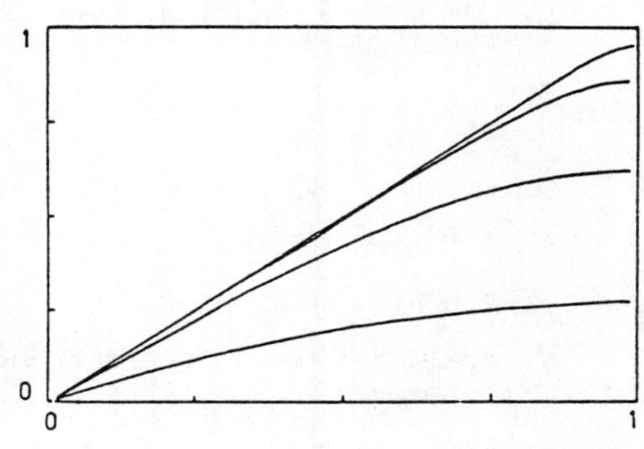

Fig. 2: Enzyme concentration as a function of the cell concentration for induction. The following parameters were used in the calculations: x_0 = 0.0003; e_0 = 0.001; k_1' (from top to bottom) = 10^4, 10^3, 10^2, 10; k_4' = 10.

Fig. 3. Enzyme concentration as a function of the cell concentration for repression. Same initial values of e and x as in Fig. 1. k_1' (from top to bottom) = 0.1, 1.0, 10.0, 100.0; k_4' = 10.

Fig. 3 (Van Dedem, Moo-Young 1973)

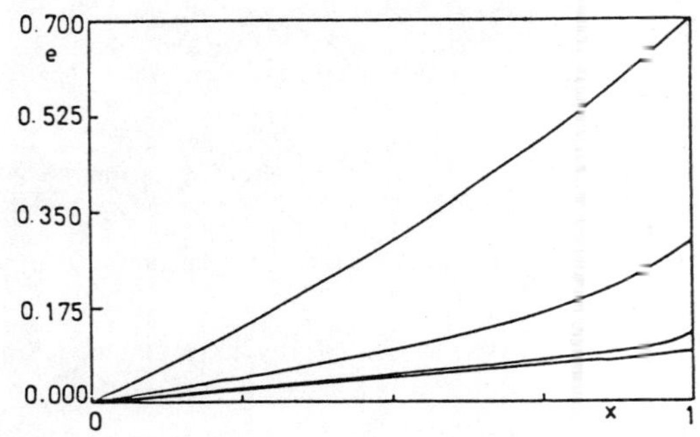

Fig. 4 (Van Dedem, Moo-Young 1975)

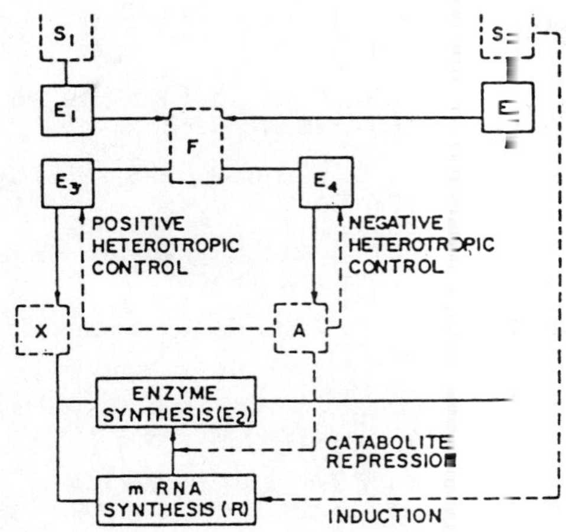

Fig. 4. Schematic representation of the model.

Table II (Lee, Bailey 1984)

Table II. Model equations for λ *dv* plasmid replication.

1) Repressor synthesis

$$\frac{d[\mathrm{mRNA}]_R}{dt} = k_m^0\eta[G] - k_{-m}[\mathrm{mRNA}]_R - \mu[\mathrm{mRNA}]_R \qquad (1)$$

$$\frac{d[R]_0}{dt} = k_p^0[\mathrm{mRNA}]_R - k_{-p}^R[R]_0 - \mu[R]_0 \qquad (2)$$

2) Initiator synthesis

$$\frac{d[\mathrm{mRNA}]_I}{dt} = k_m^0\eta(1-f)[G] - k_{-m}[\mathrm{mRNA}]_I - \mu[\mathrm{mRNA}]_I \qquad (3)$$

$$\frac{d[I]_0}{dt} = k_p^0[m\mathrm{RNA}]_I - k_{-p}^I[I]_0 - \mu[I]_0 \qquad (4)$$

3) Origin activation

$$\frac{d\theta}{dt} = k_m^0\eta(1-f)(1-\theta) - \mu\theta \qquad (5)$$

4) Replication complex formation

$$[\mathrm{REP}] = \left(\frac{\theta}{VN_A}\right)\left(\frac{K_0[\mathrm{I}]_0}{1+K_0[\mathrm{I}]_0}\right) \qquad (6)$$

5) Cell volume

$$\frac{dV}{dt} = \mu V \qquad (7)$$

Fig. 5 (Van Dedem, Moo-Young 1975)

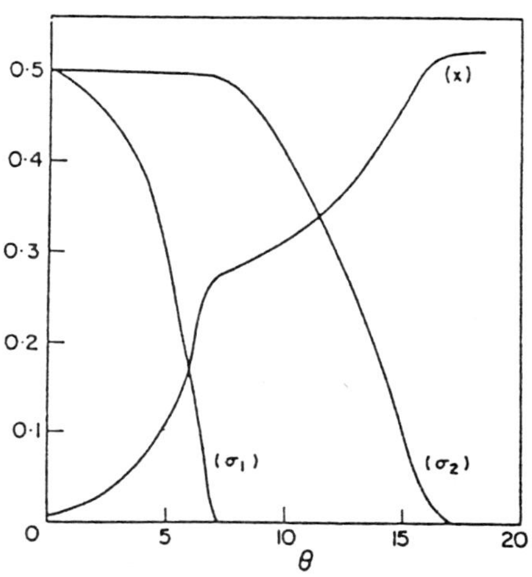

Batch growth curve for two carbon and energy sources.

Initial values

are: $x = 0.001; \sigma_1 = 0.5; \sigma_2 = 0.5; \rho = 0.001; \eta = 0.1.$

Fig. 6 (Toda 1976)

Schematic representation of the dual control mechanism.

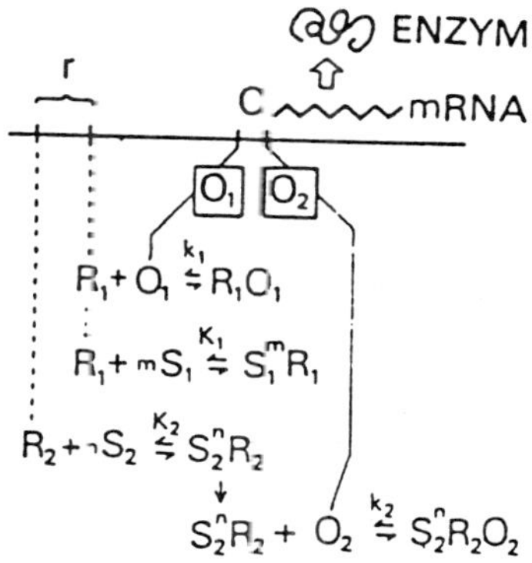

Fig. 7 (Toda 1976)

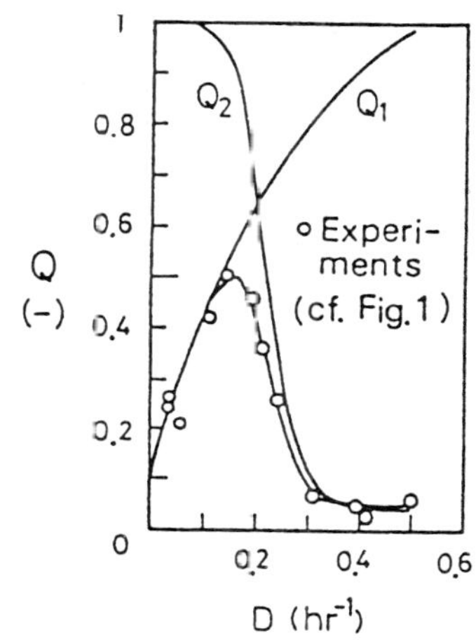

Fig 7. The extend of induction (Q_1) and repression (Q_2) estimated hypothetically from experimental results.

Fig. 8a (Toda et al. 1979)

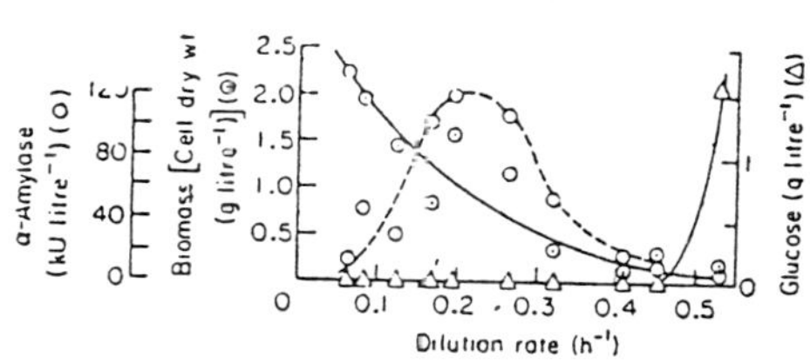

Figure 8a Effect of dilution rate on α-amylase production by *Bacillus subtilis* in a glucose-limited chemostat culture.

Fig. 8b (Toda et al. 1979)

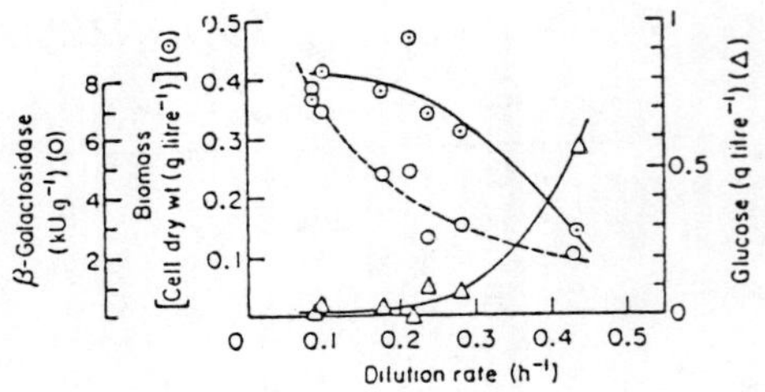

Fig. 8b Effect of dilution rate on β-galactosidase production by *Saccharomyces fragilis* in a galactose-limited chemostat culture.

Fig. 8c (Toda et al. 1979)

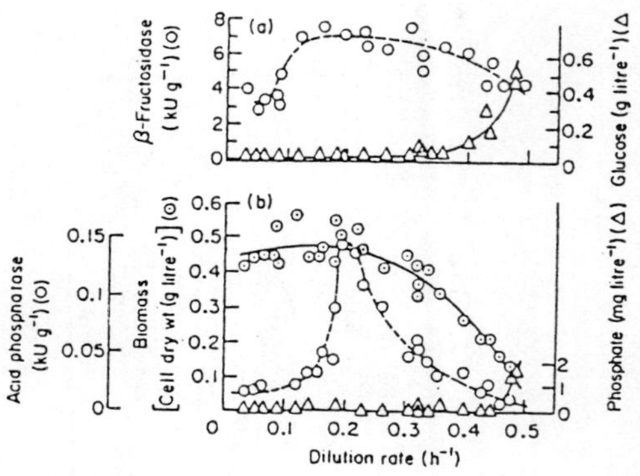

Fig. 8c Effect of dilution rate on production of acid phosphatase and β-fructosidase by *Saccharomyces carlsbergensis* in a phosphate-limited chemostat culture.

Fig. 8d (Toda et al. 1979)

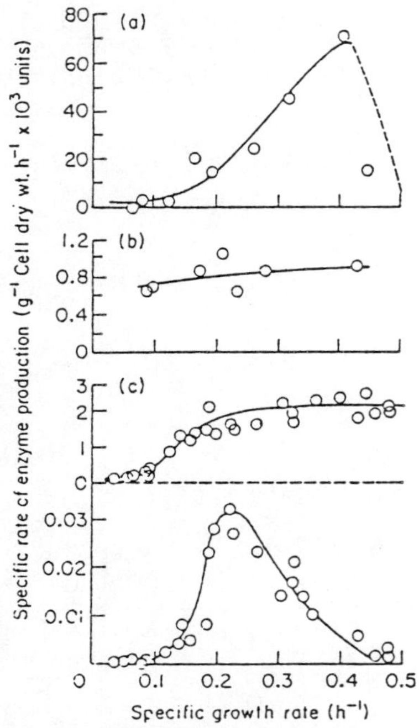

Figure 8d Relationship between specific rate of enzyme production and specific rate of growth. (a) α-Amylase production by *Bacillus subtilis*, (b) β-galactosidase production by *Saccharomyces fragilis*, (c)β-fructosidase (upper part) and acid phosphatase (lower part) production by *Saccharomyces carlsbergensis*. Solid lines are calculated according to equation (10) using parametric constants shown in Table 2.

Fig. 9 (Imanaka, Aiba 1977)

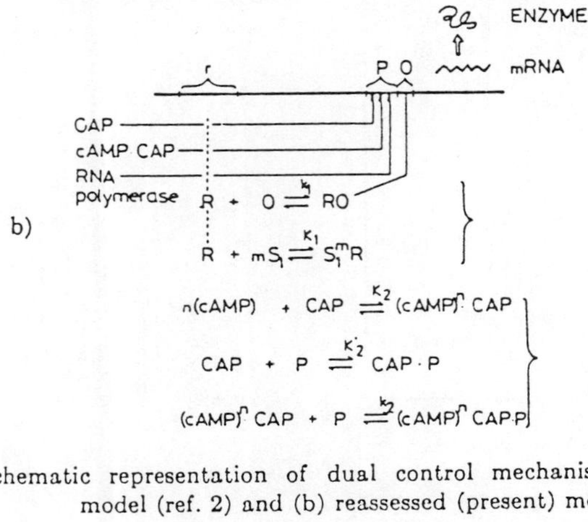

Fig 9 Schematic representation of dual control mechanism, (a) previous model (ref. 2) and (b) reassessed (present) model.

Fig. 10 (Gondo et al. 1978)

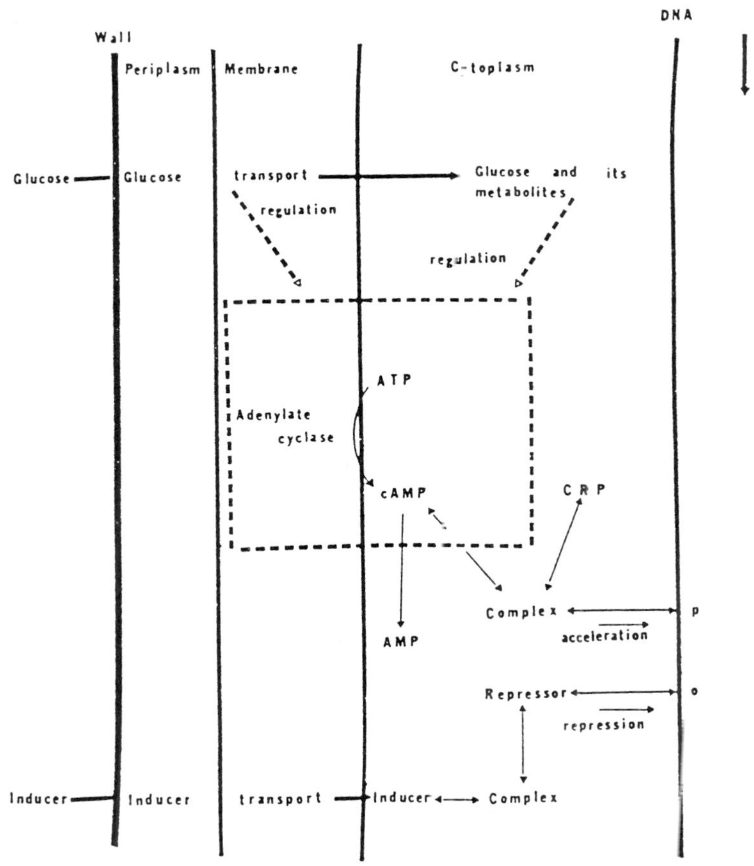

Fig. 10 Schematic diagram of induction and catabolite repression.

Fig. 11 (Harder, Roels 1982)

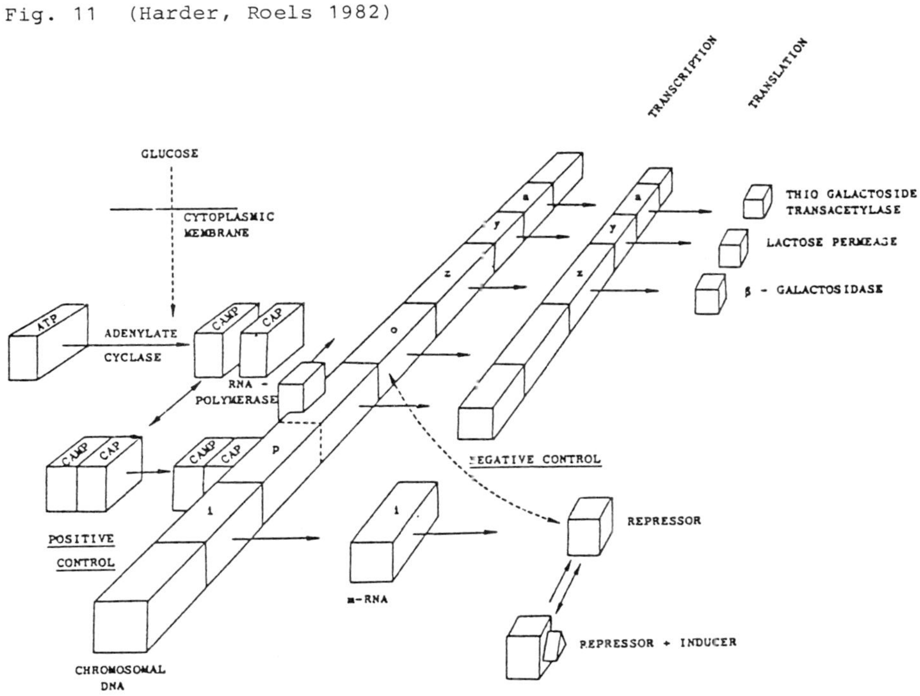

Fig. 11 Schematic representation of the lac operon; its negative and positive control units

Fig. 12 (Gondo et al. 1978)

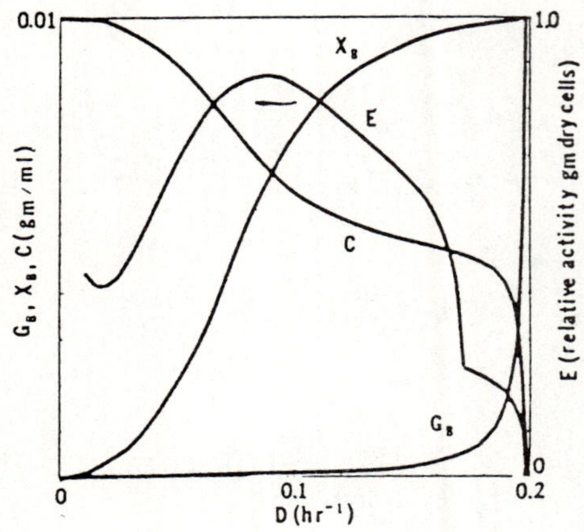

Fig. 14 (Lee, Bailey 1984)

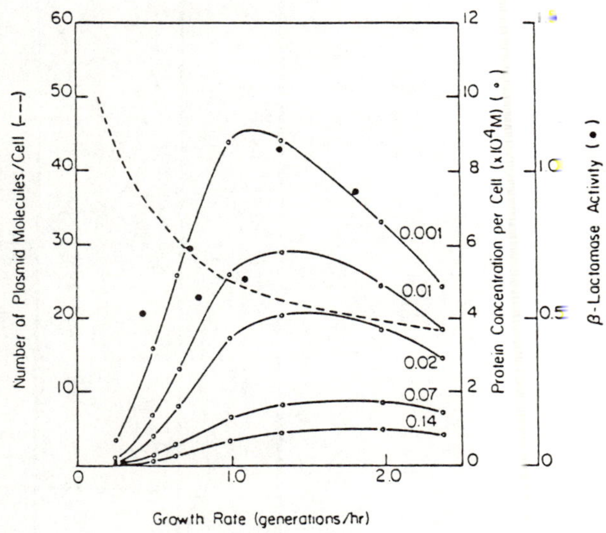

Fig.12 Effects of dilution rate on steady-state performances of continuous fermentation. Parameter values are same as those in the Appendix except for G_{BC} = 0.005 and $\alpha_{G_{BC}}$ = 0.05.

Figure 14 The effect of growth rate on productivity of cloned-gene protein. Assumed decay constants of product protein are given near each curve (units are min^{-1}). The experimental data (●) were taken from Engberg and Nordström (ref. 23). The dotted line shows plasmid copy number (curve a from Fig. 2).

Fig. 13 (Lee, Bailey 1984)

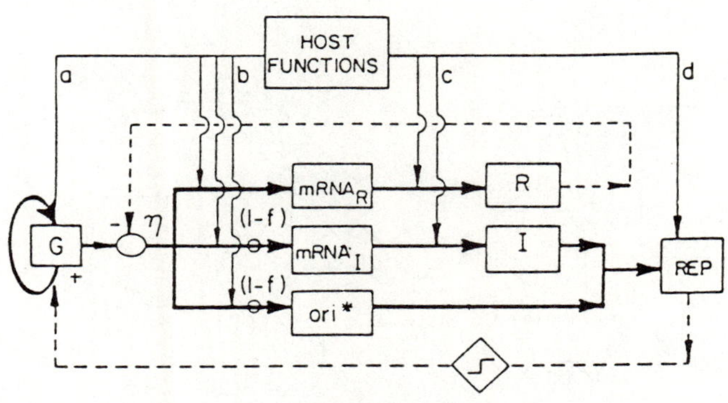

Figure 13 Schematic representation of the λdv plasmid replicon. The thick lines represent the flow of information encoded in the plasmid while thin lines represent the activities provided by host functions. The dotted line shows the regulatory action of repressor protein and initiator protein. Traces (a), (b), and (c) represent the replication, transcription, and translation processes, respectively. Line (d) indicates the involvement of replication protein of the host cell in formation of replication initiation complexes. Details of molecular mechanisms for the λdv plasmid replicon have been described in detail elsewhere (see refs. 5 and 7).

Fig. 15 (Lee, Bailey 1984)

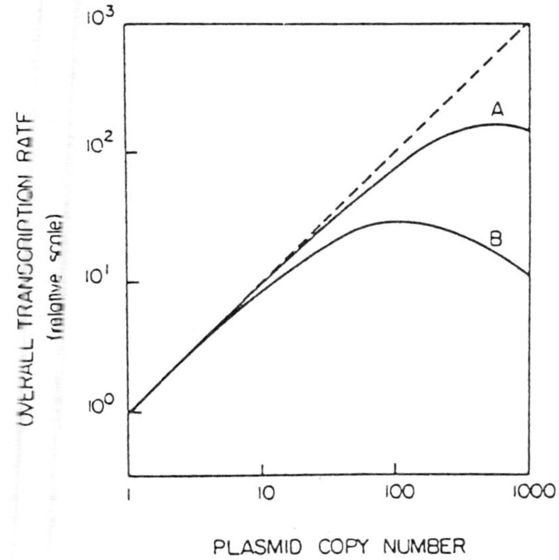

Fig. 16 (Seo, Bailey 1985)

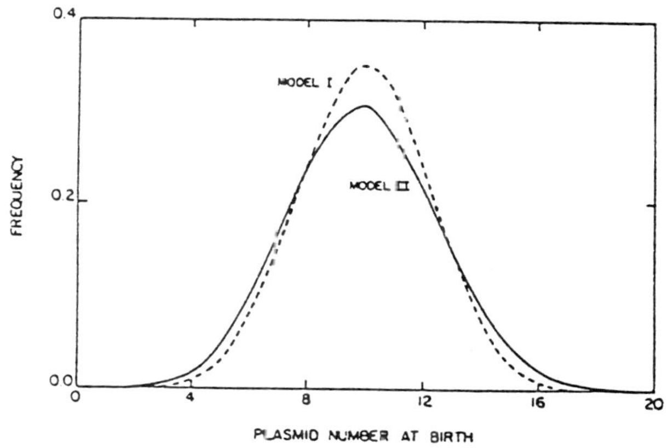

Figure 15 Overall transcription rate initiated at the cloned *lac* promoter-operator in the fully induced state as a function of plasmid copy number $\bar{N}$. Ordinate is normalized by the transcription rate for a single plasmid. Plasmid genotype is P^+O^+: (A) 5 promoters, 0.1% genome size; and (B) 20 promoters, 1% genome size.

Figure 16 Plasmid number distribution at birth for Model I (dashed curve, $N = 20$) and Model III (solid curve, $K = 10$, $N_M = 75$) for $\gamma = 0.5$ and $\nu = 1\ h^{-1}$.

Fig. 17 b (Ryder, DiBiasio 1983)

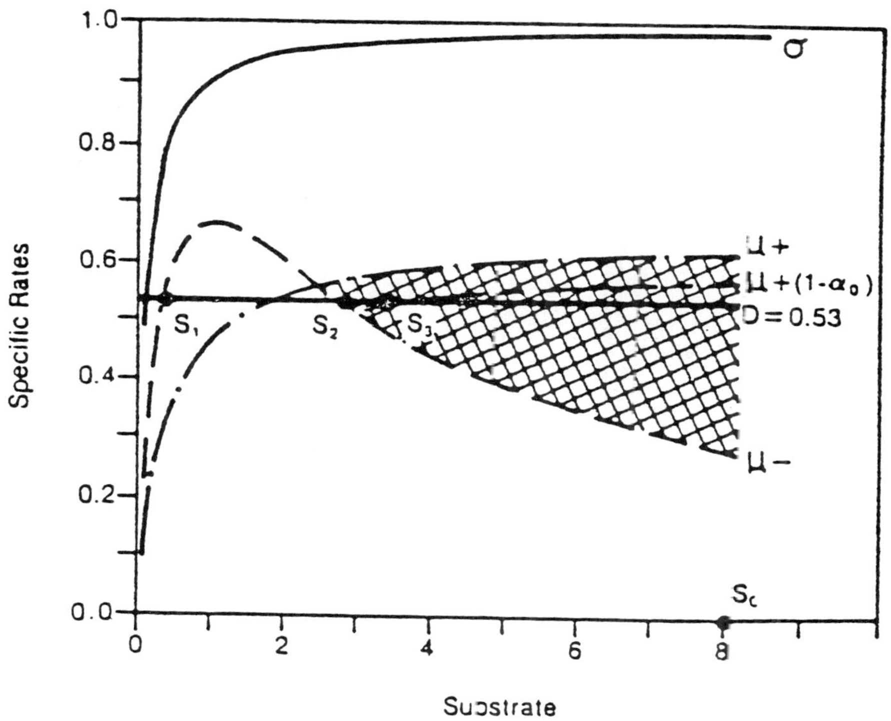

Fig. 17 b. Specific rates versus substrate concentration

Fig. 19 (Brauer 1979)

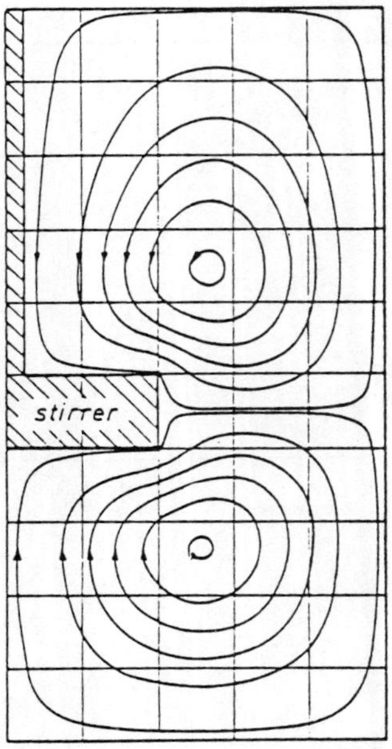

Fig. 19. Streamlines of secondary fluid movement

Fig. 21 (Zehner 1982)

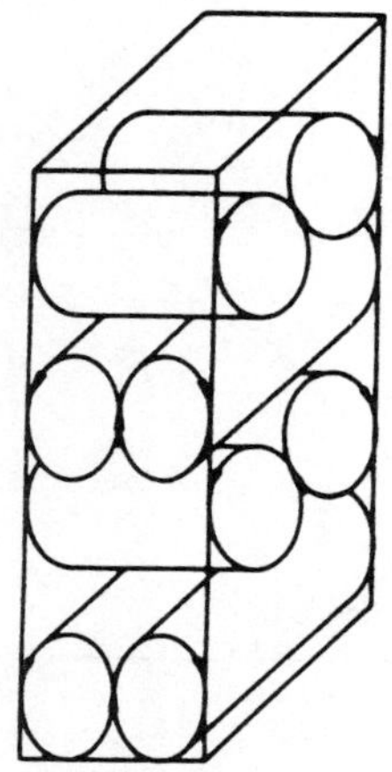

Fig. 21. Roller vortex model

Fig. 20 (Joshi 1980)

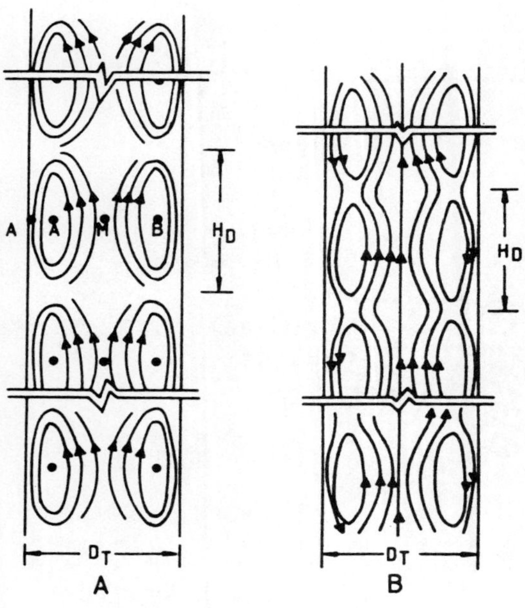

Fig. 20. Multiple vortex in tower reactors
a) without interaction of vortices
b) with interaction of vortices

Fig. 22 (Lübbert 1985)

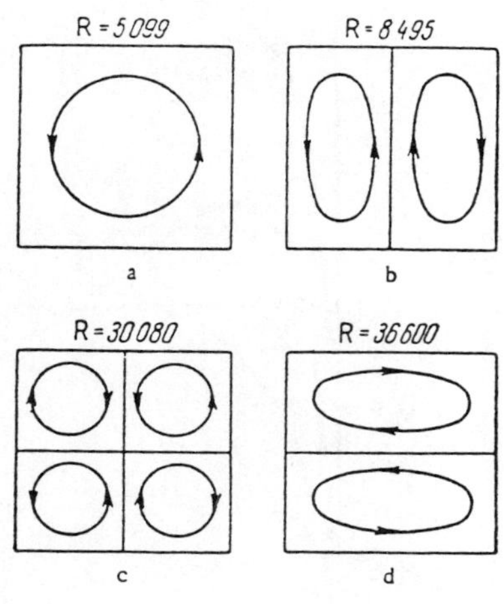

Fig. 22 Synergetic flow patterns at different Rayleigh-numbers

Fig. 22b (Lübbert 1985)

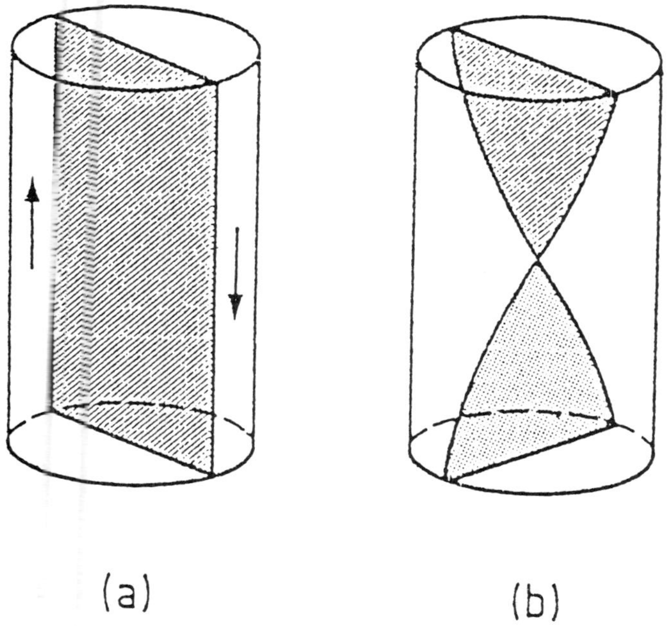

Fig. 23b (Sinclair, Brown 1970)

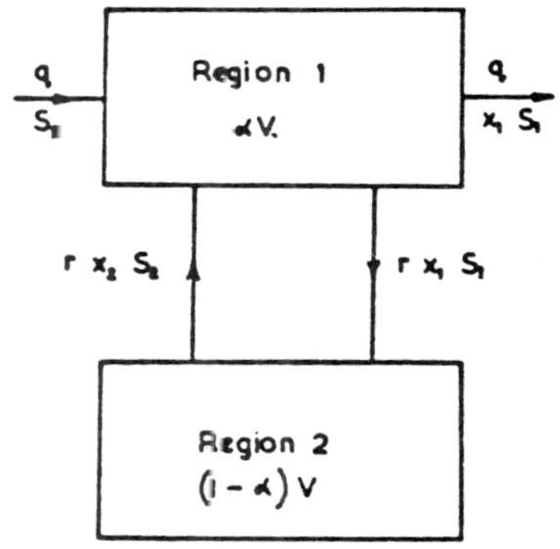

Fig. 23b. Two regions mixing model

Fig. 23a (Sinclair, Brown 1970)

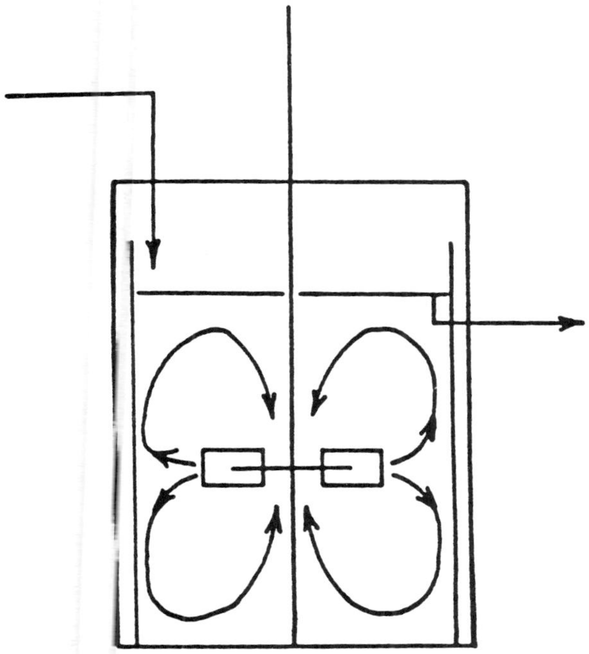

Fig 23a. Liquid flow patterns

Fig. 24 (Sinclair, Brown 1970)

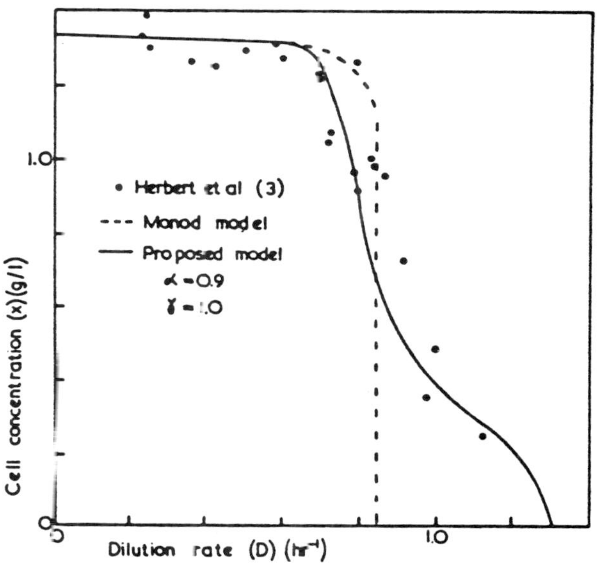

Fig. 24. Proposed model for data of Herbert et al.

Fig. 25 (Sinclair, Brown 1970)

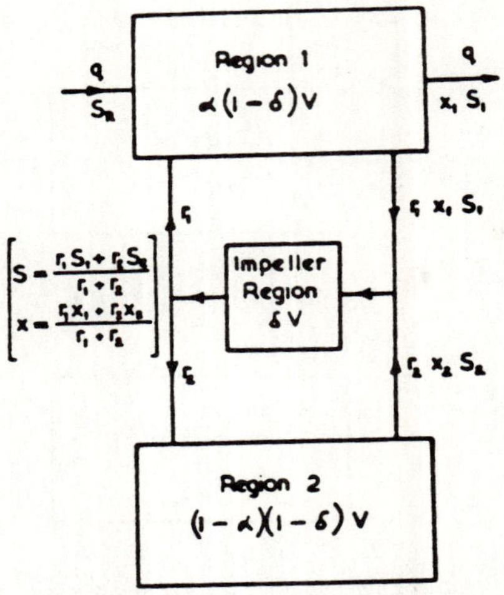

Fig. 25 Model with completely mixed region at impeller.

Fig. 26 (Sinclair, Brown 1970)

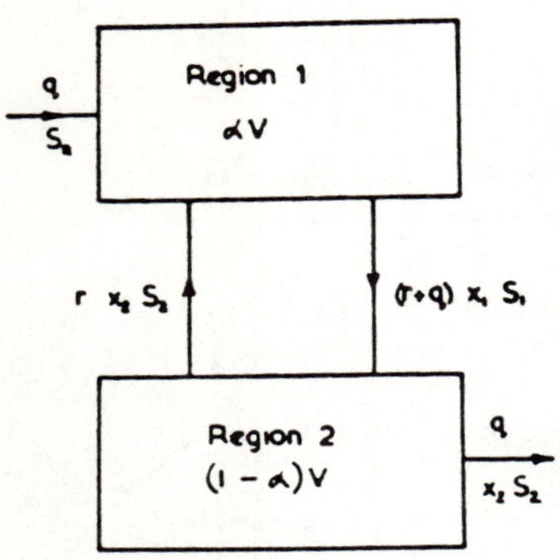

Fig. 26. Two regions mixing model with forced interchange

Fig. 27 (Sinclair, Brown 1970)

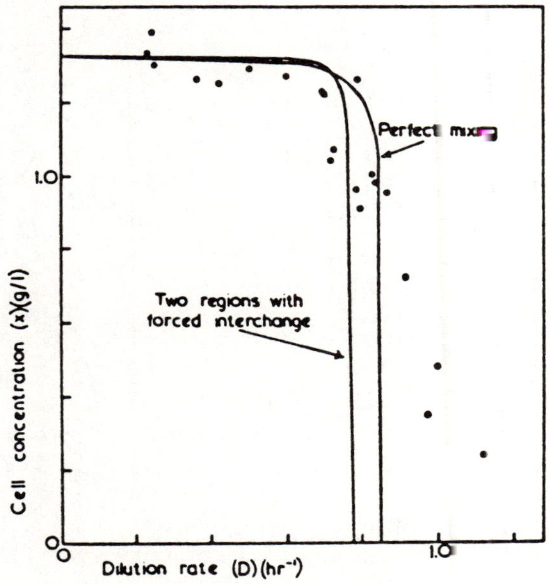

Fig. 27. Effect of forced interchange

Fig. 28 (Bajpai, Reuss 1982)

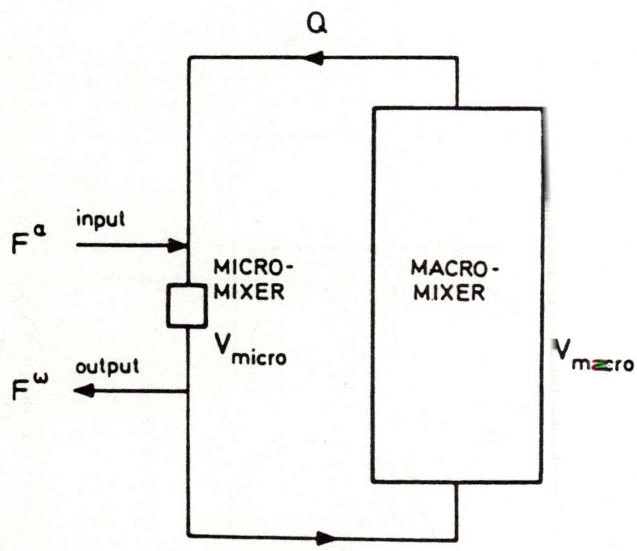

Fig. 28. A two environment model for mixing in a stirred bioreactor

Fig. 30 (Bajpai, Reuss 1982)

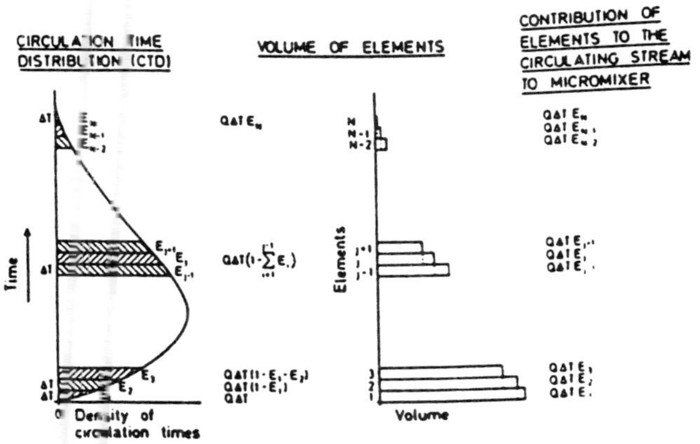

Figure 30. CTD and flow through the macromixer.

Fig. 29 (Bajpai, Reuss 1982)

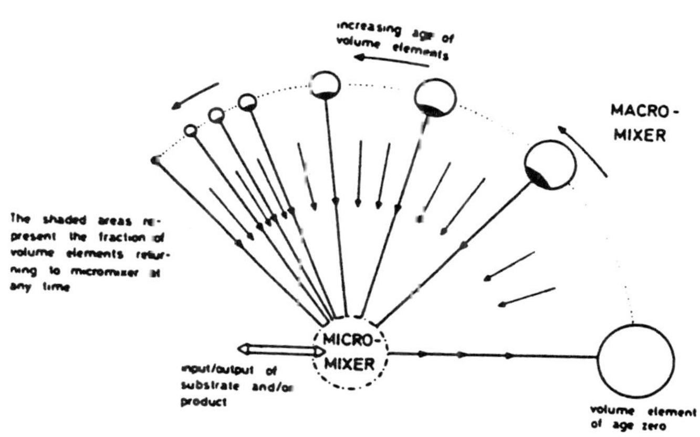

Figure 29. A schematic representation of the discrete simulation procedure.

Fig. 31 (Bajpai, Reuss 1982)

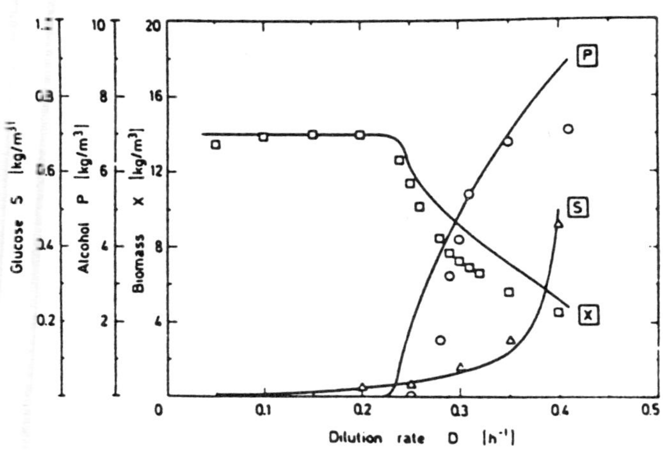

Figure 31. A comparison of predictions of the empirical model with the continuous culture data of van Meyenberg[100]. (Points represent the experimental data and solid lines the model predictions using equations (22–27) and parameters as in Table 1. Inlet stream glucose concentration, S^0 = 28 kg/m^3.

THE INDUSTRIAL APPLICATION OF MODERN MEASUREMENT AND ESTIMATION TECHNIQUES IN BIOTECHNOLOGY

A. Halme

Helsinki University of Technology, Laboratory of Automation Technology, Helsinki, Finland

INTRODUCTION

It is a well known fact that sensors available today for bioreactors and other parts of biotechnical processes do not cover all necessary measurements. In spite of recent relatively rapid progress in developing optical, silicon and so-called biosensors most of the concentration variables especially in the fermentation liquid phase cannot be still measured reliably on-line. When looking at the industrial practice the situation is almost the same as three years ago when this meeting was held. The only clear steps that have been taken are in developing commercially available sensors for glycose and ethanol. Most of the new biologically oriented sensors are still in the laboratory development phase and their commercialization seems to take a surprisingly long time. The difficulties seem to be of two kinds.

Firstly, sensors like redox or fluorescence for NAD/NADH, which are technically relatively well working, are difficult to relate in a simple way to a specific variable describing the state of cell culture, e.g. biomass (Armiger and Zabriskie, 1984). Secondly, sensors like those with enzyme or microbial coupling mechanisms, although giving in principle specific information which is close to the process itself, are technically difficult to handle because they are usually difficult to sterilize, have a short life-time and are sensitive to environmental disturbances. Those who want to get an updated view on current sensor technology could read the thorough review on the subject made recently by Wang and Stephanopoulos (1984) in Critical Reviews TM in Biotechnology.

The situation in biotechnology is thus still today much more difficult than in chemical engineering generally. This means that in industrial practice the processes have to be controlled mainly by relying in the on-line data obtained from a few conventional types of sensors and the always more or less delayed information obtained from laboratory samples.

On the other hand, due to the continuous decrease of the relative price of data handling and computing power it is in practice today always available for on-line use in biotechnical plants. Also the development in digital instrumentation makes the sensor and other on-line data easily available for the control and computing systems. This fact together with what was said before makes attractive such advanced estimation and other computing methods which can further refine the raw measurement data. It is the basic purpose of this presentation to consider how and what kind of methods can be really useful in industrial use.

However, before going so far it is a great temptation to consider somewhat the present development in measurement technology and what it is probably bringing to this field.

PROGRESS IN MEASUREMENT TECHNOLOGY

At the moment there are three interesting technologies which have great influence on the development of on-line measurements. The first one is optics and the use of optical fibers. The second one is "sensor on chip" technology and the third one biosensors, which are of special interest in biotechnology. The first two are technologies which probably have generally the greatest influence in industrial measurement and sensor technology in the near future. Biosensors are specialities in biotechnology which have an interesting but uncertain future and in any case they are developing slower than the two other technologies.

Photonic systems and fibre optic sensors have at the moment the widest application area in mechanical or elctromechanical systems but they also have many possibilities in chemical processes (see e.g. Jones, 1984, Knight, 1984). They offer in principle following advantages

- noncontact measurement
- easy transmitting of the primary measurement signal (light) to remote sensing devices
- small size, light construction and fast time constant
- very small power levels
- construction of sensing heads which are corrosion resistant, functionable in high temperature and easy to install.

In biotechnology, optical measurement technology and especially the use of fibers may result in the development of more reliable and easily sterilized small size sensors for many of the conventional variables like temperature, pressure and level. However, optical measurement principles which utilize absorption, fluorescence, reflection, colour, turbidity, luminescence etc. to detect different substances will probably have more applications because optical fibers make it possible to locate the main optical instrument far from the process area letting only a simple optical part (a lens or similar) come into contact with the process itself. This is a great advantage because it means that several of the optical techniques now used in laboratories can be essentially linked directly to the process without sample taking and transfer. Spectrophotometers and other similar complex analytical instruments can be used more easily for on-line measurements because they need not be installed for on-site operation but can be kept in a laboratory environment. By using several fiber connections and multiplexing techniques in different locations in the process, optical sensors can also measure large samples like the spatial distribution of substances in a fermentor. It may be predicted that several optical sensors that are available today e.g. refractometers, turbidometers, fluorescence and luminescence measurements are going to be modified in the near future so that optical fibers are used to transfer the light signal from the measurement site to the central unit in a more convenient location. This could help considerably the installation and asepticity problems in fermentors. It does not, however, solve the problem of relating the measurement result to the biochemical variable e.g. cell mass.

In addition to using fibers as measurement signal pathways they can also be applied as sensors. The principle is that the measured variable affects the light transmission characteristics of the fiber itself or some special material incorporated at the other end of it (for the principles see e.g. De Groot, 1983). In this way many conventional variables like temperature, pressure, distance, displacement etc. can be measured by commercially available sensors. For example, temperature measurements done in this way are interesting in biotechnology because a quick and almost pointwise measurement can be obtained.

Another interesting development in the modern sensor technique is going on in silicon sensors, i.e. sensors which are integrated on chips. This technology allows one to make a sensor in a very compact form and gives better possibilities to construct cheap, sensitive and long lasting sensors. From the biotechnical applications point of view there are some interesting developments. At the moment relatively cheap oxygen and carbon dioxide sensors have been already developed and a lot of effort is being devoted to ion selective sensors which are made by using ion selective FETs (Field Effect Transistors). Such FETs absorb ions out of the solution through an ion selective coating which changes the induced charge and hence the current through the transistor (Sansen, 1983). The dissolved oxygen sensors developed on this basis, although using the well-known polarographic or clark cell principle, give more flexibility in their use because the whole construction including necessary temperature compensation circuits can be installed in a very small size and the entire sensor is changed instead of the membrane. In the ion selective sensors this would be even more important because their use and maintenance in fermentation processes have been very difficult so far.

The enzyme and microbe coupled sensors are still in their early development phase. In spite of partial commercial success in e.g. the glucose electrode most of the reported sensor developments are laboratory enterprises (see the review by Wang and Stephanopoulos, 1983). Problems with them seem to be very difficult relating to sterilization, instability and a short life-time. In principle, however, these sensors are very suitable for biological process control because of their own biological origin. It is interesting to note that conventional sensors like pH and pO_2 are usually used in the detection part of these sensors. In this context it could be also noted that a bioreactor itself is a kind of biological sensor when the information available from conventional sensors is used to estimate biological variables like substrate or product concentrations.

THE NEED FOR ESTIMATION TECHNIQUES AND THE PRECONDITIONS FOR THEIR USE

In biotechnical processes, fermentations and certain downstream processing units are usually the most difficult ones to control. The main problem after maintaining asepticity and avoiding equipment faults is keeping the maximum production. This is done with tight process control and is difficult on one hand, because of unpredictable changes in process parameters like different yield coefficients in fermentation, filtering properties in separation etc., and on the other hand, because of lack of on-line information concerning the process state. All methods that help plant operators to specify their ideas on the present detailed state of a process are usually welcome. Computer supported methods which may be used for this purpose are

- General methods for supporting the understanding of process behaviour like
  - presentation of measured, analyzed and calculated variables in compact illustrative form
  - simulation models for steady state and dynamical analysis
  - predictive models for predicting process behaviour forward in time.
- Computational refining of sensor and laboratory data to obtain updated values for
  - derivative variables (OUR, respiration quotient etc.) describing the behaviour of the process
  - parameters (e.g. growth kinetic parameters of the models used
  - on-line estimates of variables which are only indirectly related to the on-line data.
- Expert type of interactive systems which help plant operators to analyze the present state of the process using the data available and the rules stored in the knowledge base.

In industrial practice all these kinds of methods can be classified under the name "estimation" although we in science mean by it mainly certain specific algorithmic methods like RLS, Kalman filters etc. to obtain parameter and state estimates for a process model. Whatever method is used there are some basic principles which such advanced operator aids or parts of control systems should follow.

First. It should be remembered that a method and its implementation is used by a person who is not a specialist in the method. This is important, because most estimation methods, for example, need quite a lot of initial parameter value settings, tuning and maintenance during the process run in order to work well. It is a great challenge to put all this into such a form that the designer himself is not needed to run the system.

Second. In biotechnical processes specifically unpredictable exceptional situations in the processes are very frequent. The method and its implementation should be designed so that such situations can be cleared up easily and without any risk of causing wrong actions in the control systems. For example, suppose that an estimator which utilizes gas analyzer readings to estimate substrate or product level in a bioreactor is further used to control substrate or nutrient feeding to the reactor. The gas analyzers, although quite reliable in a normal operation, may give wrong information for example when foam escapes to the outlet gas tubing system or excess antifoam agent is fed accidently into the reactor and the carbon dioxide accumulated in the broth escapes. In such cases there is a danger that wrong control actions are generated even though the estimator itself works perfectly well. However, when the operator notes the situation he probably disconnects the analyzer and turns the control on manual for a while. If he forgets to give a note to the computer systems the estimator becomes crazy because of invalid data. When he again connects the analyzer to the system and starts automatic control the system is totally out of touch. Such situations, which are also difficult in practical applications of adaptive controllers, can be avoided by a logic system which disconnects and restarts the estimator automatically whenever there is a risk that it is not working correctly. Designing such a logic system may be, however, a more complex task than designing the estimator itself.

Third. It is necessary that the plant operational personnel understand the meaning of an estimate compared to the corresponding laboratory analysis, for example. A biomass dry weight estimate may or may not give about the same reading as the laboratory analysis, such a situation being just normal. In the case where these two differ considerably a right decision is necessary. There may be several possibilities. First, there is something wrong in the laboratory analysis; make a new one and wait until it is ready. In this case it is not wise to feed the sample value to the computer, because it may disturb computations. Second, the analysis is correct and the estimator works also correctly but is just now giving somewhat biased values, perhaps because of incorrect parameters. In this case it is wise to feed the sample value to the computer to let the computations adapt to the changed situation. Third, the estimator is working incorrectly. In this case it may be wise to wait for the next laboratory analysis round to be sure before restarting the estimator. However, if the estimator is connected at the moment in closed loop control the loop should be turned manual now.

In this context I cannot help saying a couple of words about AI and expert systems. Although I don't know of any real applications this far in biotechnical processes it is my personal feeling that the field is a very favourable one for these techniques. One of the potential applications is a detailed state analysis of a bioreactor by considering all the data available. An expert system which conducts the operator to make decisions could give advice e.g. when to terminate a batch process or to make changes in current recipes, or how to control a continuous process when an unidentified disturbance seems to be present or there is a temporary shortage of substrate liquid. Such a system (or program) may also give an early alarm on contamination when this is probable because of changed characteristics of the process. According to my experience such control rules or diagnostic type of logic are difficult to put in simple algorithmic forms but can be given as a set of facts and rules by a person who knows the process well. Interactive systems which the operator can consult before taking actions in the process are also accepted more easily in the industry than the advanced algorithmic type of high level controllers which just tell how to change certain process control variables.

SOME PRACTICAL EXPERIENCE IN APPLYING ESTIMATION TECHNIQUES TO BIOREACTOR CONTROL

Most of the experiences which are available on estimation techniques concern applications in fermentor control. The area of down stream processing - although it's importance is growing all the time in industry seems to be free from any attempt as far as I know. On-line estimation techniques have usually been applied to getting on-line estimates for such variables as biomass growth rate, biomass, substrate and nutrients concentrations and in some cases product concentrations. The on-line measurement information utilized is usually the necessary flows, reactor volume, gas O_2, CO_2 or CH_4, in some cases the consumption of pH control agents and optical density. The estimation algorithms used have been based on different versions of the EKF (Swiniarski et. al. 1982; Stephanopoulos and San, 1984) or RLS in some cases (Jefferis et. al. 1977) and modelling the growth dynamics in the reactor in an appropriate way. Yeast, bacteria and also some fungi fermentations have been under test. Our group has also implemented estimators which are based on an EKF-like nonlinear filtering method in the Pekilo process (an SCP-process, see e.g. Halme, 1979) and in yeast fermentations. This method is described in more detail in another paper in this symposion (Halme et. al. 1985). In the Pekilo-process we have implemented it also for industrial use. In this case the estimator gives continuous on-line estimates on the protein production rate, biomass, sugar, phosphorous and potassium concentrations in the continuous fermentation. These readings are further used to control the fermentor volume (residual sugar level) and nutrient feedings (residual nutrient levels). The estimate of the production rate (closely related to the growth rate) is also used in production reports to give an on-line estimate of production. In this case the production can be physically measured only with a couple of days delay when the product is transported away by lorries which are weighed.

Although general experiences seem to be quite positive there are two aspects which still need development. Almost all studies agree that the parameters used in the growth/production models are changing and also that the parameters should also be estimated on-line. A problem which exists is that the number of parameters to be estimated is easily too large leading to identifiability problems (Homberg, 1982). Another problem, partly related to the previous one, is that the present practice ignores the information included in the laboratory samples. Although delayed and irregular, this information should be considered when available because it enlarges the measurement information vector, thus helping to increase the number of estimated parameters, and also gives a better possibility to maintain the consistency of the estimates. Unfortunately the well known forms of the estimation algorithms, like the EKF, cannot take into account the measurement information which is available at irregular time points and is possibly delayed. The algorithms can, however, be modified for that purpose. A possibility, which is based on a so-called windowing techniqiue, is described in our paper presented later on.

Supposing we have developed an estimation method which works well in laboratory tests, there is still quite a long way to go when implementing it in a production environment. The preconditions discussed earlier, namely user-friendliness, reliability and understanding functioning, are very important and may be quite difficult to realize in practice. When implementing and testing the estimators in the Pekilo-plant we had to keep a side terminal to the plant computer in our laboratory for about a year so that the persons who made the implementation could see how the method worked in normal use and make necessary modifications.

CONCLUSIONS

What has been said before may be summarized as follows.

Although the measurement and sensor technology is developing currently quite fast this far only very few new sensors have reached the level where industrial use is possible. Biotechnical processes have still to be controlled by relying on the quite conventional instrumentation and heavy laboratory support.

The new technologies which may lead to new breakthroughs in the near future are optics and especially the use of optical fibers, "on chip" sensors and the enzyme and microbe sensors, in that order.

Estimation methods when understood in a broad sense including different types of aids which can help plant operators to get better idea of what is going on in the fermentors and other critical parts of the plant, may be of great help provided certain preconditions concerning their use are fulfilled. In addition to those already used quite commonly in many applications, namely illustrative presentation of the measured and analyzed data and model simulations supporting the plant run, standard estimation techniques are probably soon becoming part of the commercially available software packages.

Although not yet applied in biotechnical field AI and expert systems may come very soon because of having in principle a very fruitful substrate in this area. Applications may be for example operator aids to make a detailed analysis on the state of a fermentor or to get advice on how to continue the operation in exceptional process situations, just to mention a few possibilities. Such applications may be classified in the same category as the conventional estimation methods when used to generate indirectly measured or unmeasurable process variables.

Estimation methods have been already applied succesfully in the biotechnical industry. In spite of the further development in the sensor and measurement area I personally feel that their use will increase because so many potential application possibilities exist. It should, however, be remembered that such preconditions as user-friendliness, reliability and understandable functioning are hard to fulfill in industrial applications and may be even more difficult to realize than a well-working algorithm.

REFERENCES

Armiger, W.B. and D.W. Zabriskie (1984). The use of the fluoroMeasure TM system for monitoring and controlling fermentations, First Annual Congress for Automation, Scale-up and the Economics of Biological Process Engineering, San Diego.

De Groot, P.J. (1983). Fibre sensors, Journal A, Vol. 24, No. 3.

Halme, A., R. Kuismin and M. Korteniemi (1985). A method to consider delayed laboratory analysis in state and parameter estimation of bioreactors. Preprints of the 1st IFAC Symposium on Modelling and Control of Biotechnical Processes, Pergamon Press.

Holmberg, A. (1982). On the practical identifiability of microbial growth models incorporating Michaelis-Menten type nonlinearities. Math. Biosci. 62, 23-43.

Jefferis, R.P. III, H. Winter and H. Vogelman (1977). Digital filtering for automatic analysis of cell density and productivity, in Workshop Computer Applications in Fermentation Technology, GBF Monograph Series No. 3, Verlag Chemie, New York.

Jones, B. (1984). Photonic systems meet monitoring and control, C&I, September 1984.

Knight, J., Process measurement by fibre optic sensors, C&I, September 1984.

Sansen, W. (1983). Sensors in silicon. Journal A, Vol. 24, No. 3.

Stephanopoulos, G. and K-Y San (1984). Studies on on-line bioreactor identification. I. Theory. Biotechnol. & Bioeng. 26, 1176-1188.

Swiniarski, R., A. Lesnieski, M.A.M. Dewski M.H. N[illegible] and J.R. Leigh (1983). Progress towards biomass-estimation in a batch fermentation process. Proceedings of the 1st IFAC Workshop on Modelling and Control of Biotechnical Processes, Pergamon Press.

Wang, N.S. and G.N. Stephanopoulos (1984). Computer applications to fermentation processes, in CRC Critical Reviews TM in Biotechnology, CRC Press Inc.

STABLE ADAPTIVE ALGORITHMS FOR ESTIMATION AND CONTROL OF FERMENTATION PROCESSES

D. Dochain and G. Bastin

Laboratoire d'Automatique, Dynamique, et Analyse des Systèmes, University of Louvain, Batiment Maxwell, 1348 Louvain-la-Neuve, Belgium

ABSTRACT : Fairly simple "ad hoc" adaptive estimation and control algorithms for bioreactors can be shown to be stable. We illustate our approach with four different examples of adaptive observers and adaptive regulators. For three of them a stability or convergence proof is given. The effectiveness of these algorithms is illustrated by simulation results.

1. INTRODUCTION

The aim of this paper is to present stable adaptive algorithms for parameter estimation, state estimation and regulation of fermentation processes.

We consider fermentation plants described by the following 2nd order state-space model :

$$\frac{dX}{dt} = \mu(t)X(t)-D(t)X(t) \tag{1.a}$$

$$\frac{dS}{dt} = -k_1\mu(t)X(t)+D(t)[S_{in}(t)-S(t)] \tag{1.b}$$

where X, S, S_{in} represent the biomass concentration, the limiting substrate concentration and the inlet substrate concentration respectively. The parameter k_1 represent the yield coefficient and the parameter $\mu(t)$ the specific growthrate. The dilution rate is D(t) (>0).

It is well known that model (1,a-b) is valid for continous stirred tank (CST) bioreactors (with a constant volume of culture) and for stirred tank fed batch bioreactors as well. In this later case, the volume of culture V(t) is variable and described by :

$$\frac{dV}{dt} = F(t) = D(t)V(t) \tag{2}$$

with F(t) the input flow rate.

Throughout the paper, the specific growth rate $\mu(t)$ is assumed to be a completely unknown time varying parameter, in line with a number of recent works on estimation and control of fermentation processes (Holmberg and Ranta, 1982; Dekkers, 1983; Holmberg, 1983; Stephanopoulos and San, 1983; Nihtila et al., 1984; Dochain and Bastin, 1984; Stephanopoulos and San, 1984; Bastin and Dochain, 1985).

In case of aerobic fermentation processes, an additional equation is needed to describe the dynamics of the dissolved oxygen (DO) concentration (e.g. Olsson and Hansson, 1976) :

$$\frac{dC}{dt} = k_2W(t)[C_0-C(t)]-D(t)C(t)-k_3\mu(t)X(t) -k_4X(t) \tag{3}$$

where C represents the DO concentration, C_0 the saturation DO concentration, W(t) the air flow rate and k_2, k_3, k_4 are constants.

Obviously Extended Kalman Filter algorithms can be used for adaptive parameter estimation and state estimation of non stationary and non linear systems like (1,a-b), (2), (3) (e.g. Stephanopoulos and San, 1984; Nihtila et al., 1984) but their stability is not easy to analyze.

In this paper, we show that fairly simple "ad hoc" adaptive estimation (and control as well) algorithms can be shown bo be stable. We illustrate our approach with four different examples of adaptive observers and regulators (one of them being specific to aerobic processes).

In section 2, we describe two adaptive observers :

1) on line estimation of $\mu(t)$ and S(t) from noisy measurements of X(t)

2) aerobic processes : on line estimation of $\mu(t)$, X(t), S(t) from measurements of C(t).

In section 3, we describe two adaptive regulators :

1) Adaptive regulation of unstable anaerobic digestion processes

2) Adaptive regulation of substrate concentration in fed batch processes with a liquid product

The stability and/or convergence of three of these algorithms are analyzed under the following set of (mild and realistic) assumptions :

A1.1 The specific growth rate is positive

and bounded (the maximum growth rate μ^* is unknown) :

$$0 < \mu(t) < \mu^* \tag{4}$$

A1.2 The inputs $D(t)$, $S_{in}(t)$, $W(t)$ and the state variables $X(t)$, $S(t)$, $C(t)$ are positive (in accordance with the physical situation)

A1.3 The inlet substrate concentration $S_{in}(t)$ is bounded :

$$0 \leqslant S_{in}(t) \leqslant S_{mx} \tag{5}$$

A1.4 $\mu(t)=0$ when $S(t)=0$ or $C(t)=0$ (6)

A1.5 The initial conditions fulfill the following inequalities :

$$k_1 X(0)+S(0) \leqslant S_{mx} \qquad C(0) \leqslant C_0 \tag{7}$$

A1.6 The time derivative of $\mu(t)$ is bounded :

$$\left|\frac{d\mu}{dt}\right| < M_1 \tag{7}$$

Then, we have the following preliminary result (Goodwin et al.,1983; Dochain and Bastin, 1984) :

Lemma 1 : Under assumptions A1.1 to A1.5, the state variables $X(t)$, $S(t)$, $C(t)$ are bounded for all t :

$$\begin{aligned} &0 \leqslant S(t) \leqslant S_{mx} \\ &0 \leqslant X(t) \leqslant (1/k_1)S_{mx} \\ &0 \leqslant C(t) \leqslant C_0 \end{aligned} \tag{9}$$

It must be emphasized that, from lemma 1, the states of the process are bounded without imposing any upper bound on the inputs $D(t)$ and $W(t)$.

For the readibility of the paper, the theoretical proofs are relegated in Appendix.

2. ADAPTIVE OBSERVERS

2.1. Example 1 : On line estimation of $\mu(t)$ and $S(t)$ from noisy measurements of $X(t)$.

Statement of the algorithm

Consider a bioractor described by equations (1.a-b).

Assume that :

- the inputs $D(t)$ and $S_{in}(t)$ are known.
- the yield coefficient k_1 is known
- a noisy measurement X_m of the biomass concentration X is available :

$$X_m(t) = X(t) + \epsilon(t) \tag{10}$$

with ϵ the measurement noise.

Then the following adaptive observer can be used

- to estimate on line the specific growth rate $\mu(t)$
- to estimate on line the substrate concentration $S(t)$:

$$\frac{d\hat{X}}{dt} = [\hat{\mu}(t)-D(t)+\lambda_1(X_m(t)-\hat{X}(t)]X_m(t) \tag{11.a}$$

$$\frac{d\hat{\mu}}{dt} = \lambda_2 X_m(t)[X_m(t)-\hat{X}(t)] \tag{11.b}$$

$$\frac{d\hat{S}}{dt} = -k_1\hat{\mu}(t)X_m(t) + D(t)[S_{in}(t)-\hat{S}(t)] \tag{11.c}$$

$\lambda_1 > 0$, $\lambda_2 > 0$

Convergence analysis

Defining the estimation errors:

$$\begin{aligned} ex &= X(t)-\hat{X}(t), \quad e\mu = \mu(t)-\hat{\mu}(t) \\ es &= S(t)-\hat{S}(t) \end{aligned} \tag{12}$$

we have the following convergence results:

Assumptions B2.1

a) $X_m(t)$ is strictly positive :

$$X_m(t) \geqslant \eta \geqslant 0 \tag{13}$$

b) The measurement noise is bounded :

$$|\epsilon(t)| < M_2$$

c) $\lambda_1 = \frac{\alpha}{4}(\lambda_2)^2$

d) the dilution rate is bounded:

$$0 \leqslant D(t) \leqslant D_m$$

Theorem 2.1.

Under assumptions A1.1 to A1.6 and B2.1

a) $$\lim_{t\to\infty} |ex| \leqslant \left(\frac{\beta_1}{\lambda_1} + \beta_2\right)M_2 + \frac{\beta_3}{\lambda_1^2}M_1 \tag{16}$$

b) $$\lim_{t\to\infty} |e\mu| \leqslant (\beta_4\lambda_1 + \beta_5)M_2 + \frac{\beta_6}{\lambda_1}M_1 = G \tag{17}$$

c) $$\lim_{t\to\infty} |es| \leqslant \beta_7 G \tag{18}$$

with β_i ($i=1,7$) positive constants independent of M_1, M_2, λ_1.

Comment

From expressions (16) and (17), it follows clearly that there exists an optimal value of the design parameter λ_1, minimizing the asymptotic upper bounds on $|e\mu|$ and $|es|$:

$$\lambda_1 = \sqrt{\left(\frac{\beta_4}{\beta_6}\right)}$$

Simulation results.

A simulation experiment has been performed for an anaerobic digestion process with a Monod specific growth rate and with a square wave input $S_{in}(t)$. Fig.1 illustrates the effectiveness of the method.
An application of this algorithm to real life data can be found in Bastin and Dochain (1985).

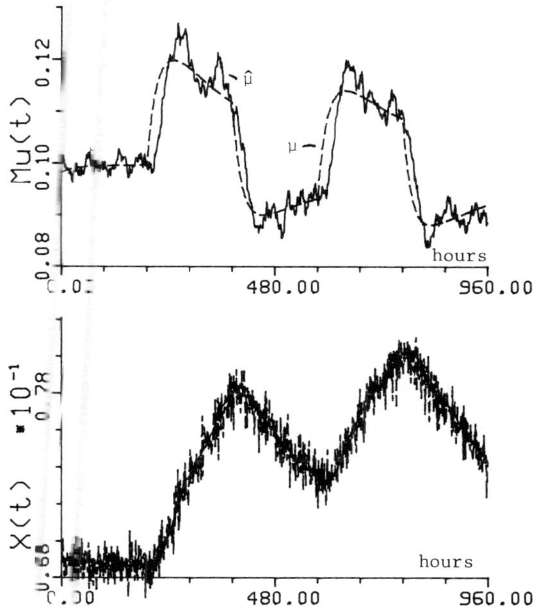

Fig. 1. Estimation of μ with noisy measurements of X(t).

2.2. Example 2 : (Aerobic processes) On line estimation of $\mu(t)$, $X(t)$, $S(t)$ from measurements of $C(t)$.

Statement of the algorithm.

Consider a CST aerobic bioreactor described by equations (1,a-b) and (2).

Assume that
- the inputs $S_{in}(t)$, $W(t)$ and $D(t)$ are known
- the parameters k_i $(i=1,4)$ and the saturation DO C_0 are known
- the DO concentration $C(t)$ is available from measurement.

Then, the following adaptive observer can be used
- to estimate on line the specific growth rate $\mu(t)$
- the state variables $X(t)$ and $S(t)$:

$$\frac{d\hat{C}}{dt} = k_2W(t)[C_0-C(t)] - D(t)C(t) - k_3\hat{\varphi}(t) - k_4\hat{X}(t) + \lambda_1[C(t)-\hat{C}(t)] \quad (19.a)$$

$$\frac{d\hat{X}}{dt} = \hat{\varphi}(t) - D(t)\hat{X}(t) - \lambda_2[C(t)-\hat{C}(t)] \quad (19.b)$$

$$\frac{d\hat{\varphi}}{dt} = - \lambda_3[C(t)-\hat{C}(t)] \quad (19.c)$$

$$\frac{d\hat{S}}{dt} = - k_1\hat{\varphi}(t) + D(t)[S_{in}(t)-\hat{S}(t)] \quad (19.d)$$

$$\hat{\mu}(t) = \hat{\varphi}(t)/\hat{X}(t) \quad (19.e)$$

$\lambda_1, \lambda_2, \lambda_3 > 0$

Clearly, the parameter $\varphi(t)$ is an on line estimate of the "biological activity" : $\varphi(t) = \mu(t)X(t)$.

Stability analysis.

We define the following estimation errors:

$$ex = X(t)-\hat{X}(t) \ , \ e\varphi = \varphi(t)-\hat{\varphi}(t)$$
$$ec = C(t)-\hat{C}(t) \ , \ es = S(t)-\hat{S}(t) \quad (20)$$

Assumptions B2.2

a) $D(t)$ is strictly positive :

$$D(t) \geq \eta > 0 \quad \text{for all } t$$

b) $\lambda_1\lambda_2 > \lambda_3$

c) $\left|\frac{d\varphi}{dt}\right| < M_3$

Theorem 2.2

Under assumptions A1.1 to A1.6 and B2.2, there exists a positive finite constant β such that the estimation errors ex, $e\varphi$, ec and es are all bounded by βM_3.

3. ADAPTIVE REGULATORS.

3.1. Adaptive regulation of unstable anaerobic digestion processes.

Statement of the algorithm.

We consider a CST anaerobic digestion process where the methanization step is rate limiting. The bacterial growth is assumed to be described by (1,a-b) while the methane gas production rate is expressed as follows:

$$Q(t) = k_5\mu(t)X(t) \quad (27)$$

It is a well known fact that this process ca be unstable, due to inhibitory effects at high substrate concentrations (e.g. Antunes and Installé, 1981). We consider the problem of regulating the reactor when it is used for waste water treatment. The control objective is to regulate the output pollution concentration $S(t)$ at a prescribed level S^*, despite the fluctuations of the input pollution level $S_{in}(t)$, by using the dilution rate $D(t)$ as control input.

From equations (1,a-b) and (27), we have:

$$\frac{dS}{dt} = -KQ(t) + D(t)[S_{in}(t)-S(t)] \quad (28)$$

$$\text{with} \quad K = \frac{k_1}{k_5} \quad (29)$$

Equation (28) can be viewed as a 1st order dynamic model of the system with a bounded time varying parameter $Q(t)$ (the boundedness of $Q(t)$ results from assumption A1.1 and Lemma 1). This equation is the basis for the derivation of the control algorithm.

Assuming that

- $S_{in}(t)$, $S(t)$, $Q(t)$ are measured on line
- the parameter K is a priori unknown;

Then, the following self-tuning algorithm can be used to regulate S(t) at the level S*:

Let

$$D^o(t) = \frac{C_1(t)[S^*-S(t)] + \hat{K}(t)Q(t)}{S_{in}(t)-S(t)} \qquad (29.a)$$

with $C_1(t)$ a strictly positive function $(C_1(t)>0)$ and $\hat{K}(t)$ an adaptive estimate of K.
The control input D(t) and the parameter estimate $\hat{K}(t)$ are calculated as follows:

$$\begin{cases} D(t) = D^o(t) & \text{if } 0 \leq D^o(t) \leq D_m \\ D(t) = 0 & \text{if } D^o(t) \leq 0 \\ D(t) = D_m & \text{if } D^o(t) \geq D_m \end{cases} \qquad (29.b)$$

$$\begin{cases} \hat{K}(0) > 0 \\ \frac{d\hat{K}}{dt} = C_2 Q(t)[S^*-S(t)] \\ \text{except if } \hat{K}(t)=0 \text{ and } S(t)>S^* \\ \text{then } \frac{d\hat{K}}{dt} = 0 \text{ (to ensure } \hat{K}(t)\geq 0\ \forall t) \end{cases} \qquad (29.c)$$

If we define

$$es = S^* - S(t) \quad \text{and} \quad ek = K - \hat{K}(t)$$

the closed-loop system (28),(29) can be written, when $D(t) = D^{\circ}(t)$:

$$\frac{d}{dt}\begin{bmatrix} es \\ ek \end{bmatrix} = \begin{bmatrix} -C_1(t) & Q(t) \\ -C_2 Q(t) & 0 \end{bmatrix}\begin{bmatrix} es \\ ek \end{bmatrix} \qquad (30)$$

i.e; as a linear time varying unforced system, with a bounded coefficient matrix.

Convergence analysis.

Assumptions B3.1.

a) The input pollution level is bouded as follows :

$S_{mn} \leq S(t) \leq S_{mx}$

b) $S^* < S_{mn}$

c) $D_m > \frac{\mu^* S_{mx}}{S_{mn}-S^*}$

Theorem 3.1.

Under assumptions A1.1 to A1.5 and B3.1,

a) all the closed-loop system variables are bounded;

b) $\lim_{t\to\infty} S(t) = S^*$;

c) $\lim_{t\to\infty} \hat{K}(t) = K$;

d) there exists t_1 such that $D(t) = D^o(t)$ for all $t \geq t_1$;

e) if $C_1(t) = \alpha Q(t)$ $(\alpha>0)$, then the closed loop system is exponentially stable.

Comments.

1) This theorem is based on the assumption that the methane flow rate Q(t) is strictly positive for all t. Within the context of our previous assumptions, it is not possible to prove that $Q(t) > 0$ since Q(t) depends on $\mu(t)$ and since we do not specify $\mu(t)$. However for simple models of $\mu(t)$, it is fairly easy to show that Q(t) is effectively > 0 for all t.

2) It is worth noting that the regulator achieves a zero steady-state error even with a varying input disturbance $S_{in}(t)$, since the algorithm includes a feedforward action (see the simulations in the next section).

Simulation results.

A simulation experiment has been performed for an unstable biomethanization plant with a Haldane specific growth rate (fig.2). A pulse of toxic substance is introduced after 1 day. Fig 2. shows clearly the stabilization of the plant by the adaptive controller (while in open loop it is led to wash-out).

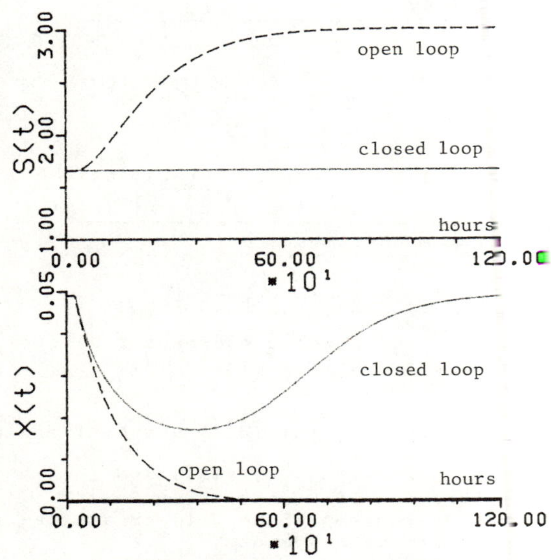

Fig. 2. Adaptive control of an unstable anaerobic digestion process.

3.2. Adaptive regulation of substrate concentration in a fed batch reactor with a liquid product.

Statement of the algorithm.

We consider a fed batch bioreactor described by equations (1,a-b) and (2), with a reaction product, in liquid phase, whose production dynamics is described by the following equation:

$$\frac{dP}{dt} = \gamma(S(t))X(t) - D(t)P(t) \qquad (31)$$

where P(t) is the concentration of the reaction product. The specific production rate $\gamma(S(t))$ is assumed to be a non linear function of S(t), inhibited at high concentrations: a typical example, taken from Takamatsu et al. (1974), is shown in fig.3.

Our conjecture is then that adaptive

regulation of S(t) at values corresponding to a maximum production rate should contribute to a maximization of the yield of the reactor (without any modelling of $\mu(t)$ or $\gamma(S(t))$ being necessary, unlike earlier optimal control studies, e.g. Ohno et al., 1975)

In accordance with assumption A1.4, we define the specific growth rate as follows:

$$\mu(t) = \rho(t)S(t) \tag{32}$$

Then, equation (1.b) is rewritten:

$$\frac{dS}{dt} = -\varphi(t)S(t) + D(t)[S_{in}(t)-S(t)] \tag{33}$$

$$\text{with } \varphi(t) = k_1\rho(t)X(t) \tag{34}$$

This equation is the basis for the derivation of the regulation algorithm.

Assuming that

- the flow rate is constant : $F(t) = F_0$ (until the tank is full, of course);
- the inlet substrate concentration $S_{in}(t)$ is the control input;
- the concentration S(t) is measured on line.

Then the following adaptive algorithm can be used to regulate S(t) at a reference level S^* :

$$S_{in}(t) = \frac{1}{D_0}\ \{C_1[S^*-S(t)]+[D_0+\hat{\varphi}(t)]S(t)\}$$
$$C_1 > 0$$

where $\hat{\varphi}(t)$ is an on line estimate of $\varphi(t)$ updated as follows :

$$\frac{d\hat{\varphi}}{dt} = C_2S(t)[S^*-S(t)]$$

except if $\hat{\varphi}(t)=0$ and $S(t)>S^*$,

then $\frac{d\hat{\varphi}}{dt} = 0$ (to ensure $\hat{\varphi}(t)\geq 0\ \forall t$).

Simulation results.

On fig.4, we compare closed loop and open loop performance of a fed batch process with a Monod specific growth rate and the specific production rate of fig.3. In both cases the reactor is fed with the same amount of substrate. In open loop, a constant input $S_{in}(t)$ is used, while in closed loop the input $S_{in}(t)$ is computed by the control algorithm. Fig.4 show that the closed loop operation increases the final product concentration by nearly 40%.

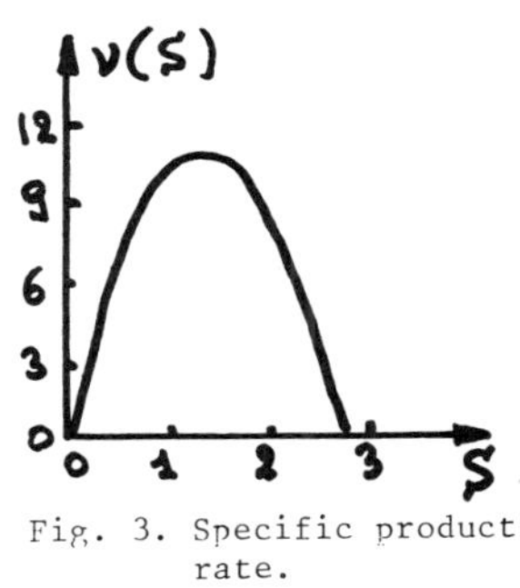

Fig. 3. Specific production rate.

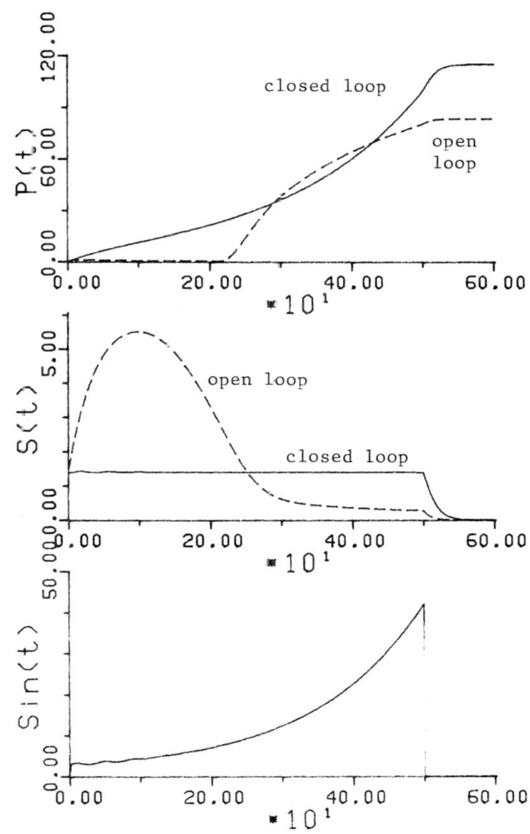

Fig. 4. Adaptive control of a fed-batch process.

APPENDIX.

PROOF OF THEOREM 2.1.

step 1: Defining $e^t = (ex, e\mu)^t$ the following error system" derives from (11.a-b) :

$$\frac{de}{dt} = X_m(t)Ae(t) + B(t)u(t) \tag{A1}$$

$$\text{with } A = \begin{bmatrix} -\lambda_1 & 1 \\ -\lambda_2 & 0 \end{bmatrix} \tag{A2}$$

$$B(t) = \begin{bmatrix} D(t) - \mu(t) - \lambda_1 X_m(t) & 0 \\ -\lambda_2 X_m(t) & 1 \end{bmatrix} \tag{A3}$$

$$u(t) = \begin{bmatrix} \epsilon(t) \\ \frac{d\mu}{dt} \end{bmatrix} \tag{A4}$$

step 2: The matrix A is stable (since λ_1,λ_2 are >0) and we denote its eigenvalues by $-p_1$, $-p_2$ (<0).

step 3: The solution of system (A1) ca be written :

$$e(t) = \exp[X_m(t)At] + \int_0^t PR^{-1}(t)R(s)P^{-1}B(s)u(s)ds \tag{A5}$$

with P = matrix of eigenvectors of A and

$R(t) = \text{diag}\ \{\exp[p_1\zeta(t)]\ ,\ \exp[p_2\zeta(t)]\}$

with $\zeta(t) = \int_0^t X_m(\tau)d\tau$

step 4: $\exp[X_m(t)At]$ tends exponentially to zero, by assumption B2.1 and step 2.

step 5: By calculating the last term of (A5), one obtains :

$$\lim_{t\to\infty} ex = \lim_{t\to\infty}\int_0^t \{\left[\frac{1-\omega}{\omega}z_1(s) - \frac{1+\omega}{\omega}z_2(s)\right][D(s)-\mu(s)$$

$$-\lambda_1 X_m(s)]\epsilon(s) + \frac{2}{\lambda_1\omega}[z_2(s)-z_1(s)][-\lambda_2 X_m(s)\epsilon(s) + \frac{d\mu}{ds}]\}ds \quad (A8)$$

$$\lim_{t\to\infty} e\mu = \lim_{t\to\infty}\int_0^t \{\frac{\alpha\lambda_1}{2\omega}[z_1(s)-z_2(s)][D(s)-\mu(s)-\lambda_1 X_m(s)]$$

$$+ \left[\frac{1+\omega}{\omega}z_1(s) - \frac{1-\omega}{\omega}z_2(s)\right][-\lambda_2 X_m(s)\epsilon(s) + \frac{d\mu}{ds}]\}ds \quad (A9)$$

with $\omega = \sqrt{1-\alpha}$

$$z_i(t) = \exp(-p_i\zeta(t)) \quad (i=1,2)$$

step 6: then by computing the asymptotic bounds of (A8) and (A9) one obtains:

$D_m < \mu^*+\lambda_1 M_0$	$D_m > \mu^*+\lambda_1 M_0$
$\beta_1 = 4\mu^*/\eta\omega$	$\beta_1 = 4D_m/\eta\omega$
$\beta_2 = 6M_0/\eta\omega$	$\beta_2 = 2M_0/\eta\omega$
$\beta_3 = 8/\eta\alpha\omega$	$\beta_3 = 8/\eta\alpha\omega$
$\beta_4 = (4-\alpha)M_0/\eta\omega$	$\beta_4 = (2-\alpha)M_0/\eta\omega$
$\beta_5 = 2\mu^*/\eta\omega$	$\beta_5 = 2M_0/\eta\omega$
$\beta_6 = 4(2-\alpha)/\eta\alpha\omega$	$\beta_6 = 4(2-\alpha)/\eta\alpha\omega$

with $M_0 = (1/k_1)S_{mx} + M_2$

Then the asymptotic upper bound on es follows immediately with $\beta_7 = D_m/k_1\eta$. ▽▽▽

PROOF OF THEOREM 2.2.

step 1: Defining $e^t = (ec,ex,e\varphi)^t$ the following "error system" derives from equations (19.a-c):

$$\frac{de}{dt} = A(t)e(t) + Bu(t)$$

$$\text{with } A(t) = \begin{bmatrix} -\lambda_1 & -k_4 & -k_3 \\ \lambda_2 & -D(t) & 1 \\ \lambda_3 & 0 & 0 \end{bmatrix} \qquad B = \begin{bmatrix} 0 \\ 0 \\ 1 \end{bmatrix}$$

$$u(t) = d\varphi/dt$$

step 2: Let the following quadratic function :

$$V = k_4\lambda_3 ex^2 + \lambda_2\lambda_3 ec^2 + (\lambda_2 k_3+\lambda_1)e\varphi^2 + 2\lambda_3 ece\varphi$$

This function is positive definite, since by assumption B2.2 we have :

$$\lambda_3 < \sqrt{\lambda_3^2+\lambda_2^2\lambda_3 k_3} < \sqrt{\lambda_1\lambda_2\lambda_3+\lambda_2^2\lambda_3 k_3} = \sqrt{\lambda_2\lambda_3(\lambda_2 k_3+\lambda_1)}$$

Now $\frac{dV}{dt} = -k_4\lambda_3 D(t)ex^2 - \lambda_3(\lambda_1\lambda_2-\lambda_3)ec^2 - k_3\lambda_3 e\varphi^2$

which is negative definite, by assumption B2.2.

Hence the free system $\frac{de}{dt}=A(t)e(t)$ is exponentially stable.

step 3: Then it is a standard result of stability theory (Willems, 1970, theorem 3.1) that there exists a positive constant β such that:

$$\lim_{t\to\infty} \|e\| < \beta M_3$$

and the boundedness of es follows readily. ▽▽▽

PROOF OF THEOREM 3.1.

a) trivial from lemma 1 and by the definition of the control algorithm.

b) see Dochain and Bastin, 1984b.

c) and d) Since S(t) is continuous and derivable, we have from b):

$$\lim_{t\to\infty}\frac{dS}{dt} = 0 \text{ and } \lim_{t\to\infty} D(t) = \lim_{t\to\infty}\frac{KQ(t)}{S_{in}(t)-S^*}$$

$$\text{Now } 0< \frac{K\epsilon}{S_{mx}-S^*} < \frac{KQ(t)}{S_{in}(t)-S^*} < \frac{KQ_{mx}}{S_{mn}-S^*} < D_m$$

$$\text{Hence } \lim_{t\to\infty} D(t) = \lim_{t\to\infty} D^0(t) = \lim_{t\to\infty}\frac{\hat{K}(t)Q(t)}{S_{in}(t)-S^*}$$

and c) and d) follows readily.

e) for $t > t_1$, the closed loop system is equivalent to th system (30). We prove that this system is exponentially stable when $C_1(t) = \alpha Q(t)$ and $\alpha > C_2 > 1$: let the following positive definite function

$$V(es,ek) = \frac{\alpha}{2(C_2-1)}(es^2+ek^2) - esek$$

then

$$\frac{dV}{dt} = -Q(t)\,[(\frac{\alpha^2}{C_2-1} - C_2)es^2 + ek^2]$$

is clearly, a negative definite function and the system (30) is exponentially stable for $t > t_1$. And, from a), the closed loop system is exponentially stable for all $t > 0$.

REFERENCES.

Antunes S. and M.Installe (1981). The use of phase plane analysis in the modelling of a biomethanization process. Proc. 8th IFAC World Cong., Kyoto, 22, pp 165-170.
Bastin G. and D.Dochain (1985). Adaptive estimation of microbial growth rates. 7th IFAC Symp. on Identification. York,U-K, 3-7 July
Dekkers R.M. (1983). State estimation of fedbatch baker's yeast fermentation. in Mod. and Cont. of Bioteh. Proc. (ed. Halme). Pergamon,pp 201-212.
Dochain D. and G.Bastin (1984). Adaptive identification and control algorithms for non linear bacterial growth systems. Automatica,20(5), pp621-634.
Dochain D. and G.Bastin (1984b). Stability and convergence analysis of a continuous time adaptive control algorithm for fermentation processes. in Anal. and Opt. of Syst. Springer Verlag, pp468-474.
Goodwin G.C., B.McInnis and R.S.Long (1982). Adaptive control algorithms for waste water treatment and pH neutralization. Opt. Cont. Appl. and Meth., 3(4), pp443-460.
Holmberg A. (1983). A micro-processor based estimation and control system for the activated sludge process. in Mod. and Cont. of Biotech. Proc. (ed. Halme), Pergamon, pp 111-120.
Holmberg A. and J.Ranta (1982). Procedures for parameter and state estimation of microbial growth process models. Automatica,13,181-193.
Nihtila M., P Harmo, M.Perttula (1984). Real time growth estimation in batch fermentation. Prc. 9th IFAC World Cong., July 2-6, 225-230
Ohno H., E.Nakanishi, T.Takamatsu (1976) Optimal control of a semi batch fermentation. Biotech. and Bioeng., 18, pp847-864.
Olsson G and O.Hansson (1976). Modelling and identification of ar activated sludge process. Proc. 4th IFAC Symp. on Ident., Tbilisi.
Stephanopoulos G. and K-Y San (1983). On line estimation of time varying parameters: application to biochemical reactors. in Mod. and Cont. of Biotech. Proc. (ed. Halme), Pergamon, pp195-199.
Stephanopoulos and K-Y San (1984). Studies on on-line bioreactor identification. Biotech. and Bioeng., 26, pp 1176-1188.
Takamatsu T. et al. (1975). Theory and practice of optimal control in continuous fermentation process. Automatica, 11, pp 141-148.
Willems J.L. (1970). Stability theory of dynamical systems. Nelson.

ADAPTIVE LQ CONTROL OF FERMENTATION PROCESSES

H. N. J. Poulisse* and C. van Helden**

*Department of Physics and Meteorology, Section Systems and Control Engineering, Agricultural University, Duivendaal 2, 6701 AP Wageningen, The Netherlands
**Department of Process Engineering, Agricultural University, Wageningen, The Netherlands

Abstract. A mathematical model is developed for a fairly large class of so-called fed-batch fermentation processes. The model consists of a set of non-linear stochastic differential equations. Measurements of a linear combination of states with additive noise are available at sampling instants. Our objective is to control the system such that it follows same prespecified trajectory. An integral criterion is formulated which should be minimized to achieve the desired system performance. The non-linear stochastic differential equations are linearized around the prespecified trajectory. The linearized continuous time system is transformed into an equivalent discrete time system (EDS) and the integral criterion is transformed into a sum criterion. Robustness of the control is achieved by applying adaptive control to the fermentation process. The EDS is rewritten as an ARMAX-representation and the sum criterion is formulated in terms of output and control rather than in terms of states and control. The relation between the parameters of the ARMAX-model and the original microbiological parameters are established. Finally an adaptive control law is formulated, which, under certain restrictive conditions, will be close to optimal with respect to the averaged quadratic sum criterion.

Keywords. Adaptive control; biocontrol; equivalent discrete time system; nonlinear control systems; parameter estimation; time varying systems.

INTRODUCTION

From usual material balance equations (Roels, 1983), a non-linear state space representation, involving biomass and substrate concentration can be obtained for fermentation processes, in particular so-called fed-batch fermentation processes. There is a tendency in the literature to leave the bacterial growth rate analytically unspecified but instead to conceive the bacterial growth rate as a slowly varying parameter (Stephanopoulos and San, 1984; Dochain and Bastin, 1984). However, if this approach is adopted for a fed-batch fermentation, the substrate state will always be unobservable. In this paper an analytical expression for the bacterial growth rate is given by Monod's law. For optimal control of the fed-batch fermentation, on-line estimates of the states, in particular the biomass are necessary. Since there are no sensors available, suitable for industrial applications to measure directly the biomass, this state variable must be measured indirectly from an analysis of the effluent gas of the fermentor. However, a measurement obtained in this way is usually not related to the biomass only, but in many cases to a - linear - combination of the biomass and the biomass production rate (BPR). Therefore a state equation for the biomass production rate has been added to the state equations resulting from the material balances. It is well known that the kinetic microbiological parameters are generally uncertain: the parameters may be slowly changing; furthermore these parameters are usually estimated off-line under circumstances quite different from the actual process conditions (Pirt, 1975) Proportional integral control action is therefore often added to the calculated optimal control to compensate for these parameter uncertainties (Alvarez Gallegos and Alvarez Gallegos, 1982). In this paper robustness of the control is achieved by applying - sub - optimal adaptive control to the fermentation process. Our objective is to control the system such that it follows some prespecified trajectory. An integral criterion is formulated, which should be minimized to achieve this goal. Since control of the fermentation process will be performed using direct digital control using observations taken at the sampling instants, the linearized state equations are transformed into an equivalent discrete time system (EDS) and the integral criterion is transformed into a sum criterion. The EDS is rewritten as an ARMAX representation and the unknown parameters are estimated using a least squares method. The sum criterion is reformulated in terms of output and control. Because the transformation of the original state equations into the ARMAX-representation is done analytically, the relationship between the parameters in the ARMAX-model and the original microbiological parameters can be established. An adaptive control law is formulated which is close to optimal with respect to the averaged sum criterion. Adaptive control of fermentation processes is also used in (Dochain and Bastin, 1984), but there minimum variance control is used instead of LQ-control.

A detailed account on adaptive control of fermentation processes can be found in (Poulisse, 1985).

THEORY

First of all we give a description of the system, leading to a set of linearized stochastic differential equations and a measurement equation. The control objective is formulated in terms of an integral criterion. Next the EDS is established together with a sum criterion. The EDS is subsequently rewritten as an ARMAX-representation and the sum criterion is adapted to this input-output description. In the last part of this section the adaptive control law is given.

Description of the system

The state space representation of bacterial growth for a fed-batch fermentation immediately follows from mass-balances (Roels, 1983).

$$\dot{M} = \mu(M.,S)M \tag{1}$$

$$\dot{S} = (\mu(M,S)/Y)S - mM + FSi \tag{2}$$

where

M: biomass (kg)
S: substrate (kg)
$\mu(M,S)$: bacterial growth rate (hour^{-1})
Y: yield coefficient (kg ciomass/kg substrate)
F: influent flow rate (m^3/hour)
Si: substrate concentration in influent (kg/m^3)
m: maintenance (kg/(kg.hour))

We note that it is in this stage of the system description not relevant to consider system noise. Stephanopoulos and San (1984) and Dochain and Bastin (1984) leave the bacterial growth rate unspecified, i.e. the bacterial growth rate is considered to be an unknown time-varying parameter. For a fed-batch fermentation this will lead to an unobservable substrate state. In this paper we adopt the Monod law (Roels, 1983) as an analytical expression for the bacterial growth rate

$$\mu(M,S) = \mu m S/(K+S) \tag{3}$$

where

μm: specific growth rate (hour^{-1})
K: Monod "constant" (kg)

It should be noted that sine we use absolute quantities for the substrate state S rather than concentrations, K in (3) changes with the working volume of the fermentor (Roels, 1983).

There are no sensors available, suitable for industrial applications to measure directly important process variables in fermentation processes. The usual approach is to observe the state variables indirectly from an analysis of the effluent gas of the fermentor. Dekkers (1982) gives a detailed analysis of this approach. It follows that a measurement obtained in this way is usually not related to the biomass only, but in many cases to a linear combination of the biomass and the biomass production rate (BPR).
Defining BPR by the equation

$$BPR = \dot{M} \tag{4}$$

we arrive at a continuous time system description of a fairly large class of fed-batch fermentation processes

$$\dot{x}(t) = f(x(t),u(t)) + v(t) \tag{5}$$

$$y(kT) = Cx(kT) + w(kT) \tag{6}$$

$k = 0, 1, 2, \ldots: N-1$; T sampling interval

where

$$x^T = (M\ S\ BPR) \tag{7}$$

$$f^T(x,u) = (f1(x,u),\ f2(x,u),\ f3(x,u)) \tag{8}$$

$$f1(x,u) = \mu m\, MS/(K+S) \tag{9}$$

$$f2(x,u) = \mu m\, MS/((K+S)Y) - mM + uSi \tag{10}$$

$$\begin{aligned} f3(x,u) = {} & \mu m^2 M(uSi-m)/(K+S) \\ & + \mu m MS(\mu m S + mM - uSi)/(K+S)^2 \\ & - \mu m^2 M^2 S/(Y(K+S)^2) \\ & + \mu m^2 M^2 S^2/(Y(K+S)^3)\ ;\ u = F \end{aligned} \tag{11}$$

$$C = (C1\ 0\ C2) \tag{12}$$

The initial condition is $x(0) = x_0$. The observation noise process $\{w(kT),\ k = 0,1,\ldots,N-1\}$ and the system noise process $\{v(t),\ t \geq 0\}$ have known constant means and covariances

$$E\{v(t)\} = 0,\ cOv(v(t),v(s)) = V\delta(t-s),\quad V \geq 0 \tag{13}$$

$$E\{w(kT)\} = 0,\ cOv(w(kT),w(jT)) = W\delta d(kj),\ W > 0 \tag{14}$$

C is assumed to be a constant vector. Actually equation (5) is defined in terms of an integral equation. However, we shall omit technical details here and refer to (de Koning, 1980; Poulisse, 1985). We assume that the continuous time system is connected to a digital computer by means of a sample - and - hold circuit, i.e. u(t) in (5) is given by

$$u(t) = u(kT),\ t\varepsilon[kT,(k+1)T)\ \ k = 0,1,\ldots,N-1 \tag{15}$$

Suppose our objective is to control the fermentation process such that the system state follows some prespecified trajectory $x^*(t)$, after application of the control $u^*(t)$. $x^*(t)$ and $u^*(t)$ generally result from a dynamic optimization, e.g. (Dekkers, 1983). This objective can be achieved if the actual control u(t) is chosen such that the integral criterion

$$J = E\{\int_0^t((x(t)-x^*(t))^T Q(x(t)-x^*(t)) + u^2(t)R)dt\} \tag{16}$$

is minimized. Q is a known constant matrix and R a known scalar. Since coping with the non-linear equation (5) is in fact impossible, equation (5) is linearized. It is common practice in control engineering to linearize the system around the prespecified trajectory by using a first order Taylor series expansion. Indeed the so obtained linear system is even locally an approximation, i.e. in not any sense equivalent to the non-linear system (Hazewinkel, 1984). But if our control efforts are successful, this approximation is certainly not any restriction for application purposes. Linearization of (5) around the prespecified trajectory gives

$$\dot{x}(t) = A(t')\Delta x(t) + B(t')\Delta u(t) + f(x^*(t'),u^*(t')) + v(t) \text{ for some } t' \tag{17}$$

where

$$\Delta x(t) = x(t) - x^*(t) \quad (18)$$

$$\Delta u(t) = u(t) - u^*(t) \quad (19)$$

$$A = \{a_{ij}\};\ a_{ij} = \partial f_j/\partial x_i|_* \ i,j = 1,2,3 \quad (20)$$

$$B = (b1,\ b2,\ b3)\ j\ b_i = \partial f_i/\partial u|_* \quad (21)$$

The partial derivatives are evaluated for the nominal trajectory and control $x^*(t)$ respectively $u^*(t)$.

Equivalent discrete time system (EDS)

We approximate equation (17) in the following way

$$\dot{x}(t) = A(kT)x(t)+B(kT)u(t)+r(kT)+v(t) \quad (22)$$

$$t\varepsilon\ [kT,(k+1)T);\ k\varepsilon\{0,1,2,..,\ N-1\}$$

where

$$r(kT) = f(x^*(kT),u^*(kT))-A(kT)x^*(kT) - B(kT)u^*(kT) \quad (23)$$

In view of (20,21) the meaning of A(kT) and B(kT) will be clear. We now make the following restrictive assumption

$$r(kT) \simeq 0 \quad \forall k \quad (24)$$

Using (24) it may be shown that the continous time system (22) with the integral criterion (16) can be transformed into the following EDS with sum criterion (de Koning, 1980; Poulisse, 1985)

$$x((k+1)T) = \Phi(kT,T)x(kT)+\Gamma(kT,T)u(kT)+v(kT) \quad (25)$$

$$J^* = E\{\sum_{k=0}^{N-1} \Delta x^T(kT)Q(kT,T)\Delta x(kT) + u^2(kT)R(kT,T)\} \quad (26)$$

The matrices $\Phi(kT,T)$, $\Gamma(kT,T)$, $Q(kT,T)$ and the scalar $R(kT,T)$ can be related to the matrices A(kT) and B(kT) in (22) and the matrix Q and scalar R in the integral criterion (16). For instance,

$$\Phi(kT,T) = \Phi'(kT,T)-\Gamma(kT,T)R^{-1}(kT,T)M^T(kT,T) \quad (27)$$

$$\Phi\ (kT,T) = \exp(A(kT)T) \quad (28)$$

$$\Gamma(kT,T) = \int_0^T \Phi'(kT,\lambda)B(kT)d\lambda \quad (29)$$

$$R(kT,T) = \int_0^T [R+\Gamma^T(kT,T)Q\Gamma(kT,T)]d\lambda \quad (30)$$

$$M(kT,T) = \int_0^T \Phi'^T(kT,\lambda)Q\Gamma(kT,\lambda)d\lambda \quad (31)$$

Details can be found in (de Koning, 1980; Poulisse, 1985). The coefficients of the matrix $\Phi(kT,T)$ and the coefficients of the vector $\Gamma(kT,T)$ can be explicitely related to the coefficients of the matrix A(kT) and of the vector B(kT). For instance if

$$\beta^2(kT) = -(a_{11}(kT)+a_{22}(kT))^2/4 + a_{11}(kT).a_{22}(kT)-a_{12}(kT)a_{21}(kT) > 0$$

then

$$\Phi_{11}(kT,T) = \exp(\alpha(kT)T)\cos(\beta(kT)T) + a_{11}(kT).\exp(\alpha(kT)T)\sin(\beta(kT)T)/\beta(kT) \quad (32)$$

where

$$\alpha(kT) = a_{11}(kT) + a_{22}(kT)$$

Note that $a_{ij}(kT)$ are known functions of the original micro-biological parameters Y, μm and K. A detailed calculation is given in (Poulisse, 1985). The system noise process $\{v(kT),k=0,1,..,N-1\}$ has known mean and covariance

$$E\{v(kT)\} = 0,\ \mathrm{cov}(v(kT),v(jT)) = V(kT,T)\delta_a(k-j) \quad (33)$$

$$V(kT,T) = \int_0^T \Phi'(kT,\lambda)V\Phi'^T(kT,\lambda)d\lambda$$

The control u(kT) is related to the original control u'(kT) in the following way

$$u(kT) = u'(kT)+R^{-1}(kT,T)M^T(kT,T) \quad (34)$$

Note that although Q and R in the integral criterion (16) are constant, the corresponding quantities Q(kT,T) and R(kT,T) in the sum criterion (26) are time-varying. It will be convenient for our further work to use the innovations representation of the state model (25, 6)

$$x((k+1)T) = \Phi(kT,T)x(kT)+\Gamma(kT,T)u(kT) + K(kT,T)W(kT) \quad (35)$$

$$y(kT) = Cx(kT) + w(kT) \quad (36)$$

This can be done without any - further - loss of generality (Anderson and Moore, 1979).

ARMAX-representation

Following (Goodwin, Ramadge and Caines, 1981) the state space system can be rewritten as an ARMAX representation

$$y(kT)+\alpha_1 y((k-1)T)+\alpha_2 y((k-2)T)+\alpha_3 y((k-3)T) = \beta_1 u((k-1)T)+\beta_2 u((k-2)T)+\beta_3 u((k-3)T) + w(kT)+\gamma_1 w((k-1)T)+\gamma_2 w((k-2)T)+\gamma_3 w((k-3)T) \quad (37)$$

For ease of notation we have dropped the time arguments from the parameters $\alpha_i,\beta_j,\gamma_l$ (i,j,l = 1,2,3).

The ARMAX-parameters can be written in terms of the parameters of the state space system, e.g.

$$\alpha_1 = -(\Phi_{11}+\Phi_{22}+\Phi_{33}) \quad (38)$$

$$\beta_1 = C_1\Gamma_1+C_3\Gamma_3 \quad (39)$$

$$\gamma_1 = C_1+C_3+\alpha_1 \quad (40)$$

where Φ_{ij}, Γ_i and C_i are elements of Φ, Γ and C respectively - again dropping the time arguments. A detailed account can be found in (Poulisse, 1985). Since the coefficients of the matrices $\Phi(kT,T)$ and $\Gamma(kT,T)$ can be given as explicit functions of the original microbiological parameters, it follows that the relationship between the ARMAX-parameters and the original micro-biological parameters is known. However, this relationship is very complicated and difficult to interpret. The sum criterion (26) must now be formulated in terms of input and output rather than in terms of states and input. Assuming that the matrix Q(kT,T) in (26) can be written as

$$Q(kT,T) = C^T\bar{Q}(kT,T)C \quad (41)$$

where $\bar{Q}(kT,T) \geq 0 \quad \forall\, k$, it follows that the appropriate sum criterion is

$$J' = E\{\sum_{k=0}^{N-1} (y(kT)-y^*(kT))\bar{Q}(kT,T)(y(kT)-y^*(kT))$$

$$+ u(kT)R(kT,T)u(kT)\} \qquad (42)$$

where

$$y^*(kT) = Cx^*(kT) \qquad (43)$$

ADAPTIVE CONTROL

We shall give a very brief account on this subject. Details can be found in (Chen and Caines, 1983, 1984; Poulisse, 1985). First we must approximate the ARMAX-representation (37):

$$y(kT)+\alpha_1 y((k-1)T)+\alpha_2 y((k-2)T)+\alpha_3 y((k-3)T) \approx$$

$$\beta_1 u((k-1)T)+\beta_2 u((k-2)T)+\beta_3 u((k-3)T)+w(kT)$$

$$+\gamma_1 w((k-1)T)+\gamma_2 w((k-2)T) \qquad (44)$$

Define

$$A(k) = \begin{pmatrix} -\alpha_1 & 1 & 0 \\ -\alpha_2 & 0 & 1 \\ -\alpha_3 & 0 & 0 \end{pmatrix} B(k) = \begin{pmatrix} \beta_1 \\ \beta_2 \\ \beta_3 \end{pmatrix} C(k) = \begin{pmatrix} 1 \\ \gamma_1 \\ \gamma_2 \end{pmatrix}$$

$$H = (1\ 0\ 0) \qquad (45)$$

Then (44) can be rewritten as

$$x(k+1) = A(k)x(k)+B(k)u(k)+C(k)w(k+1) \qquad (46)$$

$$y(k) = Hx(k);\ x^T(0) = (y(0)00) \qquad (47)$$

Here we have left out the sampling time T in the time arguments to avoid confusion with earlier parts of this paper. The following conditions are needed: $(A(k),B(k))$ is controllable and $(A(k),D(k))$ is observable, where Dk is any matrix such that $D(k)D(k)^T = H^T Q(kT,T)H$; furthermore all zero's of $(1-\alpha_1 z-\alpha_2 z^2-\alpha_3 z^3)$ lie outside the closed unit disk. Define

$$\theta^T = (\theta 1,..,\theta 8) = ((\gamma 1-\alpha 1),(\gamma 2-\alpha 2),$$

$$-\alpha 3,\beta 1,\beta 2,\beta 3,\gamma 1,\gamma 2) \qquad (48)$$

and

$$\eta^T(kT) = (y(kT),y((k-1)T),y((k-2)T),u(kT),$$

$$u((k-1)T),u((k-2)T),-\eta^T((k-1)T)\theta(kT),$$

$$-\eta^T((k-2)T)\theta((k-1)T) \qquad (49)$$

The estimates $\theta(kT)$ can be obtained from the following recursion (Chen, 1982)

$$\theta((k+1)T) = \theta(kT)+a(kT)P(kT)\eta(kT)$$

$$(y((k+1)T)-\eta^T(kT)\theta(kT)) \qquad (50)$$

$$P(kT) = P((k-1)T)-a((k-1)T)P((k-1)T)$$

$$\eta((k-1)T)\eta^T((k-1)T)P((k-1)T) \qquad (51)$$

$$a(kT) = (1+\eta^T(kT)P(kT)\eta(kT))^{-1} \qquad (52)$$

$$P(0) = sI_8;\ s > 0 \qquad (53)$$

Because the relationship between the ARMAX-parameters and the micro-biological parameters is known, we can calculate the initial estimate $\theta(0)$ by using crude estimates of the microbiological parameters - say from off-line experiments and invoking equation (48).

The class of admissible controls u consists of sequences $\{u(kT)\}$ such that:

I. $u(kT)$ is F_k^y -measurable and

$F_k^y = \sigma\{y(0),y(T),..,y(kT)\}$ and

II. $E\{||u(kT)||^2\} <$

In our fed-batch situation these conditions are satisfied. Define

$$r(kT) = trP^{-1}(kT) = 1+\sum_{k=0}^{k-1}||\eta(kT)||^2 \qquad (55)$$

en let λ_{max}^k en λ_{min}^k be the maximum respectively minimum eigenvalue of $P(kT)$. Under certain conditions for the noise polynominal $\gamma_1 z+\gamma_2 z^2$, including a strictly positive real condition and under the condition that the ratio of λ_{max}^k and λ_{min}^k is uniformly bounded, then if $r(kT) \to$ als $k \to$ our parameter estimates are strongly consistent, i.e. converge a.s. (Chen, 1983).

We comment that in Chen's algorithm it is assumed that the parametervector θ is time--invariant, while in our situation θ is time varying. However, if θ is only slowly time--varying, as is often the case in fermentation processes, we closely approximate the situation in Chen's theorem. Detailed comments and calculations are given in (Poulisse, 1985).

Let $\tau_\eta(c)$ be the stopping time associated with the parameter estimates first exceeding some specified bound c (Chen and Caines, 1984). For a sharp specification of this bound c, the relationship between the ARMAX-parameters and the original micro-biological parameters can be exploited (Poulisse, 1985). Define

$$A_k(c) = A(\theta(kT))1_{[\eta < k < \tau_\eta(c)]} \qquad (56)$$

for any $\eta \geq 0$ where $1_{[\ ]}$ is the indicatorfunction and $A(\theta(kT))$ the estimate of $A(k)$ based upon $\theta(kT)$. Likewise, for the other quantities. Than we have the following adaptive LQ control law (Chen and Caines, 1984; Poulisse, 1985)

$$u_k(n) = -(R+B_k^T(c)S_k(c)B_k(c))^{-1}B_k^T(c).$$

$$(S_k(c)A_k(c)X_k+b_k(c)) \quad k > n \qquad (57)$$

$$= 0 \qquad k \leq n$$

where we assumed that $R(kT,T)$ in (42) is constant (R) and $S_k(c)$ is given by the solution of the algebraic Ricatti-equation

$$S_k(c) = A_k^T(c)S_k(c)A_k(c)-A_k^T(c)S_k(c)B_k(c).$$

$$(R+B_k^T(c)S_k(c)B_k(c))^{-1}B_k^T(c)S_k(c)A_k(c)+H^TQH \qquad (58)$$

where we assumed that $Q(kT,T)$ in (42) is constant (Q). X_k in (57) is given by the following recursion

$$X_{k+1} = A_k(c)X_k+B_k(c)\ddot{u}_k(n)+C_k(c)(y((k+1)T)-$$

$$HA_k(c)X_n-HB_k(c)u_k(n) \qquad (59)$$

$$X_n^T = (y(nT)00)$$

$b_k(c)$ in (57) is given by

$$b_k(c) = -H^TQy^*(kT) - \sum_{j=1}^{N} [A_k(c) - B_k(c) (R + B_k^T(c) S_k(c) B_k(c))^{-1} B_k^T(c) S_k(c) A_n(c)]^{jT} H^T Qy^*((k+j)T) \quad (60)$$

N sufficiently large.

Chen and Caines (1984) have shown that the calculated control minimizes 1/N J', where J' is given by (42) for N sufficiently large. Again we have in fact a somewhat different situation here, but in case of slow time variation we closely approximate the result of Chen and Caines, see (Poulisse, 1985) where also some results are given for the time varying case using techniques given in (Moore and Anderson).

CONCLUSIONS

A mathematical model has been derived for a fairly large class of fed-batch fermentation processes. A discrete-time adaptive LQ control law has been given. The original continuous-time problem has been transformed analytically into a discrete-time problem. In this way, we have not lost contact with the original micro-biological system even when arriving at the ARMAX-representation. In the slowly time-varying situation investigated here we have only given approximate results.

REFERENCES

Alvarez Gallegos, J. and Alvarez Gallegos, J. (1982). Optimal control of a class of discrete multivariable nonlinear systems. Applications to a fermentation process. Trans. ASME, Journal of Dynamic Systems Measurement and Control, 104, 212-217.

Anderson, B.D.O. and Moore, J.B. (1979). Optimal Filtering, Prentice Hall Inc.

Chen, H.F. (1982). Strong consistency and convergence rate of least squares identification. Scientia Sinica (Series A), 25,7, 771-784.

Chen, H.F. and Caines, P.E. (1983). Adaptive linear quadratic control for stochastic discrete time systems. Research report, Department of Electrical Engineering, Mc.Gill University.

Chen, H.F. and Caines, P.E. (1984). Adaptive linear quadratic control for stochastic discrete time systems. Preprints IFAC 9th World Congres, vol. VII, 150-154.

Dekkers, R.M. (1982). State estimation of a fed-batch Baker's yeast fermentation. Proc. IFAC Workshop on Mod. and Control of Biotech. Processes. 201-211.

Dekkers, R.M. (1983). Dynamic optimization of a fed-batch fermentation process. Proc. IASTED Symp. on applied control and identification. 3.34-3.39.

Dochain, D. and Bastin, G. (1984). Adaptive identification and control algorithms for non-linear bacterial growth systems. Automatica 20,5, 621-631.

Goodwin, G.C., Ramadge, P.J. and Caines, P.E. (1981). Discrete time stochastic adaptive control, SIAM. J. Control and Optim. 19,6, 829-853.

Hazewinkel, M. (1984). Notes on (the philosophy of) linearization. Report Centre for Mathematics and Computer Science, Amsterdam.

De Koning, W.L. Equivalent discrete optimal control problem for randomly sampled digital control systems. Int. J.Syst.Sci. 11,7,841-850.

Moore, J.B. and Anderson, B.D.O. Coping with singular transition matrices in estimation and control stability theory. Int. J. Control 31,3, 571-586.

Pirt, S.J. (1975). Principles of micro and cell cultivation. Blackwell, London.

Poulisse, H.N.J. (1985). Adaptive LQ control of continuous culture and (fed)-batch fermentation processes, report department of Physics and Meteorology, Agricultural University, Wageningen.

Roels, J.A. (1983). Energetics and kinetics in biotechnology, Elsevier, Amsterdam.

Stephanopoulos, G. and San Ka-Yiu (1984). Studies on on-line bioreactor identification. I. Theory. Biotechn. and Bioeng. XXVI, 1176-1181.

COMPUTER CONTROL OF IMMOBILIZED PRODUCT FERMENTATIONS

Henry Y. Wang, Gregory F. Payne* and Frank M. Robinson**

Department of Chemical Engineering, The University of Michigan, Ann Arbor, MI 48109–2136, USA
**Present address: Department of Chemical Engineering, Cornell University, Ithaca, NY 14853, USA*
***Present address: Molecular Genetics, Inc., Minnetonka, MN, USA*

Abstract. A generalized engineering approach has been shown to improve final concentration and productivity of a fermentation process through in-situ product removal. Feedback regulation and permeability control are typical biological control mechanisms in which microbial cells prevent the over-production of the synthesized products. A computer coupled extractive fermentation system has been built to carry out automatic control and optimization of various fermentation processes. Our experimental results indicate that in-situ product removal will increase the final product yield and productivity of a fermentation. Additional advantages of this extractive fermentation include reduced processing time and induced partial purification inside the fermentor.

Keywords. Fermentation, Adsorbent resin, On-line extraction, Computer control, Biocontrol.

INTRODUCTION

The on-line use of computers has now become a well accepted technique in the chemical and petroleum industries (Evans, 1977). Computers provide extremely rapid data acquisition, reduction, and storage. They, therefore, either enable an operator to reach faster, more effective decisions or may themselves be programmed to dictate operational policy. Computers have since been applied in biochemical industries in both on-line and off-line operation. For off-line applications, a time lag exists between data acquisition and data interpretation. Reduction of manual data, the testing of fermentation models and the simulation of the fermentation under changes of various parameters have been the major application of off-line operation (Humphrey, 1973; Nyiri, 1972; Dobry and Jost, 1977; Weigand, 1978). With technical advances in both computer technology and bioengineering, the application of on-line computer monitoring and control to increase product yield and/or productivity of a fermentation becomes not only possible but also economically desirable. For the on-line application, however, it must be assumed the monitoring sensors are in existence and well characterized and so is the fermentation process. Unfortunately, these are the major bottlenecks in monitoring and control of most industrial fermentation processes (Flynn, 1982; Wang, 1984).

Besides the problem of sensing the disturbances of the fermentation processes, the other key limitation is the lack of manipulated variables to effect fermentation control. Temperature and pH can only be manipulated within a narrow range that is inherent to the biological system. Only dissolved oxygen and nutrient feed have been the major targets of control. In this paper, we will show in-situ product removal techniques such as introducing adsorbents directly to the fermenting culture can be used to effect process control. We demonstrated that such an approach can indeed enhance product synthesis and avoid product degradation.

Cycloheximide Fermentation - a model system

Cycloheximide is one of a family of glutarimide antibiotics produced by several *Streptomyces spp.* Its discovery in 1946 by Whiffen *et al.* occurred in the broth of streptomycin producing strains of *Streptomyces griseus*. The importance of this antibiotic has grown from its ability to inhibit the growth of various eucaryotic yeasts and fungi at low concentrations. Biochemical and fermentation studies of this cycloheximide fermentation had already been extensively carried out (Kominek, 1975a, 1975b). The time course for the conventional cycloheximide fermentation had been observed to be quite reproducible (Fig. 1). The cycloheximide concentration reached a peak at the 168th hour. This corresponded to the depletion of glucose in the medium. The medium pH increased and product degradation occurred (cycloheximide is less stable at higher pH's). The maximum cycloheximide concentration that we could achieve in the laboratory fermentors was about 850 mg/l.

The accumulation of cycloheximide (P) in the fermentation medium by *Streptomyces griseus* had been shown to be feedback regulated and to be a dynamic process involving both synthesis (r_P) and degradation (r_D) of the antibiotic (Kominek, 1975a):

$$\frac{dP}{dt} = r_P - r_D \qquad (1)$$

There were some evidence that the degradative reaction of the cycloheximide fermentation was part of the biosynthetic pathway but in reverse. The biosynthetic system responsible for cycloheximide formation appeared to be inducible and retain a high level of activity throughout the fermentation provided the biomass was viable and residual glucose was present. Cycloheximide, on the other hand, had been found to interact with its own production due to feedback regulation. If t_i is the time when antibiotic synthesis begins, the production of cycloheximide can be represented by:

$$r_P = 0, \qquad t < t_i \qquad (2)$$

$$r_P = r_{Po} - K\ P \qquad t \geq t_i \tag{3}$$

The system responsible for degradation of cycloheximide is strongly dependent on pH and subject to catabolite inhibition in the presence of glucose (Kominek, 1975; Payne, 1984). Threrefore, the accumulation of cycloheximide is a product of the balance between synthesis and degration (Eqn. 1). When product regulation, most likely product inhibition becomes great enough to balance the rate of synthesis to that of degradation, accumulation of the antibiotic ceases:

$$r_{Po} = r_{Pm}\ (1 - \frac{P}{Pm}) \tag{4}$$

r_{Pm} is the maximum cycloheximide synthesis rate and can be related to viable cell concentration that is capable to synthesize the antibiotic. Pm is the maximum product concentration that can be achieved when $r_{Po} \rightarrow 0$. Our experimental results indicated Pm was about 1,200 mg/l for the particular strain of Streptomyces griseus used (Payne, 1984).

In order to maximize the antibiotic synthesis rate, r_P throughout the fermentation, it is essential to keep the cycloheximide concentration in the fermentation broth low and adequate residual glucose be present to prevent product degradation.

Immobilized Product Fermentation (IPF)

End product regulation, the reduction of enzyme synthesis (repression) and/or enzyme activity (inhibition), is a common control mechanism used by biological systems including antibiotic production (Martin and Demain, 1980). Product degradation can also play a significant role in the production of antibiotics from fermentation processes. One of the engineering approaches is to remove the product continuously out of the aqueous broth during fermentation so that the inhibitory and degradation phenomena can be minimized. After reviewing and experimenting various separating agents such as non-aqueous solvents, ion-exchange resins, activated carbon and other extractants, we found the non-ionic polymeric adsorbents (XAD-2, XAD-4, Rohm and Haas, Philadelphia, PA) had the best adsorption capacity of the antibiotic and minimal toxic effect on the producing cells (Wang et al., 1981; Wang, 1983).

By adding sterilized polymeric resins such as XAD-4 directly into the fermentation broth (Fig. 2), the profiles for both glucose utilization and cycloheximide production were changed drastically when 6 wt % of XAD-4 were added to a 14 liter cycloheximide fermentation during the 48th hour (cell growth already stopped). The glucose utilization rate increased from 0.36 g/l-hr for a normal cycloheximide fermentation to 0.5 g/l-hr indicating the metabolic activities of the cells were changed in the presence of the solid adsorbents. Even though the bulk cycloheximide concentration in the fermentation broth was reduced to zero, the total antibiotic synthesis rate was increased to 17.5 mg/l-hr with a product yield of 36.5 mg CH/g of glucose. These were substantially higher than the standard control fermentation. All the antibiotic molecules produced were adsorbed onto the adsorbent resin. Conceivably, the cycloheximide loaded resin could be harvested and thus drastically reducing the volume of material that needed to be further processes and purified.

Computer Coupled Extractive Fermentation System

Since the time and the amount of resin addition were very critical to the performance of the Immobilized Product Fermentation (IPF), we decided to build a computer coupled extractive fermentation system to control and optimize this type of fermentation. The IPF has several unique characteristics. It tends to foam frequently after addition of the adsorbent resins (Wang et al. 1982). This was due to the fact that the resins reduced the surface active agents in the fermentation broth. The foam was quite unstable and can easily be controlled through both mechanical and chemical means. Since the cells of the IPF were metabolically more active, glucose feed programming was needed to prevent glucose depletion (Payne, 1984).

A highly instrumented 14 liter fermentation system (Chemapec, Woodbury, NY) is coupled to a HP 9845 computer with a HP 3497A interface (Hewlett Packard, Palo Alto, CA). The microcomputer has been employed to collect data, analyze data and effect real time control (Fig. 3). HP 3497A interface system is used for A/D and D/A conversion. Other peripherals include thermal printer, dual channel cassette tape units for data and program storage and DRT with compatible graphics for plotting graph and time history of the fermentation. Many physico-chemical sensors have been used to monitor the fermentation. As shown in Fig. 3, the computer can read data from each of the nine established sensors. Feedback control loops have been developed for pH control dissolved oxygen control, foam control and glucose feed control using this computer coupled fermentation system (Wang et al., 1982). The control of in-situ product removal can be achieved in two modes. First, addition of sterile adsorbent resin slurry (60 gm of XAD-4 in 600 ml of distilled water) through the top of the fermentor into the fermentation broth. Shots of the resin slurry can be achieved on a daily basis. Dispersing the resins in the culture broth has several engineering problems such as clogging the valves and attrition of the adsorbent resins by the impeller. The resin particles are also abrasive to the tank wall and the shaft seals. A separate extractor containing the adsorbent resins was built to accommodate these problems (Fig. 4). The fermenting broth can be recirculated between the fermentor and the extractor. In a practical situation, the innoculation tank near the fermentor is ideally suitable to be converted to the extractor (Fig. 4).

In the presence of the solid adsorbents, very low bulk cycloheximide concentration can be achieved:

$$\frac{dP}{dt} = r_P - \frac{dmQ}{dt} \tag{5}$$

Q is the loading capacity of the polymeric resin added and m is the amount of the adsorbent resin added. As long as dmQ/dt is greater than or equal to r_P, no accumulation of P can be observed. The adsorption rate can be modeled as follows:

$$\frac{dmQ}{dt} = K_s\ (P - P^*) \tag{6}$$

$$P^* = f(Q) \tag{7}$$

K_s is the adsorption rate constant of cycloheximide and P^* is the surface cycloheximide concentration which is in equilibrium to Q. K_s has been shown to vary drastically due to the characteristics of the fermenting broth and internal pore diffusion through the adsorbent particles (Fig. 5). Since Q decreased with time, m must be added to compensate this problem. When a continuous glucose feed was used in combination with direct resin addition, final cycloheximide concentration as high as 3 gm/l of broth could be obtained (Fig. 6). Over 95% of the produced cycloheximide stayed on the XAD-4 resin particles. The cycloheximide synthesis rate would eventually

The cycloheximide synthesis rate would eventually drop down and glucose correspondingly started to accumulate (Fig. 6). We are now interested in investigating this decline which may be due to nutrient depletion and intracellular enzyme degradation.

The time of resin addition had been shown to be quite important. Even though these adsorbent resins had been found to have less undesirable effects on microbial growth, the optimum time of addition was still right after microbial growth had completed and the antibiotic synthesis had just begun. The amount of adsorbent resin needed depended on the adsorption capacity of the resin used and the programming of the resin addition. By adding the entire amount of adsorbent resin in the beginning as a single shot, 6 wt % of the adsorbent resins (XAD-4) based on the total fermentation volume was needed. When the resin slurry addition to the fermentor was programmed by the computer, only half the amount of adsorbent resins were needed. When smaller amounts of adsorbent were used, the resin became saturated with cycloheximide and cycloheximide started to accumulate in the broth. If too much XAD-4 adsorbents were used, other factors such as nutrient depletion and physical shear started to play a significant role. The final cycloheximide concentration was reduced (Fig. 7).

CONCLUSION

An immobilized product fermentation (IPF) has been developed by adding sterilized polymeric adsorbents directly into the fermenting broth and reducing the product concentration inside the fermentor. A computer coupled extractive fermentation system has been developed to control this type of fermentation. Foam control and programmed glucose feed must be instituted to optimize the on-line extraction. Only half the total amount of adsorbent is needed if the adsorbent addition is controlled by the computer.

Since the cycloheximide antibiotic molecules are preferentially adsorbed onto the polymeric adsorbent. The product recovery is facilitated by separating the large adsorbent particles (> 0.5mm) from the fermentation broth and subsequent solvent extraction. The small amount of resin used indicates that a significant volume reduction occurred in this process which should simplify further purification steps.

ACKNOWLEDGMENTS

The financial support from NSF (CPE-8406686) is gratefully appreciated. We would also like to acknowledge the technical assistance of K. Sobrosky.

REFERENCES

Dobry, D. D. and J. L. Jost (1977). Computer Application to Fermentation Operations. Annual Reports on Fermentation Processes. 1, 95-114.

Evans, L. B. (1977). Impact of the Electronics Revolution on Industrial Process Control. Science, 195, 1146-1151.

Flynn, D. S. (1983). Instrumentation for Fermentation Processes. Modelling and Control of Biotechnical Processes, IFAC, Helsinki, Finland, 5-12.

Humphrey, A. E. (1973). Rationale for and Principles of Computer Coupled Fermentation Systems. The First European Conference on Computer Process Control in Fermentations. Institute National de La Recherche Agronomique, Dijon, France, 1-13.

Kominek, L. A. (1975a). Cycloheximide Production by Streptomyces griseus: Control Mechanisms of Cycloheximide Biosynthesis. Antimicrob. Chemotherap. 7, 856-860.

Kominek, L. A. (1975b). Cycloheximide Production by Streptomyces griseus: Alleviation of End-Product Inhibition by Dialysis-Extraction Fermentation. Antimicrob. Chemotherap. 7, 861-863.

Martin, J. F. and A. L. Demain (1980). Control of Antibiotic Biosynthesis. Microbiological Reviews, 44, 230-251.

Nyiri, L. K. (1972). A Philosophy of Data Acquisition, Analysis and Computer Control of Fermentation Processes. Developments in Ind. Microbiol. 13, 136-145.

Payne, G. F. (1984). Enhanced Secondary Metabolite Production through On-Line Product Removal. Ph.D. Thesis, The University of Michigan, Ann Arbor, MI.

Wang, H. Y., L. A. Kominek and J. L. Jost (1981). On-Line Extraction Fermentation Processes. Advances in Biotechnology, M. Moo-Young (Ed.), 1, 601-604.

Wang, H. Y. (1983). Integrating Biochemical Separation and Purification Steps in Fermentation Processes. Annals of the New York Academy of Sciences. 413, 313-321.

Wang, H. Y., D. Connors and F. M. Robinson (1982). Computer Coupled Extractive Fermentation System. Presented at the AIChE Annual Meeting, Los Angeles, CA Nov. 14-19.

Wang, H. Y. (1985). Sensor Development for Fermentation Monitoring and Control. Biotechnol. Bioeng. Symp. 14, in press.

Weigand, W. A. (1978). Computer Applications to Fermentation Processes. Annual Reports on Fermentation Processes 2, 43-72.

Wiffen, A. C., N. Bohonos and R. Emerson (1946). The Production of An Antifungal Antiobiotic by Streptomyces griseus. J. Bacteriol. 52, 610-611.

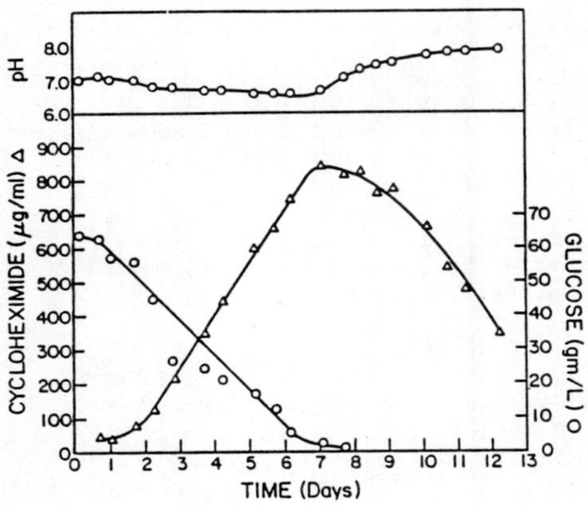

Fig. 1. Typical fermentation profile for the batch production of cycloheximide by S. griseus.

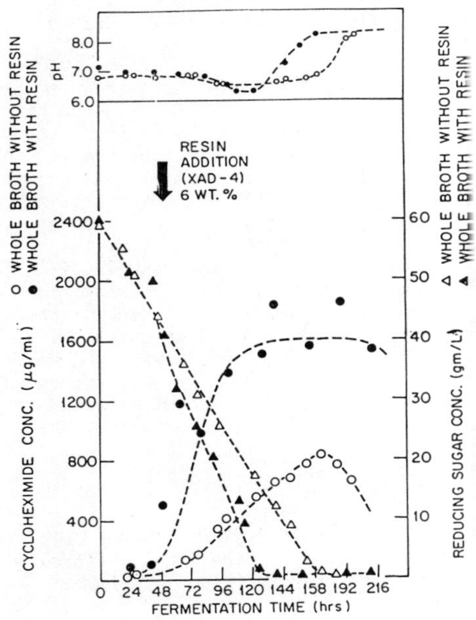

Fig. 2. Comparison between a normal cycloheximide fermentation and immobilized product fermentation using 6 wt % XAD-4 resin. The adsorbents were added at the 48th hr.

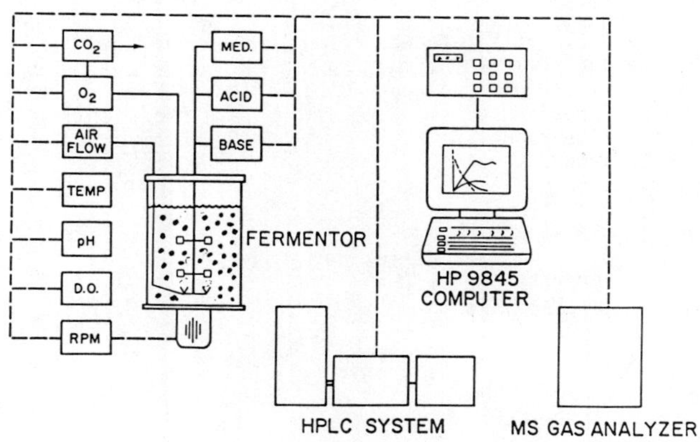

Fig. 3. Schematic diagram of the computer coupled extractive fermentation system.

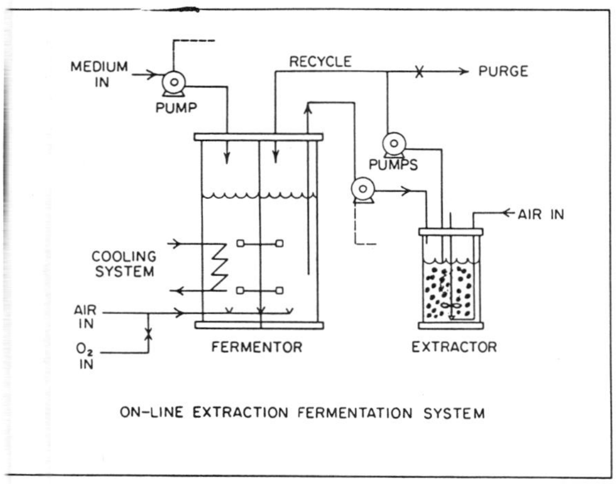

Fig. 4. Computer coupled extractive fermentation system. The on-line extraction was achieved using an external extractor. The fermentation broth was recirculated between the fermentor and the extractor.

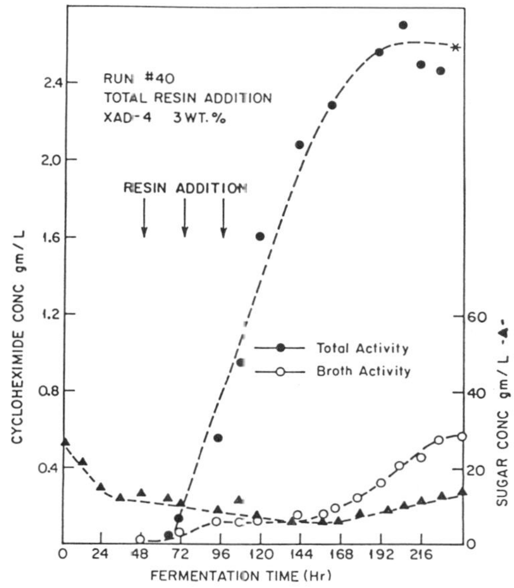

Fig. 5. The specific adsorption rate constants of the free XAD-4 resins and immobilized XAD-4 resins in hydrogel as a function of the loading capacity in aqueous cycloheximide broth.

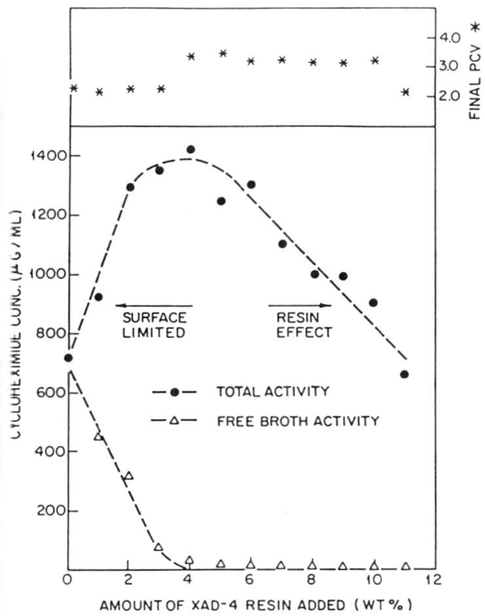

Fig. 6. Immobilized product fermentation with programmed addition of the adsorbent resins to the fermentor.

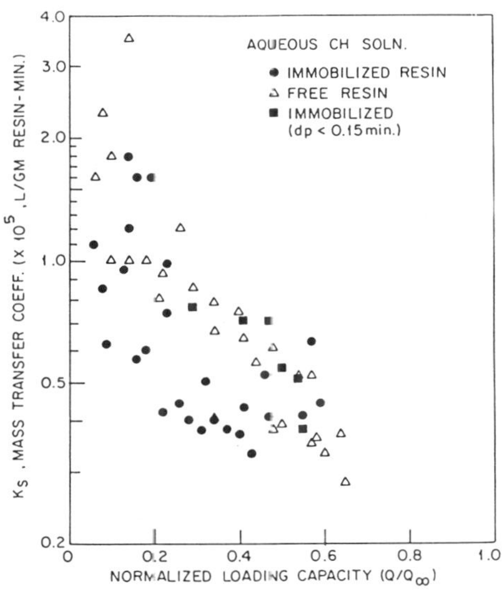

Fig. 7. Cycloheximide production as a function of the various amount of XAD-4 resin added. The optimal amount occurred between 3 - 6 wt %. The time of addition was at the 48th hour.

MODELLING AND CONTROL OF A BUBBLE FREE AERATED BIOREACTOR FOR MAMMALIAN CELL CULTIVATION

K.-H. Bellgardt, R. Luttmann, R. Schulz, G. W. Piehl, J. Nothnagel and J. Lehmann

GBF-Gesellschaft für Biotechnologische Forschung mbH, Mascheroder Weg 1, D-3000 Braunschweig, FRG

Abstract. A relativly new field in biotechnology is the cultivation of animal cells in large scale bioreactors. For this, particular requirements for reactor design, process control and careful treatment of the cells must be met.
In this paper, modelling and control strategy are presented for a 30 l and a 70 l-bioreactor with a special bubble-free membrane aeration system. The plants are used for repeated batch and continuous cultivations of modified mouse L 929 Tk-cells.

The cells are growing on micro-carriers and producing glycosilated human interferon-ß constitutively. For a mathematical description of the plant, a lumped parameter model of the reactor liquid phase and a distributed parameter model of the bubble-free aeration system was developed. The latter includes mass balances of the gas phase and of gas exchange with the fermentation broth. The control strategy is based on system analysis with this model and the design steps of the control are explained. The control algorithm was implemented on a freely programmable DDC-process-station MICON P-200. (Manufacturer: VDO, Hannover, Germany).

Keywords: Tissue culture bioreactor, membrane aeration, mathematical modelling, pH-control, pO_2-control, MICON P-200.

TISSUE CULTURE BIOREACTOR WITH MEMBRANE AERATION

In general the cultivation of human and animal cells in a pilot plant scale is a not solved problem in biotechnology. Tissue cultures are very shear sensitive and must be aerated and agitated very carefully and without stress. Figure 1 shows the design of a bioreactor, especially developed in the GBF for cultivation of human and animal cells /1/. These types of reactors are used in a scale of 1 l up to 70 l. In this system two ways of bubble- and foam-free aeration are possible. The first one is a direct gas exchange between top gas phase and liquid phase of the reactor. Another way is a mass transfer across a hydrophobic tube-like membrane. The tube is winded around a frame. The whole aeration system inside the liquid phase is excentrically suspended at a vertical drive. During one rotation each point of the rack describes a small horizontal circle. This avoids high turbulence and cell damage.
In this reactor the two variables pO_2 and pH are controlled by a proper mixture of air, oxygen, nitrogen and carbon dioxide flows into the membrane tube. The top gas phase of the reactor is aerated with a constant air flow. The increasing oxygen uptake rate of the cells has to be satisfied at a constant pO_2 in the liquid phase. This will be realized with an increasing oxygen supply from the gas phase. On the other hand, the control of pH leads to a carbon dioxide transfer from gas into liquid phase or vice versa. This depends on the actual pH-value and on the amount of carbon dioxide and acid produced by the cells.
The control of the two variables is strongly coupled via the gas phase and can lead to contradictory requirements. The increase of CO_2-supply at too high pH-values results automatically in a lower oxygen concentration in the inlet gas and therefore in a lower oxygen transfer rate.
The control strategie to supply the bioreactor with a proper gas mixture is developed with mathematical system analysis and simulation of the bioprocess.
First we well discuss the model and analyze the process. In a second step the development and practical realisation of a control system is shown.

SYSTEM CHARACTERIZATION

The mathematical description of this reactor leads to a combination of a classical stirred tank reactor model (lumped parameter system) and a plug flow reactor model (distributed parameter system). In fig. 2 the three subsystems of the membrane reactor are shown schematically in order to develop mass balance equations. It is assumed that the components in the liquid phase and in the top gas phase (index T) are ideally mixed, whereas the membrane gas phase (index M) shows a different behaviour. The first two subsystems can be modelled with only time dependent ordinary differential equations. The gas concentration profiles in the membrane tube are time (t) and space (z) dependent. For starting of the modelling, this third subsystem must be described by partial differential equations, but these are simplified later to ordinary equations.

MATHEMATICAL MODEL

Biological system equations

The liquid phase model can be divided into a biological and a physical/chemical part.

The biological model describes cell growth, oxygen uptake, CO_2-production and acid formation.

All these are influencing the control variables pH and pO_2.

The cells, a modified mouse L 929 Tk-cell line for production of human interferon-ß, are growing on micro-carriers /2/. It is assumed, that the carriers form a homogenous suspension.

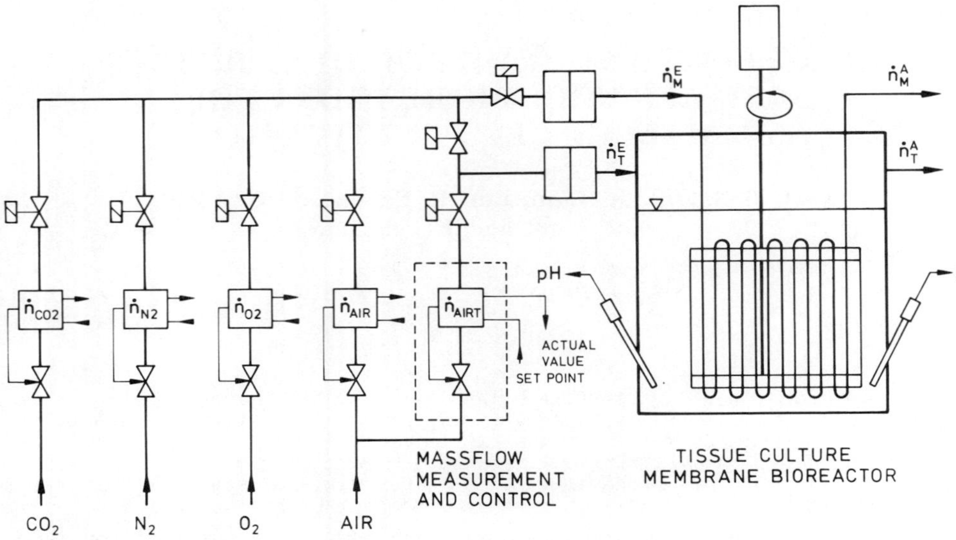

Fig. 1. Bioreactor for cultivation of human and animal cells

The growth restriction by total overlay of the micro-carrier surface can be roughly described by a growth reaction of first order. Detachment and cell desactivation is modelled by a death term of second order.

The cell balance equation yields

$$\frac{dX(t)}{dt} = Y_{X/O}\ OUR(t) - K_d\ X^2(t) \tag{1}$$

with the initial cell number

$$X(0) = X_o. \tag{1a}$$

The oxygen uptake rate OUR,

$$OUR(t) = q_{O/Xmax} \frac{pO_2(t)\ O^*_{Fo}}{K_O + pO_2(t)\ O^*_{Fo}} X(t), \tag{2}$$

as well as the carbon dioxide production rate CPR,

$$CPR(t) = K_{C/O}\ OUR(t), \tag{3}$$

are assumed to be proportional to cell growth. During the process cells are producing acid (lactate).

The acid production rate APR,

$$APR(t) = q_{A/Xmax} \frac{K_{IA}}{K_{IA} + pO_2(t)\ O^*_{Fo}} X(t), \tag{4}$$

is inhibited by a high pO_2-level.

The lactate balance equation

$$\frac{dC_A(t)}{dt} = APR(t) - D(t)\ C_A(t) \tag{5}$$

with

$$C_A(0) = C_{Ao} \tag{5a}$$

completes the simple biological model.

For the description of the dynamics of the control variables, dissolved oxygen tension pO_2 and pH-value, model equations for dissolved O_2 and CO_2 are needed. Both variables are strongly influenced by mass exchange with the two gas phases of the reactor.

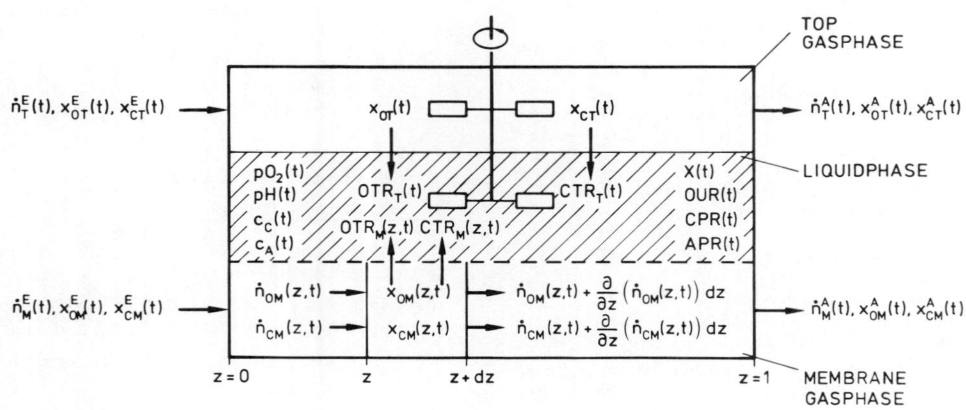

Fig. 2 Subsystems of a membrane bioreactor

Oxygen mass balance equations

The oxygen balance equation of the liquid phase is

$$\frac{dpO_2(t)}{dt} O^*_{Fo} = OTR_T(t) + \overline{OTR}_M(t) - OUR(t) + D(t)\,[O_R - pO_2(t)\,O^*_{Fo}] \quad (6)$$

with

$$pO_2(0) = pO_{2o}. \quad (6a)$$

The oxygen transfer rate from the top gas phase OTR_T,

$$OTR_T(t) = k_La_T(t)\,O^*_{Fo}\left[\frac{p(t)\,x_{OT}(t)}{p_o\,x_{OTo}} - pO_2(t)\right], \quad (7)$$

is a space independent oxygen source, whereas the overall oxygen transfer rate from the membrane gas phase $\overline{OTR}_M$,

$$\overline{OTR}_M(t) = \int_0^1 OTR_M(z,t)\,dz$$

$$= \int_0^1 k_La_M(z,t)\,O^*_{Fo}\left[\frac{p_M(z,t)\,x_{OM}(z,t)}{p_o\,x_{OMo}} - pO_2(t)\right]dz, \quad (8)$$

is a spatial integrated value.

In eq. (7) and eq. (8) the volumetric oxygen transfer coefficients k_La_G (G = T,M) are defined as

$$k_La_G(t) := k^{O2}_{LG}(t)\,\frac{A_G}{V_F(t)} \quad (9)$$

with unknown oxygen mass transfer coefficients k^{O2}_{LG}, and in this reactor well known gas/liquid interfacial areas A_G and liquid volume V_F.

For calculation of the oxygen transfer rates (eqs. (7) and (8)) the values of the oxygen mole fraction in the top gas phase x_{OT} and in the membrane gas phase x_{OM} are needed. Therefore the oxygen mass balances of the gas phases are also necessary. The oxygen mass balance of the top gas phase yields a differential equation for x_{OT},

$$\frac{dx_{OT}(t)}{dt} = \frac{R\,T(t)\,V_F(t)}{p_T(t)\,M_{O2}\,V_T(t)}\,[Q_{O2T}(t) - OTR_T(t)] \quad (10)$$

and includes the oxygen supply rate of the gas phase Q_{O2},

$$Q_{O2T}(t) = \frac{\dot{n}^E_T(t)\,M_{O2}\,[x^E_{OT}\{1-x_{CT}(t)\}-x_{OT}(t)\{1-x^E_{CT}\}]}{V_F(t)\qquad[1 - x_{OT}(t) - x_{CT}(t)]}, \quad (11)$$

and the oxygen transfer rate OTR_T (eq. (7)).

The not measureable space dependence of x_{OM} can be calculated from distributed parameter balances of all gas components /3/.

The oxygen mass balance of the membrane gas phase

$$\frac{\partial}{\partial t}(n_M(z,t)\,x_{OM}(z,t)) = \frac{\partial}{\partial z}(\dot{n}_M(z,t)\,x_{OM}(z,t)) - \frac{OTR_M(z,t)}{M_{O2}} \quad (12)$$

is a plug flow equation with mass transfer loss into the liquid phase. At stationarity and with some assumptions and simplifications an equation for the spatial course of the mole fraction x_{OM},

$$x_{OM}(z,t) = [x^E_{OM}(t) - x^*_{OM}(t)]\,e^{-\varphi_{O2}(t)\,z} + x^*_{OM}(t), \quad (13)$$

with the abbreviations

$$x^*_{OM}(t) = \frac{pO_2(t)\,p_o\,x_{OMo}}{p^E_M(t)} \quad (14)$$

and

$$\varphi_{O2}(t) = \frac{k^{O2}_{LM}(t)\,A_M\,p^E_M(t)}{H_{O2}\,\dot{n}^E_M(t)\,M_{O2}} \quad (15)$$

is found.

After integration of eq. (8) with the substitutions eqs. (13) to (15), $\overline{OTR}_M$ is analytically known. Now the pO_2 equation (eq. (6)) is solvable and the dynamics of the first control variable are known.

pH-calculation

The situation for the second control variable pH is more complicated. The premise for its calculation is a solution of the total carbon dioxide balance equation

$$\frac{dC_C(t)}{dt} = CPR(t) - CTR_T(t) - \overline{CTR}_M(t) \quad (16)$$

with

$$C_C(0) = C_{Co}\ . \quad (16a)$$

This balance equation includes CO_2 production of cells (eq. (3)) and mass transfer terms to both gas phases. The latter in eq. (16) are calculable with similiar methods as shown for eqs. (7) to (15). The hydrogen-ion concentration c_{H+} depends on five components. The (molar) dissoziated parts of pH-buffer (c_B), total carbon dioxide (c_C), produced acid (c_A), titration medium (c_T) and water lead to a quasi-stationary hydrogen-ion balance

$$c_{H+}(t) = \frac{K_B\,c_B}{K_B + c_{H+}(t)} + \frac{K_C\,c_C(t)}{K_C + c_{H+}(t)} + \frac{K_A\,c_A(t)}{K_A + c_{H+}(t)} + \frac{K_W}{c_{H+}(t)} + c_T\ . \quad (17)$$

From the iterative solution of eq. (17) the pH-value

$$pH(t) = -\log_{10} [c_{H^+}(t)] \tag{18}$$

can be calculated.
Based on the derived model, unknown biological parameters will be identified from experimental results.

PROCESS SIMULATION

For design of the control the process dynamics, biological and physical model parameters must be quantitatively known. These were identified using a long time simulation of a batch culture and a short time simulation of a shift experiment.

Fig. 3 shows the simulation results of the batch cultivation in the 30 l-reactor over a period of 90 hours. In the upper part measurements (symbols) and simulation results (lines) of cell number (X) dissolved oxygen pO_2, lactate C_A and pH are plotted. The lower part of fig. 3 shows the time courses of process input variables, the molar flow of O_2, CO_2, and air into the membrane tube, and the outlet gas mole fraction of O_2 from the top gas phase.

Fig. 4 shows the simulation results of a CO_2-aeration shift experiment. In the upper part measurements and simulation results of dissolved oxygen pO_2, O_2-mole fraction x^A_{OT} and CO_2-mole fraction x^A_{CT} at exit of the top gas phase and the calculated oxygen uptake rate OUR are plotted. The lower part shows the manipulated CO_2-flow rate $\dot{n}_{CO2}$, the CO_2-mole fraction x^A_{CM} at the exit of the membrane gas phase, the produced lactate concentration C_A, and the pH-value.

The identified parameters of the simulations in fig. 3 and fig. 4 are summarized in table 1.

The parameters of the O_2-balance, $Y_{X/O}$, k_La_M and k_La_T are very difficult to determine. Exit gas O_2-measurements of the membrane tube were not available during this time period, because of too low gas flow rates. These were below the minimum requirement of the measurement equipment. The parameter identification, only from pO_2 and top gas mole fractions, is probably not error free

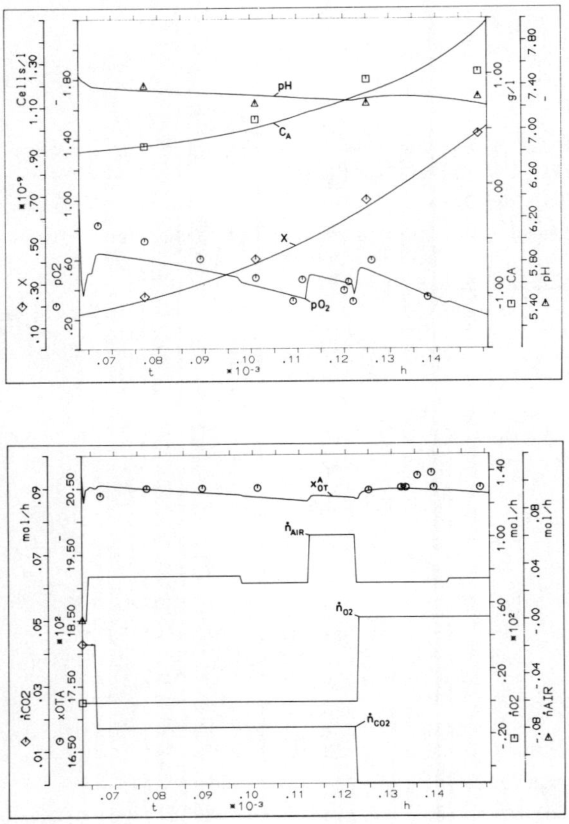

Fig. 3. Simulation of a batch cultivation

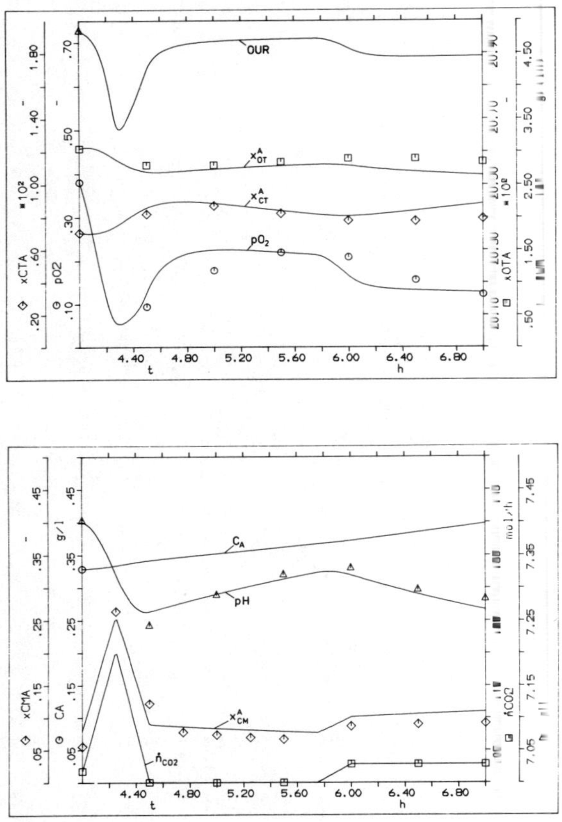

Fig. 4. Simulation of a CO_2-shift experiment
$X \approx 2.3 \cdot 10^9$ Cells/l, $\dot{n}_{O2} = 0.27$ mol/h,
$\dot{n}_{N2} = \dot{n}_{AIR} = 0.$, $\dot{n}_{AIRT} = 1.33$ mol/h

Growth of the mouse cells on the carriers is exponentially in this early period of the process. To compensate the increasing oxygen demand and the drift of pH by acid production of the cells, the oxygen supply through the membrane was shifted up at t = 107 h and 117 h and the CO_2 flow was reduced. This can also be seen in the pO_2-course.

TABLE 1 Results of Parameter Identification

μ_{max}	$= 3.9\cdot10^{-2}$ h^{-1}	$k_L^{O2}a_M$	$= 1.32$ h^{-1}
K_d	$= 6.0\cdot10^{-12}$ l $Cells^{-1}$ h^{-1}	$k_L^{O2}a_T$	$= 0.88$ h^{-1}
K_O	$= 2.3\cdot10^{-4}$ g l^{-1}	$k_L^{CO2}a_M$	$= 0.51$ h^{-1}
$Y_{X/O}$	$= 1.7\cdot10^{9}$ $Cells$ g^{-1}	$k_L^{CO2}a_T$	$= 0.15$ h^{-1}
$q_{A/Xmax}$	$= 2.1\cdot10^{-11}$ g $Cells^{-1}$ h^{-1}		
K_{IL}	$= 1.0\cdot10^{-3}$ g l^{-1}		
$K_{C/O}$	$= 0.73$ -		

All parameters must be taken as rough mean values. Especially the changing growth conditions of the cells on the microcarriers (monolayer, multilayer) influence the biological parameters very strongly. The proposed model is only the first step of mathematical description of this process and must be further improved in conjunction with measurement techniques.
In spite of a very rough model with several simplifications the dynamics of the process are described good enough to encourage us to use the model as a base for the configuration of a controller.

CONTROL CONCEPT

The aim of our work was to replace an existing control system with two seperate PID-loops for pH and pO_2 by a coupled system. Single loop control of the process has several disadvantages.

1. Oxygen transfer rate and carbon dioxide transfer rate are strong coupled via the gas mixing device.
2. Depending on the sign of the pH-setpoint deviation, the system has to be aerated with CO_2 or CO_2 has to be outgassed from the liquid phase. This requires different control strategies.
3. Exponential growth of the cells leads to wide gain variation of the plant.

Because of this, conventional control loops work unsatisfactory and could not stabilize the control variables in the entire operating area. Furthermore simple control cannot ensure a most economic use of oxygen and nitrogen.

Our intention in designing a control system is to decouple the oxygen- (pO_2) and CO_2-(pH) system. The basic idea is suggested for the special case of a top gas phase aeration with gas mixing. At stationarity and with the simplification RQ_T=1 following equations hold true for oxygen supply:

$$OUR(t) = OTR_T(t) = Q_{O2T}(t) = \frac{Q^E_{O2Tmax}(t)}{Q^E_{O2Tmax}(t) + OTR_{Tmax}(t)} (x^E_{OT}(t) - p_{O2}(t)\, x_{OAIR}) \tag{19}$$

with

$$Q^E_{O2Tmax}(t) = \frac{\dot{n}^E_T(t)\, M_{O2}}{V_F(t)} , \tag{20}$$

$$OTR_{Tmax}(t) = \frac{k_La_T(t)\, O^*_{Fo}}{x_{OAIR}} , \tag{21}$$

$$x^E_{OT}(t) = \frac{\dot{n}_{O2}(t) + \dot{n}_{AIR}(t)\, x_{OAIR}}{\dot{n}^E_T(t)} , \tag{22}$$

and

$$\dot{n}^E_T(t) = \dot{n}_{O2}(t) + \dot{n}_{N2}(t) + \dot{n}_{CO2}(t) + \dot{n}_{AIR}(t). \tag{23}$$

Similar equations could be written for CO_2-supply or desupply.

Equation (19) shows, that the magnitude of the mass transfer rates depends directly on the entrance mole fractions (x^E_{OT} or x^E_{CT}) in the gas phase.

When the pO_2-controller varies Air-, O_2- or N_2-flow rates in that manner, that the CO_2-mole fraction is kept constant, pH-control is nearly not disturbed. The corresponding holds true for varying the CO_2-flow rate by pH-control at a constant oxygen mole fraction.
In the case of CO_2-desupply the control has to be carried out by changing the total gas flow rate $\dot{n}^E_T$ without disturbing pO_2-control.
The control system is shown schematically in fig. 5. The described behaviour can be realized with two PID-controllers and a non-linear output characteristic.

With the limiting characteristic

$$R\{r\} = \begin{cases} r & r \geq 0 \\ 0 & r < 0 \end{cases} \tag{24}$$

and a common gas phase G (G=T, M), the positive parts of the controller outputs are

$$\dot{n}_{CG}(t) = R\{\dot{n}^*_{CG}(t)\} \tag{25}$$

and

$$\dot{n}_{OG}(t) = R\{\dot{n}^*_{OG}(t)\} . \tag{26}$$

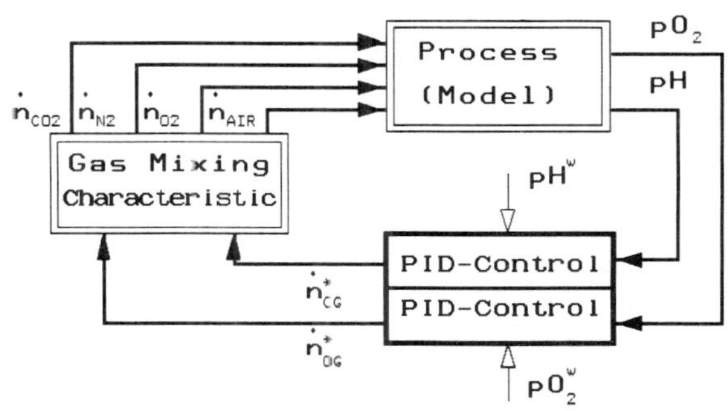

Fig. 5. Simultanuous control of pO_2 and pH

The controllers request an overall gas flow rate

$$\dot{n}^E_G(t) = \dot{n}_{Gmin} + R\{-\dot{n}^*_{CG}(t)\} + R\{\dot{n}_{OG}(t) + \dot{n}_{CG}(t) - \dot{n}_{Gmin} - R\{-\dot{n}^*_{CG}(t)\}\} \tag{27}$$

The manipulated variables are

$$\dot{n}_i(t) = R\{\dot{n}^*_i\} \quad , \quad i = N2, AIR, O2, CO2 \tag{28}$$

where

$$\dot{n}^{*}_{N2}(t) = \dot{n}_{G}(t) - \dot{n}_{CG}(t) - \frac{\dot{n}_{OG}(t)}{x_{OAIR}} \quad (29)$$

$$\dot{n}^{*}_{O2}(t) = - \frac{x_{OAIR}}{1 - x_{OAIR}} \dot{n}^{*}_{N2}(t) \quad (30)$$

$$\dot{n}^{*}_{AIR}(t) = \dot{n}_{G}(t) - \dot{n}_{O2}(t) - \dot{n}_{CG}(t) - \dot{n}_{N2}(t) \quad (31)$$

$$\dot{n}^{*}_{CO2}(t) = \dot{n}^{*}_{CG}(t) \quad (32)$$

Equation (30) ensures, that O_2-flow and N_2-flow cannot be present at the same time.

SIMULATION OF AN AUTOMATIC CONTROLLED PROCESS

Figure 6 shows the dynamic simulation of the control variables pO_2 and pH and of the four input variables for a batch experiment which is similar to that in fig. 3.
The process is controlled via the membrane gas phase by two simulated PID-controllers and the gas mixing characteristic described earlier.
The top gas phase of the reactor is aerated with a constant rate

$$\dot{n}^{E}_{T}(t) = \dot{n}_{AIRT} \quad . \quad (33)$$

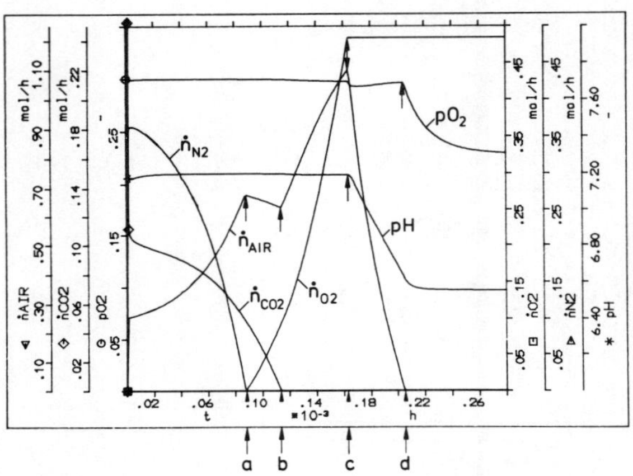

Fig. 6. Time courses of the control variables pO_2 and pH and the membrane input variables

The control action can be divided into five characteristic periods:

t = 0 : pH = pH^W = 7.2 is set with certain buffer and titration concentrations and aeration with CO_2. pO_2 = pO_2^W = 30 % air saturation is set by an additional air/nitrogen gas mixture.

0 < t < a: CO_2-production and O_2-consumption increase exponentially. The control is done by decreasing CO_2- and N_2-aeration and increasing AIR-aeration.

a < t < b: $\dot{n}_{N2}$ = 0, the growing oxygen demand is sattisfied by feeding pure oxygen (O_2) and simultanuously decreasing AIR-aeration.

b < t < c: $\dot{n}_{CO2}$ = 0; the high CO_2-production would lead to a decrease of pH. The direction of mass transfer of carbon dioxide must be changed and the total aeration has to increase. The oxygen demand is further guaranted by a proper O_2/AIR ratio.

c < t < d: The oxygen mass flow actuator is at the upper limit. The control of both variables is as from now impossible. pH-control is automatically switched off and pO_2 is controlled by increasing the oxygen mole fraction in the gas phase. This is realized by decreasing the air flow.

d < t: $\dot{n}_{AIR}$ = 0; Now pO_2 is also uncontrollable. After some time both control variables become stationary, because cell growth is also in a stationary phase.

The discussed control strategy was realized in a robust industrial microcomputer control system MICON P-200 (VDO Hannover, FRG) /4/.

The control is tested in scale up experiments in a 70 l bioreactor with membrane aeration for cultivation of mammalian cell cultures.

CONCLUSION

For cultivations of animal and human cells in large scale bioreactors a membrane tube aeration system was developed. For the control of the two variables pO_2 and pH via appropiate flows of oxygen, nitrogen, carbon dioxide and air through the membrane a special control concept was presented.

For a mathematical description of the plant, a lumped parameter model of the reactor liquid phase and a distributed parameter model of the bubble-free aeration system was developed. Two experiments were simulated to test the model and to identify unknown biological and mass transfer parameters. The first one is a batch experiment with modified mouse L 929 Tk-cells for production of human interferon-β. The second one is a short shift experiment, where a O_2-puls was supplied to the culture.

The developed control strategy for the reactor is based on system analysis with this model. In our control concept, the two strongly interdependent variables are decoupled by a nonlinear gas mixing characteristic. The control was tested by simulations of a typical batch cultivation and then implemented on a freely programmable DDC-process-station MICON P-200.

REFERENCES

/1/ Lehmann, J. et al. (1985). Verwendung von offenporigen, hydrophoben Membranen zur blasenfreie Begasung von tierischen Zellkulturen. ACHEMA 85, Frankfurt a. M.

/2/ Delzer, J. et al. (1984) Production of Human ß-Interferon in Mouse-L-Cells. 3rd European Congress on Biotechnology, Vol. I, pp. 179-181.

/3/ Luttmann, R. et al. (1983). Model Development, Parameter Identification, and Simulation of SCP Production Processes in Air Lift Tower Reactors with External Loop, I and II. Computers and Chemical Engineering Vol. 7 No. 1, pp. 43-50 and pp. 51-63.

/4/ Gêci, F., Sievers, H. (1983). Prozeßleitsystem für Fermentationsprozesse in Produktion und Forschung. Chemie Technik 12, Heft 7, S. 37-40.

NOMENCLATURE

L (length), T (time), M (mass), M_M (mole mass),
K (temperature), - (dimensionless)

A	gas/liquid interfacial area,	L^2
APR	acid production rate,	$M\ L^{-3}\ T^{-1}$
$a = A/V_F$	specific interfacial area,	L^{-1}
C_I, c_I	concentration of component I,	M, M_M
CPR	carbon dioxide production rate,	$ML^{-3}\ T^{-1}$
$D = F/V_F$	dilution rate,	T^{-1}
F	feed rate,	$L^3\ T^{-1}$
K_{O2}	(Monod-) oxygen saturation constant,	$M\ L^{-3}$
K_A, K_B, K_C	dissoziation constants,	$M\ M_M^{-1}\ L^{-3}$
K_W	dissoziation constant of water,	$M^2\ M_M^{-2}\ L^{-6}$
$K_{C/O}$	CPR/OUR ratio	-
K_d	cell death coefficient,	$L\ M^{-1}\ T^{-1}$
k_L^{O2}	oxygen mass transfer coefficient,	$L\ T^{-1}$
k_La	volumetric oxygen mass transfer coefficent,	T^{-1}
M_I	mole mass of component I,	M_M
N_S	stirrer speed,	T^{-1}
n	mole number,	$M\ M_M^{-1}$
$\dot{n}$	mole flow rate,	$M\ M_M^{-1}\ T^{-1}$
O_J	oxygen concentration in phase J,	$M\ L^{-3}$
OUR	oxygen uptake rate,	$M\ L^{-3}\ T^{-1}$
OTR	oxygen transfer rate,	$M\ L^{-3}\ T^{-1}$
p	pressure,	$M\ L^{-1}\ T^{-2}$
pH	pH-Value,	-
pO_2	dissolved oxygen tension,	-
Q_{O2}	oxygen supply rate,	$M\ L^{-3}\ T^{-1}$
q	specific reaction rate,	T^{-1}
R	gas constant,	$M_M\ L^2\ K^{-1}\ T^{-2}$
R_I	reaction rate of component I,	$M\ L^{-3}\ T^{-1}$
RQ	respiratory quotient,	-
S	substrate concentration	$M\ L^{-3}$
T	temperature,	K
t	time,	T
V	volume	L^3
$\dot{V}$	volume flow rate,	$L^3\ T^{-1}$
X	cell (biomass) concentration,	$M\ L^{-3}$
x	gas mole fraction,	-
$Y_{X/I}$	yield coefficient of substrate I	-
z	related membrane coordinate,	-

Subscripts

A	acid (lactate)
AIR	air
B	pH-buffer
C,CO2	carbon dioxide
F	liquid phase
G	gas phase
H^+	hydrogen ion
I/J	component I per component J
M	membrane gas phase
max	maximum rate
N2	nitrogen
O,O2	oxygen
P	product
R	reservoir
S	substrate
T	top gas phase
T	titration
W	water
X	cell mass
c	initial value

Superscripts

A	at the exit
d	death
E	at the entrance
gr	growth
O2	oxygen
*	saturation value

COMPUTER CONTROL OF AN AEROBIC BATCH PROCESS IN A TWO REACTOR SYSTEM

J. Van Breugel*, G. Ch. van Eybergen*, A. J. Overwater*, G. Honderd and H. B. Verbruggen****

**Computer Group, Laboratory of Bioengineering, Department of Chemical Technology Delft University of Technology, Julianalaan 67, 2628 BC Delft, The Netherlands*
***Laboratory of Control Engineering, Department of Electrical Engineering, Delft University of Technology, Delft, The Netherlands*

Abstract. Aerobic processes tend to show gradients in the oxygen concentration in the liquid, when performed on a large scale. In parts of the reactor productivity decreases, due to oxygen limitation. Controllers capable of maintaining a sufficiently high level of oxygen throughout the entire reactor volume are necessary. Some intermediate steps in the design of such controllers are described. These controllers are tested in a pseudo full scale reactor system, consisting of two separate small scale reactors: one aerated, the other reactor stirred only, without aeration. The two compartments are interconnected by a circulation flow. Two ways of control are described, both of which function satisfactory. However, they both use information available in a well defined bench scale system, but not on a full scale. Therefore a controller has been designed, combining online parameter estimation and a Luenberger observer. This combination is the subject of present research.

Keywords. Biology, chemical industry, computer control, feedback/feedforward, industrial control, modelling, observers, parameter estimation, state estimation.

INTRODUCTION

One of the main research items in the laboratory of bioengineering is the mathematical modelling of biotechnical processes. In describing these processes, empirical or mechanistic models can be used. Empirical models lack a physical foundation and are not suitable for extrapolation purposes. It is our intention to develop control algorithms applicable to bench scale as well as industrial scale reactors, therefore attention is focused on mechanistic modelling. It should be emphasized that only those mechanisms relevant to the control of the process have to be included in the model. Otherwise, a highly descriptive model, that cannot be verified and is too complex to use in a control structure, results. Secondly, all models and the controllers containing them, should be tested in practical situations with real processes. At the moment, the majority of the publications dealing with the modelling and control of biotechnical processes describe controllers containing a model, which are used to control another, often more compex, model of the same process.

An adequate process model has several applications:

- the estimation of unknown parameters, e.g. kinetic, mass transport, hydrodynamical parameters.
- the estimation of non-observable state variables. It is a well known fact, that there is a tremendous lack of suitable sensors.
- the optimization of a certain performance index defined in the model, that represents a desirable state or sequence of states (trajectory).

At the moment three processes are being studied:

- the production of gluconic acid from glucose in an aerobic batch process
- the production of ethanol from glucose in an anaerobic continuous process
- the purification of wastewater in an up-flow anaerobic sludge blanket reactor

In this paper the gluconic acid production process will be highlighted. During this process gluconic acid is produced from glucose by the bacterium *Gluconobacter oxydans*. If the dissolved oxygen concentration becomes too low, the growth rate of the bacteria and the production rate of gluconic acid are limited. On the other hand a high concentration of oxygen is useless and control at such a level will only result in high costs, either for power input for the stirrer or for expensive pure oxygen. The practical importance of this research is on the field of optimal production with respect to minimal costs for oxygen transfer.

THEORY

Process description

As an introduction to the control of this process, a single reactor set-up was used. In this reactor several strategies were tested: P, PI and PID controllers both applied to the stirrer speed and the composition of the inlet gas flow. A simple model describing the dynamics of the process combined with a simple feedback controller proved to be well satisfying (v. Breugel, 1985). On an industrial scale the stirrer speed cannot simply be changed, so oxygen enrichment of the inlet air flow is used to supply the bacteria with enough oxygen. Moreover, gradients in the oxygen concentration occur in a full-scale reactor. A maximal productivity however, only is reached when a sufficient oxygen level is realized in the entire reactor volume. On a bench scale this can be simulated in a two reactor system (fig. 1) (Oosterhuis, 1984). One reactor is being stirred and aerated, in the other reactor only stirring takes place. The two reactors are interconnected by a pump simulating the circulation flow induced by the impellers on a large scale.

Development of a model

Based upon mass balances for the two reactor system the following model equations can be derived for the process:

The biomass:

$$\frac{dC_x}{dt} = \mu_{max}\, C_x \tag{1}$$

For the aerated volume:

$$\frac{dC_1}{dt} = \frac{\phi_c}{V_1}\,(C_2 - C_1) + OTR - r_1 \tag{2}$$

For the non-aerated reactor:

$$\frac{dC_2}{dt} = \frac{\phi_c}{V_2}\,(C_1 - C_2) - r_2 \tag{3}$$

The oxygen transfer rate in the aerated reactor:

$$OTR = k_l a \left(\frac{C_g}{H} - C_1\right) \tag{4}$$

The dissolved oxygen is assumed to be at a level that permits the bacteria to grow at a maximal rate. As already stated there is a direct relation between growth and oxygen consumption:

$$r_1 = r_2 = r_{max} = \frac{\mu_{max} C_x}{Y_{ox}} = r_{max}(0)\, \exp(\mu(t - t_0)) \tag{5}$$

The sensors used are an oxygen electrode and a gasphase oxygen analyser. Assuming the gasphase in both the reactors to be mixed ideally, the following mass balance holds:

$$\frac{dC_g}{dt} = \frac{\phi_g}{V_g}\,(C_{g,in} - C_g) - OTR\,\frac{V_1}{V_g} \tag{6}$$

On the analogy of the full-scale reactor, where the dissolved oxygen concentration is measured near the stirrer, an oxygen electrode is placed in the aerated reactor. Its response is described using a first order approximation:

$$\frac{dC_e}{dt} = k_e\,(C_1 - C_e) \tag{7}$$

MATERIALS AND METHODS

The bacterium *Gluconobacter oxydans* ATCC 621H is used in all experiments. A Biolafitte/Applikon bv 15 S fermenter is used as the larger reactor, with a liquid volume of 10 l. The volume of the smaller reactor is 2.5 l. The medium in the fermenters consists of 0.4% w/w yeast extract and 10% w/w glucose. The experimental set-up is shown in Fig. 1.

The pump, simulating the circulation flow, has a capacity of 0-20 l/min. The exact amount of liquid that circulates during an experiment is determined with acid pulse-pH response measurements, using special software, that allows large numbers of pH response curves to be fitted with fluid flow models.

Temperature and pH are controlled with dedicated analog closed loop controllers at values of 30 °C and 5 respectivily. The flows of oxygen and nitrogen can be adjusted separately, using mass-flow controllers,

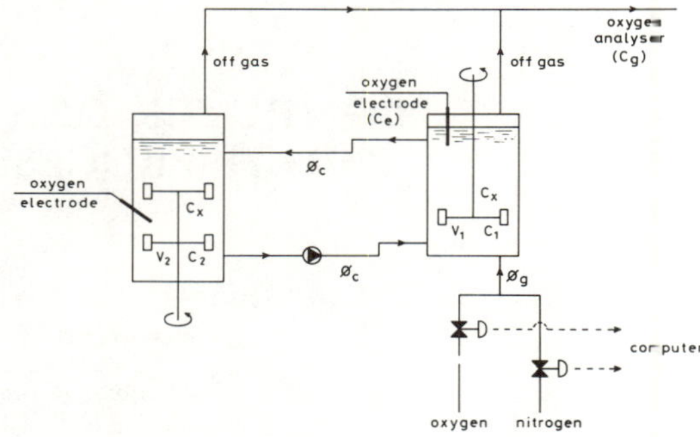

Fig. 1 Experimental set-up

(Brooks 5850). The total gasflow is kept to a value of 2 l/min. The oxygen concentration in the liquid is determined with a Schott and a Biolafitte electrode. The oxygen concentration in the off-gas is measured using a paramagnetic analyzer (Taylor Servomex OA 540).

Computer hardware of Hewlett Packard is used (v. Breugel, 1984). Data acquisition and control programs are running on the HP1000/A600 microcomputer. Online parameter/estimation and state observer software are available on the faster HP1000/F900 minicomputer. The two computers are interconnected using DS1000/IV networking software. All applications software is self-designed and written in FORTRAN 77.

DESIGN OF THE CONTROLLERS

The model presented above holds for a bench scale set-up as well as for a full-scale reactor. Since it is our final goal to design a controller suitable for application to an industrial scale, part of our knowledge of the bench scale system should not be used to develope a controller. The distribution of the reactor volume over an aerated and a non-aerated part and the size of the circulation flow are process parameters that are known in a two reactor set-up, but cannot be determined in a large scale reactor. The same holds for the oxygen concentration in the non-aerated volume.

Some intermediate steps will be necessary before such a (pseudo-) large scale controller can be realized.

Control based upon the oxygen electrode in the non-aerated reactor

Equations 1 and 2 are Laplace transformed (Dorf, 1980 and Ogata, 1970).

$$C_1(s) = \frac{C_2(s) + \tau_1 k_l a C_g(s)/H}{s\tau_1 + \tau_1 k_l a + 1} - \frac{\tau_1 r_{max}}{s\tau_1 + 1} \tag{8}$$

$$C_2(s) = \frac{C_1(s)}{s\tau_2 + 1} - \frac{\tau_2 r_{max}}{s\tau_2 + 1} \tag{9}$$

τ_1 and τ_2 are the mean residence time in V_1 and V_2 respectively.
The two last terms in eq. 8 and 9 are assumed to be a disturbance of the process and are omitted. $C_2(s)$ has to be controlled by $C_g(s)$, this transfer-function results from eq. 8 and 9:

$$\frac{C_2(s)}{C_g(s)} = \frac{\tau_1 k_l a/H}{\tau_1 \tau_2 s^2 + s(\tau_1 \tau_2 k_l a + \tau_1 + \tau_2) + \tau_1 k_l a} \tag{10}$$

To calculate the poles of the process, τ_1, τ_2 and $k_l a$ have to be known. In the experimental configuration used, τ_1 = 10 s, τ_2 = 40 s and k_1a = 0.05 s^{-1}. So, the poles of the process are P_1 = - 0.0075 s^{-1} and P_2 = -0.17 s^{-1}. The small pole makes it difficult to control the process. It is logical to design a controller compensating this pole. Furthermore, an integration is included:

$$\frac{C_g(s)}{E(S)} = K_c \frac{s+P_1}{s} \quad (11)$$

$$\text{with } E(s) = \text{Setp}(s) - C_2(s) \quad (12)$$

When pole P_1 is assumed to be compensated ideally, and the transferfunction of the sensor is not taken into account, the closed-loop roots of the characteristic equation can be calculated from:

$$s^2 + P_2 s + \frac{K_c k_1 a}{H \tau_2} = 0 \quad (13)$$

For an overshoot of 5% after a step disturbance of the setpoint K_c is:

$$K_c = 11.5 \text{ H} \quad (14)$$

Control based upon the oxygen electrode in the aerated reactor

When the control of C_2 is succesful, $\frac{dC_2}{dt}$ = 0. From equation 2 it follows:

$$C_1 = C_2 + \frac{V_2}{\phi_c} r_{max} \quad (15)$$

or

$$C_2 = C_1 - \frac{V_2}{\phi_c} r_{max}(0) \exp(\mu (t-t_0)) \quad (16)$$

C_1 has to increase exponentially to keep C_2 at its set-point value. When V_2, ϕ_c and the growth rate of the micro-organisms are known, the right trajectory for C_1 as a function of time can be calculated. However, V_2 and ϕ_c are not known exactly and it can vary due to disturbances of the bacterial growth. A pure feedforward control of the process, as eq. 16 implies, will give rise to increasing deviations from the setpoint desired. For this reason, the value of $V_2\, r_{max}(0)/\phi_c$ is adapted by measuring C_2 with a certain time interval. The growth rate, μ, is assumed to be constant and variations in it are compensated for in the estimate of $V_2\, r_{max}(0)/\phi_c$. In fact, feedback is introduced in the controller in this way.

It should be noticed that the oxygen electrodes are assumed to give the right values of C_1 and C_2. The value of C_1 is maintained on the setpoint given by equation 16, by the controller described in the previous chapter.

Control using a state observer

The equations 1 to 7 can be readily used to formulate a Luenberger observer (Luenberger, 1979). The differential equations happen to be linear and can be rewritten as:

$$\dot{\underline{C}} = A\, \underline{C} + B\underline{u} \quad (17)$$

with

$$\underline{C} = (C_x\ C_1\ C_2\ C_g\ C_e)^T \qquad \underline{u} = C_{g,in}$$

The output vector is:

$$\underline{y} = C\, \underline{C} \quad (18)$$

$$A = \begin{bmatrix} \mu_{max} & 0 & 0 & 0 & 0 \\ \frac{-\mu_{max}}{Y_{ox}} & \frac{-\phi_c}{V_1} - k_1a & \frac{\phi_c}{V_1} & \frac{k_1a}{H} & 0 \\ \frac{-\mu_{max}}{Y_{ox}} & \frac{\phi_c}{V_2} & \frac{-\phi_c}{V_2} & 0 & 0 \\ 0 & \frac{V_1}{V_g} k_1a & 0 & \frac{-\phi_g}{V_g} - \frac{V_1}{V_g}\frac{k_1a}{H} & 0 \\ 0 & k_e & 0 & 0 & -k_e \end{bmatrix}$$

$$B = \begin{bmatrix} 0 \\ 0 \\ 0 \\ \frac{\phi_g}{V_g} \\ 0 \end{bmatrix} \qquad C = \begin{bmatrix} 0 & 0 & 0 & 1 & 0 \\ 0 & 0 & 0 & 0 & 1 \end{bmatrix}$$

In this case the observer is:

$$\dot{\hat{\underline{C}}} = A\, \hat{\underline{C}} + B\, \underline{u} + G\, (\underline{y} - C\, \hat{\underline{C}}) \quad (19)$$

Every differential equation is thus extended with two terms:

$g_{i1}\ (C_g-\hat{C}_g)$ and $g_{i2}\ (C_e-\hat{C}_e)$ with i=1 to 5 (5 diff. eq.).

The magnitude of these coefficients, g_{i1} and g_{i2}, has to be determined experimentally. Their value depends on the convergence speed needed and the quality of the measurements of C_g and C_e respectively.

At the start of the experiment the oxygen consumption rate is rather low, due to low biomass concentrations. As a result, changes in C_g will be small and give little information. On the other hand, the oxygen electrode will give accurate values for C_1 in this phase of the process because it has just been calibrated. In the course of the experiment, when biomass increases rapidly, the readings of the gasphase oxygen analyser will become more accurate and the oxygen electrode tends to deviate from its calibration. So, g_{i1} and g_{i2} can be adapted during the experiment.
The value of $\hat{C}_2$ can be used in the feedback controller derived in the equations 8 to 14, or in equation 16 to replace the intermittent measurement of C_2 needed to correct the feedforward controller.

Parameter estimation

What still remains is the necessity of having estimates for the values of the ratio V_1/V_2 and ϕ_c. These estimates are needed to design the feedback controller presented in equations 8 to 14 and to calculate $\hat{C}_2$ when the state observer is used. This problem can be solved by using parameter estimation techniques. On-line parameter estimation software has been developed for this purpose. The set of differential equations is solved for several estimates of the parameters, using a direct grid search method (Beech, 1975). To solve the differential equations, values for the state variables are needed at t = t_0. $C_g(0)$ and $C_e(0)$ can be measured, $C_1(0)$ is assumed to be equal to $C_e(0)$, however, $C_x(0)$ and $C_2(0)$ are unknown. They can be calculated using the state observer in an interactive procedure. As an alternative, they can be considered to be parameters as well. When this alternative is applied, there is no need to use a state observer.

RESULTS AND DISCUSSION

Figure 2 gives the results of an experiment using the feedback controller: the oxygen concentration in the gasphase of the smaller reactor is controlled by comparing C_2 to the desired setpoint. During the first part of the experiment the controller gain, K_c, was 0.8 H, in the second part K_c was 2.4 H. In Fig. 3 the experiment is repeated with a constant gain of 2.4 H.

It can be seen in the figures that the signal in reactor 2, C_2, becomes rather noisy in the course of an experiment. This is due to gradients in the oxygen concentration in the liquid. Mixing is rather poor in this reactor to keep oxygen transfer in this volume to a minimum.

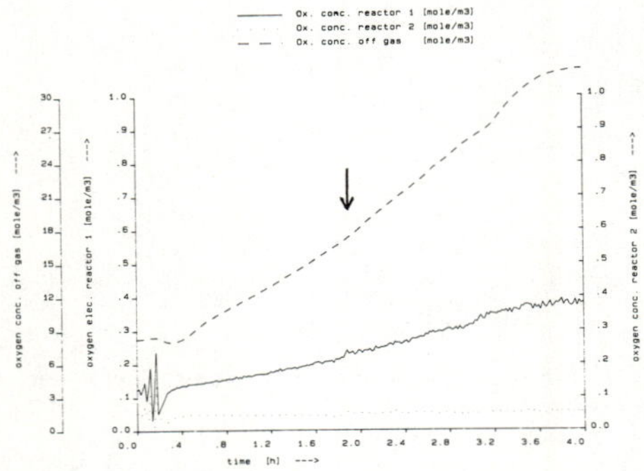

Fig. 2 Results of the feedback controller for successive values of 0.8 H and 2.4 H for K_c (Arrow indicates switching point)

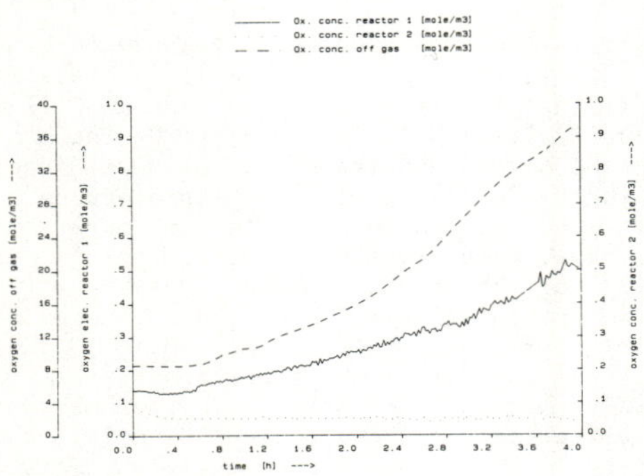

Fig. 3 Results of the feedback controller for a constant value of 2.4 H for K_c

Figure 4 shows the results of an experiment with a controller using a model describing bacterial growth as described with equation 16. During this experiment, C_2 was determined every 15 minutes and used to adjust the parameter group $V_2\ r_{max}(0)/\phi_c$. C_2 is kept well at its setpoint although in the end variations in C_2 occur due to gradients.

In Fig. 5 the same controller is applied, however, with a time interval of one hour between two succesive measurements of C_2. It can be observed, that C_2 deviates significantly from the setpoint at the end of each period.

The observer, derived in the equations 17 to 19, has been applied to data of an experiment with the feedback controller. The result has been plotted in Fig. 6. The observer is well able to reconstruct the value of C_2 from measurements of C_e and C_g. The arrow marks a specific time after which the reactor started foaming. From this time on the observer estimates show an off set. Probably,

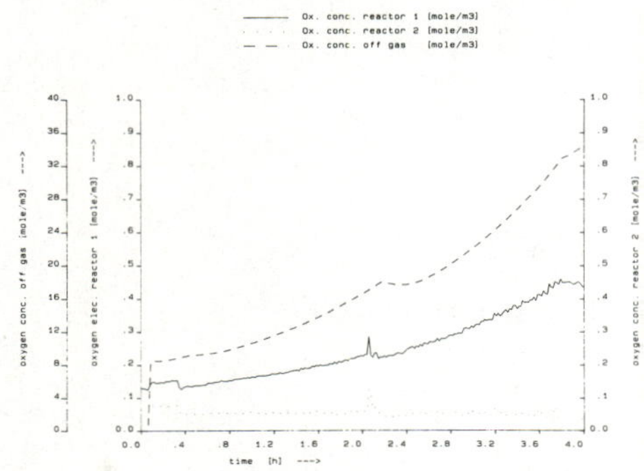

Fig. 4 Results of a controller combining feedback and a model; parameter correction every 15 minutes

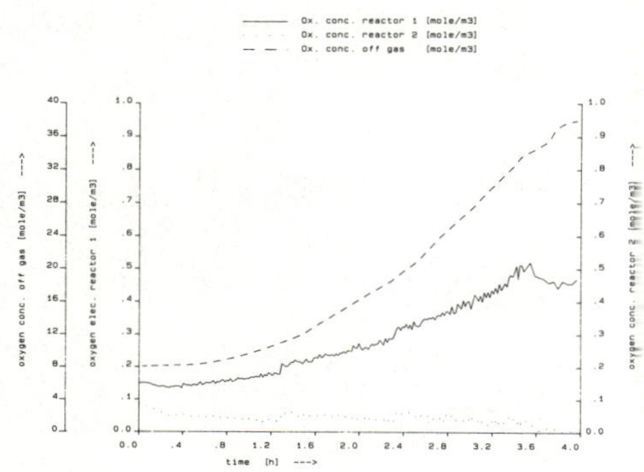

Fig. 5 Results of a controller combining feedback and a model; parameter correction every hour

the value of the kinetic parameters is no longer correct.

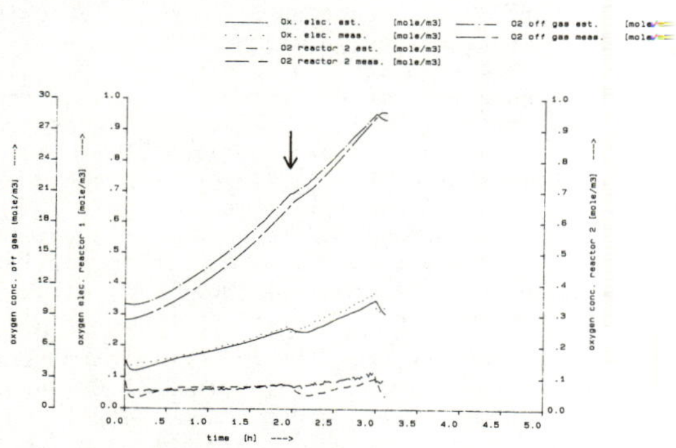

Fig. 6 State estimation using an observer. Plotted are the estimated and measured values for C_2, C_g and C_e

CONCLUSIONS

The problems encountered in the design of a controller applicable to a full scale process have been shown. Incomplete observability of the systems and varying or badly known process parameters are the main problems. The controller using a pure feedback of the oxygen concentration in the non-aerated reactor as well as the controller using a model to predict the trajectory of the oxygen concentration in the aerated reactor functioned well. The state observer derived, is able to reconstruct the state of the reactor system, unless there are disturbances not accounted for in the parameter values or the model structure. The combination of online parameter estimation and a Luenberger observer is the subject of present research.

SYMBOLS AND ABBREVIATIONS

A	state matrix	
B	input matrix	
C	output matrix	
$\underline{c}$	state vector	
$\hat{\underline{c}}$	estimated state vector	
c_x	biomass concentration	C-eq./m^3
c_1	oxygen concentration in reactor 1	mole/m^3
c_2	oxygen concentration in reactor 2	mole/m^3
c_3	oxygen concentration in gasphase	mole/m^3
c_e	oxygen electrode signal	mole/m^3
$c_{3,in}$	oxygen concentration in inlet gasphase	mole/m^3
E	error signal	
G	observer 'gain' matrix	
H	Henry coefficient	-
k_e	time constant oxygen electrode	s^{-1}
$k_l a$	specific volumetric mass transfer coefficient	s^{-1}
K_c	controller gain	
OTR	oxygen transfer rate	mole/m^3s
P	pole	s^{-1}
r	oxygen consumption rate in reactor 1	mole/m^3s
r_2	oxygen consumption rate in reactor 2	mole/m^3s
r_{max}	maximal oxygen consumption rate	mole/m^3s
t	time	s
$\underline{u}$	input vector	
V	volume of reactor 1	m^3
V_2	volume of reactor 2	m^3
V_g	total volume of gas in reactor/system	m^3
$\underline{y}$	output vector	
Y_{ox}	yield of biomass on oxygen	C-eq./mole

greek

ϕ_c	liquid circulation flow	m^3/s
ϕ_g	volumetric gas flow	m^3/s
μ	specific growth rate of micro-organisms	s^{-1}
μ_{max}	maximal specific growth rate of micro-organisms	s^{-1}
τ	mean residence time in reactor 1	s
τ_2	mean residence time in reactor 2	s

VALUES OF RELEVANT PARAMETERS

H	= 33.	
k_e	= 0.1	s^{-1}
$k_l a$	= 0.056	s^{-1}
V_1	= 2.5 10^{-3}	m^3
V_2	= 8.0 10^{-3}	m^3
V_g	= 22 10^{-3}	m^3
Y_{ox}	= 0.25	C-eq /mole
ϕ_c	= 2.7 10^{-4}	m^3/s
ϕ_g	= 3.3 10^{-5}	m^3/s
μ_{max}	= 1.4 10^{-4}	s^{-1}
τ_1	= 10	s
τ_2	= 30-40	s
Setp	= 0.05	mole/m^3

REFERENCES

Breugel, J.van, Driest, P.R.van, Eybergen, G.Ch. van, Heese, R.J.van, Ramakers, B.F.J.M. (1984). The use of a computer in a biotechnological research environment; some applications and results. Innovations in Biotechnology, 331-351. Elseviers Science Publishers bv, Amsterdam, E.H. Houwink and R.R. v.d. Meer editors.

Breugel, J.van, Overwater, A.J., Eybergen, G.Ch. van, Honderd, G. and Kossen, N.W.F. Computer control of the dissolved oxygen concentration during a gluconic acid batch fermentation. Submitted for publication, (1985).

Dorf, R.C. (1980). Modern Control Systems. Addison-Wesley Publishing Company

Luenberger, D.G. (1979). Introduction to Dynamic Systems. John Wiley, New York.

Ogata, K. (1970). Modern Control Engineering. Prentice-Hall Inc.

Oosterhuis, N.M.G. (1984). Scale-up of Bioreactors (a scale down approach). Thesis Delft University of Technology.

Beech, G. (1975). FORTRAN IV in Chemistry; an introduction to computer assisted methods. John Wiley, New York.

PARAMETER ADAPTIVE CONTROL OF THE FED-BATCH PENICILLIN FERMENTATION

G. A. Montague*, A. J. Morris*, A. R. Wright*, M. Aynsley* and A. Ward**

**Department of Chemical Engineering, University of Newcastle upon Tyne, Newcastle upon Tyne, UK*

***Department of Microbiology, University of Newcastle upon Tyne, Newcastle upon Tyne, UK*

Abstract This paper describes a study to investigate the application of parameter adaptive control to mould growth in a fed-batch fermentation for penicillin production. Two models for the penicillin fermentation are described: a new mechanistic model is outlined and compared with a model based upon the work of Bajpai with an extension to include terms for the production of carbon dioxide. The Bajpai model was initially used to represent the penicillin process for simulation purposes, with real-time experiments being carried out on a twenty litre pilot plant fermenter. The penicillin mould biomass is controlled to a reference trajectory by applying adaptive control to the on-line estimate of biomass obtained from a measurement of carbon dioxide production, use of the Bajpai fermentation model and an extended Kalman filter. The adaptive algorithm considered is the Generalised Predictive controller, a form of long range, receding horizon, predictive control. Results from the application of adaptive control to the penicillin fermentation are presented. This work forms part of an industrial collaborative project the aim of which is the optimising control of large fed-batch fermenters.

Keywords. Biocontrol; Adaptive control; Predictive control; Industrial control; Fermentation modelling

INTRODUCTION

Optimisation of antibiotic fermenter feed schedules is of considerable importance to the competitive pharmaceutical industry. Since feed costs are high, with raw materials constituting a significant proportion of the manufacturing cost, even a small increase in the yield of product may have a considerable effect on process profitability. Optimisation of the fermentation process, subject to physical, chemical and economic considerations, requires the availability of system states and the ability to control certain states to optimised trajectories. The problem can be seperated into three parts : (a) the determination of state variables, (b) calculation of optimised state variable trajectories and (c) control of the process to these trajectories. In this paper only stages (a) and (c) are considered. Stage (b) will be considered at a later date.

In order to control the process to a predefined (optimised) trajectory the following scheme has been adopted. Sugar addition to the fermentation is regulated in order to control the inferred biomass to the trajectory. Control of biomass, and hence specific growth rate, in a fermentation with sufficient supplies of dissolved oxygen enables the maximisation of antibiotic production. The controlled variable, biomass estimate, is inferred from a measurement of CO_2 production by using a mathematical model of the penicillin fermentation and an extended Kalman filter. On completion of stage (b) of the project an optimum biomass trajectory will be defined. Until then it is assumed that the optimum biomass trajectory is that of the present industrial fermentation.

The fermentation model, as well as being a major component of the extended Kalman filter, provides a simulation of the penicillin fermentation on which several control strategies have been investigated. The major problem in tuning the sugar addition / biomass control loop is that the process is of a nature such that classical design techniques cannot be applied. Applications of steps, pulses or PRBS signals to the fermenter feed addition subjects the mould to conditions which are unfavourable to growth. This gives impetus to an investigation of the applicability of parameter adaptive control. Montague (1985) has investigated the application of several adaptive algorithms to the control of estimated biomass. In this paper a new adaptive control method, developed by Clarke (1984), which overcomes many of the problems experienced with previous adaptive methods generally has been applied to the control of estimated biomass.

INDUSTRIAL PENICILLIN FERMENTATION

The original industrial process for penicillin production was a scaled up version of the laboratory equipment used by Florey and Chain in their work at Oxford in the 1930's. The process involving the growth of penicillin mould in contact with air proved to be inefficient and was soon superceded by a more efficient deep fermentation developed by Pfizer in 1943. This approach is still in use today, the mould being grown in large, aerated, stirred tank fermenters.

Initially the fermenter is charged with penicillin mould starter culture, water and an organic liquid, A, which contains many amino acids which can be used for growth. This initial charge accounts for two-thirds of the fermenter volume. As fermentation proceeds several other additions are made to the fermentation; sugar solution is added to serve as a further carbon source for growth, acid and a neutralising agent are added in order to maintain a desired pH and supply nitrogen and sulphur, an organic liquid, B, which serves both as a feed and an antifoam agent and a liquid, C, which is a precursor for penicillin production.

At the start of fermentation the carbon concentration in the substrate is high, amino acids present in the broth promoting rapid penicillin mould growth. At high rates of biomass production virtually no penicillin is produced. As the liquid, A, is consumed the mould starts to rely upon the sugar fed to the fermenter for growth, since the amino acids present in the broth are depleted. The mould is then fed in such a way as to maintain a low substrate concentration and a low biomass production rate. Under these conditions penicillin is produced in significant quantities. The production of penicillin has apparently no direct relation to the processes involved in growth and is therefore refered to as a secondary metabolite.

The aerobic process of penicillin mould fermentation is supplied with oxygen by bubbling sterilised air through the broth in order to maintain a satisfactory dissolved oxygen concentration. In the particular penicillin fermentation process of interest it was found that the dissolved oxygen concentration had many high frequency transients due to antifoam addition. However, these variations have little effect on the fermentation since dissolved oxygen concentration never drops below a value which would have a significant effect on the process.

PENICILLIN FERMENTATION MODELLING

At the outset of fermentation modelling three possible approaches were considered development of a model based upon one of the many secondary metabolite models from the literature eg. Shu (1961); development of a model based upon a mechanistic approach, or the use of a model available from the literature.

The development of a new model with the associated verification and model parameter determination is a lengthy process. Initial studies therefore concentrated on a well tested model.

A survey revealed that three models in the literature were suitable for describing an industrial penicillin fermentation. Bajpai (1980) developed a model for penicillin production which was found to give good agreement with practical results. Neestam (1983) had considered the various stages through which penicillin mould develops and has constructed a segregated model for which good agreement has been obtained with experimental results. Calam (1980) considered the rate of production of carbon dioxide during penicillin fermentation and derived a model which related the CO_2 production rate to a function of biomass.

Difficulties were foreseen in obtaining several of the parameters for the Neestam model. The segregated nature of this model (biomass is considered to consist of a mixture of three different states) causes some difficulty in the measurement of kinetic parameters. Effort was therefore concentrated on the Bajpai model with the inclusion of the carbon dioxide production terms of the Calam model since the Bajpai model does not include a description of the carbon dioxide production rate.

There are several advantages to the use of the Bajpai model for penicillin fermentation. Since the parameters of the Bajpai model available in the literature were determined from early industrial strains the model could act as a realistic test for initial investigation of control strategies. The relative simplicity of the model combined with physically realisable kinetic parameters serves to reduce the task of model fitting to the industrial penicillin strain. The model was also chosen because it had been tested experimentally by Bajpai and more recently by Thompson (1984).

In parallel with this approach however, work on the development of an alternative mechanistic model is now underway. The model considers the hyphae as tubular chemical reactors along which nutrients are absorbed and transported to a growth tip. The growth at the tip is modelled using Michaelis Menton enzyme kinetics. The results of this approach have so far been very encouraging as discussed below. Further details of the model, still in the development stages, will be presented at the conference.

THE BAJPAI AND REUSS FERMENTATION MODEL

The Bajpai and Reuss model incorporates a Contois kinetic description of growth, and substrate inhibition kinetics for product biosynthesis. The model is used with the exclusion of the oxygen limitation terms because dissolved oxygen concentration during industrial penicillin fermentation does not reach a level which would inhibit the process.

Two basic models are available to describe the growth during microbial fermentation, the Monod model (1950) and the Contois model (1959). Both models have been found to perform equally well at low microbial cell densities. However, at the high cell densities achieved during industrial penicillin fermentation, the Contois model is superior due to its ability to account for diffusional limitations. The variation of biomass with time for a fed-batch fermentation being described by:

$$\frac{dX}{dt} = \frac{\mu_x * S * X}{K_x * X + S} - \frac{X \cdot dV}{V \cdot dt} \qquad (1)$$

where S represents substrate concentration, μ_x is the maximum specific growth rate, X the biomass, V the fermenter volume, K_x the Contois saturation constant for substrate limitation of biomass production.

Bajpai and Reuss, unable to find an adequate biochemical mechanism to model penicillin production, turned to a substrate inhibition model which other workers had used to good effect, eg Hegewald (1978). This model, which accounts for the repressive effects of high substrate concentration on penicillin production is:

$$\frac{dP}{dt} = \frac{\mu_p * S * X}{K_p + S\ (1 + S / K_I)} - K * P - \frac{P \cdot dV}{V \cdot dt} \qquad (2)$$

where μ_p represents the maximum specific rate of product formation, K_p is the Monod saturation constant for substrate limitation of product formation, K_I the substrate inhibition constant for product formation and K the first order decay constant for product. The product decay has been included to account for penicillin hydrolysis, Broderson (1946).

The substrate consumption was modelled by assuming constant yield and maintenance requirements, thus:

$$\frac{dS}{dt} = \frac{-1}{Y_{x/s}} \cdot \frac{dX}{dt} \quad \frac{-1}{Y_{p/s}} \cdot \frac{dP}{dt} - X \cdot m_x + F - \frac{S \cdot dV}{V \cdot dt} \qquad (3)$$

where $Y_{x/s}$ represents the yield of biomass on substrate, $Y_{p/s}$ the yield of product on substrate, m_x the maintenance requirement of substrate and F the fermentation feed rate.

Calam (1980) modelled the carbon dioxide production rate by assuming that its production was due to (a) growth and associated energy production, (b) maintenance energy and (c) penicillin biosynthesis, hence:

$$\frac{dCO_2}{dt} = \frac{1}{k_4} \cdot \frac{dX}{dt} + m_c * X + k_5 \qquad (4)$$

where k_4 relates CO_2 production to growth, m_c relates CO_2 production to maintenance energy and k_5 relates CO_2 production to penicillin formation. Calam treated the rate of production of CO_2 due to penicillin production as a constant since in practice the rate of penicillin production is relatively constant following the initial delay period.

BIOMASS ESTIMATION

It is frequently found on process plant that certain variables are not ameanable to direct observation due to their inaccesibility or instrumentation problems and associated costs; biomass concentration in the penicillin fermentation is such a variable. This problem has been overcome with the aid of an extended Kalman filter and from a measurement of the carbon dioxide production rate, the penicillin biomass can be inferred.

The extended Kalman filter is a recursive state estimator, the states of the system being defined as X, S, V and CO_2. The penicillin concentration, P, need not be considered a state since its value does not effect the relationship between biomass and CO_2 production. Fermentation model linearisation about the previous operating condition and partial knowledge of the new states enables an estimate of the new states at the present operating condition to be obtained.

Likely problems arising from biased state estimates and no guarantee of estimate convergence due to the non-linearity of the system have been found not to be a significant problem with this particular system.

The extended Kalman filter algorithm, written in prediction correction form, is described by Litchfield (1979). The algorithm together with the fermentation model, measurements of carbon dioxide production and fermenter volume enables estimates of substrate concentration and penicillin mould biomass to be made. The estimate of penicillin mould biomass is considered to be the controlled variable modified by manipulation of the sugar addition rate.

ADAPTIVE CONTROL POLICY

In an attempt to find the most suitable adaptive control algorithm of the many available in the literature several were applied to the control of biomass in a simulation of the penicillin fermentation. In addition to those algorithms described by Montague (1985) the Generalised Predictive Controller (GPC) of Clarke (1984) has been applied. It is this algorithm which will be briefly discused.

The GPC algorithm developed by Clarke is a new form of long range predictive control which has been found to give good performance on a wide range of process plant. Those characteristics which make it particularly suitable for application to the control of biomass in the penicillin fermentation are the incremental nature of control action, low overshoot of process output above the set point and the ease of tuning.

During the controlled phase of the penicillin fermentation constant sugar feeding causes a ramping of biomass. The system can therefore be thought of as being type 1. Control of such a system to a ramping set point without offset requires a controller containing an integrator, the incremental GPC is such a controller. The usual methods for gaining insight into a process, for example step tests, pulse tests, PRBS signals etc. can not be applied to the penicillin fermentation due to the nature of the penicillin mould growth. The robust nature of the GPC algorithm combined with only a few tuning parameters helps to alleviate many of the controller tuning problems. The GPC algorithm uses input/output data (sugar addition/biomass estimate) to fit a time series model to the system using a recursive least squares technique. UD factorisation of the covariance matrix has been applied in order to improve the parameter estimate stability. The adaptive controller parameter estimation is carried out explicitly, the parameters estimated being those of the system. The controller multi-step ahead predictor parameters are then calculated by recursion of the diophantine equation.

The GPC is a minimum variance type of adaptive controller, the control action being calculated to minimise a quadratic cost function, the weighting on the control action allowing fine tuning. A further two tuning parameters are available. The assumption is made that increments in control action are zero after a finite number of samples, the control action prediction horizon. This is used for coarse tuning of the controller. The calculation of control action is based upon predictions of system output at a future instant, the number of samples into the future being the predicted output prediction horizon. Clarke found that the predicted output prediction horizon had little effect on the control action as long as it was significantly greater than the control action prediction horizon.

CONTROLLER IMPLEMENTATION

The control algorithm, programmed in Fortran to run on an IBM PC (the real-time process control computer), has system level software interfaces to the machine operating system to enable real-time operation. Control application to the process plant is achieved through an intelligent multi loop/ distributed digital control system which has the ability to monitor and regulate fermenter performance. The IBM PC serves several purposes, the calculation of sugar addition, overall fermenter supervisory control and provides full data logging facilities all via an RS422/232 data link. An IBM PC / IBM mainframe link gives the ability for data evaluation and plotting on a mainframe computer.

RESULTS

Figures 1 and 2 compare the behaviour of the industrial plant with that of the fermentation models. Industrial fermentation feed schedules are applied to both the Bajpai model and the tubular reactor penicillin mould growth model. Figure 1 shows the variations in penicillin mould biomass and Figure 2 the subsequent antibiotic production. It can be seen that both models are suitable for penicillin fermentation simulation and on-line biomass estimation since the responses are close to those observed on process plant.

Figure 3 shows results from the application of the Generalised Predictive controller to the Bajpai simulation of the penicillin fermentation. During the first forty hours of fermentation the sugar addition is under open loop control. The carbon supply for the rapid growth period is mostly that which is in the fermenter at start-up. At forty hours into the fermentation, control is taken over by the GPC.

In (a) three biomass profiles are shown, the desired biomass (set point), the 'actual' fermenter biomass and the estimated biomass. The three curves are closely aligned. The addition of sugar calculated by the control algorithm to achieve desired biomass tracking is shown in (b). In responses (d) it can be seen that the substrate concentration is high during the initial rapid growth phase falling to a low value during the antibiotic production phase.

Responses (c) show the measured and estimated carbon dioxide production rate, the extended Kalman filter serving to remove the noise on the measured signal. The actual production of penicillin is shown in (e). The delay in penicillin production, as observed on real plant is predicted by the model, the majority of penicillin being produced after the initial rapid growth phase. The final response, response (f), shows the variations in the A and B parameters estimated by the recursive least squares algorithm. The parameters exhibit wide variations during the initial growth phase due to tuning in effects and the non-linearity of the system, the parameters settling to steady values during the antibiotic production phase. The prediction error is seen to approach zero after 35 hours of the fermentation.

DISCUSSION

The work presented in this paper has summarised the steps taken in the application of control to a pilot scale fed-batch penicillin fermentation. This represents the first stage of the application of full control and optimisation to a large industrial penicillin fermentation.

The modelling of the penicillin fermentation served two purposes. Firstly, control strategy verification on a digital process simulation. Since the time taken for fermentation is long, thus prohibiting the investigation of several adaptive techniques on process plant, the digital simulation provided a means by which it could be achieved. Secondly, the fermentation model is a major component of the extended Kalman filter algorithm, the observer for the unmeasurable state - biomass. The availability of an estimate of biomass has enabled an adaptive control scheme to be applied to force the actual fermenter biomass to follow a prespecified trajectory. The development of a penicillin fermentation model based upon tubular reactor theory has produced promising results, however, work is still being carried out on model verification and parameter determination. Until the completion of this work, on-line biomass estimation will continue to be rely upon the use of the Bajpai model. Investigation of the optimisation and the ability of the process to achieve the specified optimum forms the next stage of the project.

It is intended that results further to those of the previous section, obtained from both simulation and pilot scale studies will be presented at the conference.

ACKNOWLEDGMENTS

The authors would like to acknowledge the support of the University of Newcastle upon Tyne, Tyne and Wear County Council and the collaborating industrial body. Two authors (G.A.M. and M.A.) would like to thank the U.K. Science and Engineering Research Council for financial support.

REFERENCES

Bajpai R.K. and M.Reuss (1980). A mechanistic model for penicillin production. J.Chem.Tech.Biotechnol., 30, 332

Broderson R. (1946). Acta Pharmacol, 2, 1

Calam C.T. and B.A.K.Ismail (1980). Investigation of factors in the optimisation of penicillin production. J.Chem.Tech.Biotechnol., 30,249

Clarke D.W. and L.Zang (1984). Long range predictive control using weighting sequence models. Oxford University Engineering Lab. Report No. 1542/84

Contois D.E. (1959). Kinetics of bacterial growth: relationship between population density and specific growth rate of continuous cultures. J.Gen.Microbiol. 21,40

Hegewald E. and I.Ruckbeil (1978). Model of growth and product formation of Streptomyces Hygroscopicus in batch culture and prediction of continuous culture. 7th International Symp. on continuous culture of microorganisms, Prague, Czechoslovakia

Litchfield R.J., K.S.Campbell and A.Locke (1979). The application of several Kalman filters to the control of a chemical reactor. Trans.I.Chem.E. Vol.57, 113-120

Monod J. (1980). La technique de culture continue. Theorie et applications. Ann. Inst. Pasteur 79, 390

Montague G.A., A.J.Morris, A.R.Wright and A.Ward (1985). Investigation into the application of state estimation and parameter adaptive control in fed batch fermentation for penicillin production. 7th IFAC Symposium on Identifcation and System parameter estimation, University of York, July 1985.

Neestas E. and D.I.C.Wang (1983). Computer control of the penicillin fermentation using the filtration probe in conjuction with a structured process model. Biotechnol.Bioengng, 25,781-796

Shu P.J. (1961). J. Biochem. Microbiol. Technol. Engng., 3, 95

Thompson M.L. (1984). System analysis, simulation, control and optimisation of the fed batch penicillin fermentation. M.Sc. thesis, MIT

Figure 1

Comparison of biomass against time

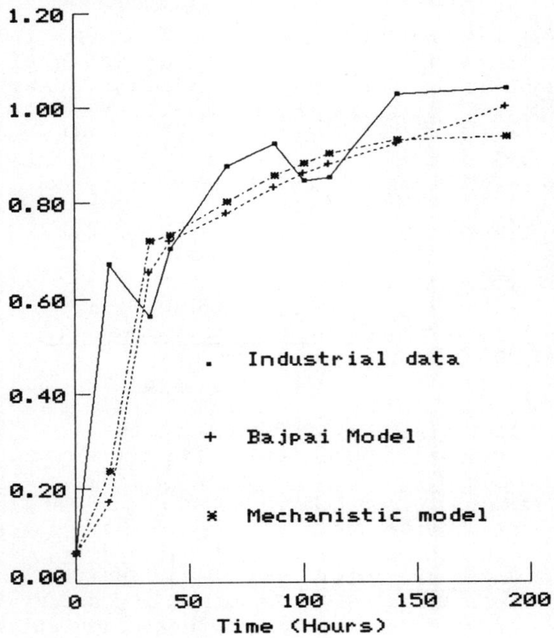

Figure 2

Comparison of Penicillin conc v time

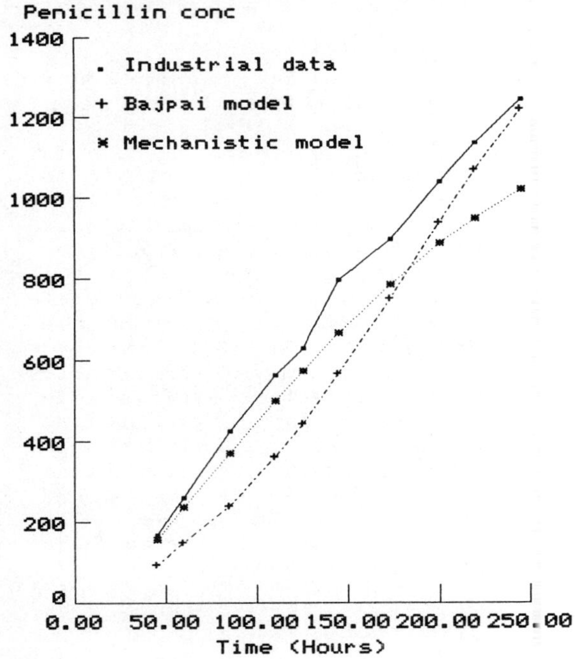

FIGURE 3: GENERALISED PREDICTIVE CONTROL OF BIOMASS

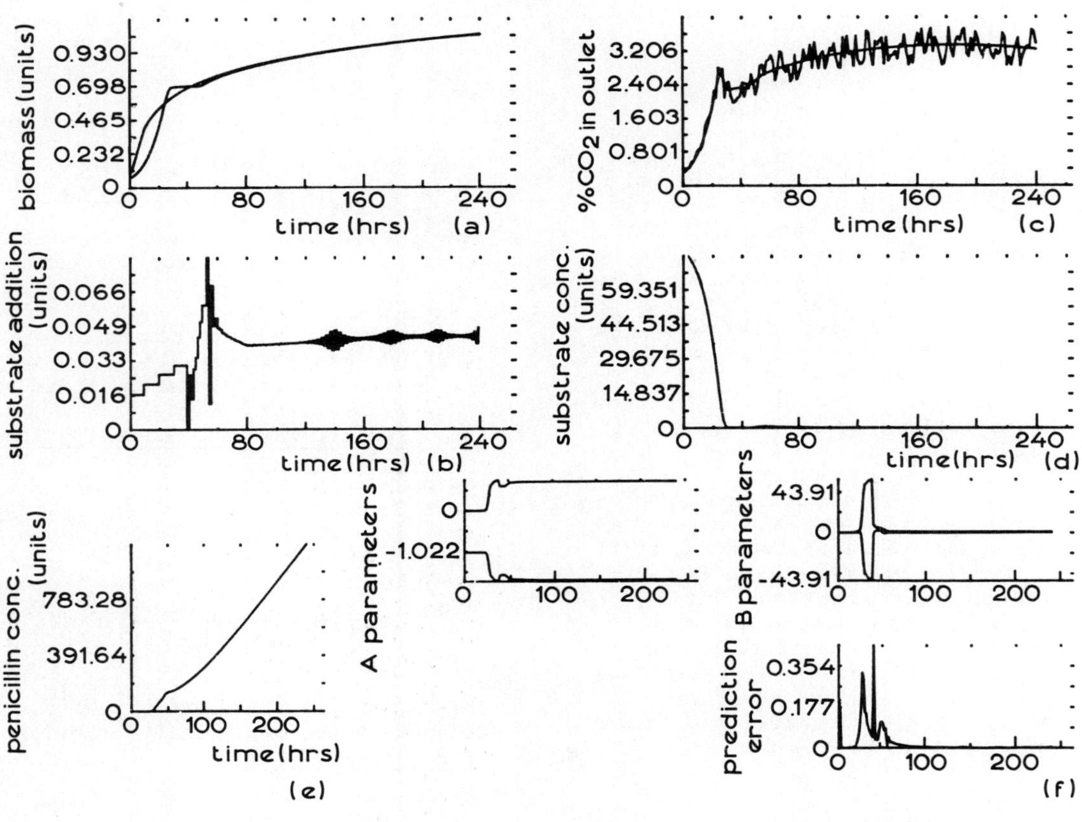

ON-LINE MEASUREMENT AND CONTROL OF PENICILLIN V PRODUCTION

K. Frueh, Th. Lorenz, J. Niehoff, J. Diekmann, R. Hiddessen and K. Schuegerl

Institut fuer Technische Chemie, Universitaet Hannover, Callinstr. 3, 3000 Hannover, FRG

Abstract. Penicillium chrysogenum was cultivated in a suspension of pellets of a suitable size (less than 0.4 mm in diameter) in a 98-l-air-lift reactor with an outer loop and a working volume of 90 l to produce Penicillin V during fed-batch fermentations.
The measurement of several key values improved the control of the rather complex fermentation process, to attain an increased productivity of antibiotics. Physico-chemical sensors as well as automatic chemical analyzers were applied.
A thermal mass flow meter was used for the control of the sole energy input by aeration to minimize the aeration rate. To achieve an optimal productivity, the dissolved oxygen concentration was held above a critical value.
The use of a PI-controller failed to maintain the dissolved oxygen concentration necessary, because the properties of the fermentation broth were changing frequently. A self-tuning controller allowed a satisfactory control of the penicillin process.

Keywords. Process control, sensors, on-line computer-control, self-tuning control, fed-batch process, tower-loop-reactor, penicillin.

INTRODUCTION

It is important to monitor several key variables of biotechnological processes to control and optimize the process for the minimization of production costs. Three steps are necessary to achieve these aims. (Bor, 1981):

1. Development of instruments to monitor the environment of cells.
2. Correlation of the different measured data.
3. Maintaining the optimal environmental conditions by continuous analysis and feed-back control of the environment.

INSTRUMENTATION AND MEASUREMENT TECHNIQUES

The instruments for measuring the process variables, e.g. pH, temperature, liquid volume, overall volume of aerated medium, aeration rate, medium recirculation rate through the outer loop, O_2-and CO_2-concentrations in the outlet gas and the dissolved oxygen concentration have been known for many years. Most of these sensors are steam-sterilizable and their signals are stable for longer periods.
In addition, it is also necessary to measure the concentrations of some substrates in the fermentation broth, to avoid undesired limitations or catabolite repressions during the different phases of the process.

The penicillin fermentation process can be devided into three phases: a growth phase with a high growth rate and a negligible product formation rate, a transition phase, and a production phase with a low growth rate and high product formation rates. During the growth phase, limitations should be avoided by monitoring the concentrations of the carbon, nitrogen and phosphor sources. Later on, the growth is limited by the carbon source (e.g. glucose). Ammonium or urea serve as nitrogen sources and sulfate as a sulphur source for the biosynthesis of penicillin. These sources as well as the product were measured.
Since no specific steam-sterilizable sensors detecting these components have been available up to now, an ultrafiltration system was developed to sample cell-free medium. It was possible to steam-sterilize this system, and it was stable for longer periods. The sample was divided to feed different automatic chemical analyzers using wet-chemical methods. If necessary, it was diluted and treated by reagents. The reactions resulted in colored complexes whose adsorbances were measured by photometers.

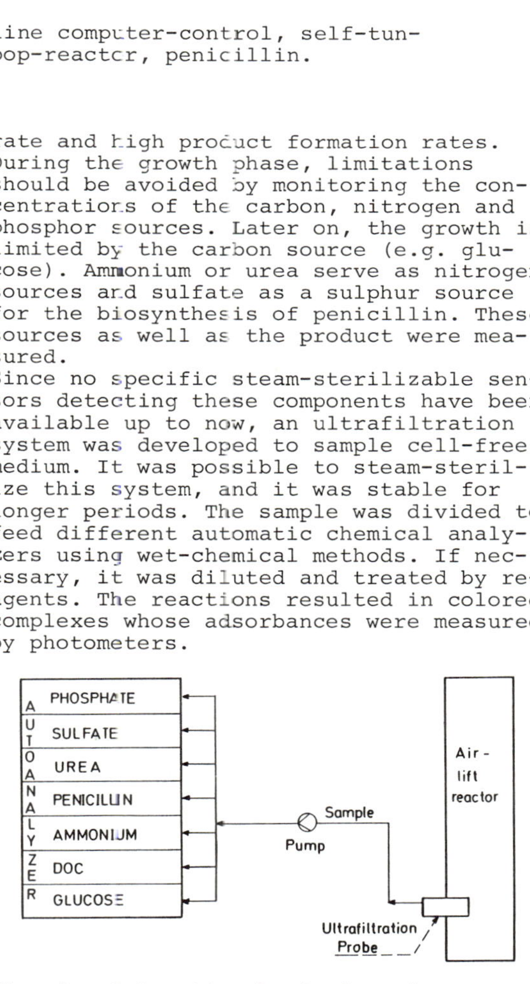

Fig. 1. Automatic chemical analyzer

The sensors which measured the ammonium and dissolved organic carbon (DOC) concentrations were based on pH-electrodes detecting ammonia and CO_2. Controlled by a computer program, the chemical analyzers were rinsed by base, acid or buffer solutions, and the signals were actualized by automatic recalibration of the instruments.

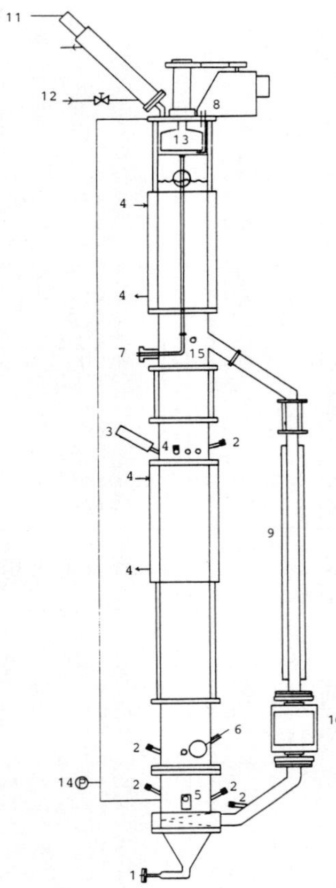

Fig. 2. Airlift-loop-reactor

1 air inlet
2 oxygen probe
3 pH-probe
4 temperature control
5 microfluorimeter
6 sampling probe
7 level meter
8 foam detectors
9 outer loop
10 velocity meter
11 air for gas analysis
12 cooling water
13 mechanical foam breaker
14 pressure meter (liquid volume)
15 inoculum connection

CORRELATION OF THE MEASUREMENTS AND MANUAL CONTROL

Some of these measurements were correlated, e.g. the urea concentration and the pH-value or the CO_2 production rate (CPR) and the penicillin productivity.

At the end of the growth phase, the proteins of the nutrient were exhausted and the pH dropped. To avoid the addition of base (KOH) by the pH control, urea was used as a nitrogen source. If the urea concentration was held at a certain level, no acid or base was needed to keep the pH at the desired value, and the pellet morphology of the mold was stable. By monitoring the ammonium concentration, nitrogen limitation was avoided.
In the production phase, the active cell mass should be kept constant. This is the main prerequisite to maintain an optimal production rate. Since it was not possible to determine the active biomass continuously, the CPR was used as a measure for the active cell concentration, because the metabolism of the cells does not change and the CPR remains constant during that time.

$$CPR = \frac{\dot{V}_o \, c^o_{CO_2} - \dot{V}_i \, c^i_{CO_2}}{V_L} \qquad (1)$$

CPR	CO_2-production rate
$\dot{V}_i$	gas inlet rate
$\dot{V}_o$	gas outlet rate
$c^i_{CO_2}$	CO_2-concentration in gas inlet stream
$c^o_{CO_2}$	VO_2-concentration in gas outlet stream
V_L	liquid volume

The CPR was kept constant by varying the feeding rate of the sugar solution manually. Automatic control will become possible, if the consumption of carbon can be predicted by a model.

SELF-TUNING CONTROL

Since energy input occurs by aeration alone, it is important for the minimization of the production costs to aerate only as much as is necessary to maintain a certain level of dissolved oxygen concentration (DOT) and to achieve a sufficient mixing of the fermentation broth.

Efforts to control DOT by means of the aeration rate with a conventional PI-controller failed, because the system parameters often changed during fermentation, and the controller had to be retuned frequently. By using a self-tuning controller, it was possible to overcome this problem. Its application also had the advantage that it was possible to manage nonlinearities of the system by using local linearization in which the model parameters change with the set-point.
It is assumed that the process can be described by the following linear difference equation of the order m:

$$y(k)+a_1 y(k-1)+\ldots+a_m y(k-m) =$$
$$= b_1 u(k-1-d)+\ldots+b_m u(k-m-d)+v(k) \qquad (2)$$

u(k) and y(k) are the sampled process input and output signals and v(k) is a discrete white noise. The process dead-time d is expressed in multiples of the sampling time T_O.
The z-transform of equation (2) is:

$$y(z) = \frac{B(z^{-1})}{A(z^{-1})} z^{-d} u(z) + \frac{D(z^{-1})}{A(z^{-1})} v(z) \quad (3)$$

with the polynomials

$$A(z^{-1}) = 1 + a_1 z^{-1} + \ldots + a_m z^{-m} \quad (4)$$

$$B(z^{-1}) = b_1 z^{-1} + \ldots + b_m z^{-m} \quad (5)$$

$$D(z^{-1}) = 1 \quad (6)$$

Now the model order m and dead-time d have to be determined. Simulations (Kurz, 1980) have shown that the process order m_p does not need to be known exactly. Parameter-adaptive controllers are insensitive to a wrong model order within:

$$m_p - 1 \leq m \leq m_p + 2 \quad (7)$$

The main problem in controlling DOT by aeration is the unknown time-varying dead-time. This however can be overcome by using dead-time estimation with an algorithm, e.g. RLSVT (Kurz, 1979).
The RLSVT algorithm basically consists of the well-known recursive least squares algorithm in which a forgetting factor λ is introduced. By using λ < 1 slowly, time varying process parameters can be tracked.

An upper limit d_{max} of the process dead-time has to be known a priori, and the number of parameters of the B polynomial (eq. 5) is increased by d_{max}:

$$\hat{B}(z^{-1}) = \hat{b}_1 z^{-1} + \ldots + \hat{b}_{m+d_{max}} z^{-(m+d_{max})}, \quad (8)$$

which results in a process model:

$$y(z) = \frac{\hat{B}(z^{-1})}{A(z^{-1})} u(z) + \frac{D(z^{-1})}{A(z^{-1})} v(z). \quad (9)$$

The impulse response of the estimated model (eq. 9) is obtained and used to calculate the dead-time and parameter values for the process model (eq. 3).
To control DOT, the RLSVT method is now combined with an extended minimum variance controller, which minimizes the criterion:

$$E\{y^2(k+d+1) + r u^2(k)\} \quad (10)$$

The weighting r on the input was proposed by Clarke and Hastings-James (1971) and prevents excessive input changes that sometimes occured with the original minimum variance controller (Astroem 1970). The controller transfer function is given by:

$$G_{MV}(z) = - \frac{L(z^{-1})}{zB(z^{-1})F(z^{-1}) + \frac{r}{b_1} D(z^{-1})} \quad (11)$$

with the polynomials:

$$L(z^{-1}) = l_o + l_1 z^{-1} + \ldots + l_{m-1} z^{-(m-1)} \quad (12)$$

$$F(z^{-1}) = 1 + f_1 z^{-1} + \ldots + f_d z^{-d} \quad (13)$$

which are obtained from

$$D(z^{-1}) = F(z^{-1})A(z^{-1}) + z^{-(d+1)} L(z^{-1}) \quad (14)$$

To avoid offsets for set-point changes, the minimum variance controller is multiplied by a proportional integral action term of the following transfer function:

$$G_{PI}(z) = \frac{1-(1-\alpha)z^{-1}}{1-z^{-1}} \quad (15)$$

The parameter-adaptive controller combined in this way was applied for DOT control using a process model of the third order. It showed good performance and stability. However, several precautions had to be taken.
When there are no disturbances or set-point changes for longer periods, the covariance matrix of the estimator will increase because of the forgetting factor. This results ultimately in excessive changes of parameter estimates (Astroem, Westerberg, Wittenmark, 1978). Therefore, whenever DOT was within one percent of the set-point, the forgetting factor was set to 1.
Sudden changes of the process - e.g. through the addition of an antifoam agent - could result in estimating a system with a negative gain and cause the controller to react the opposite way. The estimated parameters were thus checked, if they describe a system with a positive gain, otherwise parameter estimates of the preceding sample interval were taken. Finally, a level meter was used to prevent the airlift reactor from flooding. Whenever a critical upper level was reached, the set-point was reduced automatically. When the level dropped below a defined lower level, the set-point was increased by gradual steps until the original set-point was reached, or the level was in between the lower and the upper level.

RESULTS

Figure 3 shows the aeration rate and DOT during 24 hours of run 8407. The set-point was at 75 % DOT.

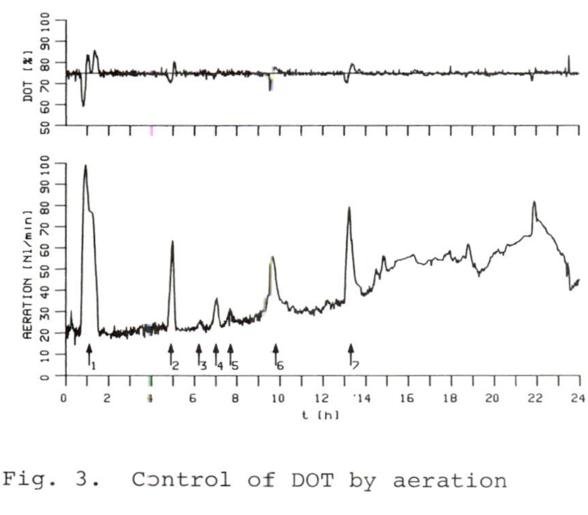

Fig. 3. Control of DOT by aeration

The disturbances were caused by addition of urea (marker 1, 2) or precurser (marker 4), or by oil (marker 3, 5, 6, 7). There is a general increase in the aeration rate in that phase of fermentation, because the oxygen transfer coefficient from the gas into the liquid phase became smaller during the transition phase, when oil was added as a substrate.

Because of these disturbances, it was not possible to control the carbondioxide production rate by means of the feeding rate of the sugar sulotion.
To overcome this problem, a way of running the fermentation has to be found were the glucose solution will be the only energy source. Furthermore, the urea- and phenoxyacetate solutions have to be added more frequently in smaller quantities.

CONCLUSIONS

A self-tuning controller was applied to control dissolved oxygen by means of the aeration rate in an airlift reactor.
During several fermentations it was very reliable in keeping the desired set-point. Its applicability to DOT-control in other fermentations is presently explored.
The additional wet-chemical measurements improved the insight in the rather complex fermentation process and an extension of the automatic control is planned.

ACKNOWLEDGEMENTS

The authors thank Prof. M. Thoma and Dr. A. Munack, Institut fuer Regelungstechnik, Universitaet Hannover, for their support.

REFERENCES

Astroem, K.J., B. Westerberg, B. Wittenmark (1978). Self-tuning Controllers Based on Pole-placement Design. Dept. of Automatic Control, Lund Institute of Technology, Lund, Sweden, Report: LUTFD2/(TFRT-3148)/1-52/(1978).

Astroem, K.J. (1970). Introduction to Stochastic Control Theory. Mathematics in Science and Engineering, Vol. 70, Academic Press.

Clarke, D.W., R. Hastings-James (1971). Design of digital controller for randomly disturbed systems. Proc. IEE 118.

Kurz, H. (1979). Digital Parameter-Adaptive Control of Processes with Unknown Constant or Time-varying Dead Time. 5th IFAC-Symposium on Identification and Parameter Estimation, Darmstadt.

Kurz, H. (1980). Erprobung und Vergleich von parameteradaptiven Regelalgorithmen bei verschiedenen Prozessen. Regelungstechnik, 28, 2-10.

Mor, J.R. (1981): A review of instrumental analysis in fermentation technology. 3rd International Conference on Computer Application in Fermentation Technology, Manchester, England.

SIMULATION AND EXPERIMENTAL STUDIES OF THE PROFILE CONTROL OF THE SPECIFIC GROWTH RATE IN A FED-BATCH CULTURE

S. Shioya, H. Shimizu, M. Ogata and T. Takamatsu

Department of Chemical Engineering, Kyoto University Kyoto 606, Japan

Abstract. The aim of this paper is to develop and discuss a practically useful system so that the specific growth rate in a baker's yeast fed-batch cultute will follow the desired profile specified in advance as accurately as possible. For this, first, the specific growth rate should be estimated using indirect measurement of the overall growth rate or using online data of cell concentration. Based on the careful analysis of the previously proposed method, an improved method was proposed which uses moving averages and dynamic mass balancing. It was shown that the specific growth rate could be estimated with sufficient accuracy using an extended Kalman filtering algorithm using the estimated overall growth rate, or using the cell concentration. Secondly, the control scheme called "Programmed-controller/Feedback-compensator(PF) system" consists of a programmed controller and a feedback compensator has been used for the profile control of μ. The PF-MRAC in which the model reference adaptive controller was used for the compensator, was shown to be effective for this purpose. Finally, the profile control of the specific growth rate in a baker's yeast fed-batch culture was realized experimentally using the proposed control scheme. The experiments showed that the estimation and control schemes proposed were sufficiently useful.

Keywords. Adaptive control; profile control; fed-batch culture, specific growth rate; Kalman filter.

INTRODUCTION

Fed-batch cultures have frequently been desired to be controlled so that a parameter or a state variable follows a desired profile. For example, the biomass concentration in a baker's yeast fed-batch culture (Nakamani et al.,1983) was controlled so as to follow the desired profile in order to attain maximum production of yeast. In a different case, it was desired that the specific growth rate follow a certain profile such as a bang-bang type one in order to obtain the maximum yield of a metabolic product(Dairaku and Shioya, 1982).

Specific growth rate μ is one of the important process parameters which represent the state of the fermenter. For example, μ is desired to be maximum during the fermentation in order to get the maximum production of cell. An optimal condition is frequently searched in laboratory scale fermenter by changing μ as a dominant parameter. As shown above, the bang-bang type pattern of μ gives the highest quality of baker's yeast. Maximum production of baker's yeast can be also achieved by the other methods using RQ(Aiba et al., 1976) and ethanol concentration(Dairaku and Shioya, 1981). However, μ is maximum but can not be changed arbitrarily by these methods. Chemostat culture can change μ according to the selected dilution rate. But it takes so long time to get a steady state, and any profile of μ except constant value cannot be made by the chemostat culture. Then, the specific growth rate μ is a important parameter to be observed and to be controlled in a fed-batch culture.

In order to construct the profile control system for the specific growth rate, first, the specific growth rate itself must be observed or estimated. In many practical cases, it is difficult to measure the specific growth rate. It should be estimated using indirect measurement of overall growth rate or on-line data of the cell concentration. It has been proposed that the overall growth rate was estimated by elemental and macroscopic balance of a fermenter utilizing other measurable rates(Cooney et al.,1977; Stepanopoulos and San,1984). By careful analysis of the method, it is shown that the selection of which rates are used for the macroscopic balance is important and that the macroscopic balance calculation scheme can produce a large error in the estimated overall growth rate in some cases.

In fed-batch cultures. the cell concentration, liquid volume and other state variables change significantly during the course of operation. In such a system with time-varying dynamic characteristics, a feedback control system with a constant controller parameter will not be adequate to follow a desired profile. The nonlinear and nonstationary characteristics of the biochemical process require a control system to adapt its parameters to values appropriate to the current conditions. For example, in a proportional-plus-integral(PI) feedback control system, the parameter values of the controller which is optimall tuned based on a dynamics should be changed correspoding to the current system dynamics. For this purpose, the Model Reference Adaptive Control (MRAC) algorithm is practical and attractive. MRAC is a well-known control technique in which the control parameters are adjusted automatically to compensate for varying process conditions(Landau, 1974; Goodwin et al., 1980; Losano and Landau, 1981). The task of the adaptation is to minimize a function of the difference between the process response and the reference model response which represents the desired response to be followed.

The aim of this paper is to develop and to discuss a practical and useful computer control scheme so that the specific growth rate $\mu(t)$ will as accurately as possible follow the desired profile specified in advance. For this, first, an improved method based on the macroscopic balances was proposed which uses moving averages and dynamic balancing. The method will be useful for the case that the cell concentration is high. When the cell concentration is low, it can be measurable using the turbidity of the sample, e.g., UV-spectrophotometer. Using an extended Kalman filter, the specific growth rate was estimated with sufficient accuracy based on the calculated overall growth rate or cell concentration.

Next, the control scheme called "Programmed-controller /Feedback-compensator(PF) system", was proposed. This control system consists of a programmed controller which should follow the desired profile unless there is noise or disturbance, and a feedback compensator which should compensate the disturbance and the noise. As the feedback compensator, the Model Reference Adaptive Control(MRAC) algorithm(Lozano and Landau, 1981) was also proposed. The PF system with MRAC was named PF-MRAC in which the second term, MRAC means that the MRAC was used for the feedback compensator. A classical PI feedback controller was also adopted as the feedback compensator, and the whole control scheme with PI was named PF-PI. Numerous computer simulations verified the usefulness of the proposed control system.

Finally the profile control of the specific growth rate in a baker's yeast fed-batch culture was realized experimentally using the proposed estimation and control schemes. The experiments were performed using a microcomputer coupled laboratory scale fermenter. Till now, not so many experiments was done, but the results showed that the estimation and control schemes were sufficientrly useful for the profile control of the specific growth rate.

ESTIMATION OF THE SPECIFIC GROWTH RATE

The specific growth rate μ is a coefficient of the rate equation (1) as

$$dVX/dt = \mu VX \qquad (1)$$

$$dV/dt = F \qquad (2)$$

where X, V, F are the biomass concentration, the liquid volume, the input flow rate, respectively. The specific growth rate is a function of the environmental conditions, such as the substrate concentration, temperature, etc. At now stage, there is no sensor which directly measures the specific growth rate. Then, μ should be estimated using other observed variables.

For the range of low cell concentration, the cell concentration can be measured by a type of turbidity meters, such as a UV-photospectrometer, which was used here in the experiment. However, for the range of high cell concentration, there is no practically available sensor without sampling and diluting.

Estimation of the Overall Growth Rate

The overall growth rate, R_X can be estimated by the method(Cooney et al.,1977; Stephanopoulos and San, 1984) which utilizes the elemental and macroscopic balance of a fermenter. By the method, the stoichiometric equation (3) is assumed.

$$aC_nH_mO_\ell + bO_2 + cNH_3 \longrightarrow C_\alpha H_\beta O_\gamma N_\delta + dH_2O + eCO_2 + f\cdot\text{Product} \qquad (3)$$

where $n, m, \ell, \alpha, \beta, \gamma$ and δ are known parameters. If parameters a∿f are known, the overall growth rate R_X can be calculated using a measurable rate, e.g., oxygen uptake rate R_{O2} as

$$R_x = (MW/b)\cdot R_{O2} \qquad (4)$$

where MW is molecular weight of the cell. Then the problem is reduced to how to obtain the parameters a∿f. In case of baker's yeast considered here, [Product] is ethanol, and gulcose is adopted as a cabon source($n=\ell=6$, $m=12$). Eq.(3) is derived from the assumption that intermediate products from the substrate to the cell can be neglected and that nutrients can be represented by the simple materials as in Eq.(3). In some cases, these assumptions may not be appropriate.

To estimate 6 unknown parameters, six equations are required. First, four balance equations of elements, C, H, O and N can be used. Then, the more two equations or relationships are necessary. From Eq.(2), three of production rates and consumption rates can be utilized for this condition. The rate of oxygen consumption and carbon di-oxyside production rate can be relatively easily obtained using off-gas analyzers. For one more rate required here, the following two cases; i)ethanol production rate R_{EtOH}, ii)consumed rate of ammonia R_{NH3}, are considered. In thses cases, parameters a∿f can be obtained by solving the following linear algebraic equation,

$$\begin{bmatrix} 6 & 0 & 0 & 0 & -1 & -2 \\ 12 & 0 & 3 & -2 & 0 & -6 \\ 6 & 2 & 0 & -1 & -2 & -1 \\ 0 & 0 & 1 & 0 & 0 & 0 \\ & & & A & & \end{bmatrix} \begin{bmatrix} a \\ b \\ c \\ d \\ e \\ f \end{bmatrix} = \begin{bmatrix} \alpha \\ \beta \\ \gamma \\ \delta \\ b \end{bmatrix} \qquad (5)$$

where for case i),

$$A = \begin{bmatrix} 0 & R_{CO2} & 0 & 0 & -R_{O2} & 0 \\ 0 & R_{EtOH} & 0 & 0 & 0 & -R_{O2} \end{bmatrix} \quad b = \begin{bmatrix} 0 \\ 0 \end{bmatrix} \qquad (6)$$

and for case ii),

$$A = \begin{bmatrix} 0 & 0 & 0 & 1 & 0 & 0 \\ 0 & 1 & 0 & 0 & 0 & 0 \end{bmatrix} \quad b = \begin{bmatrix} R_{CO2}/R_{NH3} \\ R_{O2}/R_{NH3} \end{bmatrix} \qquad (7)$$

Now the required overall growth rate R_x can be written as

$$R_x = (MW/b)\cdot R_{O2} \qquad \text{for case i)} \qquad (8a)$$

$$R_x = (MW/\delta)\cdot R_{NH3} \qquad \text{for case ii).} \qquad (9)$$

Eq.(8a)can be rewritten using the solution of Eq.(5), such as

$$R_x = \frac{4\ MW}{\beta-2\gamma-3\delta}(R_{CO2}-R_{O2}-R_{EtOH}) \qquad (8b)$$

If the data of these rates have measurement error, the solution of Eq.(5) for two cases, that is, R_X obtained from Eq.(9) and one obtained from Eq.(8b) are different each other. And it is shown that the effect of the measurement error on the obtained R_x depends on the employed equations and the amount of measurement error of instrument. The estimated R_x for case i) will have larger error than for case ii), because the denominator in Eq.(8b) is small.

As easily understood from Eq.(8b), if the following condition

$$\beta - 2\gamma - 3\delta = 0 \qquad (10)$$

is satisfied, parameters a∿f can not be obtained for case i). This condition is coincides with the condition that the respiratory quotient(RQ) is unity (RQ= e/b=1) when ethanol is not produced. Because RQ is written as

$$RQ = 1 + (\beta-2\gamma-3\delta)/4b \qquad (11)$$

by solving Eqs.(5) and (6) with f=0. In practical case, Eq.(10) is not necessarily satisfied. For example, using the previous data(Stephanopolos and San, 1984); α=1, β=1.666, γ=0.511, δ=0.168, left-hand side of Eq.(10) is not equal to zero but equal to 0.14. Then, RQ is not unity but should be close to 1. Therefore, the left-hand side of Eq.(10), that is, the denominator of Eq.(8b) is small compared with R_{O2}. Afterall, both the denominator and the numerator of Eq.(8b) are small.

Hereafter, to demonstrate that the filtering technique employed here can reduce the measurement noise and gives good estimate of μ, simulation studies were performed especially for case i). To eliminate the measurement error, the data are filtered by taking the moving average and using the dynamic mass balancing method taking into account for the gradient of the O_2 and CO_2 concentration. It should be stressed again that the rate employed for solving Eq.(5) affects the amount of error included in the estimated R_x.

Estimation of the Specific Growth Rate

If the overall growth rate μX or the cell concentration X is measured, for the next step, μ should be estimated using the obtained data. For this, system equation which describes the behaviour of μ (including μ=constant) should be introduced. With less informations about the system, μ is desired to be estimated in any cases. From this viewpoint, the following equations of μ,

$$d\mu/dt = C\mu+\eta_1 \quad (12)$$

$$dC/dt = -(F/V)C + \eta_1 \quad (13)$$

have been proposed (Stephanopoulos and San, 1984), where C is a new variables introduced here, and η_1 is white noise. However, the physical background of Eq.(13) was not given. Here, it is shown that the same type equations

$$d\mu/dt = A\mu+\eta_1 \quad (14)$$

$$dA/dt = -2(F/V)A - aX(\mu^2+A) +\eta_1 \quad (15)$$

$$da/dt = \eta_1 \quad (16)$$

can be derived from the assumption that $\partial\mu/\partial S=0$. Using these equations and output data,

$$y = R_x + \eta_2 = \mu X + \eta_2 \quad (17)$$

or

$$y = X + \eta_3, \quad (18)$$

the extended adaptive Kalman filter can be applied to the estimation of the specific growth rate μ, where η_i (i=1,2,3) are white noises. Many numerical simulations verify that the estimation scheme can reproduce successively well. Eq.(1), (2) and (8b) based on the off-gas analysis and Eq.(17) are used for numerical simualtion for this section, and the set of Eqs.(12) and (13)(case A) and the set of Eqs.(14), (15) and (16)(case B) are used for the equations related to μ.

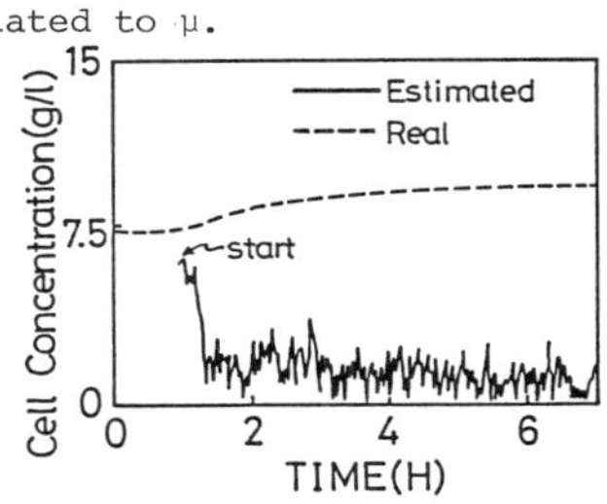

Fig.1 State estimation by the Kalman filter.

With no modification of the estimation scheme such as taking the moving average, one of the results by the ordinary Kalman filter for case A is shown in Fig.1. The estimated value of X and the data of X in a fed-batch culture are shown in the figure. The estimated X doesn't coincide with the data at all. Fig.2 shows the result for case A by the estimation scheme taking into account for the adptive filter by which the system noise covariance matrix is adjusted from the prediction error. Estimated cell concentration gradually approaches to the data. However, estimated μ is very noisy. By taking the moving average, both estimated values of the cell concentration X and the specific growth rate μ become to be very close to the data as shown in Fig.3a for case A and in Fig.3b for case B. Comparing with Fig.3a and Fig.3b, it is shown that there is no siginificant difference between using Eqs.(12)∿(13) and using Eqs.(14)∿(16). Hereafter, Eqs.(12)∿(13) are used for system equation of μ.

In Fig.3a and 3b, initial estimates of X and μ are calculated by the least square estimation of the data during initial 1 hr based on Eqs.(1) and (2), where μ is assumed to be constant during the time. Initial estimate by this method was shown to be successively good as shown in Fig.3a and 3b. And it is also verified that by the extended adaptive Kalman filter, the specific growth rate μ can be estimated successively well. As shown in Fig.4, μ can be estimated well when it changes so quickly and frequently.

PROFILE CONTROL SCHEME IN A FED-BATCH CULTURE

Structure of the Profile Control Scheme

A fed-batch culture can be mathematically described as follows:

$$\begin{aligned} dX/dt &= (\mu - \frac{F}{V})X \triangleq f_1 \\ dS/dt &= -\nu X + \frac{F}{V}(S_0 - S) \triangleq f_2 \\ dV/dt &= F \triangleq f_3 \\ dy/dt &= g(X,S,V,F,y) \triangleq f_4 \end{aligned} \quad (19)$$

where X, S, V, F, S_0 are the biomass concentration, the substrate concentration, the liquid volume, the input flow rate, and the substrate concentration in the feed, respectively. Of course, Eq.(19) includes Eqs.(1) and (2). Parameters μ and ν are the specific growth rate and substrate uptake rate which are functions of S or occasionally y and X. y represents all other state variables except X, S and V. Such variables could be, for example, an inhibitory substance, an intermediate product in the medium, or an inorganic substance needed for growth.

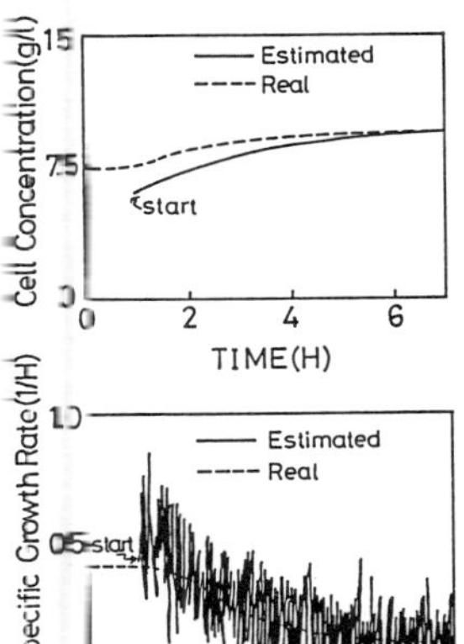

Fig.2 Effect of moving average of data on state estimation.

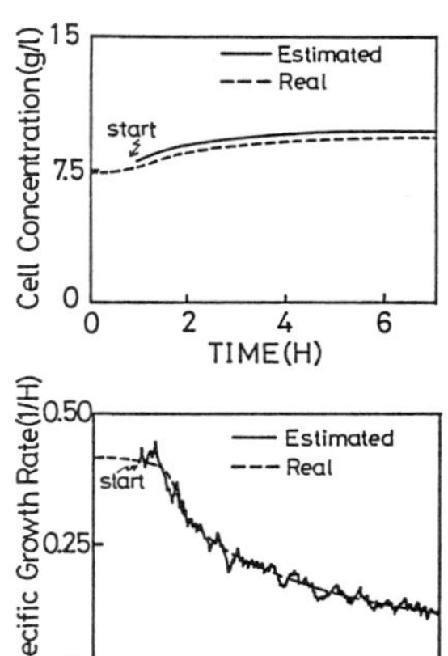

Fig.3a State estimation by Eq.(12)∿(13).

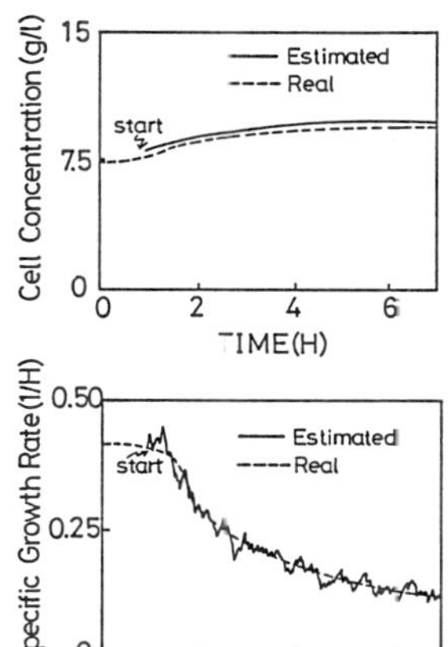

Fig.3b State estimation by Eq.(14)∿(16).

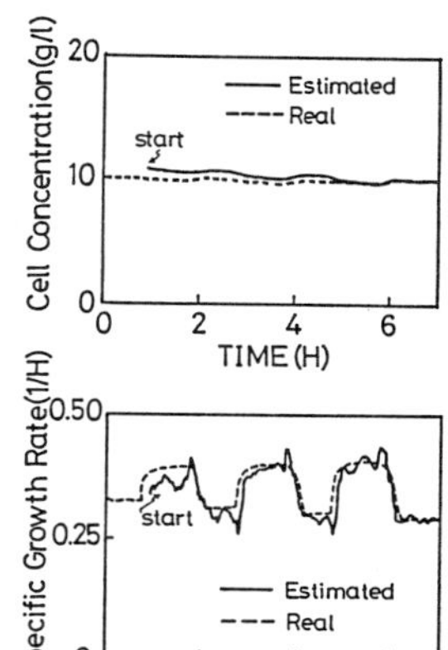

Fig.4 State estimation by extended adaptive filter.

The objective of the control system for this fed-batch process is to make the actual trajectory $\mu(t)$ follow the desired $\mu^*(t)$. The problem considered here is "to control the specific growth rate so as to follow the desired pattern $\mu^*(t)$ by manipulating the feed rate of the substrate F in the fed-batch culture, where (t) is assumed to be observed in this section".

Profile control scheme : The proposed profile control scheme, the PF system, for the above problem is shown in Fig.5. Mathematical justification of the scheme can be described as follows. The linearized version of Eq.(19) around the nominal operating condition $\bar{F}(t)$ at each time along the desired curve, is considered. It is assumed that X(t), S(t), V(t), y(t) and F(t) deviate from X*(t), S*(t), V*(t), y*(t) and $\bar{F}(t)$ as ΔX, ΔS, ΔV, Δy and ΔF. The linearization of Eq.(19) gives Eq.(20) shown as,

$$\frac{d}{dt}\begin{bmatrix}\Delta X\\ \Delta S\\ \Delta V\\ \Delta y\end{bmatrix} = \begin{bmatrix}\mu-F/V, & \mu_s X\ , & FX/V^2, & \mu_y\\ -\nu, & -\nu_s X-F/V, & -F(S_0-S)/V, & -\nu_y\\ 0, & 0, & 0, & 0\\ g_x, & g_s, & g_v, & g_y\end{bmatrix}_{X^*,\bar{F}}\begin{bmatrix}\Delta X\\ \Delta S\\ \Delta V\\ \Delta y\end{bmatrix} + \frac{\partial}{\partial F}\begin{bmatrix}f_1\\ f_2\\ f_3\\ f_4\end{bmatrix}\Delta F + \begin{bmatrix}f_1(\bar{F}) - X^*\\ f_2(\bar{F}) - S^*\\ f_3(\bar{F}) - V^*\\ f_4(\bar{F}) - y^*\end{bmatrix}$$

$$= A(t)\begin{bmatrix}\Delta X\\ \Delta S\\ \Delta V\\ \Delta y\end{bmatrix} + b(t)\Delta F + d(t) \qquad (20)$$

where A(t) and b(t) are defined as the coefficient matrices or vectors of $(\Delta X,\Delta S,\Delta V,\Delta y)^t$ and ΔF, respectively, which are evaluated using X*(t) and $\bar{F}(t)$. For example, the first element of b(t) is ; $\partial f_1/\partial F = -X^*(t)/V^*(t)$. Then, A(t) and b(t) are time-varing parameters. Superscript t means transposing a matrix or a vector. And superscript * means the desired value or profile, and superscript — means the nominal value which is not obtained from the desired profile but given arbitrarily. If $\bar{F}(t)$ gives X*(t), S*(t), V*(t) and y*(t), the distubance vector d(t) becomes zero. For such a special case, $\bar{F}(t)$ is rewritten as F*(t). However, d(t) is not zero for a $\bar{F}(t)$ given arbitrarily.

Now, F should be manipulated in such a way that ΔX(t) becomes zero. To do this, a control system which combines the programmed-control to follow the desired profile or one close to the desired curve with the feedback-compensator (abbreviated as PF), can be proposed as shown in Fig.5(Shioya et al., 1984). The feedback compensator will compensate the disturbances. The manipulated variable, ΔF, can be obtained by requiring the error of the controlled variable $\mu(t)$ from the desired biomass profile $\mu^*(t)$, that is, ε, to be zero. ε is defined as,

$$\varepsilon = \mu(\bar{F}+\Delta F,t) - \mu^*(t) = \Delta\mu(t) \qquad (21)$$

As seen in Eq.(20), the parameters A(t) and b(t) are time-varying. Thus, the introduction of the adaptive

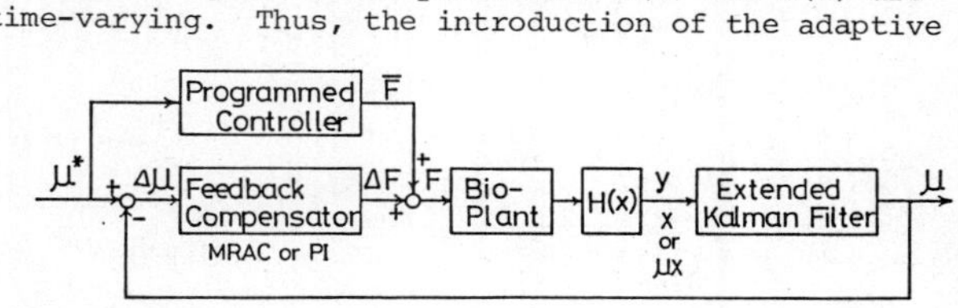

Fig.5 Structure of the PF system for a profile control of μ in the fed-batch culture.

controller as the feedback compensator in Fig.5 was proposed. The discrete-type Model Reference Adaptive Control(MRAC) algorithm was adopted here. The algorithm of MRAC proposed by Lozano and Landau(1981) is employed here. The PF system including MRAC as the feedback compensator named PF-MRAC was considered for the profile control of $\mu(t)$ in the fed-batch culture. This control system was applied to simulation and experimental studies of a profile control problem of the baker's yeast fed-batch culture.

Profile Control of μ in a Baker's Yeast Fed-Batch Culture

The control scheme proposed here was applied to profile control in a baker's yeast fed-batch culture. First, the usefulness and the validity of the proposed scheme was discussed using the computer simulation based on the assumption that the mathematical model was a real system. However, when the control scheme proposed here is applied to a profile control in an experiment, it should be stressed that by using the MRAC algorithm only the order of the system equation should be given a priori.

A profile control of baker's yeast growth on molasses has been reported by Nakatani et al.(1983). Here the same culture system was employed and was represented by a mathematical model. The relationships describing baker's yeast cultivation are assumed to be the same as the literature(Fukuda et al., 1978; Okada et al., 1981).

A bang-bang type curve of $\mu(t)$ is considered here to be the desirable profile to be followed. In this case, it is not easy to calculate the feeding rate $\bar{F}(t)$ which produces the desired bang-bang type specific growth rate $\mu^*(t)$, that is, $\mu(\bar{F}(t))\neq\mu^*(t)$, even if all system parameters and initial conditions are known. The reference value of the controlled variable is given as in Eq.(21), where $\bar{F}(t)$ is calculated as the feeding rate of an exponential fed-batch culture based on a constant μ.

A numerical example of PF-MRAC applied to the profile control of μ is shown in Fig.6a. In this example the system has a 10% disturbance of substrate concentration in the feed. The desired profile of μ ie, $\mu^*(t)$, is bang-bang type one shown in the figure. The actual value of $\mu(t)$ and the substrate feeding rate F(t) are also shown in the figure. Fig.6b shows the time courses of other state variables, C_e, P and S. At the switching point from μ_{MIN} to μ_{MAX} or from μ_{MAX} to μ_{MIN}, the parameters of the controller were not quickly adapted. At those points, controlled variable μ has a discrepancy from μ*, and the manipulated variable F should be greatly changed comparing with $\bar{F}(t)$ in Fig.6a. C_e and S also changes largely at those points. However, as shown in Fig.6a, PF-MRAC works well, and the control algorithm was shown to be superior.

It is assumed that μ can be observed with no random noise. One of the numerical examples of the effect of measurement noise is shown in Fig.7. In this example, the estimated value of μ using $(\dot{V}X)/VX)$ has additive random noise which has normal distribution function with zero mean and covariance σ^2. σ is taken as 0.03 which corresponds to more than 10% of the observed value. As shown in Fig.7, the controlled $\mu(t)$ becomes a little bit noisy compared with one in Fig.6a. However, the result will be acceptable for practical cases. If the data has larger noise than this example, the controller performance will not be adequate. But if one of the various kinds of sophisticated techniques for averaging the random noise such as by taking moving averages or filtering the noisy data are used as shown previously, the level of the measurement noise will be reduced to be in the order of the noise in this example. Anyway,

PF-MRAC or PF-PI should be used in an actual experiment as the next step in showing the validity of the proposed control scheme.

Fig.8 shows the comparison of the results obtained using a 1st order adaptive model and one using a 5-th order adaptive model(Fig.6a). In this case, during the latter half of the step-change of $\mu^*(t)$ the error of the controlled variable, $\Delta\mu(t)$ ($=\mu(t)-\mu^*(t)$), is not controlled more accurately than that by the higher order model as seen in Fig.8. However, during the first half of step-change of $\mu^*(t)$, the response of $\mu(t)$ by the 1st order model is faster than that by the 5-th order model. This is because the model parameter is adjusted or adapted so quickly in the low order model, and because the low order model, which is not real, requires a long settling time for the error to become zero. The higher order model which is the real one compensates the error of the controlled variable in the latter half of stepchange of $\mu^*(t)$. The selection of the system order depends on the characteristics of the controlled variable required for the controller, that is, depending on the quickness of the response and the amount of the off-set of the controlled variable. However, in any real problem, the 1st or 2nd order model may be sufficient to follow the desired profile of the fed-batch culture as shown in Fig.8.

In the system employed here, MRAC algorithm was basically applied for the feedback compensator. However, PI feedback controller is applicable instead of MRAC algorithm. This system was named as PF-PI. It was shown that MRAC algorithm performed better than the PI-feedback controller. It may be said that PF-PI will be sufficient for apply a practical problem if the error between the desired and the actual curves is not limited too strictly. However, PF-MRAC is favorable if the inital conditions or process paramaters change largely.

EXPERIMENTAL STUDY FOR THE CONTROL OF μ

In order to certify that the estimation scheme and control scheme of μ are practically useful for the actual plant, experimental studies were performed using a laboratory scale fermenter. Schematic diagram of the experimental apparatus used is shown in Fig.9. The main facilities used are ; 1ℓ jar fermenter with temperature control, Shimazu UV-Spectrophotometer, micro-tube-pump used for feeding of substrate, PH controller connected to the computer, micro-processor μMAC-5000(Analog Devices Co. Ltd.) which was used for A/D and D/A converter and first step data processing, and personal computer NEC-PC-9801(384KB) with two floppy drivers(1MB, each) which is a main computer used for estimation of μ, calculating the input flow rate to be manipulated, data filtering and graphic diplaying. Glucose concentration was off-line measured by the sugar analyzer(YSI model-27). At now stage, the experiment were performed only for the range of low cell concntration. Then the off-gas analyzer was not used this time, but the Gilconia-type oxygen analyzer and an infrared CO_2 analyzer are attached to the fermenter. Ethanol concnetration could be also measured by the tubing method(Dairaku and Shioya, 1981) with FID detector.

Liquid sample is auto-sampled every 5 minutes and cell concentration is measured using the UV-Spectrophotometer with auto-flow-sell unit which is controlled by the μMAC-5000. The amount of NH_4OH solution used for PH adjustment is calculated in μMAC-5000. DO concentration is only recorded by the computer but is possible to be controlled by changing the agitation speed. Thecommunication between μMAC and PC9801 was done through the RS-232C cable. PC-9801 was used for calcualtion of the control action and the estimation of μ by the extended adaptive Kalman filter. Sampling period is 5 minutes and it is enough for estimation and control of μ.

Some of the experimental results are explained here.

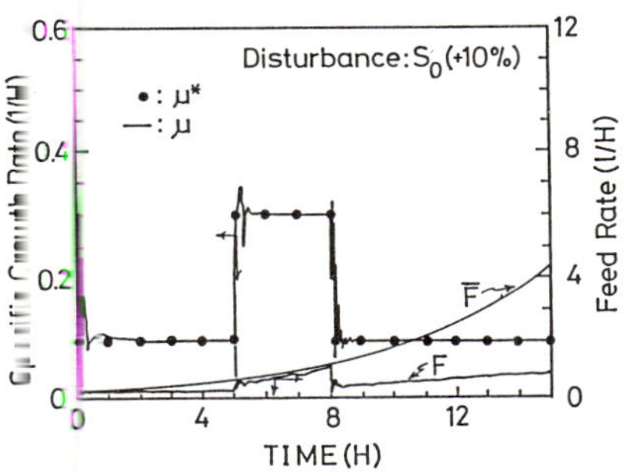

Fig 6a Profile control of μ by PF-MRAC; 5th order model.

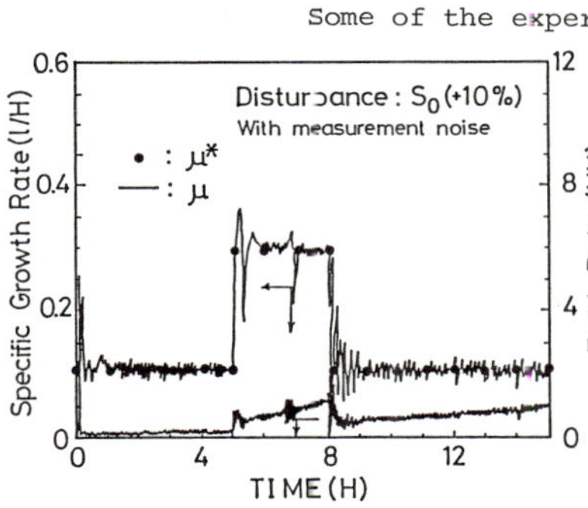

Fig.7 Profile control of μ by PF-MRAC with measurement noise.

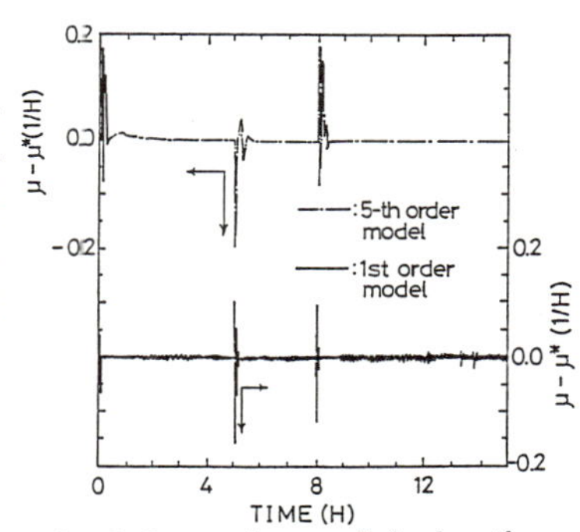

Fig.8 Comparison of $\Delta\mu$ by the 5-th order PF-MRAC and 1st order PF-MRAC.

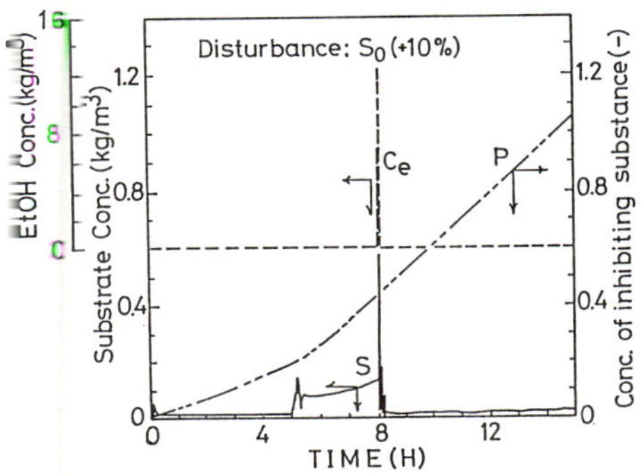

Fig.6b Profile control of μ by PF-MRAC; 5th order model.

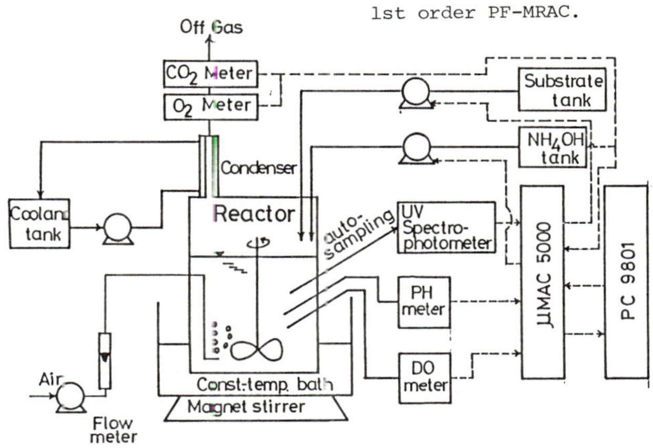

Fig.9 Schematic diagram of experimental apparatus.

Here, at the first stage of the profile control, the specific growth rate μ was desired to be kept at a constant value. Fig.10 shows one of results of the exponential fed-batch culture where the feeding flow rate was calculated so as to keep $\mu^*=0.2[1/hr]$. The exponential fed-batch culture doesn't include the feedback control scheme. Then, if the initial condition and parameters required for calculating the feeding flow rate have error, the specific growth rate is not equal to the desired value at all as shown in Fig.10. Feeding rate was too fast, then μ became large and ethanol was produced. In the latter half of the experiment, the growth of the microorganisms was inhibited by ethanol and μ decreased. This exponential fed-batch culture had failed to keep μ constant.

Fig.11 and Fig.12 show the experimental results by PF-PI and PF-MRAC, respectively. Experimental conditions for both experiments were; PH=5, DO=4∿6[ppm]. For Fig.11, initial liquid volume;V(0)=1.1[ℓ], and the substrate glucose concentration in the feed; $S_0=4.94[kg/m^3]$. For Fig.12, V(0)=1.05[ℓ] and $S_0=4.58[kg/m^3]$. Fig.11 shows the result by PF-PI. Parameters of the PI controller were; proportional gain;Kp=0.1[ℓ]and reset time;Ti=0.5[hr]. Estimated value of μ, cell concentration X, nominal flow rate $\bar{F}$ and actual flow rate $\underline{F}$ are shown in the figure. The difference between $\bar{F}$ and F is the corrected amount of flow rate by the compensator(=PI controller). $\bar{F}$ is the feeding rate calculated by the exponential fed-batch culture of $\mu^*=0.2$. If the difference of $\bar{F}$ and F is large, it means that the exponential fed-batch culture can not control μ at the desired constant value, $\mu^*=0.2$.

Fig.12 shows the result by PF-MRAC in which the order of plant model is taken as 1st order one as a simplest model. Parameter adjustment rule is taken as one proposed by Lozano and Landau(1981). The desired parameter μ is controlled fairly well compared with Fig.10. However, it was not shown that the controller performance by PF-MRAC was better than PF-PI from this experimental data. At now stage, not so many experimental results have not been obtained. Then the more experimental data than ever is required for judging the superiority of PF-MRAC compared with PF-PI. Anyway, the control scheme proposed here, the PF system, PF-MRAC and PF-PI can effectively used for the estimation and control of the specific growth rate. For other cases, e.g., high cell concentration and bang-bang type profile of μ, experiments should be performed in the near future.

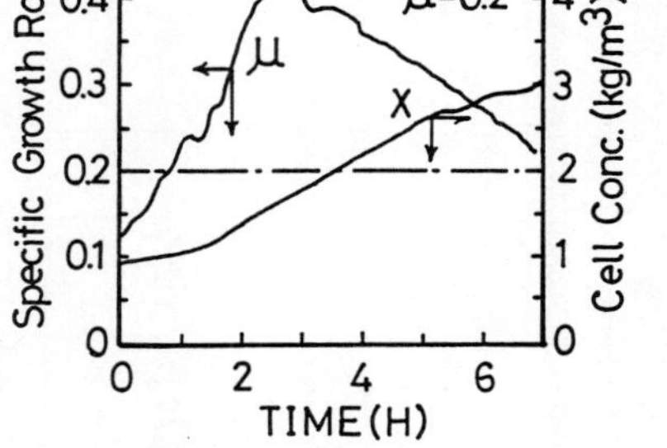

Fig.10 Exponential fed-batch culture.

CONCLUSION

In this paper, the estimation and control schemes for the profile control of the specific growth rate are proposed. The schemes are shown to be useful by many simualtion works. And some of experimental works also show that they work successively. In the near future, more complicated profile of μ should be experimentally tested using the scheme proposed here.

REFERENCES

Aiba, S., S. Nagai and Y. Nishizawa (1976). Biotechnol. & Bioeng., vol. 18, 1001.

Cooney, C.L., H.Y. Wang and D.I.C. Wang (1977). Biotechnol. & Bioeng., vol. 19, 55/86.

Dairaku K., Y. Yamasaki, K. Kuki, S. Shioya and T Takamatsu (1981). Biotechnol. & Bioeng., vol. 23, 2069.

Dairaku, K., E. Izumoto, H. Morikawa, S. Shioya and T. Takamatsu (1982). Biotechnol. & Bioeng., vol. 24, 2661.

Fukuda, H., T. Shiotani, W. Okada and H. Morikawa (1978). J. of Ferment. Technol., vol. 56, 36 .

Goodwin G.C., P.J. Ramadge and P.E. Cains (1980). IEEE Trans., vol. AC-25, 449.

Landau, I.D. (1974). Automatica, vol. 10, 353.

Lozano and I.D. Landau (1981). Intern. J. of Control, vol. 33, 247.

Nakatani, H., K. Dairaku, H. Morikawa, E. Nakanishi and K. Nanbara (1983). Proc. of 3rd Int. Conf. on Computer & Chem. Eng., C25.

Okada, W., H. Fukuda and H. Morikawa (1981). J. of Ferment. Technol., vol. 59, 103.

Shioya, S., Y. Okada and T. Takamatsu (1984). Proc. of 7th Int. Biotechnol. Symp., New-Delhi 55_.

Stephanopoulos, G. and Y.K. San (1984). Biotechnol & Bioeng., vol. 26, 1176/1197.

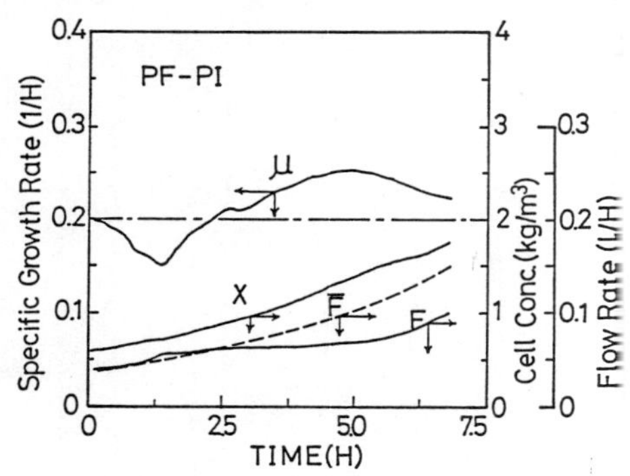

Fig.11 Control of the specific growth rate by PF-PI.

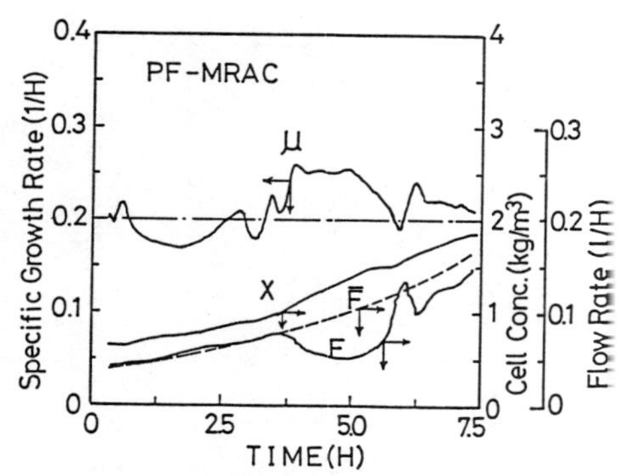

Fig.12 Control of the specific growth rate by PF-MRAC.

A NEW AUTOMATIC SAMPLING DEVICE FOR THE DETERMINATION OF FILTRATION CHARACTERISTICS AND THE COUPLING OF AN HPLC TO FERMENTORS

R. Lenz, C. Boelcke, U. Peckmann and M. Reuss

Technical University Berlin, Seestr. 13, D-1000 Berlin 65, FRG

Abstract. The major bottleneck for computer-coupling of fermentation processes is the lack of reliable,sterilisable and robust sensors for process key variables like biomass, substrate and product concentrations. Principally there are two ways to overcome this difficulty: a) construction of new sensors like product specific electrodes for in situ application or b) automatic coupling of powerful analyser systems like HPLC.

This work is aimed to present a new system for coupling automatically an HPLC-system to fermentors. Furthermore this system is able to measure automatically filtration characteristics of the fermentation broth. The relationship between filtration characteristics, the height of the filter cake and its density was applied to the determ nation of morphology parameters of Penicillium by Nestaas and Wang (1981).

Similar to other attempts for coupling autoanalyser systems, their concept was based on a filter unit with backflushed filter media. Problems in the use of such concepts for industrial media had been described several times in literature.

Therefore in this work the concept of renewable filters is chosen.The filter media is a band of filter paper or membrane filter (depending on the process), which lies between the media support and the filter body.The sequence of the different operations : sterilisation, sampling, filtration, HPLC-analysis of the filtrate cleaning and finally transportation of the filter media is computer controlled. During filtration the computer registers the data for weight of the filtrate (load cell) filtration time and height of the filter cake from a couple of reflection light barriers. Using this information the computer predicts filtration and morphology parameters.

Examples for the application of this automatic system will be discussed with the aid of fermentations with a mycelial organism and yeasts. In all of these examples the emphasis of discussion is placed on both filtration parameters and HPLC-coupling.

Keywords. Computer control; on-line analysis; automatic sampling; filtration characteristics of bio-suspensions;estimation of biomass.

INTRODUCTION

Application of computer controlled fermentation processes has been considerably hampered by the lack of reliable sensors for the most important process variables like concentrations of substrate, biomass, precursors and products. A first possibility to overcome this difficulty is the use of indirect estimations for these variables from mass balancing techniques. The application of these well known methods is restricted to only few fermentation processes in which the stoichiometric relationships are exactly known, time invariant and reproducible between different batch processes. Particularly the last two prerequisites most often cannot be guaranteed due to the well known changes in microbial behaviour from run to run even if operating conditions are exactly the same. The effect of these changes on the reliability of indirect biomass estimation can be rather drastically as shown by Nelligan and Calam (1983) for the penicillin fermentation in which carbon dioxide balancing was used for the estimation of biomass concentration. Also Mou and Cooney (1983) clearly demonstrated that maintenance coefficients during the same process are not constant.These observations are in agreement with our own experiences with computer coupled penicillin fermentations.

A second possibility to improve the strategies for computer coupled fermentation processes is the development of in situ

not required for estimation of biomass concentration.

A particular problem in the computer aided analysis of the filtration characteristics in case of yeast suspensions can be observed from Fig. 10., in which t_f/V_F is plotted against V_F for two biomass concentrations. According to the theory of cake filtration this plot should result in a straight line with a slope proportional to the specific cake resistance. The diviation from ideal cake filtration is due to various interactions between the biosuspension and the filter media during the early stages of cake buildup (Orr,1977). The length of the initial period characterised by a slope $\alpha' = 1$ in the filtration characteristics depends on the organism, its concentration and also on the material of the filter media. It is therefore difficult to decide in advance the critical filtration time t_c beyond which the data can be analysed with the aid of the cake filtration theory.

Bryszewski (1979) has suggested a simple but useful analysis of the filtration characteristics including this initial period. This analysis leads to a parabolic equation of the following form:

$$t_f = D\,V_F^2 + B\,V_F + C \qquad (9)$$

where coefficients B and C are parameters depending on the filter media resistance and the initial period of cake buildup. The most important variable is D, defined by

$$D = \frac{\eta\,\alpha'}{2\,A^2\,\Delta P} \qquad (10)$$

Thus, because of Eq. (8) a plot of 1/D vs. 1/X should result in a straight line, the slope of which is proportional to the biomass concentration X. Using the raw data of the filtrate volume as a function of time the process computer predicts the coefficients D,B and C in Eq. (9) with aid of the Gaussian elimination method. In Fig. 11 results from three different bakers yeast fed-batch fermentations are shown, in which the computer controlled substrate feeding was applied (Wang, Cooney, Wang, 1977). The scattering of the data is in the same order of magnitude as the dry weight laboratory measurements shown in this figure.

CONCLUSION

In this work a computer aided fermenter-coupled filtration unit is proposed which solves the automation of sampling, indirect biomass estimation and sample preparation for on-line analysis of key concentrations in the fermentation broth with aid of automatic analysers. In spite of the present limitations in the on-line analysis of important key variables of fermentation processes the proposed method may offer interesting possibilities to achieve greater succes in the computer-aided operation of biotechnical processes.

REFERENCES

Bryszewski, C. (1979).Einfluß der Anfangsperiode auf die Dynamik der Kuchenfiltration.Chem. Techn.,31,6,300-302.

Ferrari, A., J.R. Gerke, R.W. Watson and W.W. Umbreit (1965). Application of automated analysis to the study of bacterial growth: Part I - The instrumental system. Ann. N.Y. Acad. Scienc.,Vol.130,pp.704-721

Glaxo Laboratories, Ltd., U.S. Patent 3830106 (1974).

Imming, G., K. Schaller and M. Meiners (1982). The use of automated analysers to control microbial processes. In A. Halme (Ed.), Modelling and control of biotechnical processes, Proceedings of the first IFAC workshop. Pergamon Press, London.pp 27-32.

Mou, D.G. and C.L. Cooney (1983). Growth monitoring and control through computer-aided on-line mass balancing in a fed-batch penicillin fermentation. Biotechnol. Bioeng.,25,225-255

Nelligan, I. and C.T. Calam (1983). Optimal control of penicillin production, using a mini-computer. Biotechnology Letters,5, 561-566.

Nestaas, E. (1980). Quantitative characterisation of myc lial biomass and morphology using the filtration probe in the penicillin fermentation. Ph.D. thesis, Massachusetts Institut of Technology, Cambridge, Massachusetts.

Nestaas, E. and D.I.C. Wang (1981). A new sensor, the "filtration probe" for quantitative characterisation of the penicillin fermentation.I. Mycelial morphology and culture activity. Biotechnol. Bioeng.,23, 2803-2813.

Nestaas, E., D.I.C. Wang, H. Suzuki and L.B. Evans (1981). A new sensor, the "filtration probe" for quantitative characterisation of the penicillin fermentation. II. The monitor of mycelial growth. Biotechnol. Bioeng.,23,2815-2824.

Orr, C. (1977). Filtration.M. Dekker Inc., New York.pp. 288-302.

Silvennoinen, E. and H.N. Koivo (1982). Estimation of the biomass concentration in the Pekilo-Process from the filtrate flow rate of the concentrator. in A. Halme (Ed.), Modelling and control of biotechnical processes, Proceeding of the first IFAC workshop. Pergamon Press, London. pp. 219-258.

Valentini, L. and G. Razzano (1982). The real time analysis of broth constituents in the fermentation processes: on/off-line requirements. In A. Halme (Ed.), Modelling and control of biotechnical processes, Proceedings of the first IFAC workshop. Pergamon Press, London. pp. 253-258.

Vogelmann, H., M. Reuss, J. Grieser and F. Wagner (1973). Kontinuierliche sterile Probenentnahme bei mikrobiellen Prozessen zur automatischen chemischen Analyse. In H. Dellweg (Ed.), 3. Symposium fuer technische Mikrobiologie,Berlin 1973. VLSF, Berlin. pp. 215-220.

Wang, H.Y., C.L. Cooney and D.I.C. Wang (1977). Computer-aided baker's yeast fermentations. Biotechn. Bioeng.,21 67-86.

Zabriskie,D.W. and A.E. Humphrey (1978). Continuous dialysis for the on-line analysis of diffusible components in fermentation broth. Biotechnol. Bioeng., 20, 1295-1301.

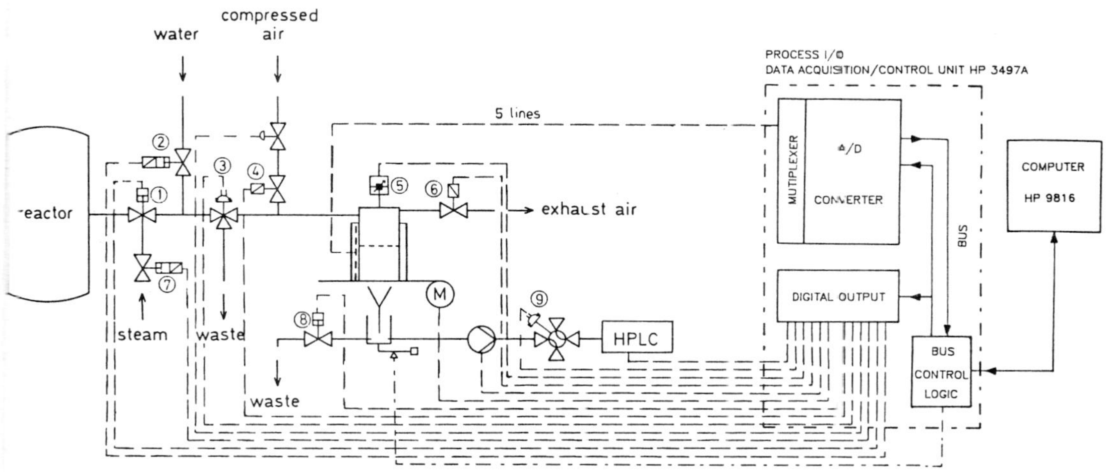

Fig. 1. Process computer coupled filtration unit.

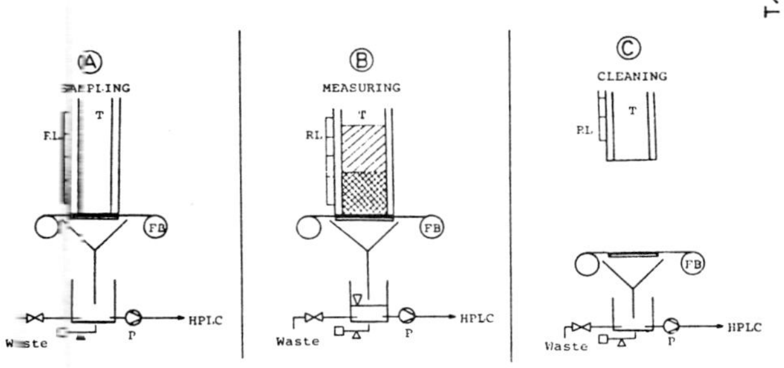

Fig. 2. Operation modes of filtration unit.

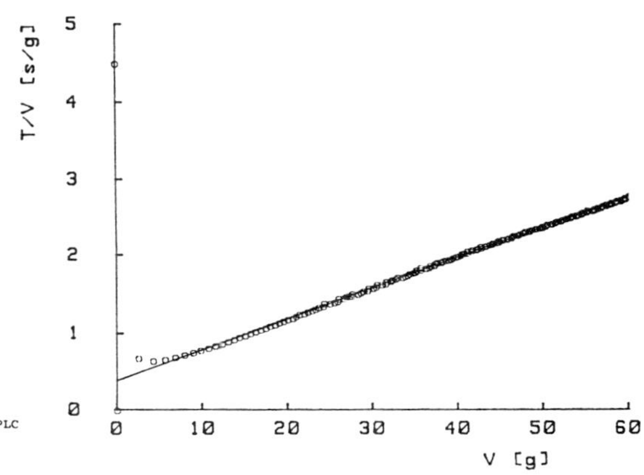

Fig. 4. Filtration characteristics.

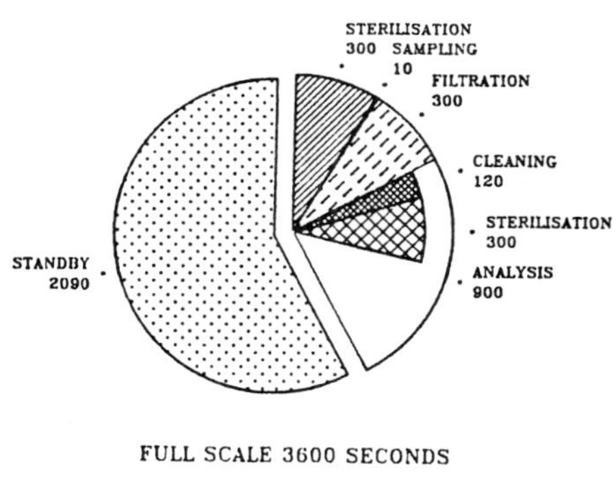

Fig. 3. Cycle of filtration unit.

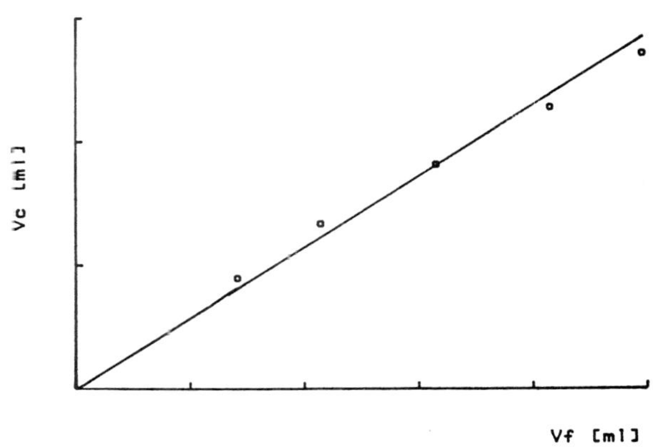

Fig. 5. Regression of filter cake vs. filtrate volume.

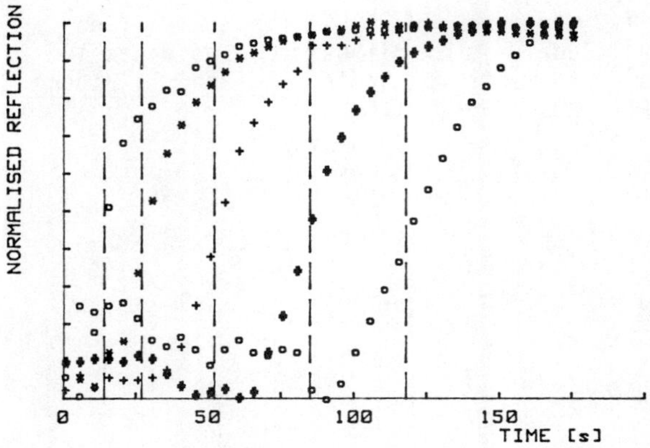

Fig. 6. Reflection light barrier data.

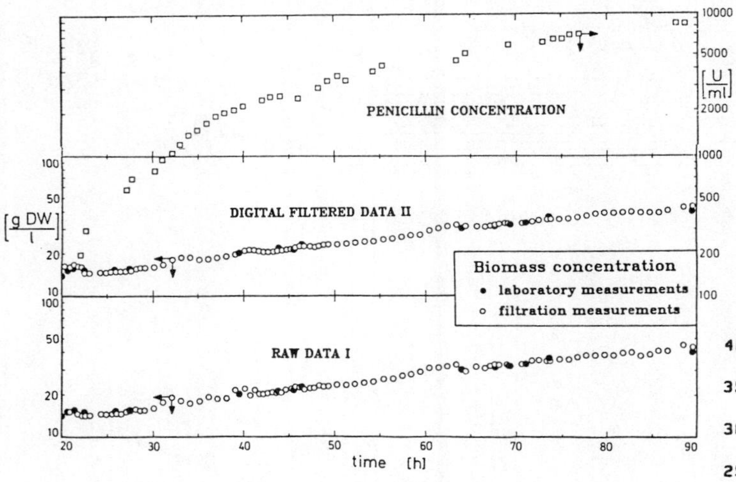

Fig. 7. Penicillin fermentation.

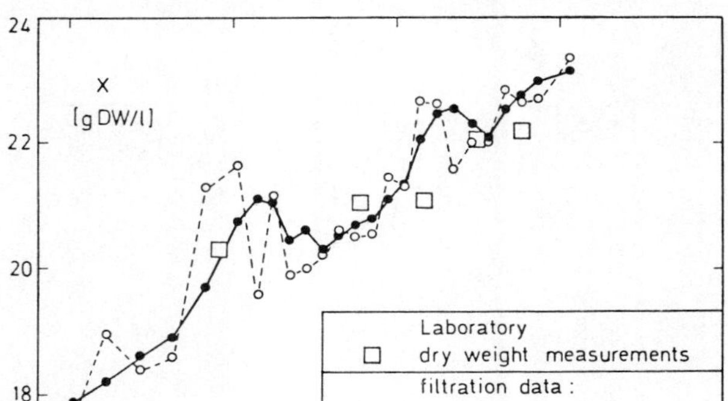

Fig. 8. Performance of digital filter.

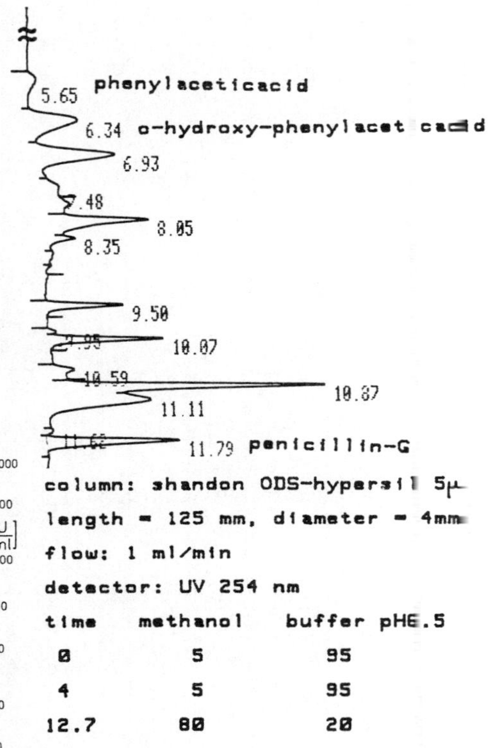

Fig. 9. On-line HLPC-chromatogramm

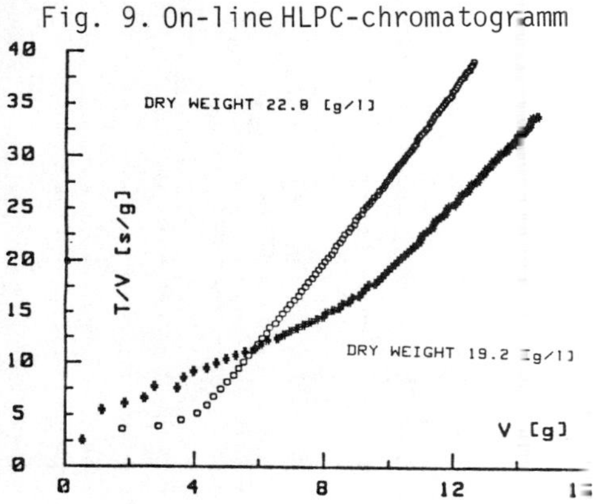

Fig. 10. Filtration characteristics of yeast suspensions.

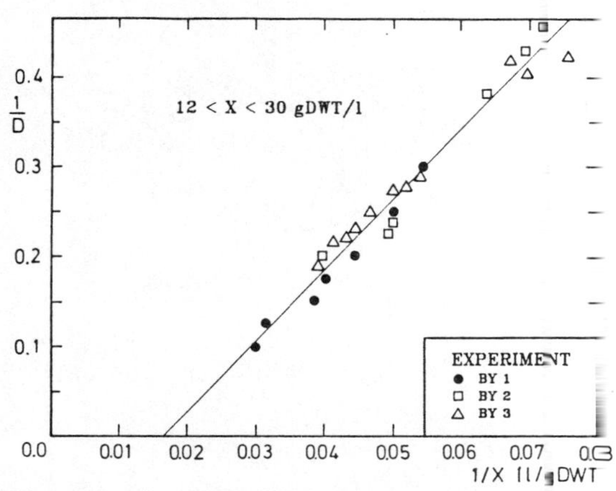

Fig. 11. Correlation between specific cake resistance and biomass (baker's yeast).

ANALYSIS OF BIOMASS AND METABOLITES USING PYROLYSIS MASS SPECTROMETRY

E. Heinzle, M. Lunden and I. J. Dunn

Chemical Engineering Laboratory, ETH, CH-8092 Zurich, Switzerland

Abstract. Pyrolysis mass spectrometry was used to analyze fermentation samples. It is shown that the analysis of poly-3-hydroxybutyric acid (PHB) in aculture of *Alcaligenes latus* is possible using mass spectrometry with a simple batch oven for pyrolysis. Tests showed that practically any organic matter gives detectable volatiles when heated to pyrolysis temperatures under helium atmosphere.

Keywords. Mass spectrometry, fermentation, pyrolysis, poly-3-hydroxybutyric acid.

INTRODUCTION

The lack of suitable sensors and of fast analytical methods are the limitations for efficient on-line fermentation analysis and control. Many attempts are presently made to solve these problems, and a large part of a recent conference was dedicated to this (Heinzle, 1984). Mass spectrometers are amongst the most versatile detectors and have successfully been applied to on-line fermentation analysis problems (Reuss and co-workers, 1983; Pungor and co-workers, 1981; Heinzle and co-workers, 1983, 1985). All these contributions dealt with the analysis of gases and volatiles.

As most of the biologically important materials are not sufficiently volatile to be easily introduced into the high vacuum of a mass spectrometer (MS), methods have to be developed to volatilize such compounds. Pungor and co-workers (1984) used continuous esterification to produce volatile products of 2-oxoglutaric acid. Such derivatizations presumably would have to be developed for each process. Reuss (1984) tried to apply an HPLC-MS interface to directly introduce liquid samples into the high vacuum. This only worked satisfactorily with clean solution.

It would be desirable to use a more versatile method for volatilization. Possibly the most versatile method is pyrolysis in a non-oxidative environment. This method has been successfully applied for a number of analytical problems (Meuzelaar and co-workers, 1982). The most successful method seems to be Curie-point pyrolysis in the vacuum of the MS, but it would not likely be useful for fermentation samples because less than 100 µg of sample can be treated, making it very difficult to get a representative sample from a fermentation broth. This method requires a specially constructed MS. Additionally it would be desirable to have the possibility to introduce samples into the pyrolysis MS automatically or even continuously.

MATERIALS AND METHODS

Alcaligenes latus (DSM 1123) was cultivated in a mineral medium containing saccharose as carbon and energy source. Analysis of biomass and poly-3-hydroxybutyric acid (PHB) was done as described elsewhere (Heinzle, 1978). Reagens were from Fluka, PHB for reference was produced in our laboratory.

The MS equipment (Balzers) used is described elsewhere in detail (Heinzle and co-workers, 1983, 1985). Ion intensities were normalized to the sum of all detected intensities. The oven design is shown in Fig.1 and is similar to the one used by Chan and co-workers (1982). This oven allows the automatic introduction of samples. Because of the temperature control, whose set point is changed by a program which allows successive drying and pyrolysis of a sample, a fermentation broth can directly be analyzed without any further pre-treatment. The oven is made of glass, the sample beaker of quartz. The heating system was made of coaxial stainless steel heating wire. The temperature was measured with a stainless steel coated thermoelement, whose signal was used to control the temperature within the oven. The oven was constantly flushed with He to avoid oxidation of organic materials at higher temperatures. The connection to the MS was made using the heated capillary inlet. The speed of data aquisition was limited by the available quadrupole data processor and the computer (Apple IIe). To get reasonable results which were not too much distrubed by noise, 70 s were needed to process a spectrum with 150 mass units. For good quantitative work, the speed can be drastically increased using selected ion monitoring.

Spectra were collected during the whole period of sample treatment, stored and further processed using factor analysis (Heinzle and co-workers, 1985; Malinowski and Howery, 1980). The calculations were done with BMDP (Biomedical Computer Programs P-series; Dixon, 1981). The number of factors was determined using the empirical factor indicator function from Malinowski and Howery (1980).

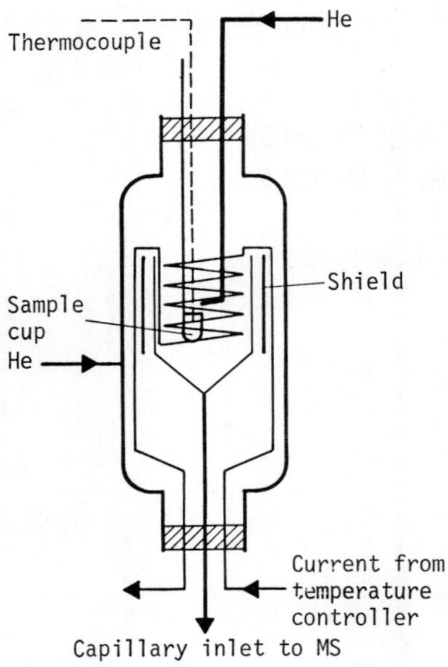

Fig. 1 Pyrolysis oven.

ANALYSIS OF PURE COMPONENTS

Spectra from a number of pure compounds were analyzed to test the whole equipment. Fig.2 shows the spectrum of yeast RNA where measurement of the spectrum was started at the temperature jump from ambient temperature to 400°C. A second spectrum measured afterwards (70 s) showed lower intensities for the peaks m/z = 82, 92, 96 and 100. The kinetics of production of pyrolysis products is very much dependent on the chemical nature of the compounds.

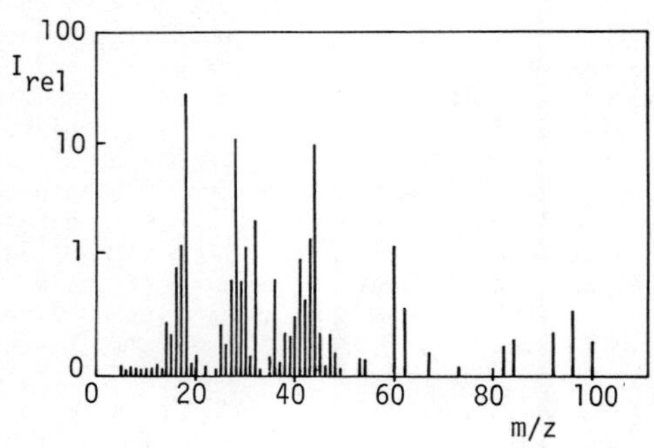

Fig. 2 Spectrum of yeast RNA taken just after changing the temperature set point stepwise to 400°C.

Cellulose was found to quickly change its product patterns and pyrolysis was found to be finished after about 60 s as can be seen from the experiment of Fig. 3. In this experiment, selected ion monitoring was used to measure selected ion currents. Also shown is a typical temperature profile. Obviously for a sample with such a rapid pyrolysis, the measurement time for a complete spectrum is too long. In this case selected ion monitoring would be preferable.

With many compounds this problem is less significant, e.g. poly-3-hydroxybutyric acid. A spectrum measured with this pure compound is shown in Fig.4. In Fig.5 a spectrum of bovine serum albumin is

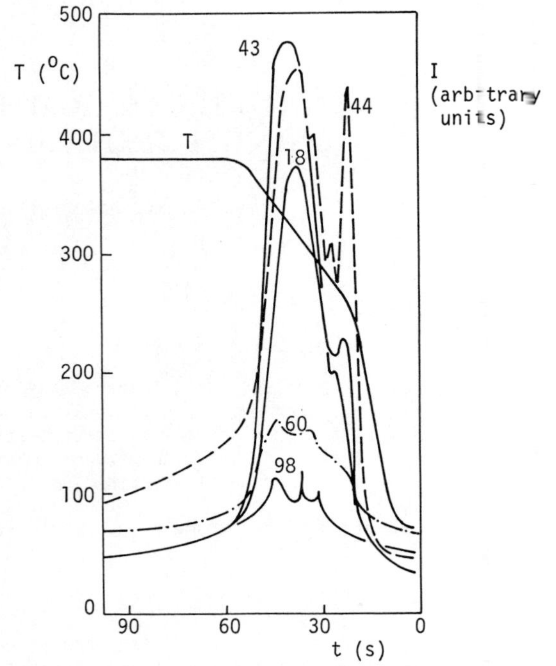

Fig. 3 Intensity-time profiles for various masses (m/z = 18, 43, 44, 60, 98) during the pyrolysis of cellulose.

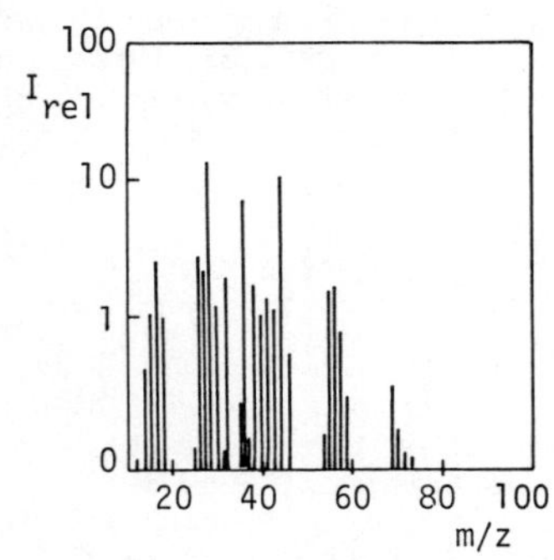

Fig. 4 Spectrum of PHB measured 10 s after the set point temperature jump to 300°C.

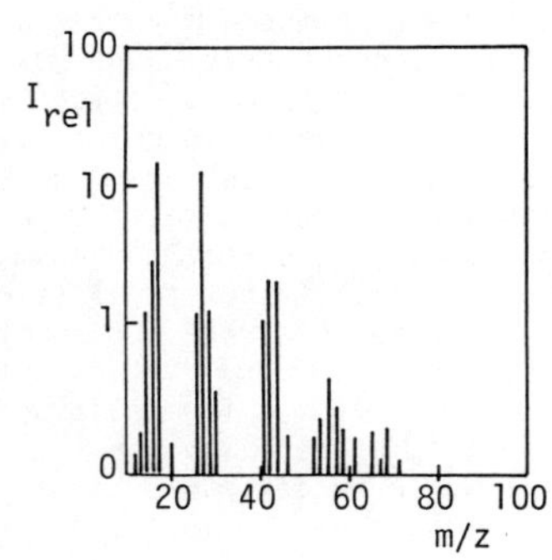

Fig. 5 Spectrum of albumin taken 11 s after the temperature set point jump to 390°C.

shown. A number of other compounds such as amino acids, penicillin, organic acids, sugars etc , were found to give detectable spectra. Thus it appears that the necessary pre-requisites for the analysis of fermentation samples are fulfilled. The more significant problem is the quantitative analysis of a fermentation broth.

ANALYSIS OF FERMENTATION SAMPLES CONTAINING PHB

To four fermentation whole broth samples (each 0.2 ml) varying amounts of PHB were added (A = 0 mg, B = 0.17 mg, C = 0.34 mg, D = 0 mg). First the water was evaporated at 95°C for 10 min, then the temperature was increased to 400°C. During each 70 s, a pyrolysis spectrum over the whole amplification range of the MS was measured. The results of measurements and subsequent factor analysis (principal component analysis) are shown in Fig. 6.

In an additional experiment PHB and albumin were added (A = reference, B = 2 mg PHB, C = 4.3 mg albumin, D = 4.2 mg albumin and 2 mg PHB). After the same data treatment and analysis as above, no significant correlation of any factor with PHB or albumin could be identified.

Samples from a A.latus batch fermentation were analyzed. For each measurement a 0.15 ml sample of whole fermentation broth was introduced into the quartz sample beaker and analyzed using the same procedure as for the previous experiments. The results are given in Fig. 8. During the batch fermentation both biomass (optical density and protein concentration) and PHB increased. Scores for Factor 4 resulting from principal component analysis without rotation and with varimax rotation

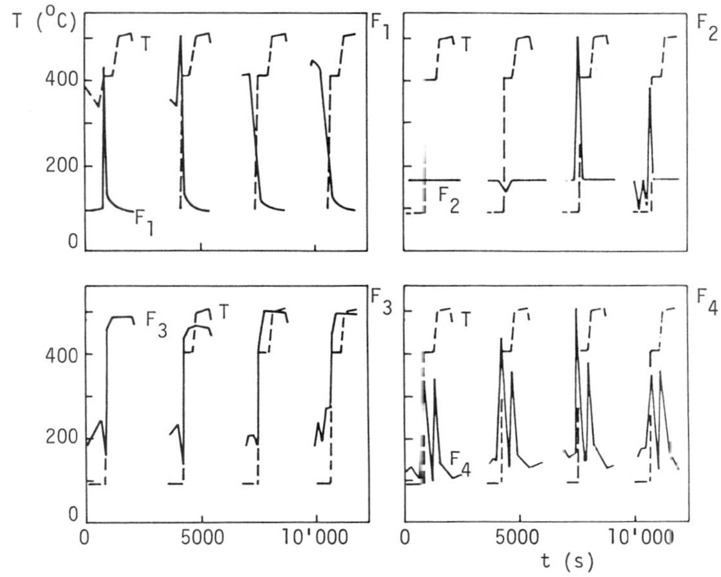

Fig. 6 Change of Factors 1 to 4 (solid lines) in 4 samples during the pyrolysis which lasted 2000 s. The Factor 4 exhibits definite peaks which are correlated with PHB. The temperature (dashed line) was varied in each case as shown with Factor 2.

Using the empirical factor indicator function, it was found that 5 factors are characteristic for these measurements. The first 4 factors together with the temperature profile indicate a complex behaviour. Factor 2 seems to characterize a disturbance in the sample handling, whereas Factor 1 and similarily Factor 3 do not show systematic differences among the 4 samples. Factor 4, however, shows an interesting behaviour. Right at the temperature jump to 400°C, this factor reaches a maximum which seems to correlate with the amount of PHB added to the fermentation sample. After going through a minimum which is very similar for all samples, it again reaches a maximum when increasing the temperature further to 500°C. This second maximum only differs slightly for the 4 samples. The factor score value of Factor 4 at the first maximum was plotted versus the added amount of PHB (Fig.7). In this figure, a correlation is shown for the unrotated factor and for the same factor after a varimax rotation with the added amount of PHB. This rotation results in a factor which is loaded with as few variables (masses) as possible to give simple spectra. Both values give a good linear correlation and the close similarity between the factor score of the first and the last sample both containing no additional PHB indicates that cleaning of the quartz reaction beaker is not very critical.

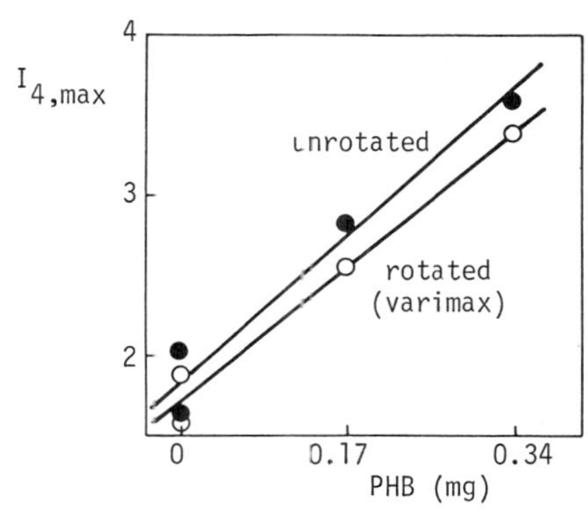

Fig. 7 Linear correlation between the maxima of Factor 4 (rotated and unrotated) and the quantity of PHB in the sample.

obviously correlated to a certain extent with fermentation product concentrations as seen by correlations of the maximum Factor 4 score with PHB and protein concentration (Fig. 9).

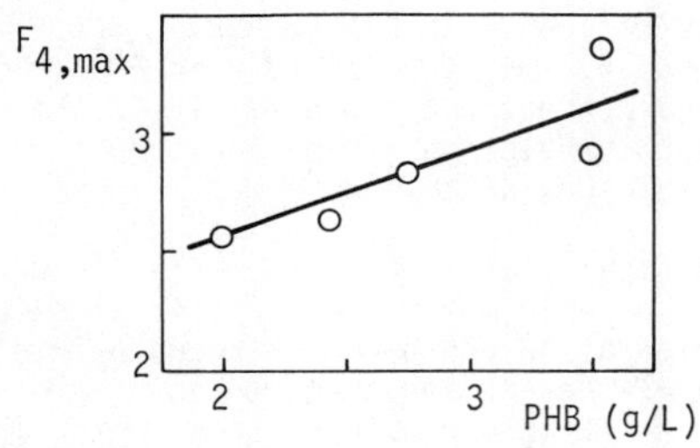

Fig. 8 Correlation between the maximum of Factor 4 and the PHB concentration.

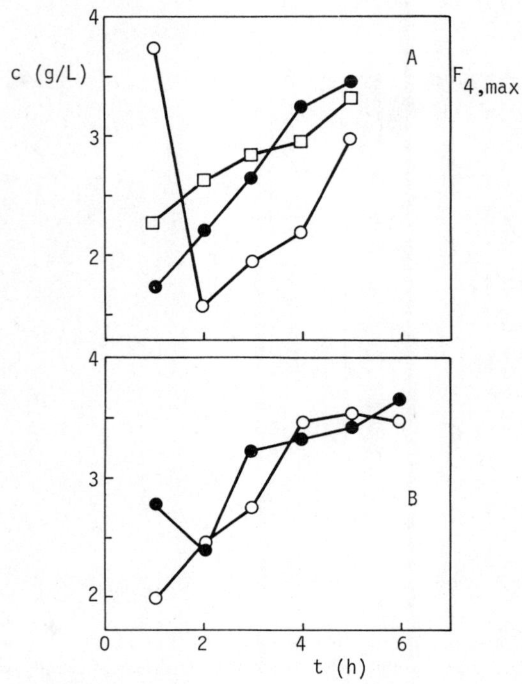

Fig. 9 Profiles of the fermentation parameters and maxima of Factor 4 versus time. Each point represents a sample during the batch culture and a pyrolysis measurement.

A: ○ Principal component analysis, $F_{4,max}$
□ Same as ○ but after varimax rotation
● Optical density

B: ○ PHB concentration
● Protein concentration

CONCLUSIONS

The first series of pyrolysis experiments using a simple batch pyrolysis oven coupled to a quadrupole mass spectrometer via a stainless steel capillary revealed that practically any organic material can be detected. The resulting spectra are considerably different from those measured by Curie-point pyrolysis (Meuzelaar and co-workers, 1982). The oven design and temperature programming allows drying of samples and subsequent pyrolysis. This makes it possible to directly use fermentation samples without any pre-treatment. Though the mass spectrometer equipment available did not allow fast spectra collection, it was possible to get useful results for the analysis of PHB when added to a fermentation sample using factor analysis to treat the data. Analysis of a series of fermentation samples indicated that fermentation parameters can be detected. Batch pyrolysis in this reactor is not ideally suited for fast measurements of fermentation samples (< 15 min) because of the long evaporation time (approximately 10 min for 0.2 ml) and because of the necessary cooling time. An additional problem is the rather slow pyrolysis procedure which gives enough time for unpredictable side reactions and therefore increasing matrix affects. A continuous pyrolysis oven, now being designed, will dry and pyrolyse the samples very quickly. It will not, however, allow reduction of the water peak intensity because its operation is continuous and not sequential.

REFERENCES

Chan K.C., R.S.Tse and S.C.Wong (1982). Anal.Chem. 54, 1238.

Dixon W.J. (1981). BMDP Statistical Software, Univ. of California Press, Berkeley.

Heinzle E. (1978). Wachstum und Synthese von Poly-β-hydroxy-buttersäure (PHB) in Alcaligenes eutrophus: Extraktion des Polymeren sowie deren mathematische Modellbildungen. Doctoral Thesis, TU Graz, Austria.

Heinzle E., K.Furukawa, I.J.Dunn and J.R.Bourne (1983). Bio/technol. 1. 181.

Heinzle E. (1984). Bio/technol. 2, 786.

Heinzle E., H.Kramer and I.J.Dunn (1985). Biotechnol.Bioeng. (in press).

Malinowski E.R. and D.G.Howery (1980). Factor Analysis in Chemistry. Wiley, New York.

Meuzelaar H.L.C., J.Haverkamp and F.D.Hileman (1982). Pyrolysis Mass Spectrometry of Recent and Fossil Biomaterials. Elsevier, Amsterdam.

Pungor E., E.Schaefer, J.C.Weaver and C.L.Cooney (1981). Direct monitoring of a fermentation in a computer-mass spectrometer-fermenter system. In M.Moo-Young, C.W.Robinson and C. Vezina (Eds.), Advances in Biotechnology, Vol. 1. Pergamon Press, Toronto.

Pungor E., M.Pecs, L.Szigeti and L.Nyeste (1984). Analyt.Chim.Acta 163, 185.

Reuss M., H.Piehl and F.Wagner (1975). J.Appl. Microbiol.Biotechnol. 1, 323.

Reuss M. (1984). Personal communication.

RAPID BIOELECTROCHEMICAL METHODS FOR THE DETECTION OF LIVING MICROORGANISMS

G. Ramsay, A. P. F. Turner, A. Franklin and I. J. Higgins

Biotechnology Centre, Cranfield Institute of Technology, Cranfield, Bedfordshire MK43 OAL, UK

Abstract. Methods suitable for the estimation of microbial biomass are reviewed. Rapid bioelectrochemical techniques for detecting viable bacteria and yeasts were investigated using fuel cell and poised potential amperometric detectors. Escherichia coli in a fuel cell was detected over the range of $4 \times 10^{6} - 4 \times 10^{9}$ cells ml^{-1}, with an analysis time of approximately 30min. The three electrode amperometric detector produced a linear calibration curve for Escherichia coli in the range $6 \times 10^{6} - 6 \times 10^{8}$ cells ml^{-1}. The correlation coefficient was 0.996 ($P < 0.001$) and the analysis time < 15 min. The two electrode amperometric system gave linear calibration for $5 \times 10^{6} - 9 \times 10^{7}$ E.coli cells ml^{-1}. The correlation coefficient was 0.999 ($P < 0.001$) and the analysis time < 15 min. Batch growth of several bacteria and a yeast was successfully monitored bioelectrochemically. It was concluded that the two electrode poised potential amperometric method provided rapid, convenient, inexpensive detection of a wide range of microorganisms.

Keywords. Biomass detector; bioelectrochemical detection; microbial fuel cell; amperometric method; biosensor; bioactivity monitor.

INTRODUCTION

In order to control biotechnological processes and to detect microbial contamination, efficient detectors of viable cells are required. Generally, the microbial concentrations in contaminated samples are much less than those found in commercial fermenters i.e. less than 10^{6} cellsml^{-1}. At present there is no available satisfactory biomass probe for either concentration range. However, current research has yielded devices that go a considerable way towards achieving the goal of an effective biomass detector (Table 1).

TABLE 1 Properties of an Ideal Biomass Probe

1) Accuracy
2) Reproducibility of response
3) Rapid response
4) Sensitivity (resolution of 10^{7} viable cells ml^{-1})
5) Large linear range
6) On-line capability
7) Autoclavability
8) Low cost
9) Ease of operation
10) Durability
11) Biological inertness
12) Harmless to the biomass
13) No added reagents
14) No sample pre-treatment
15) Operation in clear or opaque samples
16) Differentiation between viable and non-viable biomass

A recent review of biomass detection (Harris and Kell, 1985) discusses most of the current methods. Table 2 lists some of the analytical parameters of the non-bioelectrochemical biomass detection methods. Acoustic resonance densitometry (ARD) and the piezoelectric membrane methods are described in detail because of their promising potential. Tables 3 and 4 list selected analytical parameters of the bioelectrochemical biomass detection methods. Following these tables some applications of the techniques are described. The following definition of a bioelectrochemical method was used for technique classification: a method in which the analytical signal is generated by the biomass producing or consuming electrical charge, this change being detected at electrodes.

NON-BIOELECTROCHEMICAL METHODS OF BIOMASS DETECTION

See Table 2.

Piezoelectric Membrane

A novel method for total biomass detection was reported by Ishimori, Karube and Suzuki (1981). In this procedure an ultrasonic signal was generated and directed through the microbial sample onto a piezoelectric membrane. This membrane provided a measure of the intensity of the ultrasonic signal. Both yeast and bacterial cells attenuated this signal and could therefore be detected at concentrations above 10^{6} cells ml^{-1}. The upper limit of detection was 10^{9} Saccharomyces cerevisiae cells ml^{-1} and 10^{8} cells ml^{-1} for Bacillus subtilis or a Klebsiella species. In addition to providing real time analyses the detection cell could be autoclaved. However, after autoclaving several times, the epoxy resin attaching the piezoelectric membrane in position cracked necessitating replacement.

A dilution step would be required for monitoring industrial bacterial fermentations and some yeast fermentations. In addition since there is no discrimination between biomass and non-biomass particles serious errors would occur if the fermentation broth contained a suspension of substrate material. However, this method promises inexpensive real time analysis of total fermenter

Table 2. Analytical Parameters of Non-bioelectrochemical Biomass Detection Methods

Method	Detection Limit (cells ml^{-1})	Analysis Time(h)	Sample Pre-Treatment	Source of Analytical Signal	Reference
Bioluminescence	10^5	0.2	ATP Extraction	Viable Cells	Chappelle (1973)
Radiometry	1	8-16	Incubation	Viable Cells	Waters (1972)
Direct Microscopic Counts	$5x10^7$	1	Staining	Viable Cells	Pettipher (1981)
Epifluoresence	$5x10^3$	0.5	Staining	Viable Cells	Pettipher (198[illegible])
Viable Counts	10	24-72	Incubation	Viable Cells	Postgate (1969)
Dry Weights	0.02mg ml^{-1}	8	Drying	Solid Particles	Neufeld (1982)
Nephelometry	$2x10^5$	Real time	None	Suspended solids	Mallette (1969)
Turbidimetry	$2x10^6$	Real time	None	Suspended solids	Mallette (1969)
Fluorimetry	0.4 mg ml^{-1}	Real time	None	Fluorophores	Zabriskie (1978)
Microcalorimetry	10^5	2	Incubation	Viable Cells	Beezer (1978)
Filtration	-	0.5	Filtration	Solid Particles	Nestaas (1983)
Electrical Counting	-	0.1	None	Suspended solids	Dow (1979)
Piezoelectric Membrane	10^6	Real time	None	Suspended solids	Ishimori (198[illegible])
Acoustic Resonance Densitometry	$1.0x10^8$	Real time	None	Suspended solids	Clarke (1985)

Table 3. Examples of Impedimetric Detection of Microbial Contamination

Initial Concentration (cells ml^{-1})	Sample	Detection time(h)	Reference
10^4	cerebrospinal fluid	5.8	Kahn and colleagues (1975)
10^5	raw milk	8.9	Gnan and Luedecke (1982)
10^5	frozen vegetables	5.0	Hardy and colleagues (1977)
10^5	urine	2.5	Zafari and Martin (1977)
10^6	blood	6-8	Hadley (1976)
10^6	blood	8.5	Kahn and colleagues (1975)
$10^6 - 10^7$	sewage effluent	1.5	Munoz and Silverman (1979)

All experiments carried out using a Bactometer 32 instrument.

Table 4. Analytical Parameters of some Bioelectrochemical Biomass Detection Methods

Method	Detection range (cells ml^{-1})	Analysis time(h)	Sample Pre-Treatment	Reference
Potentiometric	$1x10^0 - 1x10^6$	7.5-1.0	Incubation	Wilkins (1978)
Fuel Cell	$10^7 - 10^9$	0.2	None	Matsunaga (1979)
Fuel Cell	$6x10^6 - 4x10^9$	0.5	None	Turner (1982)
Fuel Cell	$10^4 - 10^6$	0.3	Preconcentration	Nishikawa (1982)
Amperometric	$3x10^8 - 2x10^9$	0.1	None	Matsunaga (1980)
Amperometric	$7x10^5 - 4x10^7$	0.2	Preconcentration	Matsunaga (1981)
Amperometric	$6x10^6 - 6x10^8$	0.2	None	Ramsay (1983)
Amperometric	$3x10^6 - 2x10^8$	0.1	None	Matsunaga (1984)

biomass.

Acoustic Resonance Densitometry (ARD)

The principle of ARD is that when a mass is at its resonant frequency the frequency is inversely proportional to the mass. Hence, if a fixed volume of fermenter culture is made to resonate, the resonant frequency will be proportional to the specific gravity of the fluid. After calibrating the system the specific gravity can be calculated. The weight per ml of suspended solids is then obtained by subtracting the specific gravity of the culture medium (Clarke and co-workers, 1985; Krakty and Leopald, 1968). This device is extremely robust and autoclavable. Analysis is rapid but not in real time, as the specific gravity of the growth medium containing no particulate matter has to be measured. The linear range of this method 1.0 x 10^8 - 1.2 x 10^{12} E.coli ml^{-1} is sufficient to monitor fermentations and downstream processing (Blake-Coleman and colleagues, 1984; Blake-Coleman and Clarke, 1984b, 1984c). In addition, the price of the ARD instrumentation could be quite low.

BIOELECTROCHEMICAL METHODS OF BIOMASS DETECTION

Impedimetric Methods of Biomass Detection

This technique uses the change in electrical conductivity that results from microbial metabolism. To achieve reproducible results the composition of the growth medium and its temperature have to be carefully controlled. Instrumentation costs are high but many samples can be assayed simultaneously. Typically the analyte sample is placed in sterile medium and incubated for several hours until the impedance changes by a pre-set amount. The most successful applications have occurred where large numbers of samples containing quite low levels of microbial contamination need to be analysed. Hence the analysis of foodstuffs and clinical samples is often carried out impedimetrically (see Table 3).

Microbial contamination in frozen vegetables was investigated by Hardy and co-workers (1977). Extracts containing 10^5 viable cells per ml were analysed in 5h. This was far more rapid than the conventional viable count method which took several days and there was good agreement. Martins and colleagues (1982) established that pasteurised milk with a detection time of 13.5h or more had a shelf life of at least 12 days. Samples with a shorter detection time spoiled within 11 days. This impedimetric method was considerably faster than the viable count assay which took 2 days.

Several studies have been carried out on impedimetric detection of microbial contamination in urine (Cady and colleagues, 1978; Specter and co-workers, 1976; Zafari and Martin, 1977). Samples containing 10^5 cells ml^{-1} were assayed in approximately 3h. This compared favourably with the 48h conventionally required and there was good correlation with the standard method. Hadly (1976) reported that positive blood samples, i.e. those containing 10^6 cells ml^{-1} or more, were detected in 6-8h and there was good agreement with the plate count. Kahn and colleagues (1976) assayed samples of cerebrospinal fluid to detect bacterial meningitis. They found that the threshold level of this condition, 10^4 cells ml^{-1} was detected in 5.8h. This was far more rapid than the plate count method.

Potentiometric Methods of Biomass Detection

Table 4 summarises potentiometric and amperometric methods of biomass detection. The basis of potentiometric analysis is the measurement of the potential developed at an electrode with respect to a reference electrode. No external poising potential is applied. Wilkins (Wilkins, 1974; Wilkins and colleagues, 1978) measured the potential developed at a Pt electrode with respect to a saturated calomel electrode (SCE) by Enterobacter. A linear relationship was obtained between the initial viable microbial concentration and the time required to reach a potential of 30mV. Escherichia coli, with an initial concentration of 1×10^6 cells ml^{-1} was detected in 1h. One cell ml^{-1} could be detected in 7.5h. In later work, Wilkins (1978) replaced the SCE with a Pt electrode. Results for E. coli were similar to those obtained previously. The method was also shown to be applicable to a wide range of bacteria. At a concentration of 1×10^6 cells ml^{-1} the range of detection times for the microorganisms used was 1.5-3.4h; at 1 cell ml^{-1}, the range was 8.4-16.9h.

The potentiometric method is simple to operate, has low capital cost and detects a wide range of viable bacteria with a large dynamic range. However, the lengthy analysis time is a disadvantage.

Fuel Cell

Matsunaga, Karube and Suzuki (1979) reported the use of a non-mediated fuel cell to detect microbial cells (a mediator is a small redox compound used to transfer electrons from biological material to an electrode or vice-versa). Two fuel cells were used with only the anode of each in contact with the analyte samples. To provide a background signal the anode of one of the fuel cells was covered with dialysis membrane, hence excluding the microorganisms but allowing the oxidation of any interfering chemical species. Thus the analyte signal was the difference between the electrical currents from the two fuel cells. Although not a sensitive method the batch growth of Saccharomyces cerevisiae was successfully followed with the biomass probe in-situ. Lactobacillus fermentum gave a linear calibration curve in the range 9×10^8 - 3×10^9 cells ml^{-1}. Only viable cells produced a response.

Turner and co-workers (1982) increased the analytical sensitivity of the fuel cell method by using the mediator phenazine ethosulphate (PES) to detect E.coli. Analysis time was approximately 30 min and the limit of detection 4×10^6 cells ml^{-1}.

Nishikawa and co-workers (1982) used filter pre-concentration as well as a mediator (2,6-dichlorophenol indophenol) to increase analytical sensitivity. Linear calibration curves over the range 10^4 - 10^6 cells $^{-1}$ were obtained from the four bacterial species tested. The analytical sensitivity of the method differed between species. However, when the viable counts in two water samples were determined, using E. coli calibration curve, good correlation was obtained. Analysis time was about 20 min and only viable cells gave a response.

Amperometric Methods

These methods are based on measuring the current produced on reaction of the analyte species at a working electrode. They can be classified into two or three electrode systems. Three electrode systems consist of a working electrode poised at a pre-set potential, a reference electrode and a counter electrode which completes the electrical circuit. In two electrode systems the counter and reference electrodes are combined. This system is simpler to operate but the potential at the working electrode is not maintained as accurately as in the three electrode system (Sawyer, 1974; Turner, 1985).

Matsunaga and colleagues (1980) reported the detection of Bacillus subtilis in a non-mediated three electrode system. Two sets of electrodes were used, one with the working electrode covered in dialysis tubing to provide a background signal. Since no mediator was used the detection limit was high approximately 3×10^8 cells ml^{-1}. However, the absence of a soluble mediator and the stability of the system to autoclaving, allowed the successful in-situ monitoring of batch growth. Response time was less than 5 min and only viable cells responded. Ramsay and co-workers (1983) reported a three electrode method in which a mediator (PES) was used. A linear calibration curve over the range 6×10^6 - 6×10^8 cells ml^{-1} was obtained for Escherichia coli with an analysis time of less than 15 min. Only viable cells gave a response.

Recently Matsunaga and Namba (1984) have reported a cyclic voltammetric method for the detection of yeast and bacterial species. Linear calibration was obtained over the range 3×10^6 - 2×10^8 cells of a Saccharomyces cerevisiae ml^{-1}. It was shown that modification of a graphite electrode, with 4,4' -bipyridine, enhanced the current from *Saccharomyces cerevisiae* by 40%. If the 4,4' -bipyridine remained immobilised on the electrode and the electrodes were stable to sterilisation this probe could theoretically be used in a fermenter for on-line analysis.

Matsunaga and colleagues (1981) reported a two electrode biomass sensor based on a Clark oxygen electrode with a filter pre-concentration step to increase analytical sensitivity. The analysis time was approximately 10 min. Aerobic bacteria and yeasts were detected at low concentrations. The limit of detection for the yeasts was approximately 10^5 cells ml^{-1} and for the bacteria 10^6 cells ml^{-1}. Linear calibration curves were obtained over a one decade concentration range.

Both the fuel cell and the amperometric methods have in common rapid response, applicability to both bacteria and yeasts, low cost, medium sensitivity and ease of operation. For these reasons it is likely that these methods will be further developed.

Possible Developments Leading to Practical Biomass Detecters

All the biomass estimation methods reviewed have some disadvantages. However, acoustic resonance densitometry, the piezoelectric membrane method and the amperometric methods merit further attention. In particular in ARD the successful development of a real time measure of the specific gravity of the cell free culture is necessary for real time continuous monitoring of fermenter biomass. The piezoelectric membrane method requires the development of a membrane attachment system that is stable to repeated autoclaving. The amperometric methods need to be made more sensitive and if possible based on an immobilised mediator. An immobilised mediator could permit on-line analysis of fermenter biomass.

THE APPLICATION OF AMPEROMETRIC SYSTEMS TO THE DETECTION OF MICRO-ORGANISMS

Our early work showed that the current generated by bacteria in a phenazine ethosulphate mediated fuel cell was proportional to the bacterial concentration (Turner and co-workers, 1982). The anolyte was deoxygenated by bubbling with nitrogen. This prevented electron loss to oxygen and mediator autooxidation. The mediator increased the rate of charge transfer from the bacterial cells to the anode. From the anode, the electrons flowed through a load resistor to the cathode where they reacted with oxygen and protons to form water. A potentiometric chart recorder measured the potential difference across the resistor thus providing a measure of the current. The mediator concentration was optimised at 4.2mM and the temperature of the electrolyte maintained at 30°C. Using reticulated carbon electrodes *Escherichia coli* was detected at concentrations over the range $6 \times 10^7 - 4 \times 10^9$ bacteria ml^{-1}. The analysis time was approximately 30 min. When Pt gauze was used in place of reticulated carbon the detection limit was decreased to 4×10^6 cells ml^{-1}. These results were encouraging but the fuel cell was tedious to assemble and operate. Due to the short working life of the ion-exchange membrane, the fuel cell had to be dismantled frequently, hence increasing analysis time.

The experimental procedure was simplified by using a three electrode poised potential system (Fig.1). This required no ion-exchange membrane as there was only one reaction compartment. Only one solution had to be prepared and this was sparged with a single gas, nitrogen. The mediator was phenazine ethosulphate at an optimum concentration of 5.8mM. The analysis time (measuring the initial slope of the analyte peak) was less than 15 min. Linear calibration for *Escherichia coli* was obtained from the detection limit of 6×10^6 cells ml^{-1} to 6×10^8 cells ml^{-1} (see Fig.2). A correlation coefficient of 0.996 (n=6, $P < 0.001$) was obtained showing that there was little data scatter.

Development of a Two Electrode Poised Potential Amperometric Biomass Detector

The three electrode system was simplified and made easier to use by replacing the counter and reference electrodes with a single electrode. This consisted of an electrochemically deposited layer of silver chloride on a silver foil support. A robust probe was formed by attaching the Pt working electrode and the reference electrode to opposite sides of a plastic support (see Fig.3). To increase sample throughput a multichannel system was built, based on the BBC Microcomputer.

Typical experimental conditions were:
Pt working electrode 0.5cm x 1.5cm
Ag/AgCl reference electrode 0.5cm x 1.5cm
Electrolyte de-oxygenated with oxygen free nitrogen for five minutes and then sparged continuously.
Electrolyte temperature 30°C
Working electrode poised at 400mV versus Ag/AgCl reference electrode.
Sodium phosphate buffer (8.55ml, 50mM, pH 7.0)
Potassium chloride (1.00ml, 1000mM)
Glucose (0.15ml, 25mM)
Potassium ferricyanide (0.15ml, 250mM)
Microbial cells (0.15ml)
E.coli cells were grown in continuous culture on defined medium under glucose limitation. The microbial cells were added after the mediator oxidation peak had plateaued. The analysis time when the maximum slope of the analyte peak was measured was less than 15 min.

Potassium ferricyanide was normally used as the mediator in place of PES because it was more chemically stable. The potassium chloride was required to stabilise the potential provided by the reference electrode.

Two electrode amperometric studies. The relationship between respiration rate and BEC[1] response of *Escherichia coli* cells was investigated. Firstly the respiration rate over a range of substrate (glucose) concentrations was measured in a Clark oxygen electrode. The rate increased non-linearly from zero at zero substrate concentration to a plateau at 30uM glucose (130nmol oxygen min^{-1} mg dry wt^{-1}). In the BEC a similar response was obtained (see Fig.4). At zero glucose concentration a small peak was obtained (maximum slope 0.3uA min^{-1}). The maximum slope then increased non-linearly up to a plateau (1.8uA min^{-1}) at 125uM. Probably the small peak obtained at zero glucose concentration was caused by the bacteria using endogenous reserves. These experiments demonstrated that for maximum BEC[1] response *E.coli* cells needed to be respiring at their maximum rate. Having established the saturating glucose concentration for the BEC[1] response of a high concentration of *E.coli* cells (1×10^9 cells ml^{-1}) the standard glucose concentration was fixed at 0.4mM.

BEC[1] : Bioelectrochemical Cell.

Calibration of E.coli BEC response. Calibration standards were prepared by diluting fresh fermenter samples with 0.1% peptone. Three replicate analyses were carried out for each calibration standard. A linear calibration curve was obtained from the detection limit 5×10^6 to 9×10^7 cells ml^{-1})(see Fig.5). The upper limit of detection was determined by the current density that the reference could conduct before the reference potential began to drift. This occurred at approximately 100uA. There was little data scatter (correlation co-efficient of 0.999, n=8, P< 0.001).

The equipment and running costs of all three bio-electrochemical methods were low, response was rapid, and analytical sensitivity was quite good. Also the lower detection limits were very similar for all three techniques. In practice the method chosen would be the two electrode amperometric system as it is the simplest to use. If, however, the microbial concentration was sufficient to generate enough electrical current to destabilise this system then the three electrode system could be used.

Detection of Bacteria Naturally Occurring in Milk by the Two Electrode Amperometric Method

There is considerable commercial interest in rapidly detecting and quantifying microorganisms contaminating milk. One of the bacterial species that routinely occurs in pasteurised milk was assayed in the BEC. This species was a Gram negative aerobic coccus, provisionally named Milk Isolate Number 3. The batch growth of this isolate in nutrient broth was followed (30°C, 150rpm on orbital shaker). During growth the optical density, BEC response, viable count and respiration rate were measured. Fig.6 shows the variation of maximum slope and the viable count during the batch growth. There is close correlation between the maximum slope and the viable count. Similarly there was good agreement between the maximum peak height and viable count.

Range of Microorganisms Detected in BEC

In order to assess the applicability of this technique to microorganisms other than E.coli a range of organisms has been tested in the BEC. These organisms were chosen because they occur in one or more of the following matrices: lubricating oil, milk or cooling tower water. From the many microbial contaminants found in lubricating oil, Pseudomonas aeruginosa, Alcaligenes faecalis, Nocardia NCIB 8863 and Achromobacter album grown in batch culture and successfully monitored bioelectrochemically. Similarly Enterobacter aerogenes a common cooling water contaminant and Lactobacillus bulgaricus a milk contaminant, gave responses in the BEC which were proportional to their viable counts. In addition, the response of the yeast Saccharomyces cerevisiae has been measured in the BEC. These results indicate that this technique will be applicable to a wide range of microorganisms.

CONCLUSION

The results in this paper have shown that both the poised potential and fuel cell bioelectrochemical techniques provide simple, rapid, inexpensive and moderately sensitive means of detecting viable biomass. Such techniques could be cost-effective in detecting microbial contamination in samples such as foodstuffs, industrial waste waters and clinical samples. It could also be useful in monitoring the viable biomass concentrations in fermenters.

REFERENCES

Beezer, A.E., Bettleheim, K.A., Al-Salihi, S. and Shaw, E.J. (1978). The enumeration of bacteria in culture media and clinical specimens of urine by microcalorimetry. Sci.Tools, 25, 6-8.

Blake-Coleman, B.C., Calder, M.R., Carr, R.J.G., Moody, S.C. and Clarke, D.J. (1984). Direct monitoring of reactor biomass in fermentation control. Anal. Chim. Acta. (in press)

Blake-Coleman, B.C., and Clarke, D.J. (1984b). Concentration measurement of biological particles Brit.Pat.App.No.84085/28.

Blake-Coleman, B.C., and Clarke, D.J. (1984c). Fluid density measurement. Brit.Pat.App.No. 84085/27.

Cady, P. (1978). Progress in impedance measurements in microbiology. In A.Sharpe and D.Clarke (Eds). Mechanising microbiology. Charles C. Thomas, Springfield, pp.199-239.

Chappelle, E.W., Piccioli, G.L., and Deming, J.W. (1978). Determination of bacterial content in fluids. Meth. Enzymol., 57, 656-72.

Clarke. D.J., Calder, M.R., Carr, R.J.G., Blake-Coleman, B.C., and Moody, S.C (1985). The development and application of biosensing devices for bioreactor monitoring and control. Biosensors (in press).

Dow, C.S., France, A.D., Khan, M.S., and Johnson, T. (1979). Particle size distribution analysis for the rapid detection of microbial infection of urine. J.Clin.Pathol., 32, 386-90.

Gnan, S., and Luedecke, L.O. (1982). Impedance methods in raw milk as an alternative to the standard plate count. J.Food Protection, 45, 4-7.

Hadley, W.K.(1976). The use of electrical impedance measurements for the detection of microorganisms in blood cultures. In H.H.Johnston and S.W.B. Newsom (Eds), Proc. 2nd Int. Symp. Rapid Methods and Automation in Microbiology. Learned Information (Europe) Ltd. Oxford, p.14.

Hardy, D., Kraeger, S., Dufour, S., and Cady, P. (1977). Rapid detection of microbial contamination in frozen vegetables by automated impedance measurements. App.Env.Microbiol., 34, 14-17.

Harris, C.M., and Kell, D.B (1981). The estimation of microbial biomass. Biosensors, 1, 17-84.

Ishimori, Y., Karube, I., and Suzuki, S. (1981). Determination of microbial populations with piezoelectric membranes. Appl.Env.Microbiol., 42. 632-7.

Kahn, W., Friedman, G., Rodriguez., W.Coutroni, C., and Ross, S. (1976). Rapid detection and isolation of bacteria in blood and cerebrospinal fluids in children by the electrical impedance methods. In H.H.Johnston and S.W.B.Newsom (Eds) 2nd Int.Symp. on Rapid Methods and Automation in Microbiology. Learned Information (Europe) Ltd, Oxford, pp.14-15.

Kratky, O., and Leopold, H. (1968). Acoustic resonance densitometry. Austrian Patent No. 2704.

Mallette, M.F (1969). Evaluation of growth by physical and chemical means. In J.R.Norris and D.W.Ribbons (Eds), Methods in Microbiology, Vol. 1. Academic Press, London, pp.531-66.

Martins, B.B., Hodapp, S., Dufour, S.W. and Kraeger S.J. (1982). Evaluation of a rapid impedimetric method for determining the keeping quality of milk. J.Food Prot., 45, 1221-6.

Matsunaga, T., Karube, I., and Suzuki, S. (1979). Electrode system for the determination of microbial populations. Appl. Environ. Microbiol 37, 117-121.

Matsunaga, T., Karube, I., and Suzuki, S. (1980) Electrochemical determinations of cell populations. Eur.J.Appl.Microbiol.Biotechnol., 10, 125-32.

Matsunaga, Y., Karube, I., Nakahara, T. and Suzuki, S. (1981). Eur. J. Appl. Microbiol. Biotechnol., 12, 97-101.

Matsunaga, T. and Namba, Y. (1984). Selective determination of microbial cells by a graphite electrode modified with 4,4'-bipyridine. Anal. Chim. Acta, 159, 84-87.
Munoz, E. and Silverman, M. (1979). Rapid, single step most-probable-number method for enumerating faecal coliforms in effluents from sewage treatment. Appl. Env. Microbiol., 37, 527-30.
Nestaas, E. and Wang, D.I.C. (1983). A new sensor, the "filtration probe" for quantitative characterisation of the penicillin fermentation. 111: An automatically operating probe. Biotechnol. Bioeng., 25, 1981-7.
Neufeld, H. A., Zajic, J.E. (1982). A technique for monitoring biomass levels and cell adhesion in hexadecane fermentations. Biotechnol. Lett., 4, 701-704.
Nishikawa, S., Sakai, S. Karube, I., Matsunaga, T. and Suzuki, S. (1982). Dye-coupled electrode system for the rapid determination of cell populations in polluted water. Appl. Env. Microbiol., 43, 814-18.
Pettipher, G.L. (1951). The Direct Epifluorescent Filter Technique, p.55. Research studies press: Letchworth, England.
Pettipher, G.L., Mansell, R., Mackinnon, C.H., and Cousins, C.M. (1980). Rapid membrane filtration-epifluorescence microscopy for direct enumeration of bacteria in raw milk. Appl. Env. Microbiol., 39, 423-429.
Postage, J.R. (1969). Viable counts and viability. In J.R. Norris and D.W. Ribbons (Eds.), Methods in Microbiology, Vol.1. Academic Press, London, pp. 611-28.
Ramsay, G., Turner, A.P.F. and Higgins I.J. (1983). The use of bioelectrochemical cells for biomass and respiratory activity estimations. Soc. Gen. Microbiol. Quarterly, 10, (1)M14.
Sawyer, D.T. and Roberts, Jnr, J.L. (1974). Experimental Electrochemistry for Chemists. John Wiley and Sons Inc, New York.
Spector, S., Throm, R. and Friedman, H. (1976). Screening for bacteriuria in the clinical microbiology laboratory by electrical impedance monitoring. In H.H. Johnston and S.W.B. Newsom (Eds.), Proc. 2nd Int. Symp. Rapid Methods and Automation in Microbiology. Learned information (Europe) Ltd. Oxford, pp.13-14.
Turner, A.P.F., Ramsay, G. and Higgins, I.J. (1982). Applications of electron transfer between biological systems and electrodes. Biochem. Soc. Trans., 11, 445-8.
Turner, A.P.F. (1983). Applications of direct electron transfer bioelectrochemistry in sensors and fuel cells. Biotech. '83. Online Publications, Northwood, pp.643-654.
Turner, A.P.F. (1985). In K. Mosbach (Ed.), Methods in Enzymology: Immobilised Enzymes and Cells. Academic Press, New York. (in press)
Waters, J. (1972). Sensitivity of the $^{14}CO_2$ radiometric method for bacterial detection. Appl. Microbiol., 23, 198-9.
Wilkins, J.R. (1974). Pressure transducer method for measuring gas production by microorganisms. Appl. Microbiol., 27, 135-40.
Wilkins, J.R. (1978). Use of platinum electrodes for the electrochemical detection of bacteria. Appl. Env. Microbiol., 36, 683-7.
Wilkins, J.R., Young, R. and Boykin, E. (1978). Multichannel electrochemical microbial detection unit. Appl. Env. Microbiol., 35, 214-15.
Zabriskie, D.W. and Humphrey, A.E. (1978). Estimation of fermentation biomass concentration by measuring culture fluorescence. Appl. Env. Microbiol., 35, 337-43.

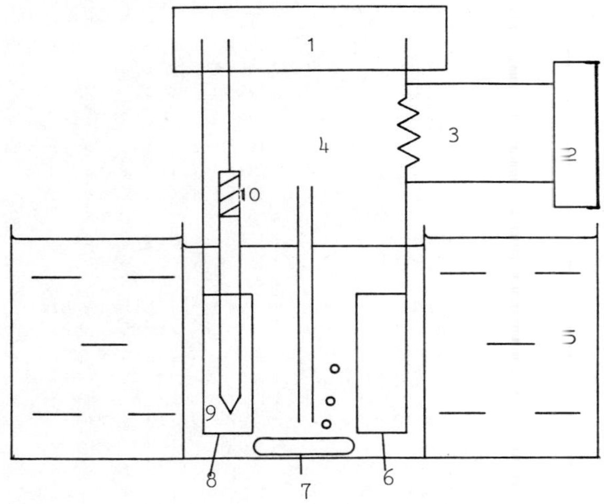

Fig. 1. Three electrode poised potential amperometric system

1, Potentiostat; 2, chart recorder; 3, load resistance; 4, air inlet; 5, water bath; 6, counter electrode; 7, magnetic stirring bar; 8, working elecrode; 9, Lugin capillary; and 10, saturated-calomel reference electrode

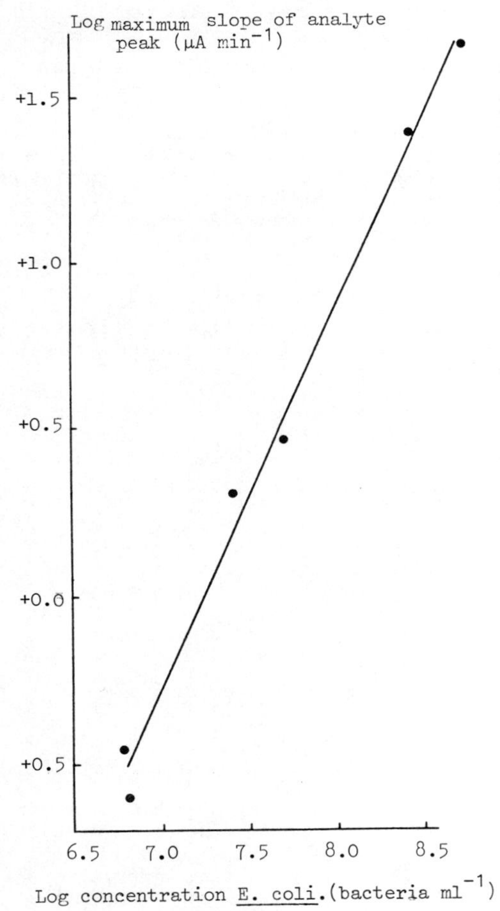

Fig. 2. Variation of maximum slope of analyte peak with concentration of *Escherichia coli* in the three electrode amperometric cell

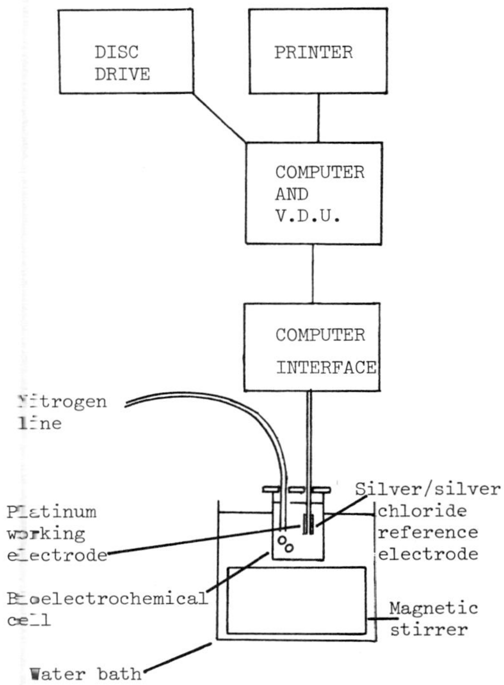

Fig. 3. Two electrode poised potential amperometric biomass detector

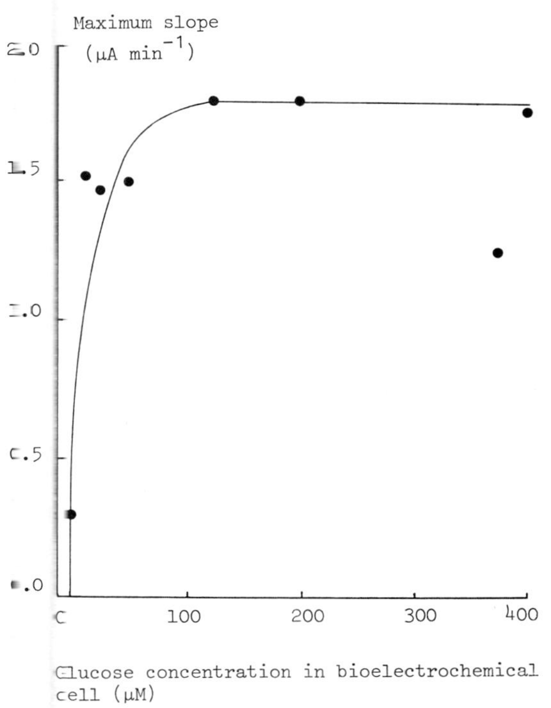

Fig. 4. Maximum slope versus glucose concentration for Escherichia coli cells.

Experimental conditions as in text. Escherichia coli concentration 4×10^8 cells ml^{-1}

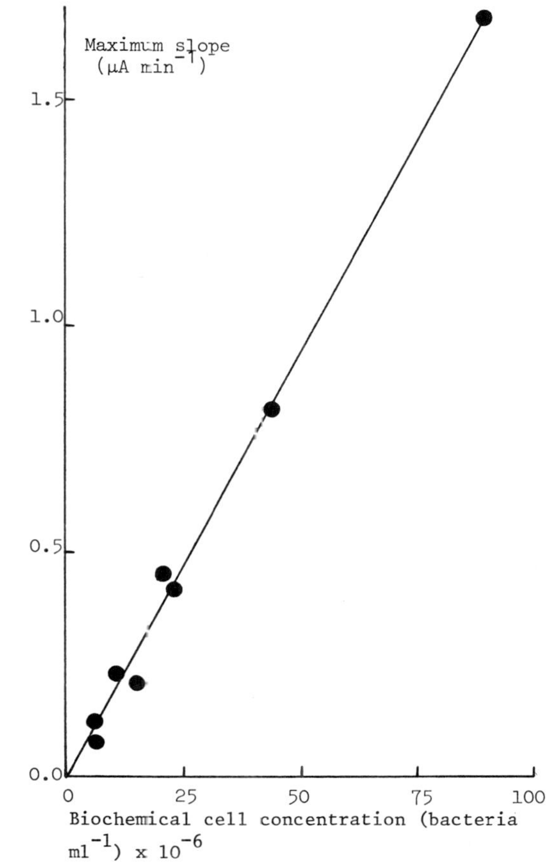

Fig. 5. Maximum slope versus Escherichia coli bioelectrochemical cell concentration: two electrode amperometric system.
Experimental conditions as in text.

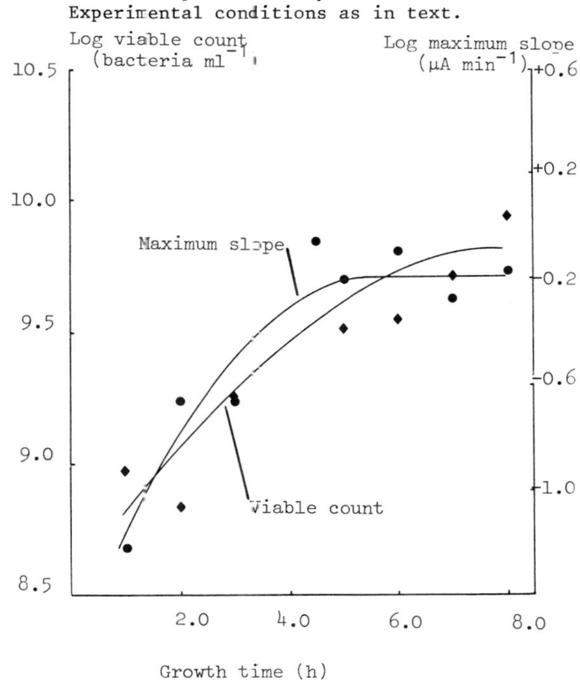

Fig. 6. Log viable count and log maximum slope versus growth of Milk Isolate No. 3 in batch culture. Experimental conditions as in text;
● : log maximum slope; ◆ : log viable count;

ADAPTIVE CONTROL OF A FED-BATCH BAKER'S YEAST FERMENTATION

R. M. Dekkers* and M. Voetter**

**Gist Brocades N.V., Research & Development, PO Box 1, 2600 MA Delft, The Netherlands*
***Department of Applied Physics, Delft University of Technology, Prins Bernhardlaan 6, 2628 BW Delft, The Netherlands*

Abstract. Two adaptive controllers for the regulation of the respiratory quotient of a fed-batch baker's yeast fermentation are presented. The first one is a feedforward adaptive controller based on a second order model from which the parameters are determined with auxiliary measurements. The second one is a self-tuning controller. Both controllers are based on a newly developed sub-optimal control design method. Experiments with a 100 liter pilot fermentor have shown the accuracy of the two controllers. The self-tuning controller is preferable for general application because less a priori knowledge is required.

Keywords. Adaptive control; gain scheduling; self-tuning controller; baker's yeast fermentation; fed-batch; respiratory quotient.

1. INTRODUCTION

During a fed-batch baker's yeast fermentation, it is important that the respiratory quotient (RQ) is accurately regulated at one desired value or trajectory. Such a requirement stems from optimization considerations like optimal economic production and/or optimal quality. On-line computer control of the glucose feed rate (GFR) has the potential to solve the control problem. However, standard fixed parameter control algorithms (PI or PID) do not work satisfactory (see e.g. Pons, Bordet and Engasser, 1982). This is inherent to the nonlinear character of a fed-batch fermentation and the fact that the state variables are not steady but dynamically varying. Nonlinear controllers are proposed by Aiba, Nagai and Nishizawa (1976) and by Wang, Cooney and Wang (1979). Although these controllers give reasonable performances, they are not guaranteed to be accurate during all stages of the process. In fact the different stages in a fed-batch baker's yeast fermentation require different controllers having parameters that are not easily determined a priori.

In this paper two adaptive control algorithms for the regulation of the RQ are presented. In section 2 a process description and a control philosophy from the biotechnical viewpoint will be given. In section 3 a new controller design will be given for a given discrete time model. Although the treatment of this design method is somewhat general and besides the principal topics of this article, we find it essential for a good understanding of the adaptive controller presented in section 4. Experiments with the application of the controllers are given in section 5. Finally some conclusions are discussed in section 6.

2. PROCESS DESCRIPTION AND CONTROL PHILOSOPHY

Those phenomena of baker's yeast fermentation that are of importance for the control problem are described in this section. Most of the given properties are well known from the literature (von Meyenburg, 1969; Aiba, Nagai and Nishizawa, 1976; Wang, Cooney and Wang, 1979; Geurts, de Kok and Roels, 1980; Dekkers, 1982, 1983, 1984), but some of them are new.

The Importance of the Respiratory Quotient

The baker's yeast production fermentation is a fed-batch process with a continuous supply of glucose and air. Generally, it may be supposed that the growth rate of the yeast is limited by the available amount of glucose. The growth is given by

$$\dot{M} = BPR = \mu M \tag{2.1}$$

where M is the total amount of biomass in the fermentor (kg), BPR is the biomass production rate (kg/hr) and μ the specific growth rate (1/hr). μ depends on a complexity of factors, but is actually controlled by the glucose concentration S in the fermentor which in turn is controlled by the glucose feed rate GFR (mol/hr). A higher glucose concentration gives a higher specific growth rate so that a big GFR is desired. However at high glucose concentrations, the so-called glucose effect occurs and ethanol is produced causing a lower efficiency i.e. the yield factor Y diminishes with

$$Y = BPR/GUR \tag{2.2}$$

where GUR is the glucose uptake rate. The effect is sketched in Fig. 1. In a wide range of $\mu < \mu_c$, Y is nearly constant and independent of μ. μ_c is the critical specific growth rate above which ethanol production starts and Y falls. With fed-batch production fermentations the glucose accumulation is small so that GUR ≈ GFR. Theoretically the fairly ideal situation $\mu = \mu_c$ could be achieved with the feedforward control

$$GFR = \mu_c M(t)/Y_m = \mu_c M(o)\exp(\mu_c t)/Y_m \tag{2.3}$$

where Y_m is the maximum of Y. A drawback of this feedforward control is that an error in the initial biomass estimate will not be damped out. However even if M(t) is accurately known all the time then difficulties remain because Y_m and especially μ_c are not constant but time varying. Therefore it is better to use a feedback controller with the respiratory quotient

$$RQ = CER/OUR \tag{2.4}$$

as the controlled variable; CER is the CO_2 evolution rate (mol/hr) and OUR is the O_2 uptake rate (mol/hr). With $\mu > \mu_c$, CER increases while OUR remains constant (provided there is no catabolic repression of respiration) and so RQ increases. From Fig. 1 it is obvious that RQ-control can only work with a setpoint $RQ_s > RQ_c$ because with $RQ_s = RQ_c$, μ is not uniquely determined and biomass production is not optimal. However if ethanol is present in the medium, $RQ_s = RQ_c$ is possible, because with $\mu < \mu_c$ the RQ will diminish as result of ethanol uptake (see dashed lines in Fig. 1). For this reason and for optimal production it is desired to take $RQ_s > RQ_c$ during start-up so that some ethanol is formed. Taking $RQ_s < RQ_c$ in later stages has no sense because the indirect way of glucose utilization by means of ethanol production/ethanol consumption, respectively, is less efficient than the direct complete oxidation of glucose.

The Importance of the Oxygen Uptake Rate

In the preceding it was shown that the RQ gives valuable information for the optimal control of a fed-batch baker's yeast production. However the control problem is not easily to solve. Suppose that $RQ_s = RQ_c$, ethanol is present in the fermentor, and that μ_c and Y_m are constant. Then the feedback controller has to realise a control signal like (2.3). Such an exponential input is hard to realise with feedback of the RQ measurement only. Therefore additional information gained via the measured OUR signal will be used. OUR is closely related with the growth and is determined by

$$OUR(t) = \min(OTR_m(t), OUR_r(t), OUR_f(t)) \tag{2.5}$$

where OTR_m is the maximum possible oxygen transport rate under the given conditions determined by the physical transport phenomena; OUR_r is the maximum possible oxygen uptake rate determined by the actual enzyme capacity of the respiratory chain; OUR_f is the maximum oxygen uptake rate determined by the available amount of carbon in the form of glucose and ethanol. As example of the last, see the curve OUR/M for $\mu < \mu_c$ and no ethanol present in Fig. 1. It is obvious that for an optimal fermentation OUR must never be determined by OUR_f because then the production is not exploited. In an optimal situation $OUR = OTR_m$ or $OUR = OUR_r$; in other words OUR is determined by the system itself and not directly by the input GFR! This is a very important property which can be used for the control problem. From an elemental and ATP balance, it can be shown that

$$GUR = \delta_1 CER + \delta_2 OUR - \varepsilon M \quad (\delta_1>0, \delta_2<0, \varepsilon>0) \tag{2.6}$$

or

$$\frac{GUR}{OUR} = \delta_1 RQ + \delta_2 - \varepsilon \frac{M}{OUR} \tag{2.7}$$

The last term in the right hand side of (2.7) is small with respect to the two other terms.

The Controller Structure

According to (2.7) define a control variable u with

$$GFR = uOUR \tag{2.8}$$

Then in a (pseudo-) steady state, a constant u will produce a constant RQ e.g. if μ_c is constant then M/OUR is constant and an exponential growth rate is achieved with a constant u. Nominal variations in u can be recovered with an integral action from RQ to u and for perturbations from the (pseudo) steady state a normal proportional-derivative type feedback is required. So we propose the control structure as given in Fig. 2. The (adaptive) RQ-controller has to regulate the encircled block by manipulating u so that RQ is equal to the desired RQ_s. The inner loop feedback of OUR is possible because the dynamics of OUR is determined by a slow mode i.e. it is growth associated (provided OUR is not determined by OUR_f). On the other hand the dynamics of RQ is mainly determined by a fast mode i.e. it is associated with the rapid uptake of fed glucose. An illustration of the fast and slow mode is given in Fig. 3. Here the RQ was controlled with $RQ_s = 1.2$. An external pseudo-random binary test signal was added to the controller output. It is obvious that the step changes in GFR are only followed by CER and thus RQ, while OUR follows an autonomous trend because OUR is not determined by OUR_f.

A Simple Model

Knowledge about the transfer function from u to RQ is required for the design of the RQ-controller. A simple model can be derived from Monod kinetics and the fermentor residence time distribution for CO_2 (de Koning, 1984). Linearization of the kinetics around a (pseudo) steady-state gives the process transfer function

$$G_p(s) = \frac{\Delta RQ(s)}{\Delta u(s)} = \frac{L\exp(-sT_d)}{(1+\tau_1 s)(1+\tau_2 s)} \tag{2.9}$$

where

$$\tau_1 = \alpha/X, \quad \tau_2 = \beta_1 + \gamma_1/F, \quad T_d = \beta_2 + \gamma_2/F \tag{2.10}$$

with α, β_1, β_2, γ_1 and γ_2 constants; X is the biomass concentration and F is the air flow rate. The gain L follows from the steady state relation (2.6)

$$L = 1/\delta_1 \tag{2.11}$$

However, it appeared from the identification experiment as shown in Fig. 3 that the model (2.9) could not be confirmed i.e. the poles are complex and the gain L is clearly less than $1/\delta_1$. These results indicate that the Monod model is not valid during dynamic transients and the smaller gain would indicate some buffering of glucose in the yeast cell. In spite of these identification results the model (2.9), (2.10) was used for controller design purposes.

3. CONTROLLER DESIGN

This section gives a brief description of the calculation of a control law for a single input - single output process with known structure and known parameters. A detailed explanation is given by Dekkers (1985).

Discrete Time Model

We consider the ARMAX-model

$$\begin{aligned} A(z^{-1})y(i) &= z^{-k}B(z^{-1})u(i) + C(z^{-1})e(i) \\ A(z^{-1}) &= 1+a_1z^{-1}+\ldots+a_{na}z^{-na}, \quad a_{na} \neq 0 \\ B(z^{-1}) &= b_0+b_1z^{-1}+\ldots+b_{nb}z^{-nb}, \quad b_0 \neq 0 \\ C(z^{-1}) &= 1+c_1z^{-1}+\ldots+c_{nc}z^{-nc}, \quad nc \leqslant na \end{aligned} \tag{3. }$$

where u and y are the input and output respectively, i is the discrete time index, z^{-1} the backward shift operator, integer k the discrete time delay and e uncorrelated noise $(0, \sigma_e^2)$. Without loss of generality C(z) can be chosen to have its zeros outside the unit circle (Åström, 1970). It is assumed that

A and B are coprime, nb $\leq$ na (3.2)

These two assumptions are fairly mild.

LQG Controller

As the control law we take the linear controller that minimizes the criterion

$$J = \lim_{\substack{i_1 \to \infty \\ i_0 \to -\infty}} \frac{1}{i_1 - i_0} E \sum_{i=i_0}^{i_1} \{y^2(i) + \rho u^2(i)\}, \quad \rho > 0 \tag{3.3}$$

for the system (3.1), (3.2). The solution is given by the steady-state LQG controller

$$R(z^{-1})u(i) = -S(z^{-1})y(i) \tag{3.4}$$

$$R(z^{-1}) = 1 + r_1 z^{-1} + \ldots + r_{nr} z^{-nr}, \quad nr = \max(nb + k - 1, nc) \tag{3.5}$$

$$S(z^{-1}) = s_0 + s_1 z^{-1} + \ldots + s_{ns} z^{-ns}, \quad ns = na - 1 \tag{3.6}$$

where R and S follow from the pole assignment identity

$$A(z^{-1})R(z^{-1}) + z^{-k}B(z^{-1})S(z^{-1}) = C(z^{-1})P(z^{-1}) \tag{3.7}$$

where on the right hand side $C(z^{-1})$ prescribes the Kalman filter poles and $P(z^{-1})$ the regulator poles according to the spectral factorization

$$\eta P(z)P(z^{-1}) = \rho A(z)A(z^{-1}) + B(z)B(z^{-1})$$
$$P(z^{-1}) = 1 + p_1 z^{-1} + \ldots p_{na} z^{-na} \tag{3.8}$$

where P(z) has its zeros strictly outside the unit circle; η is a scaling factor because $P(z^{-1})$ is defined as a monic polynomial. Tuning of the LQG controller viz. selection of ρ may be simplified by the tuning of another parameter ρ' defined by

$$\rho = \rho' K^2, \quad \rho' > 0 \tag{3.9}$$

and

$$K = B(1)/A(1) \text{ if } A(1) \neq 0$$
$$K = \lim_{z \to 1} (1 - z^{-1})B(z^{-1})/A(z^{-1}) \text{ if } A(1) = 0 \tag{3.10}$$

where it is assumed that A contains at most one zero at 1 if A(1) = 0. In our experience a good starting point is to take $\rho' = 1$.

A Sub-Optimal Controller

In practice the model (3.1) is only valid for perturbation variables around a steady-state or nominal trajectory. More often than not the nominal values are not accurately known and the consequence is that steady-state errors remain with the controller (3.4). It is possible to eliminate such errors by extending the model (3.1) with an offset and applying an adaptive controller that estimates this offset and compensates for it. The occurrence of steady-state errors only, however, is no reason for adaptive control and it is well known that such errors can be eliminated by integral feedback. Therefore we propose a sub-optimal controller whose design and performance is close to the optimal controller but which also has no steady state error viz.

$$R(z^{-1})(1 - z^{-1})u(i) = -S^*(z^{-1})y(i) \tag{3.11}$$

with the structure of $R(z^{-1})$ unchanged equal to (3.5) but with

$$ns^* = ns + 1 = na \tag{3.12}$$

The parameters of R and S^* are now computed with

$$A(z^{-1})R(z^{-1})(1 - z^{-1}) + z^{-k}B(z^{-1})S^*(z^{-1}) = P(z^{-1})C(z^{-1})(1 - qz^{-1}), \quad 0 \leq q < 1 \tag{3.13}$$

Comparing this with (3.7), it is obvious that an extra pole z = q is introduced, which determines the proportion of the integral feedback. The pole z = q can be interpreted as an observer pole for constant disturbances; q = o will give a controller with quick suppression of load disturbances but also an undesirable deterioration of the control behaviour with respect to criterion (3.3), while $q \uparrow 1$ gives an approximation of the LQG controller as closely as desired but with practically no load disturbance rejection. In fact some compromise must be made for the determination of q, but it is advisable to tune q between

$$q_0 \leq q < 1 \tag{3.14}$$

where q_0 is "the slowest mode" of P and C. The influence of q on the control performance is illustrated in the following example.

Example

Fig. 4 shows the output variance Ey^2 vs. the input variance Eu^2 for the system

$$(1 - 1.54z^{-1} + 0.59z^{-2})y(i) = z^{-1}(0.28 + 0.23z^{-1})u(i) + (1 - 0.4z^{-1})e(i) \tag{3.15}$$

controlled with the LQG controller and sub-optimal controller with q = 0.95,0.9,0.8,0.7,0.5 respectively for $0 < \rho < \infty$ ($\sigma_e^2 = 1$). Fig. 5 shows the input and output responses of the noiseless system (3.15) ($\sigma_e^2 = 0$) with the LQG controller and the suboptimal controllers with $\rho = 1$, to a step-wise load disturbance applied at i = 5. An offset remains for the LQG controller.

Servo Extension

In the preceding only the regulator problem has been discussed. The inclusion of the servo problem leads for the sub-optimal controller to a control scheme as given in Fig. 6. Note that a prefilter $G_F(z^{-1})$ is included in the forward path defined as

$$G_F(z^{-1}) = \xi P(z^{-1})/Ad(z^{-1})$$
$$Ad(z^{-1}) = 1 + ad_1 z^{-1} + \ldots ad_{nad} z^{-nad}, \quad \xi = \frac{\sum_{j=1}^{nad} ad(j)}{\sum_{j=o}^{nb} b(j)} \tag{3.16}$$

The reason is that the regulator (3.11) (also regulator (3.4)) is only suitable for constant or slowly varying set-points. When the set-point undergoes large step changes which often occur at the start-up of batch processes, the input becomes too large and large transients result, especially if the input meets physical constraints. A remedy for these effects may be a relative small value for ρ, but then the disturbance suppression of the system is deteriorated. In fact R and S are completely determined by the desired properties of the system with respect to stability and disturbance rejection. So the correct tuning sequence is first to select a suitable value of ρ and q for a desired system behaviour with respect to the regulator problem and then the prefilter is selected in order to get the desired transmittance from the set-point to the output (servo problem). Hence the regulator design is independent of the servo design, but the converse is not true. A favourable choice for $Ad(z^{-1})$ would be in general (Aström, 1979)

$$Ad(z^{-1}) = 1 + ad_1 z^{-1} + ad_2 z^{-2}$$

$$ad_1 = -2\exp(-\alpha)\cos(\alpha) \qquad (3.17)$$

$$ad_2 = \exp(-2\alpha) \;, \; \alpha \approx 0.3 - 0.7$$

Computation

The first step in determining the parameters of the proposed controller is solving the spectral factorization (3.8). An algorithm based on Newton iterations is given by Kučera (1979). However, from our experience with the control of bioreactors it is sufficient to choose na = 1 or na = 2 (provided that the sample time is not too small). For these cases a simple purely algebraic solution is possible as follows:
Define

$$v_0 = (\rho \sum_{j=o}^{na} a_j^2 + \sum_{j=o}^{nb} b_j^2)/2$$

$$v_1 = \rho \sum_{j=o}^{na-1} a_j a_{j+1} + \sum_{j=o}^{nb-1} b_j b_{j+1}$$

$$v_2 = \rho \sum_{j=o}^{na-2} a_j a_{j+2} + \sum_{j=o}^{nb-2} b_j b_{j+2} \qquad (3.18)$$

$$\mu_1 = v_o - v_2 + \surd[(v_o+v_2)^2 - v_1^2]$$

$$\mu_2 = (\mu_1 + \surd[\mu_1^2 - 4v_2^2])/2$$

Then

$$p_1 = v_1/(\mu_2+v_2)$$
$$p_2 = v_2/\mu_2 \qquad (3.19)$$

The next step is solving the polynomial identity (3.13) by equating the coefficients of equal powers of z^{-1} which leads to a set of (nr + ns + 1) linear equations with the coefficients of $R(z^{-1})$ and $S(z^{-1})$ as unknown parameters. Finally we mention a 'structural property' which is of importance with the self-tuning controller. Suppose we know that (3.1) and (3.2) are valid but we do not know k exactly, only a lower bound k_{min} and an upper bound k_{max}. Then the controller parameters are computed with an enlarged B polynomial viz:
- solution of (3.18) with nb = na + k_{max} - k_{min}
- solution of (3.13) with nb = na + k_{max} - k_{min}, k = k_{min}, nr = max(na + k_{max} - 1,nc)

where for (3.18) we have implicitly used the fact that maximal na coefficients of $B(z^{-1})$ are nonzero without explicitly indicating which coefficients are zero.

4. ADAPTIVE CONTROL

In section 2 the basic ideas were given for the solution of the control problem. The remaining task is the specification of the adaptive RQ-controller of Fig. 2. Two parameter adaptive controllers will be presented in this section viz. a feedforward adaptive controller based on a 'gain scheduling approach' and an explicit self-tuning adaptive controller (Aström, 1983). Both methods use the control design method of section 3.

Feedforward Adaptive Control

The feedforward adaptive controller makes use of the transfer function (2.9). For given τ_1, τ_2, T_d and –, the deterministic part of the ARMAX model (3.1) is given by the discrete transfer function

$$H(z^{-1}) = z^{-k}A^{-1}(z^{-1})B(z^{-1}) = (1-z^{-1})Z\{G_p(s)/s\} \qquad (4.1)$$

with

$$T_d = (k-1)T_s + T_{df} \qquad 0 \leqslant T_{df} < T_s \qquad (4.2)$$

$$na = 2,\; nb = 1 \text{ if } T_{df} = 0,\; nb = 2 \text{ otherwise}$$

where T_s is the sampling time and Z is the well known ordinary or modified Z-transform if T_{df} is zero or nonzero respectively (Jury, 1958). The parameters of $H(z^{-1})$ are updated at every sampling with the aid of (2.10) and (2.11). This updating uses a directly measured air flow rate F and an on-line estimated biomass concentration X according to an elemental balancing method (Dekkers, 1984). Now the adaption of the controller is coupled to the changing model parameters by substituting (4.1) into (3.8) and (3.13). However solution of (3.13) requires besides k, A and B also C. Because information about the stochastic part of the ARMAX model is not a priori available, we do not use a Kalman filter but a reduced order observer specified with

$$C(z^{-1}) = 1+c_1 z^{-1},\; c_1 = p_2^{1.5} \qquad (4.3)$$

Finally we take $q = p_2^{1/12}$, $\rho' = 1$ and $\alpha = 0.3$. With these specifications the feedforward adaptive controller is completely tuned.

Self-Tuning Adaptive Control

The self-tuning adaptive controller does not make use of the auxiliarly measurements of X and F. The idea of self-tuning adaptive control is well known The explicit self-tuning controller can be thought as composed of three parts: a recursive parameter estimator, a linear controller and a block which determines the controller parameters from the estimated process parameters. The last two parts follow immediately from section 3. A Recursive Prediction Error Method (Ljung and Söderström, 198–) is proposed as the parameter estimator for the ARMAX model. Bearing in mind (2.9) and (2.10) the structural parameters of the model should be taken as

$$na = 2,\; nb = 2 + k_{max} - k_{min},\; nc = 0,1 \text{ or } 2,\quad k = k_{min} \qquad (4.4)$$

where k_{max} and k_{min} depend on the maximum and minimum time delay respectively and the sample time. The set-point RQ_s is (nearly) constant so that with the suboptimal controller the closed loop system identifiability condition (Söderström, Gustavsson and Ljung, 1975) becomes

$$k \geqslant nc + 1 \qquad (4.5)$$

In our application we guess $0 \leqslant T_d \leqslant 2T_s$ and thus $k_{min} = 1$, $k_{max} = 2$. So with (4.4) and (4.5) the structural parameters are taken as

$$na = 2,\; nb = 3,\; nc = 0,\; k = 1 \qquad (4.6)$$

Now with nc = 0 the RPEM becomes the ordinary RLS identification method and hence the parameter estimate $\hat{\theta}^T \triangleq (\hat{a}_1 \ldots \hat{a}_{na}, \hat{b}_o \ldots \hat{b}_{nb})$ is generated from the equations

$$\hat{\theta}(i) = \hat{\theta}(i-1) + P(i)\phi(i)[y(i)-\phi^T(i)\hat{\theta}(i-1)] \tag{4.7}$$

$$P(i) = \frac{1}{\lambda(i)}[P(i-1) - \frac{P(i-1)\phi(i)\phi^T(i)P(i-1)}{\lambda(i)+\phi^T(i)P(i-1)\phi(i)}]$$

where $\phi^T(i)=(-y(i-1)...-y(i-na),u(i-k)...u(i-nb-k))$ and $\lambda(i)$ the exponential forgetting factor. The ARMAX model (3.1) can only be a good description of the process dynamics with aspect to perturbation variables. Therefore the d.c. values u_0 and y_0 must be taken into account. The correct model is given by

$$A(z^{-1})(y(i)-y_0)=z^{-k}B(z^{-1})(u(i)-u_0)+C(z^{-1})e(i) \tag{4.8}$$

or

$$A(z^{-1})y(i) = z^{-k}B(z^{-1})(u(i)-\bar{u})+C(z^{-1})e(i) \tag{4.9}$$

with

$$\bar{u} = u_0 - A(1)/B(1)y_0 \tag{4.10}$$

$\bar{u}$ is estimated by recursive averaging with exponential forgetting

$$\bar{u}(i) = q\,\bar{u}(i-1) + (1-q)(u(i)-x(i)) \tag{4.11}$$

$$B(z^{-1})x(i) = \xi A(z^{-1})y_s(i)$$

and hence u(i)must be replaced with $\nabla u(i)$ in (4.7), where $\nabla u(i) = u(i) - \bar{u}(i)$.

Tuning the Self-Tuner

In the preceding paragraph the principle of our self-tuning controller has been given. A number of parameters of the self-tuning controller must be tuned by the user. Some have been given such as the integers na, nb, nc, k and thus nr, ns*; the real numbers q, ρ, α are discussed in section 3. Other parameters, not yet discussed, refer to the identification algorithm. Tuning of these parameters has been considered by Ljung and Söderstrom (1983). But there also are other issues such as start-up, bumpless transfer, prefiltering, numerical problems, false measurements etc. which must be reckoned with. It is beyond the scope of the article to discuss all these items, though two remarks will be made with respect to the start-up and model order reduction.

The start-up of the self-tuner. It is a common practice to start the parameter estimator with

$$\hat{\theta}(0) = (0,...,0)^T, \; P(0) = rI \tag{4.12}$$

where I is a unit matrix and r a large number e.g. r = 1000. In the controller precautions must be taken with respect to the solution of (3.11) because of the initial value of $\hat{\theta}$. Besides this the control signal must be limited during the first period of time

$$u_{min} = \bar{u} - \Delta u \leq u \leq u_{max} = \bar{u} + \Delta u \tag{4.13}$$

where $\bar{u}$ is the output before start-up determined by the manual control of the process operator and Δu the allowable magnitude of the initial test signal. With the self-tuning LQG controller it is normally found that in such a situation the control signal jumps a small number of times between u_{min} and u_{max}, then the parameters estimates become useful values and at the same time a desireable control behaviour is obtained. However with the self-tuning sub-optimal controller we found by simulations that often the control signal is not rich enough for useful parameter estimation. If the restriction (4.13) is relaxed then a large error in y may result which causes an offset for a long time especially if q is close to 1. Therefore a brief open loop identification by applying one period of a maximum length sequence (e.g. 4 bits thus a period of 15 sampling intervals) is suggested. In this way a better starting model for the self-tuning controller is obtained and the mentioned problems are avoided.

Model order reduction. If the polynomials A and B have an (almost) common factor the solution of (3.13) will be poorly conditioned and unacceptable excursions of the input signal may result. The roots of A are easily computed because in our design philosophy na = 1 or na = 2, but the determination of the roots of B may be difficult because $nb = na + k_{max} - k_{min}$ and thus B may be of a high order. Therefore the following test is suggested. Given the roots z_i (i = 1,2) of A. Then z_i is a root of B if

$$|z_i' - z_i| < \varepsilon \min(|z_i|, |z_i'|) \tag{4.14}$$

where ε is a small number e.g. ε = 0.02. z_i' is the result of one Newton-Raphson step for the B polynomial with z_i as starting value. If (4.14) is true for z_i, the factor $z - z_i$ is cancelled from A and B and thus reduced polynomials A and B are obtained. As a consequence, the order of the controller polynomials is then also reduced.

5. EXPERIMENTS

This section gives the results of three experiments which were performed in a computer controlled pilot fermentor with a working volume of 30 - 80 l. A description of the equipment and instrumentation has been given by Dekkers (1982, 1984).

Summary

TABLE 1. Overview of the Experimental Results

item	batch 1	batch 2	batch 3
control	feedf.adapt.	stc	stc
T_s(min)	1	1	1
ρ	2	1	1
q	$p_2^{1/12}$	0.9	0.9
λ(∞)	-	0.995	0.99
X(o)[g/l]	5.83	3.75	3.89
$X(t_f)$[g/l]	48.7	47.3	51.0
M(o)[g]	239	213	204
$M(t_f)$[g]	2960	3880	4020
E_m[g/l]	3.7	0.5	0.7
$E(t_f)$[g/l]	3.7	0.4	0.6
$\bar{Y}$[g/g]	0.438	0.472	0.488
t_f[hr]	15.7	16.8	12.7

An overview of the three fed-batch fermentations is given in TABLE 1. The first one is RQ-controlled by the feedforward adaptive controller and the others by the self-tuning controller. E_m is the maximum reached ethanol concentration, $E(t_f)$ the ethanol concentration at the end of the fermentation, t_f the batch time and $\bar{Y}$ the overall yield. All of the batches were inoculated with unadepted yeast. This means that the initial specific oxygen uptake rate is low because of limitation by OUR_r. For every fermentation we can distinguish three stages viz.

- an adaption stage where OUR/M goes from a small initial value to a maximum of about 7.5 mmol/g-hr
- a stage with maximum specific growth rate under the constraint of no or a little ethanol production i.e. $\mu_c \approx \mu_{c,max}$
- a stage where OUR/M is decreasing because $OUR = OTR_m$.

Batch 1. Feedforward Adaptive Control

The results of this batch are shown in Fig. 7. The setpoint RQ_s was 1.08 during the whole fermentation. Because the RQ behaves rather unpredictably during start-up, the GFR has been taken constant till t = .20 min and then the controller has been switched on. The offset from 1.7 untill 3.3 hr is the result of a too slow integrator pole q which was approximately 0.95, corresponding to a time constant of 20 min. The two 'large' deviations at 11.8 hr and 12.8 hr are the consequences of the addition of anti-foam at that time, causing a temporary drop in the oxygen uptake rate.

Regarding the OUR/M it can be seen that this batch had a fairly long adaptation phase. After 6 hr the GFR remains constant till 9 hr. It is not likely that the OUR is already limited by OTR because the GFR still increases considerably during 9 - 12 hr. Also the stirrer speed has been increased at 6 hr, which increases the oxygen transport rate and should have a marked effect on OUR/M if the OUR is transport limited. We conclude therefore that OUR is only determined by OTR from 12 hr onwards, and that the diminishing specific growth rate during 6 - 8 hr is caused by a growth synchronisation effect.

Batch 2. Self-tuning Control

Fig. 8 and Fig. 9 give the results of this batch. The setpoint RQ_s was 1.2, 1.1, 1.05 and 1.03 respectively (see the solid line in Fig. 8). The start-up procedure was as follows. After a first few minutes of constant GFR a fixed controller has been used. At 1 hr a p.r.b.s. was added to the input signal in order to get a reasonable initial estimate for the process parameter vector. Then at 1.25 hr the self-tuner control loop was closed. With this batch μ_c did not match $\mu_{c,max}$ or only during a very short time $\mu_c = \mu_{max}$ at about 5 hr. After this time $OUR = OTR_m$ as it appeared from the step in OUR/M at 7.9 as response to the step in the stirrer speed.

The RQ controller works quite satisfactory except from 8.5 hr till 9.5 hr. The reason for this is the drift of the CO_2 measurements. From a calibration procedure it appeared that $RQ_{measured} > RQ_{real}$. Therefore the controller tries to keep RQ_{real} below RQ_c. This is feasible only as long as ethanol can be consumed. At 8.5 hr all ethanol has been consumed (which has been confirmed by analyzing samples taken from the process) and RQ must increase. At 9.5 hr the meter was recalibrated causing a drop of 0.02 in the $RQ_{measured}$ and thus correct control was reestablished.

In Fig. 9 $\hat{b}_0(i)$ and $P_{b_0,b_0}(i)$ are shown,whose shape is typical for other parameters and elements of $\hat{P}$. At the first few samples the uncertainty of $\hat{b}_0$, P_{b_0,b_0} rapidly decreases as a result of the p.b.r.s. test signal. In the following hours, P_{b_0,b_0} increases considerably, although the system identifiability condition (4.5) is fulfilled. This poor identifiability may be due to the following factors. Firstly, a forgetting factor < 1 is used, together with closed loop identification and with lack of setpoint variations. At those instances where large input excitations occur, new information is gathered and consequently P_{b_0,b_0} drops. Secondly, the large value of the integrator time constant has a poor effect, as the closed loop system will become unidentifiable as $q \uparrow 1$. Also shown in fig. 9 are the poles of the process. As they are complex for some time, the model (2.9) is not confirmed by this experiment.

Batch 3. Self Tuning Control

The results of this batch are shown in figs. 10, 11, 12. Again the setpoint RQ_s was 1.2, 1.1, 1.05 and 1.03 respectively. The principle of the start-up procedure was the same as that of batch 2. The main difference between this batch and batch 2 is that here we have a plain stage with $\mu = \mu_{c,max}$, attained with higher stirrer speed rates and thus higher oxygen transfer rates. During time 5 till 8 hr $\mu \approx 0.3$ hr^{-1}. After time 8 hr OUR/M is descending while the measured dissolved oxygen concentration DO (near the stirrer and air inlet) is clearly above zero. This probably indicates that there are zones in the fermentor where the DO is zero while the measured DO is not zero.

The RQ controller is very accurate except from 9.7 till 11 hr. The reason for the failure is an estimator wind up. Due to a bad parameter estimate a control law is calculated which gives an unstable closed loop system. The resulting peak in the input at 9.7 hr allows much information to be gathered by the parameter estimator (see figure 11), so the regulator recovers. At 10.1 hr the RQ again diverges from the setpoint. From fig. 12, which shows the poles of $P(z^{-1})$, it is clear that this is due to an unstable pole of $P(z^{-1})$. The correct calculation of $P(z^{-1})$ by (3.18) and (3.19) requires the aforementioned assumption that only n coefficients of $B(z^{-1})$ be nonzero, an assumption that is not met in this region. At 10.5 hr a common factor in $A(z^{-1})$ and $B(z^{-1})$ is detected by (4.14) (see open loop poles and zeros in fig. 11 and 12 respectively). As a result of the model reduction the unstable pole of $P(z^{-1})$ is removed.

6. CONCLUSIONS

It has been shown how two types of adaptive controllers can regulate the respiratory quotient accurately. For both controllers the redefinition of the system to be controlled by the inner loop feedback of the OUR, is essential. In this way all stages of the process can be controlled with one controller structure. The sub-optimal control design method has the advantage that integral action needed for offset elimination can be included with only a slight deterioration of the stochastic control behaviour. The feedforward adaptive controller has the advantage that no excitation of the process is required for the updating of the parameters. The disadvantage is of course that a priori knowledge must be available e.g. about the transport phenomena. This would be fairly difficult to obtain for large scale fermentors. The self-tuning controller is more generally applicable because of its black box approach. The weak point is the parameter estimator with constant forgetting factor, which can cause covariance wind-up. An improvement would be attained by making the parameter estimator only adaptive if information is available e.g. with the usage of a variable forgetting factor.

Acknowledgement. This work was supported by the Department of Applied Physics of the Delft University of Technology.

7. REFERENCES

Aiba, S., S. Nagai and Y. Nishizawa (1976). Fed-batch culture of *Saccharomyces cerevisiae*: a perspective of computer control to enhance the productivity in baker's yeast cultivation. Biotechnol. Bioeng., 18, 1001-1016.

Aström, K.J. (1970). Introduction to Stochastic Control Theory. Academic Press, New York.

Aström, K.J. (1979). Simple self-tuners I. Report CODEN: LUTFD2/(TFRT-7184)/(1979), Department of Automatic Control, Lund Institute of Technology.

Aström, K.J. (1983). Theory and applications of adaptive control - a survey. Automatica, 19, 471-486.

Dekkers, R.M. (1982). State estimation of a fed-batch baker's yeast fermentation. In A. Halme (Ed.), Proceedings 1st IFAC Workshop on Modelling and Control of Biotechnical Processes, Helsinki, 201-212.

Dekkers, R.M. (1983). Dynamic optimization of a fed-batch fermentation process. In F. Conrad (Ed.), Proceedings 1st IASTED International Symposium on Applied Control and Identification, Copenhagen, 3.34-3.39

Dekkers, R.M. (1984). Optimal control of a fed-batch fermentation. In E.H. Houwink, R.R. van der Meer (Ed.), Progress in Industrial Microbiology, 20, Elsevier, Amsterdam, 313-330.

Dekkers, R.M. (1985). Theory and Practice of the Optimal Control of a Fed-Batch Fermentation. Ph.D. thesis, to be published.

Geurts, T.G.E., H.E. de Kok and J.A. Roels (1980). A quantitative description of the growth of Saccharomyces Cerevisae CBS 426 on a Mixed Substrate of Glucose and Ethanol. Biotechnol. Bioeng., 22, 2031-2043.

Jury, E.I. (1958). Sampled Data Control Systems. Wiley, New york.

Koning de, C.J. (1984). Modelvorming van een bakkers gist fermentatie (in Dutch). Thesis, Department of Applied Physics, Delft University of Technology.

Kučera, V. (1979). Discrete Linear Control: The Polynomial Equation Approach. Wiley, London.

Ljung, L. and T. Söderström (1983). Theory and Practice of Recursive Identification. MIT Press, Cambridge.

Meyenburg von, H.K. (1969). Energetics of the budding cycle of Saccharomyces Cerevisiae during glucose limited aerobic growth. Arch. Mikrobiol., 66, 289-303.

Pons, M.N., J. Bordet and J.M. Engasser (1982). A micro-mini computer hierarchical control system for a laboratory scale fermentor. In A. Halme (Ed.), Proceedings 1st IFAC Workshop on Modelling and Control of Biotechnical Processes, Helsinki, 259-266.

Söderström, T., I. Gustavsson and L. Ljung (1975). Identifiability conditions for linear systems operating in closed loop. Int. J. Control, 21, 243-255.

Wang, H.Y., C.L. Cooney and D.I.C. Wang (1979). Computer control of baker's yeast production. Biotechnol. Bioeng., 21, 975-995.

8. CAPTIONS.

Fig. 1. Schematic diagram of respiratory quotient, yield, and specific oxygen uptake rate vs. specific growth rate in steady-state.

Fig. 2. Inner loop feedback structure for RQ control.

Fig. 3. Response curves of oxygen uptake rate and carbon evolution rate to glucose feed rate.

Fig. 4. Plot of the variances Eu^2 and Ey^2 with ρ as parameter for LQG and different sub-optimal controllers for system (3.15).

Fig. 5. Responses of the noiseless system (3.15) with LQG-optimal and sub-optimal control on a stepwise load disturbance with magnitude 1 applied at t=5.

Fig. 6. Structure of the sub-optimal controller with servo extension.

Fig. 7. Respiratory quotient with setpoint, specific oxygen uptake rate and glucose feed rate for batch 1, with feedforward adaptive control.

Fig. 8. Respiratory quotient with setpoint, specific oxygen uptake rate, glucose feed rate and dissolved oxygen of batch 2, with self-tuning control.

Fig. 9. Estimated parameter $\hat{b}_0$ with estimated covariance, and estimated open loop poles of batch 2.

Fig. 10. Respiratory quotient with setpoint, specific oxygen uptake rate, glucose feed rate and dissolved oxygen of batch 3, with self-tuning control.

Fig. 11. Estimated parameter $\hat{b}_0$ with estimated covariance, and estimated open loop poles for batch 3.

Fig. 12. Estimated zeros and closed loop poles (poles of $P(z^{-1})$) for batch 3. Branches marked with '2' consist of a pair of complex zeros.

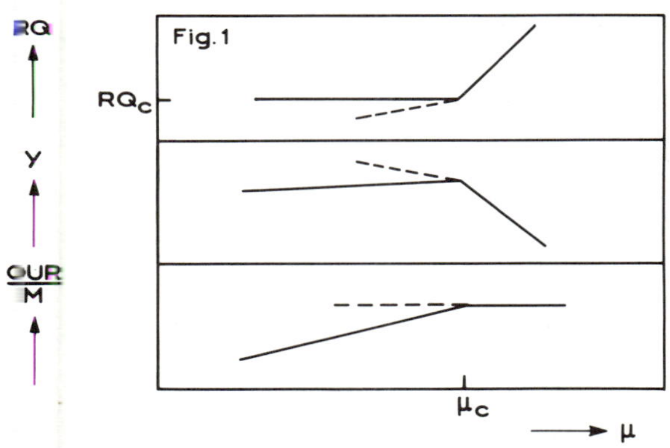

Fig. 2

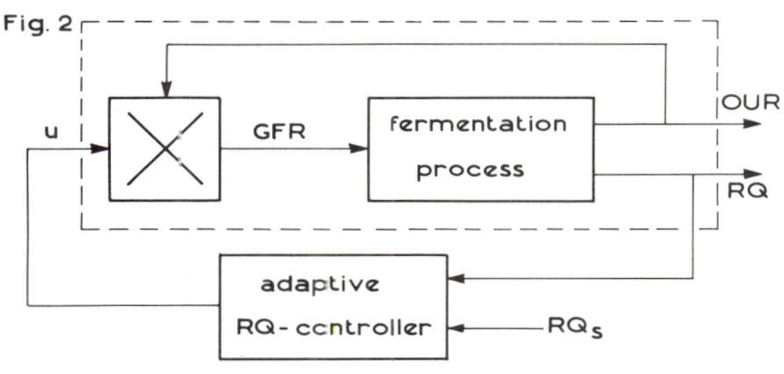

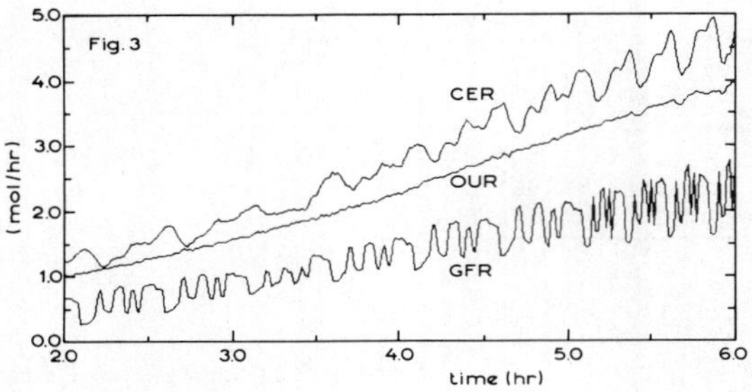
Fig.3
CER
OUR
GFR
(mol/hr)
time (hr)

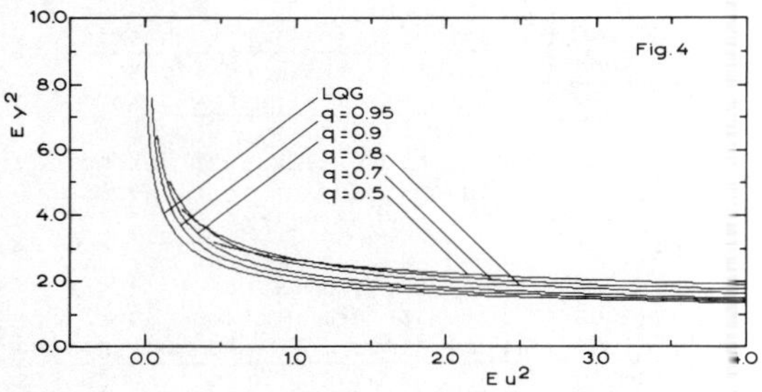
Fig.4
LQG
q=0.95
q=0.9
q=0.8
q=0.7
q=0.5
E y²
E u²

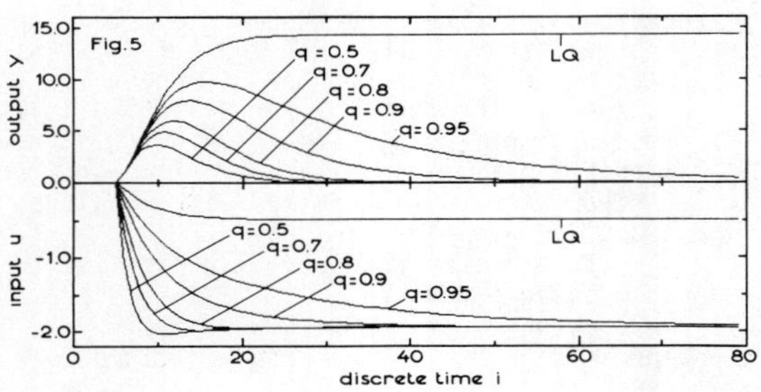
Fig.5
q=0.5
q=0.7
q=0.8
q=0.9
q=0.95
LQ
output y
input u
discrete time i

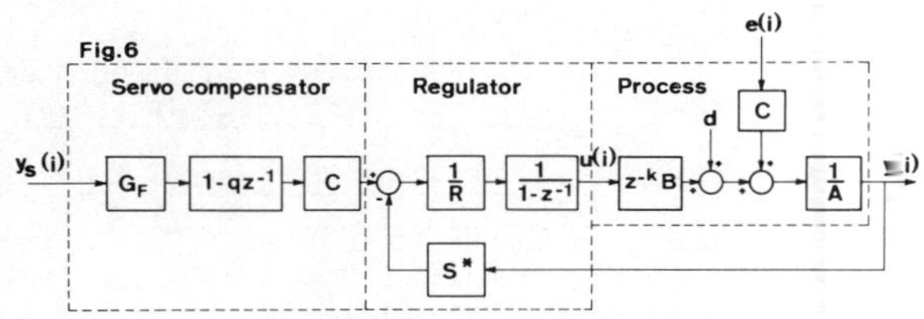
Fig.6
Servo compensator
Regulator
Process
e(i)
d
u(i)
y_s(i)
G_F
1-qz⁻¹
C
1/R
1/(1-z⁻¹)
z⁻ᵏB
1/A
S*

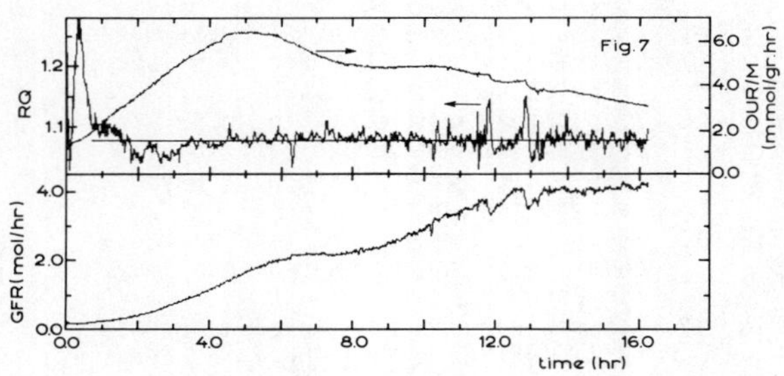
Fig.7
RQ
OUR/M (mmol/gr.hr)
GFR (mol/hr)
time (hr)

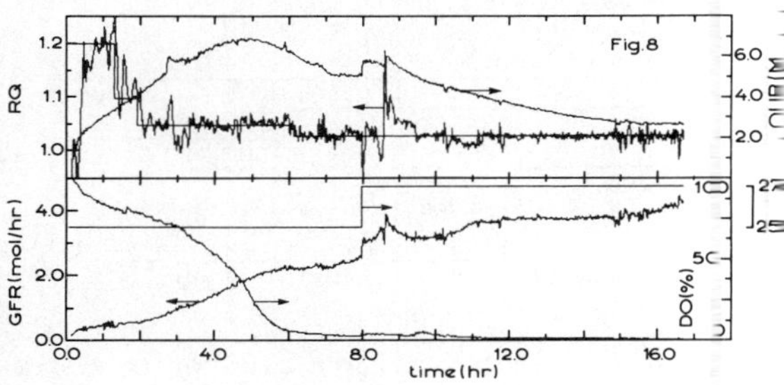
Fig.8
RQ
GFR (mol/hr)
DO(%)
N(RPM)
time (hr)

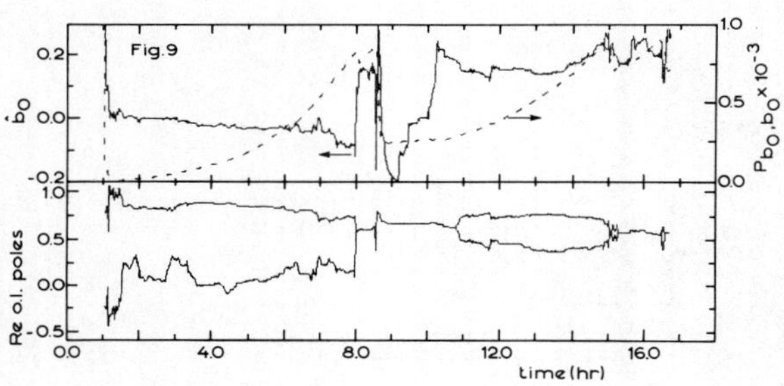
Fig.9
b̂₀
P_b0,b0 x10⁻³
Re o.l. poles
time (hr)

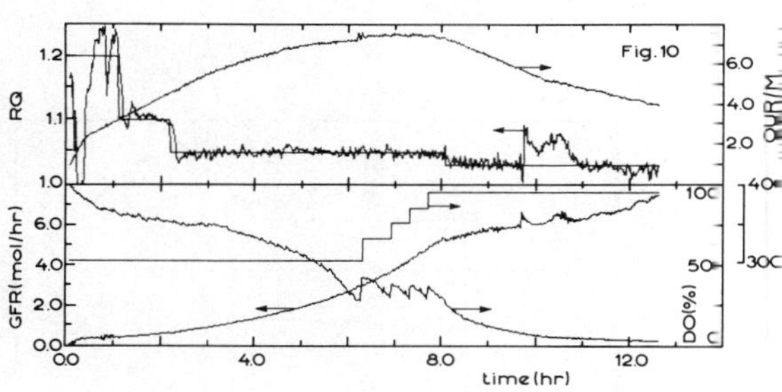
Fig.10
RQ
GFR (mol/hr)
DO(%)
N(RPM)
time (hr)

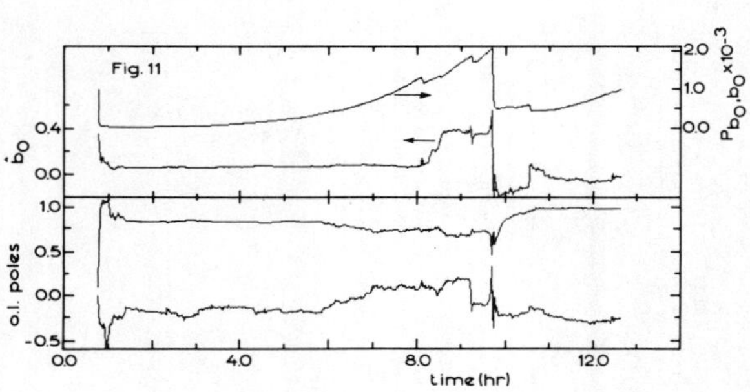
Fig.11
b̂₀
P_b0,b0 x10⁻³
o.l. poles
time (hr)

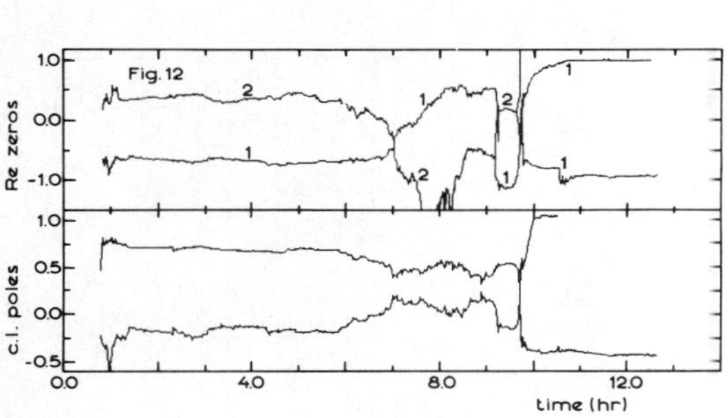
Fig.12
Re zeros
c.l. poles
time (hr)

CONTROL OF FERMENTATION PROCESS BY AN ON-LINE ADAPTIVE TECHNIQUE

P. A. Montgomery, D. Williams and B. H. Swanick

Department of Electrical and Electronic Engineering, Liverpool Polytechnic, UK

Abstract. Control of a fermentation process is a complex problem due to inherent non-linearities and time varying characteristics of the process. The application of conventional analogue controllers provides poor control due to problems in tuning individual loops. The application of standard optimal tecniques is complicated both by the complexity of the process and the lack of adequate models as a result of poorly understood dynamics. The problem is further complicated by the lack of important transducers for product measurement. Hence, the process provides a natural test for adaptive control techniques.

The paper includes details of the multivariable parameter estimation, by recursive least squares as part of an adaptive control technique. A description of the operation of the adaptive controller is given in relation to its real-time application to a fermentation plant.

Details of the pilot-plant, instrumentation and computer system are described highlighting the practical problems in these areas and the means by which these problems have been overcome.

Results are presented to show the successful performance of the adaptive technique on the pilot-plant.

The work indicates that the application of the adaptive technique could provide great opportunities for enhancement of conventional control of fermentation processes.

Keywords. Adaptive control, Biotechnology, Fermentation processes, Multi-variable control, Parameter estimations, Performance index, Three term control.

INTRODUCTION

The feasibility of using a microcomputer supporting a multi-variable adaptive control algorithm is of considerable interest to many manufacturing companies. The low cost, flexibility and improved performance offered by such a configuration can easily justify its application for the replacement of ageing controllers and for application on new process plant. The work described in this paper is directly concerned with the development and application of an industrial based microcomputer system to support a robust adaptive controller specifically for application in the fermentation industries.

The ability to control a fermentation process accurately and automatically in the production of micro-organisms has been the subject of considerable interest for a number of years (1) as the optimisation of such plants can have a major impact on the quality, cost and reproducibility of the final product.

The main disadvantage, with conventional control techniques is that they have no on-line optimisation or adaptive features and cannot therefore provide an optimum control performance throughout a complete fermentation phase, due to the non-linear and time variant properties of the system.

In recent years computers have started to make a limited impact upon the monitoring and control of fermentation processes. It is only with their implementation that optimal and adaptive techniques can be applied to industrial processes. In many practical applications it may not be economical, convenient or even possible to measure all of the system variables accurately on-line. The problem of inadequate or non-existent measurement techniques is particularly evident in biotechnology where the on-line measurement of cell concentration, cellular activity, nutrient concentration and enzyme activity are not available. The additional problems of sterility mean that the design of probes is difficult and their active life can be severely diminished.

Despite these problems Waterworth and Swanick (2) have successfully shown the feasibility of using a multivariable adaptive control technique (3) for the optimisation and control of a biological model. Yousefpour and Williams (4) have further demonstrated the on-line application of this multivariable adaptive controller using a minicomputer for the optimisation and control of a small fed-batch fermentation process. The studies are now continuing in collaboration with APV International towards the development of an industrial microprocessor based adaptive control package.

The microcomputer system, is based on an industrial computer system manufactured by APV International Ltd. A high level language has been chosen to implement the adaptive algorithm and suitable interfaces constructed.

MICROCOMPUTER SYSTEM

Many current process computers in biotechnology have been used, primarily to meet the need for extensive process monitoring with only limited conventional feedback or sequential control. In addition, many computer manufacturers have, in developing their systems, included specially written software and programming routines designed for use by process personnel with little or no computing experience. These executive routines enable an otherwise complex programme to be written as a series of simple routine calls but at the expense of losing general flexibility and numerical computational facilities. The result has meant that modern process control and optimisation techniques requiring the use of floating point arithmetic, multi-dimensional data arrays and parameter passing procedures cannot be used directly.

The APV computer system, (an Accos 2 series microcomputer) is the latest in a range of computers specifically designed to provide a flexible modualised system based on a distributed microcompressor structure.

In its basic form the microcomputer consists of three independent modules however within a small self-contained cabinet. The modules consist of a central controller, a programmable process interface and a mains power supply, Fig. 1.

The central control module consists of a number of standard Mostek plug in cards interconnected through an STD power and communication bus. In its standard form the system incorporates

A Zilog Z80 central processing card,
a 64K dynamic random access memory (RAM) card,
a diagnostic and real-time clock card
a video display and keyboard communication card and
a 64K permanent memory and programmable read only memory (PROM) card.

Pro-Pascal, a relatively new programming language based on standard Pascal with additional facilities for efficient real-time application was chosen because it provided the required numerical accuracy and supported matrix algebra. In addition the segmentation facilities of the language alllowed economy to be gained at all stages in the software's development by permitting independent compilation of the source modules.

Prior to the on-line data acquisition and control of the process, it was necessary to design signal conditioning interfaces to provide impedance matching, filtering and amplification of the low level signals prior to processing by the Accos 2 microcomputer.

FERMENTATION INSTRUMENTATION

The fermentation instrumentation and equipment shown in Figure 2 is based on a Biostat V laboratory fermentor fitted with a 5-litre culture vessel and baffled Chicane type heat exchanger. A 0.7 HP agitator drive motor complete with motor speed control is used to provide a constant stirrer speed in the range (0-1000 r.p.m.).

The process temperature is measured with a resistance thermometer probe and Uniprobe model 104 temperature monitor reading 0-100°C. The temperature is controlled by a Churchill Thermocirculator (model LTCV), consisting of a reservoir pump, ON/OFF thyristor controller, and heater suitable for maintaining temperatures in the range 20-80°C; pH is measured and automatically controlled by a Uniprobe pH electrode and controller. The instrument measures and controls pH in the range 0-14 by the use of an analogue ON/OFF controller and a dual Peristaltic pump to supply additions of either acid or alkali. The measurement of dissolved oxygen is by a sterilisable Uniprobe oxygen electrode (50). The electrode measuring the partial pressure of oxygen in the medium in the range 0-100%. The concentration of carbon dioxide (CO_2) and oxygen (O_2) in the exit gas is measured using an infrared (LIRA Model 303) CO_2 analyser and a Peramagnetic (MSA Model TMB) O_2 analyser connected in series. The instruments have been calibrated to measure CO_2 in the range 0-10% and O_2 in the range 0-2%.

The continuous measurement of alcohol is achieved by using a Radio Spares Gas Sensor (type 307) and amplifier. The unit was calibrated by progressively adding known quantities of alcohol into a 3-litre medium of pure water. Currently the instrument will measure alcohol concentrations from 0-25 g/lit. of medium. The air flow is supplied and measured by a Brooks air flow pump and controller at up to 5 lit./min. Oxygen is supplied to the medium on the basis that air contains 20.9% oxygen. The substrate (glucose feed) and ammonia additions for pH control are measured using (Cantilever Beam) load cells and high performance SIL SGA 300 series strain gauge amplifiers. An anti-foaming conductance probe together with a shot and relay controller (FT Instruments) are used to control the production and level of foam by the automatic addition of polypropylene glycol into the medium. A Watson Marlow programmable pump (Model 501U) is used to supply a 30% concentration of glucose feed to the medium either manually or via computer control.

THE CONTROL PROBLEM

Although systematic techniques exist (5) for tuning conventional three-term controllers, the problem of finding an optimum set of controller constants can occasionally be impossible or at best difficult and time consuming. In general with any complex process its satisfactory use is limited due to the process or actuators having non-linear and/or time varying dynamics. The additional problems of environmental disturbances, quantisation or output measurement noise and possible interactions between variables may result in a poor closed loop performance or even instability. The problem is even more complex for a multivariable process such as fermentation due to considerable state/state, input/input and state/input interaction.

Analysis of the process reveals that each output variable is a function of all the input variables and as such cannot be adequately controlled using conventional three-term controllers.

THE ADAPTIVE TECHNIQUE

The object of an adaptive control scheme however, is to maintain a desired process performance under varying operational and random external conditions.

The term "adaptive" signifies that the controller performs the simultaneous functions of learning about the process whilst at the same time controlling its behaviour. The adaptive scheme considered (2,3,4) is based on a multi-variable deterministic estimation procedure permitting the estimation of process parameters without the requirements for special statistical input disturbing functions. The estimated parameters being generated from the observations of the standard dynamic variations of the process as a result of a controlling action and noise within the system using the statistical method of recursive least squares.

Assuming that the controller works on a fixed sampling period, T seconds, and uses a sample and hold output, then the discrete time behaviour of the system will be given by the solution of the linear vector difference equation

$$\underline{X}\,[(K-1)]T = \phi\,(T)\,\underline{X}\,(KT) + \Delta\,(T)\,\underline{U}\,(KT) \quad (1)$$

where $\underline{X}$ (KT) is an (nx1) process state vector
$\underline{U}$ (KT) is an (mx1) process input vector
$\underline{K}$ = 0,1,2.....

$$\text{and} \quad \underline{\phi}(T) = e^{-\underline{A}T} \text{ (nxn) transition matrix} \quad (2)$$

$$\underline{\Delta}(T) = \underline{A}^{-1}(\underline{\phi}-I)\ .\ B \text{ (nxm) driving matrix.}$$

With reference to Figure 3, the process to be controlled is assumed to be represented by a linear multivariable state space equation, which can be modelled in discrete form by equation (1). A parameter estimation algorithm is initially applied to the process to estimate the driving and transition matrices $\underline{\phi}$ and $\underline{\Delta}$ or in the case of a nonlinear/time variant process their quasi-linear representations. These parameter estimates are then used by a linear suboptimal feedback control law, together with the current state vector to control the process in a manner specified by a quadratic performance index.

The feedback control law can be designed to minimise the error between a process output and an input reference, or alternatively to transfer the process from some initial state to a required state with a desired performance defined in terms of a minimum error function. This is achieved in nonlinear and/or time variance cases by the automatic adjustment of the feedback gains to counteract the changing dynamics of the process or plant due to changing quasi-linear estimates of the process parameters.

<u>Recursive Least Squares Parameter Estimation</u>

The recursive least squares estimation is a technique used extensively in data fitting applications in which it is desired to obtain a function that minimises the sum of the errors squared between the actual measurement data and a proposed function. The main advantage of the approach is the degree of noise filtering given to the measured data before the best estimated function is obtained. It can be shown that the recursive least squares estimate is given by the equation set (3).

$$\begin{aligned} \hat{\underline{\Theta}}_{K-1} &= \hat{\underline{\Theta}}_K + (\underline{V}_{K+1} - \hat{\underline{\Theta}}_K.\underline{W}_K).\underline{W}_K^t.\underline{P}_{K-1}.d_{K-1}^{-1} \\ d_{K-1} &= \rho + \underline{W}_K^t.\underline{P}_{K-1}.\underline{W}_K \qquad (3) \\ \underline{P}_K &= \frac{\underline{P}_{K-1}}{\rho}(\underline{I} - \underline{W}_K^t.\underline{W}_K.\underline{P}_{K-1}.d_{K-1}^{-1}). \end{aligned}$$

where $\underline{V}_K$ is a vector whose column represents the value of the state vector at each sampling interval.
$[X_1(K+1)T \ \ldots\ X_n(K+1)T]$

$\underline{W}_{K-1}$ is a vector whose column represents the value of the input and state vector at each sampling interval.
$[X_1(KT)\ \ldots\ X_n(KT),\ U_1(KT)\ \ldots\ U_m(KT)]$

$\underline{P}$ is a covariance matrix
$\overline{P}$ is the forgetting factor

$\underline{\Theta}_K$ is a matrix of estimated system parameters

$$\begin{bmatrix} \circ_{11} \ \ldots\ \phi_{1n}\ , & \Delta_{11}\ \ldots\ \Delta_{1m} \\ \circ_{n1}\ \ldots\ \phi_{nn}\ , & \Delta_{n1}\ \ldots\ \Delta_{nm} \end{bmatrix}$$

The equation set (3) represents the sequential algorithm for obtaining parameter estimates assuming process noise is white, with a zero mean. The estimate at stage K+1 being based on the previous K stage estimate, the current stage covariance matrix $\underline{P}_{K+1}$ and the additional data vector $\underline{V}_{K+1}$. The matrix pair $(\underline{Q},\underline{P})$ representing the only overhead in the algorithm for characterising the input-output data vectors for each estimation update. A continuous measure of the confidence and convergence of the estimated parameters is also given by the value of the trace elements of the covariance matrix and the scalar quantity d. The smaller the trace and the closer d is to 1, the more satisfactory is the convergence and the closer the estimated parameters approach their true values.

The use of the parameter ρ is to provide the recursive algorithm with a parameter estimation tracking facility by reducing the significance of past events and allowing more emphasis to be placed on the current system conditions.

The idea behind the technique is to use the fact that the covariance matrix P_K reflects the degree of confidence in the estimated parameters. By setting the forgetting factor between 0 and 1 ($0 < \rho \leqslant 1$) the speed at which the estimates adapt to changes in the process can be controlled. As ρ is decreased, the speed of adaptation increases, but this adaptive capability is gained at the expense of an increased statistical uncertainty. For noise contaminated data measurements, ρ will have to be close to 1, in order to obtain an appreciable observation length and noise filtering performance. This constraint is obviously undesirable when the estimation and tracking of (slowly) varying parameters is required.

Ideally a dynamic balance condition is sought in which the information lost from the estimator at each recursion is matched by the new information entering with the covariance matrix settling into a dynamic equilibrium condition.

<u>Control Law Formulation</u>

For on-line operation, it is essential that the control procedure operates efficiently and in a manner suitable for computer implementation. The technique presented (2,3,4) is based on a linear quadratic performance index whose solution for a single stage minimisation is particularly suitable for real-time application. Its operation can be considered as a balance between reducing any errors which exist between a desired state, which may either be constant or changing as a function of time, and the actual state of the system with the minimum of control effort.

The performance index is expressed as a quadratic, weighted function for each individual state and control variable (4).

$$J_K = (\underline{X}_{K-1}^t.\underline{Q}.\underline{X}_{K+1} + \underline{U}_K^t R \underline{U}_K) \qquad (4)$$

where $\underline{Q}$ and $\underline{R}$ are positive, definite, symmetric and represent weighting factors applied to the input and output states respectively; J_K

represents the scalar performance index at the Kth sampling interval.

The object of the control law is to compute the suboptimal (single stage minimisation) input vector $\underline{U}_K$, in order to minimise J_k (5).

$$\underline{U}_K = -\left(\underline{\hat{\Delta}}_K^t . \underline{Q} . \underline{\hat{\Delta}}_K + \underline{R}\right)^{-1} . \underline{\hat{\Delta}}_K^t . \underline{Q} . \underline{\hat{\phi}}_K \underline{X}_K \quad (5)$$

Weighting Matrices Q and R

As the system states will be required to track demand values or set points, it is necessary to introduce dummy state variables into the system. This is achieved by extending the state vector X_K by the introduction of a demand reference input vector X_d. Matrix $\underline{Q}$ is then chosen to minimise the square of the error between the reference states and the required system states, or alternatively if overall system stability is the primary objective then $\underline{Q}$ may act to weight each individual state's relative contribution in maintaining an acceptable stability level.

Weighting matrix $\underline{R}$ is chosen so that the relative importance of the state variables performance is contrasted with the importance of the expenditure of the system energy. For this reason the choice of matrix $\underline{R}$ can only be determined from a close examination of the controllers and actuators performance in maintaining an acceptable minimum variance, whilst minimising energy expenditure. As no wholly satisfactory theory exists for determining the weighting of the elements of matrix R, their values can only be obtained on a trial and error basis. However, experience has shown that the element values are generally in the range $1 < \underline{R} > 0$. If there are no input constraints then the elements of matrix R equal zero and equation (5) takes the more simplified form of a minimum variance controller.

OBJECTIVES FOR CONTROL

In the commercial process, yeast is normally grown in large aerated fermentors with a glucose or sucrose feed added to give roughly an exponential cell growth. The process is normally timed for completion in 15-18 hours when harvesting takes place and the process efficiency can be assessed. The optimum process performance being obtained when the yield coefficient equals or approaches 0.5, and a high specific growth rate usually ($\mu = 0.2$) can be maintained to provide maximum production.

On this basis the objective for control is to produce the maximum quantity of yeast with the minimum of feed, and in the minimum of time.

Prior to the application of the adaptive control package attempts were made to tune a three-term controller to regulate RQ at a set point of 1.05. A Zeigler and Nichols (5) open loop tuning procedure was investigated by applying a step input in nutrient feed to the process at the rate of 20 gm/hr. The RQ response is illustrated in Fig. 4 from which values for the three-term controller constants were estimated using the relationships

$$Kc = \frac{1.2M}{NL} \; ; \quad Ti = 2L \quad \text{and} \quad Td = 0.5L$$

where Kc is the proportional gain
Ti is the integral action time
Td is the derivative action time
and the controller transfer function is

$$Kc\left[1 + \frac{1}{Tis} + Tds\right]$$

The performance of the tuned three-term algorithm is illustrated in Fig. 5, where the process variable RQ is controlled to a setpoint of 1.05. Although RQ was maintained within 35% of its setpoint value the cellular growth was not optimum due to the large oscillations in the actuating signal.

Clearly, many results would be necessary to ascertain reasonable values for the tuning constants. The additional problem of process non-linearity would necessitate the tuning constants being continuously updated throughout the complete fermentation period which in essence is the basis of an adapive controller.

On-Line Operation of the Adaptive Controller

There are several decisions to be made when initialising an adaptive controller for optimal on-line performance.

1. The sampling time. This is generally the most critical parameter and is normally based on between 1/10 and 1/20 of the dominant process time constant.

2. The initial parameters $\phi(o)$ and $\Delta(o)$. The most popular procedure involves setting the estimated parameters to any initial value say 0.1 and Po to 10^5I where I is a unit matrix. However, to obtain reasonable control whilst the estimated parameters are allowed to converge a simple low gain proportional controller can be used.

3. For most practical applications the value of ρ is usually chosen by trial and error to provide a reasonable degree of parameter adaptation with a good noise filtering performance. However, if the approximate degree of parameter variation is known then a value of ρ can be obtained by considering the relationship

$$\alpha = \frac{1}{1-\rho} \quad \text{where } \alpha \text{ represents a data window}$$

over which the parameter estimates are assumed linear.

Analysis of the major process variables revealed that the dominant time constant (the variable DOT or dissolved oxygen) was approximately 2 hours giving a sampling time of between 12 and 6 minutes. However, as the remaining variables CO_2; O_2 and alcohol have significantly shorter time constants ranging from between 5 and 2 minutes an overall sampling time of 5 minutes was chosen.

Table 1 shows the hourly values of the estimated parameters $\underline{\phi}$ and $\underline{\hat{\Delta}}$ when the process is exponentially fed in accordance with the relationship:-

$$SFR = 2\,\mu\, X_o\, e^{\mu t}$$

Where μ is the specific growth rate (hr^{-1})
X_o is the initial value of cell concentration (g/lit)
t is the time (hr)

The degree of non-linearity can clearly be seen by the changing parameter estimates as the process evolves. A detailed analysis of the estimated parameters revealed that the process remained reasonably linear over a period of 50 minutes ie. 10 sampling intervals giving a value for the forgetting factor (ρ) of 0.9.

In addition for successful control it is essential that the estimated parameters accurately represent the dynamics of the process over each sampling interval. If the structure of the estimator does not include all the process dynamics then the estimates will not be a true representation of the process and adaptive control would not be successful. To illustrate the importance of this requirement the normal operating level of foam within the fermenter was controlled by excessive inputs of the anti-foaming agent Polypropylene Glycol. The reaction resulted in the dynamics of the oxygen transfer rate and subsequently the carbon dioxide evolution rate changing. Hence, two important output variables were being affected by an input variable that was not included in the estimation structure.

Fig. 6 and 7 demonstrate the performance of the parameter estimator with and without excessive Polypropylene Glycol additions. The divergence between the estimated and measured process variables can clearly be seen. This is further demonstrated in Fig. 8 which shows the performance of the adaptive technique without the correct process representation.

In the case of RQ control, the estimated value is much greater than the actual value Fig. 9. Hence the adaptive controller is given incorrect data, leading to a low driving signal which is insufficient to maintain RQ at its set point value. However, when RQ is accurately represented by the parameter estimator adaptive sub-optimal control can be achieved, Fig. 10.

Conclusions

On-line experiments with the microprocessor based adaptive control algorithm show that the process can be sub-optimally controlled. However, it is essential that care is taken in the design of the parameter estimator to ensure that an accurate structure of the process is available for the adaptive algorithm to refine.

Further work is required to investigate the behaviour of the covariance matrix over long time periods and the possibility of numerical instability due to the limited numerical accuracy of using 8 bit microprocessors. Following the commissioning of the adaptive system upon the commercial plant, it is expected to uncover more specilised interpretations of the generalized analytical approach to the control of fermentation processes.

ACKNOWLEDGEMENTS

The authors gratefully acknowledge the financial support given by the UK Science and Engineering Research Council and APV International Ltd. Also to M. Senior in the preparation of the Figures for this paper.

REFERENCES

Humphrey, A.E. (1973). The first European conference on computer process control in fermentations, Dijon.

Sandoz, D.J. and Swanick, B.H. (1970). Real-time hybrid simulation of an adaptive control technique. Proc. IEE, Vol. 117, No. 11. (Nov.).

Waterworth, G. and Swanick, B.H. (1981). Investigation into the adaptive control of a fermentation process. Trans. Inst. M.C., Vol. 3, No. 1. (Jan-Mar).

Williams, D. and Yousefpour, P. (1984). On-line adaptive control of a fermentation process. Proc. I.E.E., Vol. 131, No. 4. (July).

Ziegler, J.G. and Nichols, N.B. (1942). Optimum setting for automatic controllers. Trans A.S.M.E., Vol. 64, page 759.

TIME (hr)	$\phi 11$	$\phi 12$	$\phi 13$	$\phi 14$	$\Delta 11$	$\phi 21$	$\phi 22$	$\phi 23$	$\phi 24$	$\Delta 21$
0	0.1	0.1	0.1	0.1	0.1	0.1	0.1	0.1	0.1	0.1
1	1.07	0.13	-0.16	0.000	-0.30	0.004	0.03	0.02	0.000	0.007
2	1.15	0.09	-0.06	0.000	-0.37	0.005	0.006	0.005	0.000	0.008
3	1.21	0.15	-0.18	0.000	-0.39	0.009	0.005	0.009	0.000	0.006
4	1.28	0.22	-0.15	0.000	-0.40	0.017	0.013	0.017	0.000	0.005
5	1.68	0.24	-0.19	0.007	-0.34	0.029	0.022	0.029	-0.007	0.009
6	2.21	0.32	-0.22	0.13	-0.31	0.126	0.134	0.117	-0.03	0.02
7	2.98	0.30	-0.38	0.09	-0.26	0.487	0.280	0.237	-0.04	0.021

TABLE 1. Estimated process Parameters

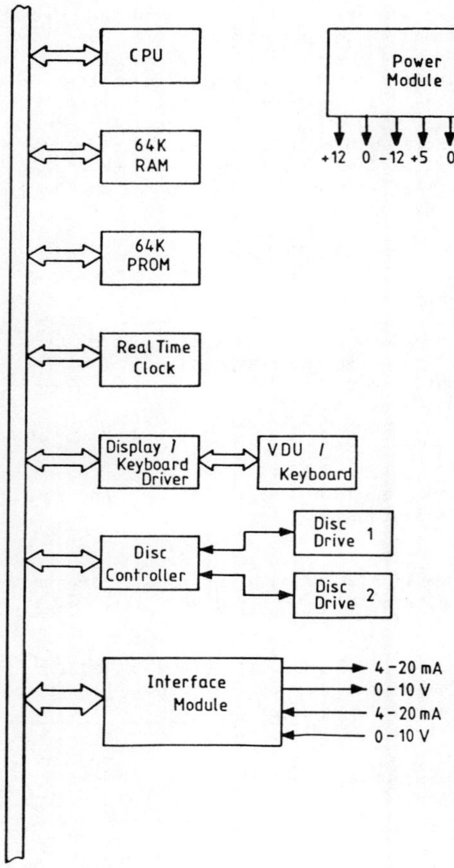

Fig.1. Schematic diagram of microprocessor system.

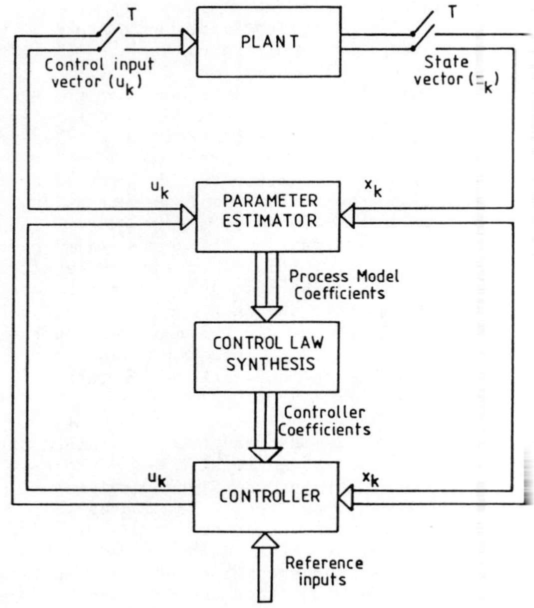

Fig.3. Structure of adaptive control seq■enc■.

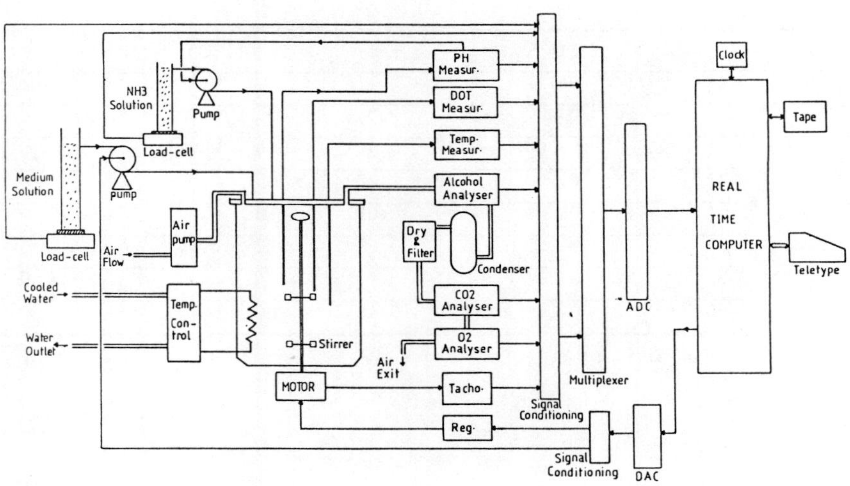

Fig.2. Schematic diagram of computer coupled fermentation process for adaptive control.

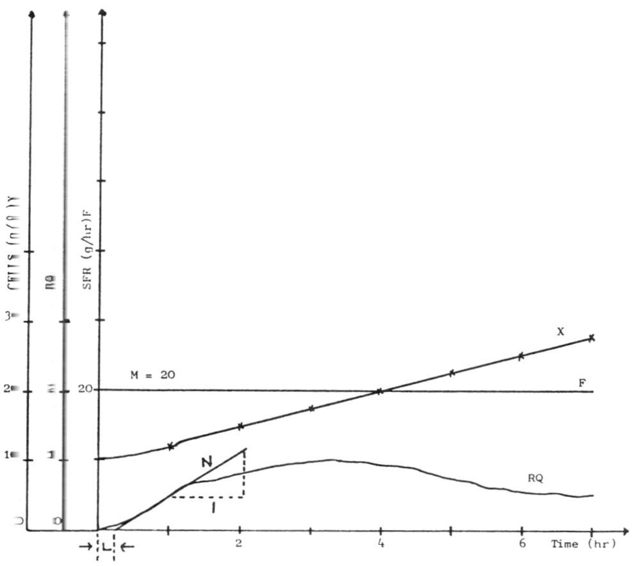

Fig.4. Open-loop fed-batch fermentation showing RQ response for a constant feed.

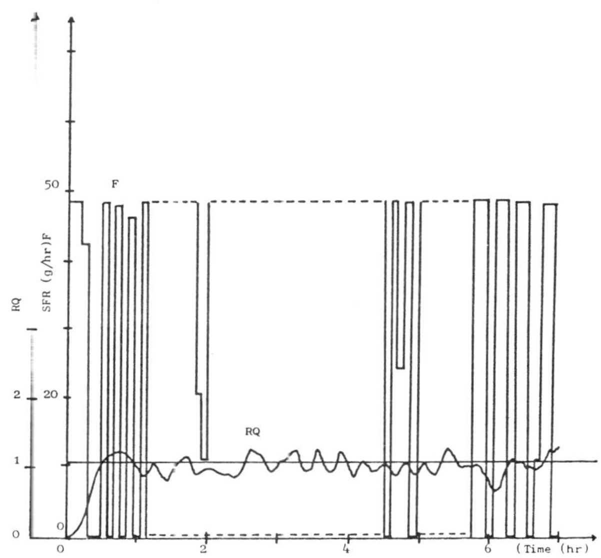

Fig.5. Proportional control of RQ in a fed-batch fermentation.

Fig.6. Comparison of estimated and actual process variables with correct representation of the fed-batch process.

MULTILOOP CONTROLLED FED-BATCH FERMENTATION PROCESS USING A SELFTUNING CONTROLLER

H. B. Verbruggen*, G. H. B. Eelderink** and P. M. van den Broecke**

*Delft University of Technology, Department of Electrical Engineering, Control Laboratory, PO Box 5031, 2600 GA Delft, The Netherlands
**Gist-Brocades NV, PO Box 1, 2600 MA Delft, The Netherlands

Abstract. A fed-batch fermentation process is controlled by a multiloop control structure, which contains a selftuning controller in order to meet the demands of a process with time-varying parameters and noisy conditions. It is shown that a biochemical multi-input multi-output process model can be rearranged into a simple ARMAX structured model. This model proved to be accurate enough to predict the biomass output within a few percent. The selftuning controller is based on the pole-assignment principle, and provided at the same time the opportunity to verify the estimated process parameters with those calculated from the biochemical modelling and from test results under ideal conditions. The set point used will increase steadily because of its dependence on the internal state of the process. This leads to the design of a multiloop control structure, where the selftuner takes care of a well-defined inner loop and a conventional controller is part of the outer loop. Some experiments carried out on a 10 litre vessel proved that the designed multiloop structure behaved rather well. The results of these experiments will be presented in the paper. This work has been performed as part of a close cooperation between the Delft University of Technology and Gist-Brocades, a fermentation company in The Netherlands.

Keywords. adaptive control, biocontrol, cascade control, direct digital control, modelling, parameter estimation.

INTRODUCTION

A cooperative study between the Delft University of Technology and the fermentation company Gist-Brocades NV has been carried out on the modelling and control of a baker's yeast fermentation process. In this paper attention will be focussed on the control of the process. Basically, the fermentation process is a multi-input multi-output fed-batch process. The modelling started with an existing model, which was developed from the biochemical point of view. Moreover, the dynamic behaviour of the process due to the physical dimensions of the fermentor and the configuration of the measurement equipment was derived under ideal conditions. Simplifications and omissions which were allowed under certain circumstances have resulted in a model that was suitable for the controller designed by Visser (1983). The main goal which the controller has to serve can be formulated as follows:
The process must be controlled such that in a minimal amount of time the maximal amount of biomass (baker's yeast) is produced using a minimal amount of raw material (molasses and air), in spite of disturbances, a not-completely- known initial state of the process and an unknown time-varying process behaviour. In order to reach this goal it was decided to use the simple selftuning controller as described by Lammers (1982). This controller, which is based on the indirect self-tuning pole-assignment principle, has the following advantages:

- It is adaptive with respect to the varying process parameters.
- The system behaviour with respect to noise reduction can be independently chosen from the system behaviour with regard to varying set points.
- Having a separation between the parameter estimation and the adjustment of the controller settings provides the opportunity to freely choose the type of controller. This approach will also bring forth the possibility to verify the theoretical model parameters a posteriori with those achieved by estimation.
- Last but not least, the assignment of poles in regions where this would lead to instable or oscillatory controller behaviour can be prevented.

The results of two fed-batch cultures are shown and discussed.

PROCESS DESCRIPTION AND PROCESS MODEL

The fermentor on which the experiments are carried out is a vessel with a maximal content of 10 litres. The process itself is a fed-batch culture, which means that the fermentation is started with a medium and seed yeast. During the fermentation process the substrate (molasses) and air, is fed into the vessel. This yields a time limitation on the fed-batch process which depends on the maximal contents and input flows. Other limitations are set by the maximal aeration capacity and the cooling capacity of the vessel.
The inputs and outputs of the biochemical model description are indicated in Fig. 1. The base flow and acid flow are used to control the pH; the anti-foam addition is necessary to prevent the aerated wort from rising above a distinct level; the sugar flow, ammonia flow and oxygen flow are the ingredients on which the biomass grows. The internal state of the process is described by pH, temperature, ammonia concentration, amount of biomass and amount of alcohol.
The measured outputs during the fermentation are the concentration of CO_2 and O_2, both determined in the outlet gas flow.

It is assumed that the pH and the temperature are well controlled. The ammonia feed is not considered in this paper. These assumptions allow us to simplify the biochemical model to the model with a structure as indicated in Fig. 2.

This model consists of two parts:

- a part describing the fermentor dynamics, related to the dimensions and operating conditions of the fermentor and to the measurement set-up. Parameters describing the dynamics of this part of the model will be called 'external parameters'. The outputs of this part of the model are defined as the oxygen uptake rate (OUR) and the carbon dioxide production rate (CPR). These outputs are calculated from the concentration measurements of (CO_2) and (O_2) and the gas flow in Fig. 1.
- a part describing the internal behaviour of the fermentor with fictitious outputs OUR_{cell} and CPR_{cell}, describing the OUR and CPR at the cell level.

The quotient of the outputs, defined as

$$RQ = \frac{CPR}{OUR} ,$$

is called the 'respiratory quotient'. This RQ indicates the behaviour of the process according to biomass growth and alcohol production. The static relationship between the inputs (oxygen and sugar) and RQ_{cell} (representing the internal behaviour) is indicated in Fig. 3.

In this figure the origin O, the intersection of the two axes, indicates that there is just enough sugar and oxygen fed into the fermentor in relation to the existing biomass in order to maintain maximal growth. The unshaded area indicates that there is a lack of sugar, forcing the biomass to consume alcohol, if present, in addition to sugar. The growth rate will therefore diminish. In this area the RQ will be smaller than the optimal value of RQ for maximal growth.
The shaded area indicates that alcohol production takes place. This may be due to a lack of oxygen (left part of the shaded area) which will also cause a smaller growth rate.
The upper right quadrant of the plane indicates the area where the biomass growth rate is maximum. However, the amount of sugar fed into the vessel cannot be used completely for growing. The extra amount of sugar will therefore be converted by the biomass into alcohol.

From this description it may be clear that the internal behaviour is indicated by the value of RQ_{cell}. However, when we want to use RQ_{cell} as the controlled variable, we must be aware of the fact that RQ_{cell} is well defined only in the right half of the plane with respect to the sugar feed. This means that the situation in which there is a lack of oxygen must be avoided. From the control point of view maximal growth has to be maintained. This also means that there must be no lack of oxygen. Therefore, the gas flow is controlled separately, by a conventional controller, in relation to the sugar flow, such that the process will remain in the right half of the plane. The theoretical modelling carried out under the above-mentioned conditions showed that the sugar flow does not influence the OUR_{cell} signal.
Moreover, the influence of variations in the gas flow (Fig. 4) can be taken into account when we consider the OUR as a disturbance upon the CPR signal.
After determining the existing relationships in the internal part of the model a simple dynamic relationship of the external part of the model is derived. As stated by Levenspel (1972) the transient response of a fermentor in relation to a set-point change of the concentration of CO_2 can be approximately described by one major time constant $\tau = V/\phi_g$ in which V is the fermentor volume.
Moreover, transportation lags of approximately 1 to 2 minutes can be calculated, due to the physical arrangement of the measuring equipment and the gas flow.

This simple model is verified under ideal test conditions (pH in the range of the real process conditions). It was found that the amount of foam, the speed of the stirrer and the amount of liquid in the fermentor hardly influenced the dynamic behaviour. The transportation lags were found to be nearly constant and independent of the value of ϕ_g, which could be expected given the arrangement of the measurement set-up. Part of the time delay is also caused by the effect of small time constants which were neglected.
The above analysis and tests resulted in a simple model of the fermentor described by a varying time constant of 3 to 10 minutes (depending on the value of ϕ_g and transportation lags of 1 minute and 2 minutes for the measurement of the concentration of CO_2 and O_2 respectively.

The above considerations and verifications allow us to set up the complete model of the process for control design purposes.
The internal part of the model provides us information about the main static influences of ϕ_s and ϕ_g on OUR_{cell} and CPR_{cell}. Simulations with parameters calculated from the biochemical model proved this model to be accurate within 5% of the responses found by various real fermentations with respect to predicting the amount of biomass and the growth rate. Time constants in this model were in the order of 75 hrs. The d.c. gain of the relation CPR/ϕ_s will vary over a wide range and is one of the main varying parameters during fermentation. By rearranging the structure of Fig. 4 the simple single-input single-output model of Fig. 5 comes into being, where the OUR signal is a measurable disturbance (because of the nearly static relationship between ϕ_g and OUR). The relationship between CPR and ϕ_g is described by fast as well as slow dynamics and a changing gain factor.

CONTROL STRATEGY

The objectives which the control must reach can be stated as follows:

- Maximize the biomass output as a function of time;
- Minimize the required amount of sugar and air;
- Achieve these objectives despite uncertain (varying) conditions using the RQ signal as the controlled variable.

Furthermore, the controller has to deal with a molasses feed with an inaccurately known sugar concentration. This concentration may even vary during the fermentation process. Moreover, the process parameters may vary due to the increase of biomass and flows. Also, the condition of the seed yeast and its exact amount is unknown at the start of the fermentation.

In order to deal with these uncertainties it was decided to use an adaptive controller, as already stated in the introduction. The controller needs an ARMA-structured model for estimation purposes. This leads to the model structure in Fig. 6, which is the result of the theoretical modelling with:

ϕ_s= the molasses flow

OUR = the measured oxygen uptake rate used as a disturbance

W = a possible constant offset,

and where e(k) is supposed to be zero-mean white noise.

It is assumed that the process can be described sufficiently well by a second order model. The polynomials are therefore of the form:

$$A = 1 + a_1z^{-1} + a_2z^{-2}$$

$$B = b_0 + b_1z^{-1}$$

$$C = 1 + c_1z^{-1} + c_2z^{-2}$$

$$D = d_0 + d_1z^{-1}$$

where z^{-1} is the backward shift operator.

For estimation purposes we define the following vectors:

$$\hat{\theta}^T = \{\hat{a}_1,\hat{a}_2,\hat{b}_0,\hat{b}_1,\hat{c}_1,\hat{c}_2,\hat{d}_0,\hat{d}_1,\hat{w}\}$$

$$\psi^T = \{-y(t-1),-y(t-2),u(t-1),u(t-2), n(t-1),n(t-2),q(t-1),q(t-2),1\}$$

where y equals the CPR signal, u equals ϕ_s and q equals the OUR signal.
The values of the parameters of the various polynomials are strongly influenced by the sampling period T_s chosen during modelling and control experiments. Because in all experiments T_s was chosen larger than the transportation lags of the fermentor, and because of the implementation of the control algorithm in a digital computer, a time lag of one sampling period could be chosen.

Following the description given in Visser (1982) this results in the set of recursive estimation equations using λ as a forgetting factor.

$$\varepsilon(t) = y(t) - \psi^T(t)\,\hat{\theta}(t-1)$$

$$\hat{\theta}(t) = \hat{\theta}(t-1) + P(t)\,\psi(t)\,\varepsilon(t)$$

$$P(t) = \frac{1}{\lambda}\left\{P(t-1) - \frac{P(t-1)\psi(t)\psi^T(t)P(t-1)}{\lambda + \psi^T(t)P(t-1)\psi(t)}\right\}$$

$$n(t) = \hat{e}(t) = \frac{\lambda}{\lambda+\psi^T(t)P(t-1)\psi(t)}\,\varepsilon(t)$$

The control structure is shown in Fig. 7 in which R and S follow from AR + z^{-1}BS = B(C+μA). The parameter μ can be tuned to achieve a reasonable noise reduction against an allowed variation in the control action ϕ_s. The polynomials R and S which define the loop behaviour are of the form:

$$R = r_0 + r_1z^{-1}$$

$$S = s_0 + s_1z^{-1}$$

Now we are able to estimate the parameters for a model describing the relationship between ϕ_s (molasses flow) and CPR (carbon dioxide production rate).

The coefficients of the controller follow from the estimated A, B and C coefficients by:

$$R : r_0 = (1 + \mu)b_0 \quad ; \; r_1 = (1+\mu)b_1$$

$$S : s_0 = c_1 - a_1 \quad ; \; s_1 = c_2 - a_2.$$

The set-point behaviour is taken care of by the transfer function T(z). Therefore T(z) is defined as $T = T_m.T_c$, in which T_m is chosen as the transfer function describing the desired set-point behaviour and T_c as a compensation network (prefilter) in order to get an overall transfer function from CPR_{set} to CPR as $z^{-1}T_m$. This can be realised with the given control strategy by choosing
T_c= C + μ A.
In case T_c may be taken equal to one

$$T = T_m = \frac{1+p_1+p_2}{1+p_1z^{-1}+p_2z^{-2}}$$

in which p_1 and p_2 follow from

$$p_1 = -2e^{-\zeta\omega T_s}\cos\omega T_s\sqrt{1-\zeta^2}$$

$$p_2 = e^{-2\zeta\omega T_s}$$

In this way, after a stepwise set-point change the system will follow rather well a second-order step response with a damping ratio ζ and a natural frequency ω_n. This holds in case $2\pi/T_s$ is sufficiently small, compared to the natural frequency. This basic controller is extended with provisions as described by Lammers (1982, 1984) for the following situations:

- No cancellation of the process zero is allowed. This provision is used when the following conditions do not hold:

$$-\frac{b_1}{b_2} < -0.1 \quad \text{or} \quad -\frac{b_1}{b_2} > 0.9$$

- The process appears to be a first-order process. The selection criterion is

$$N \leq 0.01\,\{\max(b^2_0,\ b^2_1)\}\ \text{with}$$

$$N = b^2_1 - a_1b_0b_1 + a_2b^2_0$$

- The process is noise-free, which is detected by a large variance in the c-parameters.

For compensation of the measurable disturbance OUR the controller is extended with the U polynomial. U can be found by solving

$$(U.B + D.R)z^{-1}.OUR=0,\ \text{when B and R are known.}$$

This leads to the following settings of U:

$$u_0 = -(1+\mu)d_0$$

$$u_1 = -(1+\mu)d_1$$

From Fig. 3 we learn that the RQ signal indicates the internal behaviour of the process. With this knowledge we can calculate from the measured OUR signal and a chosen RQ set point a set point for the CPR by $CPR_{set} = RQ_{set}$ x OUR. It must be taken into account that the measured OUR signal follows from events in the past. In addition, we must be aware of the fact that the relationship in Fig. 3 is a static one normalised to the existing biomass. In order to maintain a maximal growth rate, in the light of the above remarks, the RQ set point should be chosen higher than the optimal RQ.
In order to realise a maximum yield of biomass the growth must not be limited by an insufficient supply of sugar. That requirement results in a straight line for RQ_{set} parallel to the x-axis in the first quadrant of Fig. 3.

By applying the controller described above the following results can be expected:
Part T_c of the transfer function T together with the choice of R and S will take care of an overall dc gain of 1 for stepwise set-point changes. Part T_m of T describes the dynamic behaviour upon stepwise set point changes.
These results are only valid in case the orders of the polynomials A, B and C are correctly chosen a priori and after the parameters of these polynomials converge to the right values, which can take a long time especially for the parameters of C.

Moreover, the effectiveness of the control algorithm of Fig. 7 depends on an exact, updated knowledge of the value of W in order to compensate for a possible offset. It can be shown that X has to be adjusted to

$$X = -\frac{R(1)}{B(1)}\hat{W}$$

with R(1) and B(1) as the values of R and B for z=1 respectively. Consequently, besides having to have a good estimate of W,B(1) also has to be known exactly. An additional problem is that the influence of OUR on CPR is compensated for by a feedforward control action, when U is chosen as a function of D. This also requires an accurate knowledge of the order and the parameters of D.

The majority of the problems indicated above can be solved by one of the following methods (Verbruggen, Lammers and Van den Bosch, 1985):

- a forced integral action in the feedforward path of the controller;
- a cascade configuration (multiloop controller) in which the selftuning control algorithm is part of the inner loop and an integrating controller is part of the outer loop.

We have chosen the second method because this configuration is also favourable for handling the problem introduced by the time varying set point CPR_{set}. Because $CPR_{set} = RQ_{set}.OUR$ (with RQ_{set} constant and OUR increasing in time during fermentation), the value of CPR_{set} will steadily increase due to the growing biomass. The final control structure is given in Fig. 8, where the cascade controller is indicated together with a feedforward path with transfer function H_m.

For stepwise set-point changes it is shown in (Verbruggen, Lammers and Van den Bosch, 1985) that $H_m = T_m z^{-1}$ is the optimal choice, resulting in a step response as fast as that obtained by a control structure as given in Fig. 7, however, with the benefits of the cascade controller configuration (elimination of feedforward and offset errors).

In order to be able to follow a steadily increasing set point (ramp function) the above choice of H_m is not optimal because in the steady state CPR will always lag behind CPR_{set}. A better choice is to take $H_m = 1$. In that case it can be shown that

$E = CPR_{set} - CPR$ can be written as:

$$E = \frac{(1-z^{-1})(1-T_m z^{-1})}{1-z^{-1}+T_m z^{-1}} CPR_{set}$$

By applying the final value theorem, given $OUR_{set} = a+bt$, it can be shown that

$$e_{ss} = \lim_{n\to\infty} e(nT_s) = 0, \text{ with } \lim_{z\to 1} T_m = 1.$$

Defining CPR_{set} by the measured OUR provides one with the opportunity to avoid compensation via the U polynomial.

RESULT AND DISCUSSION

Two fed-batch cultures are taken as an example. In both cases the settings of the controller part were chosen as follows:

For the part of T describing the set-point behaviour.

$$\zeta = 0.7,\ \omega_n = 13.5\ 1/\text{hrs},\ T_s = 2\ \text{min}$$

For the pole assignment part:

$$\mu = 2$$

For the estimator:

$$\lambda = 0.98$$

The set point of RQ is chosen to be 5% higher than the theoretical optimal value of RQ.

Culture I has been controlled by the single-loop control structure shown in Fig. 7, and culture II by the multiloop cascaded control structure shown in Fig. 8.

The parameter estimation in both situations behaves rather well, as can be seen in Fig. 9a, where the behaviour of the parameters of the A-polynomial is shown. The dotted lines represent the parameters for culture II and the solid lines the parameters for culture I. It could be expected that the parameters of both experiments do not vary in exactly the same way due to differences in process conditions. The trend in time is, however, similar. The same can be said about the parameters of the B-polynomial.

The estimation results show that over the fermentation time constants are found in the range of 8 to 3 minutes. These values could be expected from the modelling of the process.

In open-loop experiments the large time constants in the range of 75 hrs were found using a larger sampling period.

From Fig. 9b it is clear that the multiloop controller performs very well in controlling RQ near RQ_{set}, while the single-loop controller yields values of RQ beneath the theoretically optimal value of RQ, resulting in a lower production of biomass.

The reasons for this poor behaviour are twofold:

- CPR_{set} does not change stepwise, but is calculated from the measured OUR-signal. Because of the even more than linear increase of the theoretical value of OUR over the fermentation (see Fig. 9c) the delay of CPR will increase over the fermentation.
- The chosen value of ω_n results in a rather slow system performance.

A considerable improvement is found by introducing the multiloop structure which takes into account the time-varying character of CPR_{set}. Further improvement can be obtained by increasing the value of $_n$ and at the same time decreasing the value of T_s.

Moreover, by introducing an extra integration in the outer loop even a quadratic increase in the time of CPR_{set} could be followed, although this would at the same time introduce stabilising problems in the control loop.

CONCLUSIONS

It has been shown that, despite the complex relationship of the various parameters and variables in the fermentation process, a relatively simple process model can be derived for control purposes. This model consists of a part related to the biochemical process on the cell level and a part related to the fermenter and the measurement set-up.

The model parameters were verified by various test results of real fermentations.

It was observed from the results of culture II that an adaptive control structure for controlling this process could be favorable given the changing static and dynamic parameters of the process during fermentation.

The simple selftuning controller proved to work satisfactorily for this kind of process, especially when extended with an extra control loop.

Based on the results of the parameter estimation further research will be carried out upon the model structure, which can be used for the implementation of various controller structures and algorithms.

REFERENCES

Visser, C.A. (1982). **Modeling and optimal control of baker's yeast production in a repeated batch-fed culture,** M.Sc. Thesis A83.002(305), Delft University of Technology, Dept. of Electrical

Engineering, Control Laboratory, (Dutch)

Lammers, H.C. (1982). A simple self-tuning controller. **Proceedings IEEE Conference on Application of Adaptive and Multivariable Control,** Hull, England, 19-21 July 1982, pp. 90-95

Levenspel, O. (1972). **Chemical Reaction Engineering,** 2nd ed., John Wiley & Sons, Inc. New York

Lammers, H.C. (1984). **Towards the practical application of self-tuning adaptive controllers,** Ph.D. Thesis, Delft University of Technology, pp. 287

Verbruggen, H.B., H.C. Lammers and P.P.J. van den Bosch (1985). A simple selftuning adaptive controller, **Proceedings Fourth Yale Workshop on Application of adaptive systems theory,** New Haven, Conn., USA

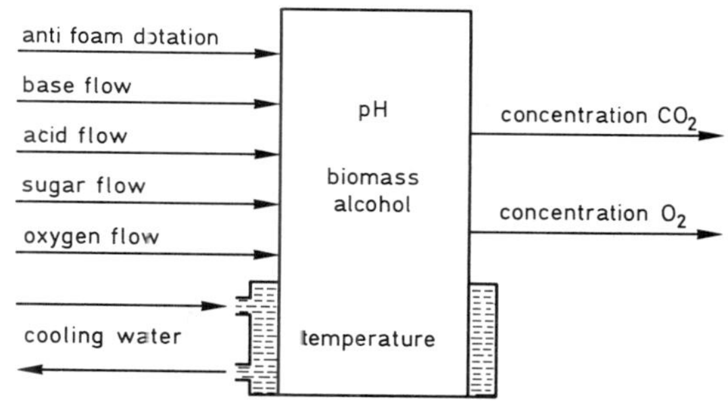

Fig. 1. Inputs, outputs and internal state of the process

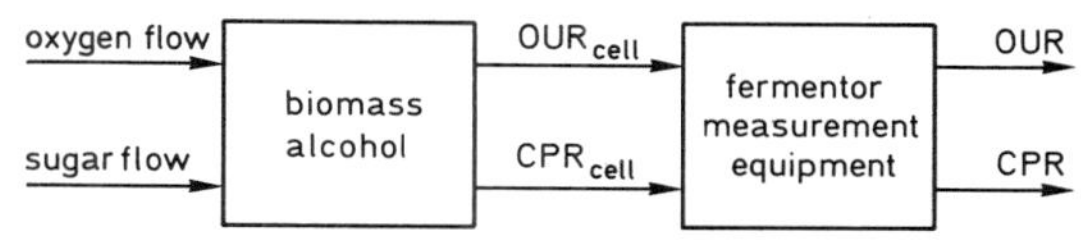

Fig. 2. Control engineering model split up

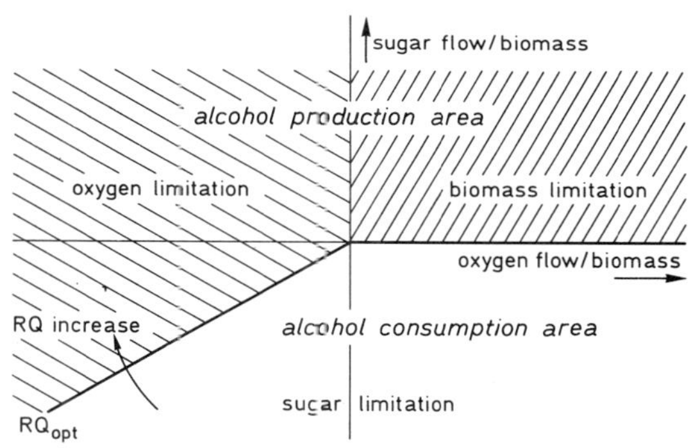

Fig. 3. Static process relations

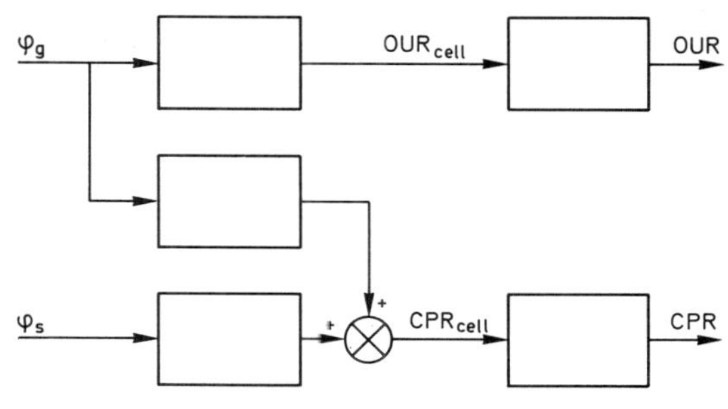

Fig. 4. Control engineering model

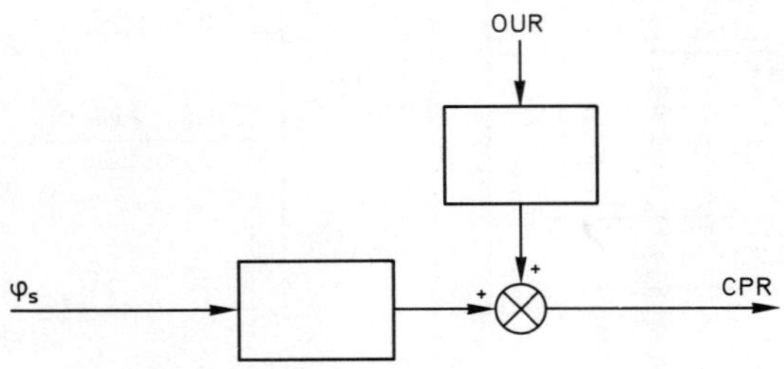

Fig. 5. Restructured model

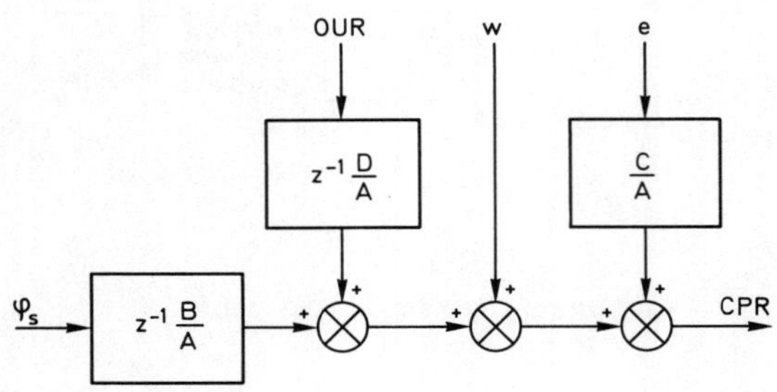

Fig. 6. ARMA structured model

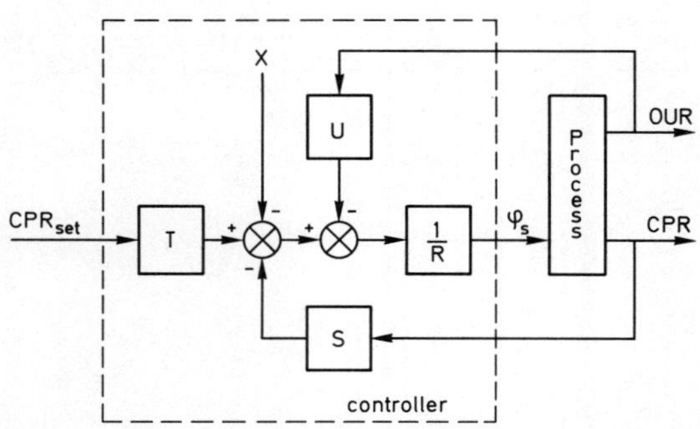

Fig. 7. Single-loop controller structure

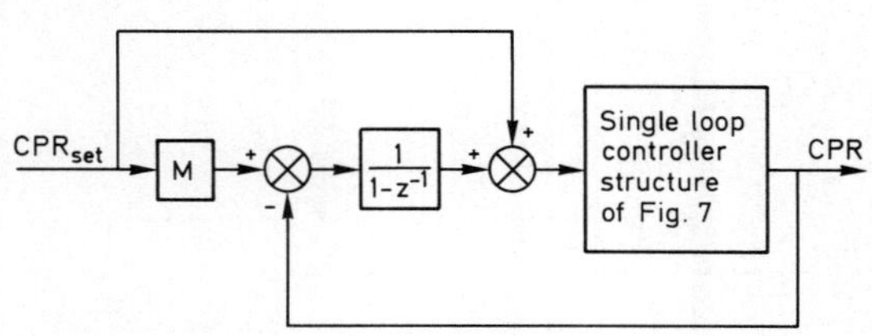

Fig. 8. Multi-loop controller structure

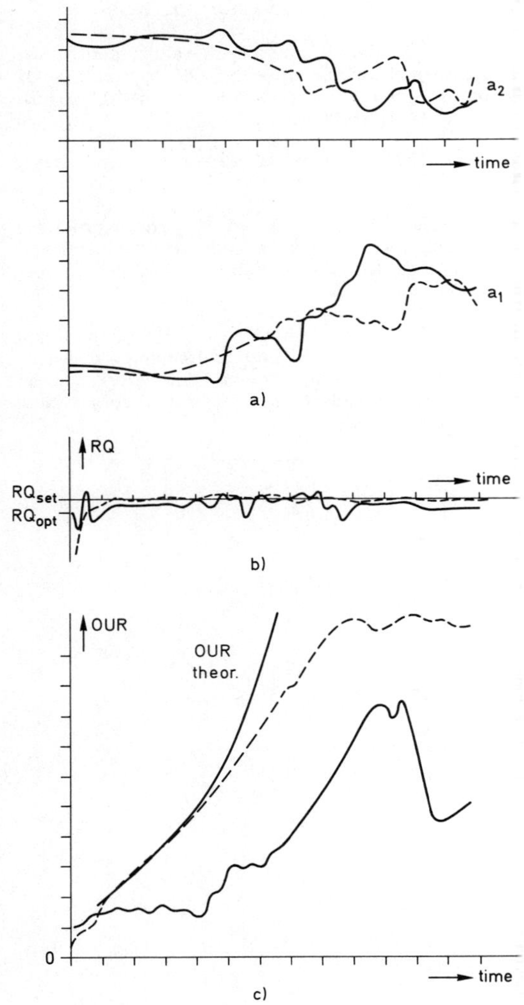

Fig. 9. Results of experiments (drawn lines: culture I, broken lines: culture II)
a. Parameter estimation
b. RQ responses
c. OUR responses

SELF-CORRECTING MEASUREMENT OF DISSOLVED OXYGEN

R. Kässer

Department of Engineering, Section for Automation, Sandoz AG, Basel, Switzerland

Abstract. The usual method of monitoring oxygen content during biotechnological reactions is by means of a pO_2 electrode, as described by Clark (1953). The accuracy of this technique, however, is often severely diminished by contamination in or on the membrane. A new method has now been developed for the continuous monitoring of the electrode sensitivity. It relies on periodic pressure changes of known magnitude yielding a well-defined reference signal in the culture broth. The new method has been confirmed in tests performed with living cultures in laboratory fermenters.

Keywords. Sensors; dissolved oxygen measurement; parameter estimation; biocontrol; error compensation.

INTRODUCTION

In biotechnology, measurement of dissolved oxygen and the maintenance of a specified oxygen concentration are one of the most important parameters for ensuring optimum conditions for microorganisms and their productivity. Accurate measurement of the oxygen concentration is conducive to improved reproducibility in experiments or production runs and provides important information for special events.

Oxygen content is usually measured by means of a pO_2 electrode originally described by Clark (1953). The accuracy of this method depends on various technological and structural parameters (Bühler, 1976; Krebs, 1975; Lucero, 1968) which affect flow dependence, response time, long-term stability and mechanical strength in different ways. A further factor that can have a relatively great effect on the accuracy is the contamination of the pO_2 electrode that can occur later in the course of a fermentation.

A new method has been developed which makes it possible to compensate substantially for the contamination of the pO_2 electrode. This consists in applying a constant, periodic change to the oxygen partial pressure in the culture broth, so that the sensitivity of the pO_2 electrode can be followed and the measured value can be corrected appropriately.

THE THEORY OF THE pO_2 ELECTRODE

The Clark pO_2 electrode operates on the amperometric principle. The oxygen reaching the cathode through the membrane is reduced selectively, and the resulting electric current is a measure of the oxygen content of the material under measurement.

Cathode reaction: $O_2 + 2H_2O + 4e^- \rightarrow 4\ OH^-$

Anode reaction : $4\ Ag + 4\ Cl^- \rightarrow 4\ AgCl + 4e^-$

The fundamental physical process in the diffusion of oxygen from the material under measurement through the membrane to the cathode (see Fig. 1). On a onedimensional basis, the diffusion current is given by

$$q = A\ D\ \frac{C_m - C_o}{x} \qquad (1)$$

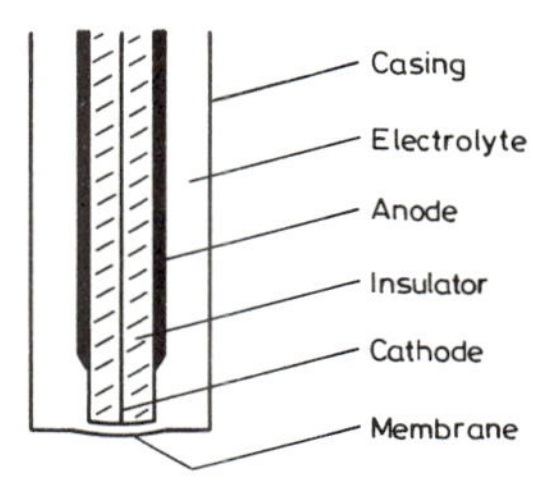

Fig. 1. Principle of the construction of a pO_2 electrode.

For an ideal pO_2 electrode, where the oxygen is immediately reduced at the cathode, the following condition applies:

$$C_o = 0 \qquad (2)$$

A further physical law which is fundamental for the functioning of the electrode is Henry's law of absorption. This gives the relationship between the oxygen concentration and the oxygen partial pressure.

$$C = p\ s \qquad (3)$$

Henry's law states that the concentration of gas absorbed by a liquid (or by the membrane material) is proportional to the partial pressure of the gas over this liquid and to its solubility in the material concerned.

For an ideal pO_2 electrode, the oxygen concentration at the surface of the membrane C_m obeys Henry's law, as long as equilibrium state is assumed. This state is virtually reached, since the electrode absorbs only very little oxygen. Real pO_2 electrodes are moreover affected by diffusion and convection processes in the liquid, which result in a certain flow dependence. In the ideal case, Eq. (3) leads to the following equation for the membrane

$$C_m = p\ s_m \tag{4}$$

which, combined with Eq. (1) and Eq. (2) yields

$$q = A\ D\ s_m \frac{p}{x} \tag{5}$$

The most important conclusion to be drawn from this equation is that the diffusion current q is proportional to the oxygen partial pressure p. All the other symbols in Eq. (5) are constant and depend only on the properties of the material and the design of the electrode. This shows that the value measured amperometrically is proportional to the oxygen partial pressure and not to the oxygen concentration. The term pO_2 electrode is therefore justified.

Temperature affects the behaviour of a pO_2 electrode by varying the diffusion constant and the solubility coefficient. The product of these two factors gives the permeability coefficient, which increases by about 2% per ^{o}C (Krebs, 1975; Lucero, 1968).

$$P = D\ s_m \tag{6}$$

Compensation for this dependence on temperature is usually provided by electronic means, but slow, i.e. within minutes.

THE OXYGEN PARTIAL PRESSURE

According to Dalton's law, the pressure in the gas phase is the sum of the various partial pressures. This also applies to the vapour pressures of liquids, the latter depending only on the temperature.

$$p_{tot} = \Sigma\ p_{vapour} + \Sigma\ p_{gases} \tag{7}$$

An alteration in pressure in a fermenter is wholly distributed over the partial pressures of the gases. Each partial pressure changes by the same percentage. If it is assumed that the partial pressure of the oxygen in the gas phase remains constant during the whole fermentation, then a periodic variation of constant amplitude in the pressure will produce a constant periodic variation in the oxygen partial pressure.

These partial pressure variations of constant amplitude can be used as a reference signal, provided that this signal is transmitted at the same amplitude from the gas phase to the liquid phase, i.e. the culture broth. This is true if there is enough time for the stationary state to be established, as shown in the next section.

THE OXYGEN PARTIAL PRESSURE CORRESPONDING TO THE OXYGEN IN THE CULTURE BROTH

The oxygen partial pressure measured in the culture broth is lower than in the gas phase, if oxygen is consumed by living microorganisms. Oxygen transfer from the gas phase to the culture broth is limited by the oxygen transfer coefficient and can be shown for the stationary case to depend on the oxygen concentration as follows

$$Q = k_L a\ (C_{sat} - C) \tag{8}$$

Applying Henry's law as given in Eq. (3) to the cases of saturation concentration and oxygen concentration in the culture broth, yields

$$C_{sat} = p\ s \tag{9}$$

$$C = p_b\ s \tag{10}$$

and Eq. (8) can hence be rewritten

$$Q = k_L a\ s\ (p - p_b) \tag{11}$$

An ideal uncontaminated pO_2 electrode would therefore measure the value p_b in the culture broth. As long as no oxygen is consumed by the microorganisms, Q = 0 and therefore $p_b = p$. This measured value corresponds to the partial pressure in the gas phase which is adjusted to 100% at the start of a fermentation. An increasing consumption of oxygen, yields a difference between p and p_b, leading to a correspondingly lower measured value for p_b.

The following conclusions can be drawn from equation (11): A change in the oxygen partial pressure in the gas phase causes the same change to the ideal measured value p_b, as long as the consumption of oxygen Q remains constant. Periodic changes in partial pressure of constant amplitude would therefore be transmitted to the culture broth.

In fact, the oxygen consumption Q of most microorganisms does not depend on the partial pressure, as long as the latter is above about 10% (Harrison, 1969; Weinfurtner, 1960). Less oxygen is only taken up below a value of 10%, as a result of limited viability. Hence, periodic pressure changes of constant amplitude in the fermenter will also produce a reference signal of constant amplitude in the culture broth during the whole fermentation, as long as partial pressure is above about 10%.

MEASUREMENT OF pO_2 WHEN THE MEMBRANE IS CONTAMINATED

The surface of the membrane of a pO_2 electrode can become more and more contaminated during a fermentation of prolonged duration. This results in a reduced accuracy of measurement, which may be broken down into two components as shown in Fig. 2. There are firstly, a reduced sensitivity of the electrode and secondly, a possible zero shift.

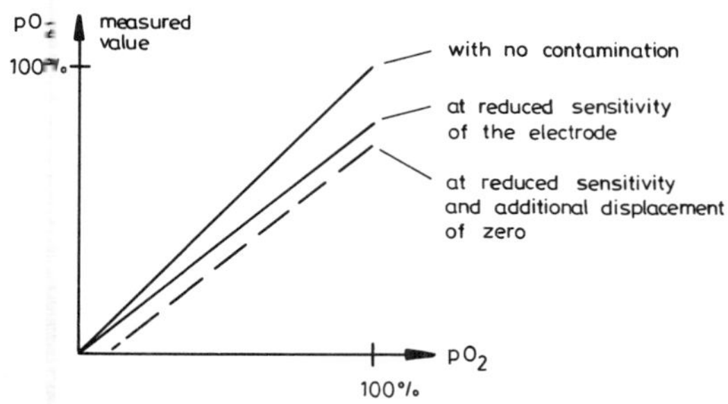

Fig. 2. Components in errors of measurement for a contaminated pO_2 electrode.

Lower sensitivity is caused by infiltration into the surface of the membrane which alters its diffusion constant or its capacity for dissolving oxygen, both factors affecting the permeability of the membrane. A shift in the zero point is caused by the infiltration of oxygen-consuming microorganisms into the membrane.

rience shows that this is relatively small in comparison with the first component of the errors in measurement. Both components result in a reduction of the measured signal and it is certainly advantageous to correct at least one of these.

EXPERIMENTAL PROCEDURE AND RESULTS

A 30-litre stainless steel stirred kettle fermenter was used for the test. The equipment includes controls for temperature, pH, pressure and aeration rate and also automatic addition of anti-foaming agents. A commercial Clark pO_2 electrode (Ingold AG model IL 235) was used for measuring the oxygen partial pressure. The test was carried out at a temperature of 30°C, an aeration rate of 16 l/minute and a pH $\geq$ 5.

The nutrient solution contained 0.75% of glucose, 2% of peptone and 2% of yeast extract, and its pH was adjusted to 5 with sodium hydroxide solution before sterilization. The anti-foaming agent used was Dow Corning's M30.

After sterilization (20 minutes) had been carried out at 120°C, the pO_2 electrode was calibrated to 100% at a pressure of 1.5 bar absolute. The nutrient solution was then inoculated with Candida tropicalis, and periodic pressure reduction was started at the same time. This consisted in alternatively keeping the pressure at

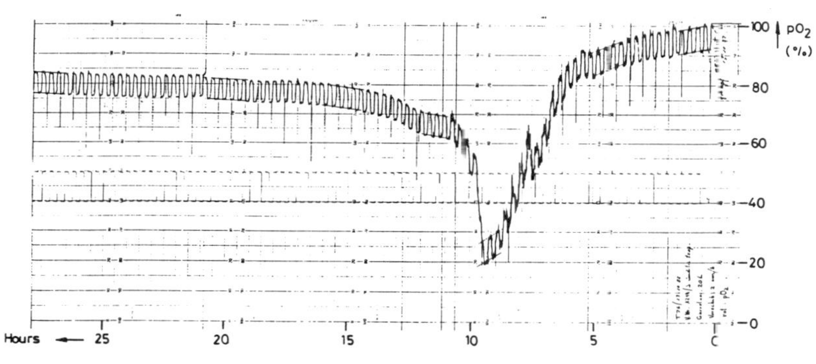

Fig. 3. The recorded curve of oxygen partial pressure.

The new method consists of the continuous monitoring of the electrode sensitivity. Periodic changes in pressure of constant amplitude produce a measurement signal with superimposed signal also of constant amplitude. A reduced measured reference signal hence leads to the conclusion, that the sensitivity of the electrode has been reduced correspondingly.

In other words, if the periodic fluctuations in the measured value diminish by a certain factor, the oxygen partial pressure measured must be corrected by means of the reciprocal of this factor.

Obviously however, this will only compensate for a loss in sensitivity of the pO_2 electrode. There will be no correction for a shift in the zero point, but expe-

1.5 bar for 10 minutes and then reducing it to 1.4 bar for 10 minutes. The oxygen partial pressure measured by means of the oxygen electrode varies with the same period, as shown in Fig. 3. The amplitude of these variations, was measured during the test, and indicated a decreasing sensitivity in the oxygen electrode, which is recorded in Fig. 4. This figure also shows the effective measured values of partial oxygen pressure obtained during the periods when the pressure was kept at 1.5 bar, as well as the compensated values obtained by multiplying the measured values with the reciprocal of the sensitivity.

The test was continued well beyond the duration of microbiological activity. At the end of the test, the glucose has for

a long time been completely consumed and the oxygen partial pressure should therefore theoretically again be 100%. The compensated measured values do actually give 100% at the end of the test and thus confirm the applicability of this method.

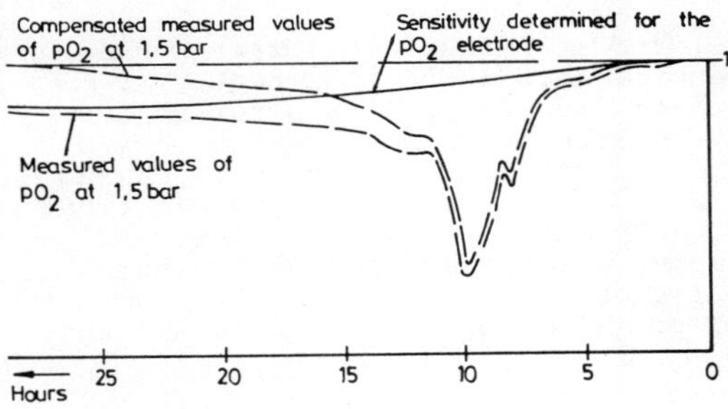

Fig. 4. The sensitivity pattern of the pO_2 electrode determined from a superimposed reference signal, together with the effective and appropriately compensated measured value.

The sensitivity of the pO_2 electrode decreased for a short time after each addition of anti-foaming agent. This is not shown in Fig. 4, because this curve shows an average figure of sensitivity over several periods of pressure variations.

CONCLUSIONS

The accuracy of the pO_2 determination can be improved by means of the method presented. A loss of sensitivity in the pO_2 electrode can be detected and compensated for, as long as the oxygen partial pressure does not fall below about 10%. No correction can be made for a shift in the zero point of the measured value, but experience shows that this is small compared with the component in the error of measurement for which compensation can be made.

Implementation of the method requires controlling the pressure in the fermenter and computer-assisted evaluation of the pO_2 measured values. It must be considered that other measurements, for example pCO_2, and also the addition of fermentation agents will also be affected by a reduction in the pressure. It is possible therefore that pressure reductions can only be carried out for certain short periods.

NOMENCLATURE

A	Membrane surface (cm^2)
C	Oxygen concentration ($mole/cm^3$)
C_o	Oxygen concentration at the surface of the membrane on the side facing the cathode
C_m	Oxygen concentration at the surface of the membrane on the side facing the material under measurement
C_{sat}	Saturation concentration of oxygen in the culture broth
D	Diffusion coefficient (cm^2/sec)
k_La	Oxygen transfer coefficient (sec^{-1})
P	Permeability coefficient of the membrane (mole/cm · sec · mm Hg)
p	Oxygen partial pressure (mm Hg)
p_b	Oxygen partial pressure corresponding to the oxygen in the culture broth
Ω	Consumption of oxygen by the microorganisms in the culture broth ($mole/cm^3 \cdot sec$)
q	Diffusion flow (mole/sec)
s	Solubility coefficient of oxygen in the relevant medium ($mole/cm^3 \cdot mm\ Hg$)
s_m	Solubility coefficient of oxygen in the membrane
x	Thickness of membrane (cm)

REFERENCES

Bühler, H., W. Ingold (1976). Measuring pH and oxygen in fermenters. Process Biochemistry, 11, 19-24.

Clark, L.C., R. Wolf, D. Granger, and Z. Taylor (1953). Continuous recording of blood oxygen tensions by polarography. J. Applied Physiol., 6, 189-193

Harrison, D.E.F., D.G. MacLenann, and S.J. Pirt (1969). Responses of bacteria to dissolved oxygen tension. D. Perlman (Ed.), Fermentation Advances. Academic Press, New York, pp. 117-143.

Krebs, W.M. (1975). Dissolved oxygen measurement in brewery systems. MBAA Technical Quarterly, 12, 977-985.

Lucero, D.P. (1968). Performance characteristics of membrane covered polarographic gas detectors. Analytical Chemistry, 40, 707-711.

Weinfurther, F., A. Uhl, and H. Ott (1960). Die Bedeutung des gelösten Sauerstoffs für die Vermehrung der Mikroorganismen. Archiv für Mikrobiologie, 36, 1-22.

AUTOMATED MEASURING SYSTEM FOR BIOLOGICAL WASTEWATER TREATMENT PLANT WITH BIOLOGICAL REMOVAL OF PHOSPHATE

K. M. Pedersen, P. Haagensen, M. Kümmel and H. Søeberg

Department of Chemical Engineering, The Technical University of Denmark, DK-2800 Lyngby, Denmark

Abstract. An automated measuring system for monitoring the concentrations of phosphate, nitrate and ammonia in a biological wastewater treatment plant of the activated sludge type with biological removal of phosphorous, nitrate and organic matter is described. The sampling method is based on dialysis, and the analyzer consists of a triple flow injection system. Pumps, valves and analyzer are controlled by a microcomputer, which also handles and stores the data from the analyzer. The dialysis-unit has been tested at a sewage water plant for a period of two weeks without loss of performance. The dialysate has been analysed with the phosphate analyzer and 0.4 mg-P/l was found.

Keywords. Water pollution, computer control, environment control, flow injection analysis.

INTRODUCTION

Increasing concern about the effects of nutrients in the effluents from wastewater treatment plants has led to investigations in biological removal of nitrate, phosphate and organic matter from the wastewater. These investigations have been restricted by the absence of a simple and robust measuring system for monitoring phosphate, nitrate and ammonia. Solid state ion electrodes for nitrate and ammonia have been investigated by Kümmel and Nielsen (1982) without satisfactory results due to drifting electrode potentials and clogging. Phosphate was not measured by these investigators, since no phosphate electrodes are so far available.

Colourimetric analysis methods offer possibilities for measuring a great variety of compounds, but these methods all require samples free from turbidities. The key element in systems using colourimetric methods is therefore the sampling and filtering system. Numerous designs with filter systems have been investigated by the authors and others so far without satisfactory results.

In the solution presented here a dialysis-unit replaces the filter and makes it possible to use colourimetric methods.

The analyzing system is based on flow injection analysis as described by Ruzicka and Hansen (1981, 1984). The method has several features that makes it suitable for this application: 1) it is simple, 2) it requires only small sample volumes, and 3) the consumption of reagents is low. Application notes from Tecator (1983) describe flow injection systems for the analysis for several compounds, including phosphate, nitrate and ammonia. Slania et al. (1980) describe a flow injection system for ammonia, and Anderson (1979) has developed a system for nitrate. A flow injection system for phosphate is described by Stewart et al. (1976).

In the system to be presented here, peristaltic pumps transport the water from the plant to the dialysis-unit. Salts from the water diffuse through a membrane into distilled water. This dialysate is pumped to the analyzing system, which consists of three parallel flow injection systems, - one for each of the component phosphate, nitrate and ammonia.

The new dialysis method is a combination of analysis methods adapted from well known laboratory standards. The system will only require limited maintenance. All of the elements in the system have been tested, but the combined system still needs testing. Results from this will be presented at the conference.

THE MEASURING SYSTEM

Figure 1 shows the system for monitoring the plant at three points. This number is easily expandable. The peristaltic pumps transport the water from the relevant point in the plant through the dialysis-unit and back to the plant. Distilled water is pumped under the membrane in the dialysis-unit through one of the valves V4-V6 to the analyzer by pump P7. By switching one of the valves V1-V3, a standard with known concentration of the relevant salts can be fed through for calibration. All valves are pinch solenoid valves from Sirai and employ silicon rubber tube.

The Dialysis-Unit

In Figure 2 the unit is sketched. In the upper part a channel for the water from the plant is cut, and in the lower part a microchannel for the acceptor stream of distilled water is formed by pressing a wire bended in sinusoidal form down in the material, which is PVC. The depth and width of the microchannel is appr. 0.7 mm, resulting in a channel volume of appr. 0.08 ml. The ends of the channel are connected to PVC tubes with ID = 0.5 mm.

Over the microchannel a polypropylene-membrane (Celanese, no. 3501) is applied with glue. The pore size in the membrane is 0.04 microns, ensuring that no turbidities contaminate the acceptor stream.

In order to minimize the axial dispersion, the microchannel is curved. At an acceptor flow rate of 0.044 ml/min the concentration of salts reaches 30% of equilibrium. If the acceptor stream is stopped for one minute, the concentration in dialysate reaches appr. 59% of the equilibrium level.

In order to achieve reproducable results, it is important to maintain a small and constant pressure drop over the membrane. This ensures that the main transport phenomenon is diffusion, which in this application is preferable as to minimize the risks for clogging of the membrane.

The dilution caused by the process of dialysation can easily be compensated through calibration with one or more standards. In this system one standard is available.

The Analyzers

Figure 3 shows the principles of the analyzers. There are three parallel systems, one for each of the components. Dialysate from the dialysis-unit is pumped by pump P7 through the sample loops formed by the three pairs of valves V7-V8, V9-V10 and V11-V12. In this period of the analysis cycle pumps P8-P10 are stopped. In the next period the valves V7-V12 are opened in the straight direction, P7 stopped and P8-P10 started, which causes reagents to be mixed with the contents of the sample loops by dispersion during the passage through the tubes. Chemical reactions take place and form coloured compounds, which are detected in the photometers. The analysis-cycle is completed in less than one minute.

The Phosphate Analysis

The reagents are:

R1 Ammoniummolybdate and sulfuric acid in distilled water

R2 Stannous chloride, hydroxylammoniumchloride and sulfuric acid in distilled water.

Flow rate of reagents: 0.4 ml/min:

The phosphate in the sample reacts with molybdate forming an acid, which is reduced to the blue molybdenum. The intensity of the colour is measured with the photometer at 660 nm.

The Nitrate Analysis

Reagents:

R3 Ammoniumchloride and sodiumchloride in distilled water

R4 Sulphanilamide, hydrochloric acid in distilled water

R5 N-(1-naphtyl)-ethylenediamine dihydrochloride and sodiumchloride in distilled water

Flow rates of reagents: 0.2 ml/min:

The sample with nitrate is injected into a carrier and passed through a reduction column containing cadmium granules covered with copper. By the passage, nitrate is reduced to nitrite, which then reacts with the sulphanilamide of R4, forming a diazonium salt. When this is mixed with reagents of R5, a diazo dye is formed and detected at 540 nm, Anderson, L. (1979).

The Ammonia Analysis

Reagents:

R6 pH-buffer, pH = 11.6

R7 hypochlorite and sulfuric acid in distilled water

R8 Nitroprusside and sodiumhydroxide in distilled water

R9 Phenol, methanol and sodiumhydroxide in distilled water

Flow rate of reagents: 0.4 ml/min:

The employed reaction is known as the Berthelot reaction (1859). The ammonia is oxidized by hypochlorite to form chloramine, which then reacts with phenol forming an indophenol dye detectable in the photometer at 625 nm.

The Photometers

The principles of the photometers are sketched in Figure 4. The flow cell is constructed with a light emitting diode (LED) as a light source and a photo diode as detector. The newest integrated circuits have been used for the amplifier, and it is contained in a printed circuit measuring 50x70 mm. The price of the components for the photometer is less than US$ 50.-. Further details about the photometer are given by Haagensen (1985).

RESULTS

The dialysis have been plant tested for a period of two weeks at a wastewater treatment plant with activated sludge. The membrane showed no sign of loss of performance.

The phosphate analysis has been performed, and Figure 5 shows a test run with four phosphate standards and a sample from dialysis-unit, which has a phosphate content at 0.4 mg-P/l. No interferences from other ions in the wastewater were observed. Results from the complete system will be presented at the conference.

DISCUSSION

It is our belief that this system will be valuable in gaining more insight in the processes taking place in biological wastewater treatment plants. Furthermore, it will enhance possibilities for automatic control of sewage plants and improve effluent quality. After more extensive testing we expect to prove the difficult sampling process for analysis of sewage water and similar fluids can be solved with a dialysis-unit.

CONCLUSION

We have described a new method for sampling from wastewater treatment plants using dialysis. The unit is simple and robust and is well suited for FIA. A test period of two weeks at a sewage plant has shown that the unit works without loss of performance.

A triple flow injection system has been designed, and the phosphate-analyzer has been tested with samples from the dialysis-unit. The analyzer is able to measure the concentration of phosphate with adequate accuracy and sensitivity in the dialysate.

ACKNOWLEDGEMENT

E. Arvin at the Laboratory for Technical Hygiejne, J. Ruzicka and E.H. Hansen at Chemical Laboratory A, all at the Technical University of Denmark, contributed with consultance, which is greatly appreciated.

REFERENCES

Anderson, L. (1979). Anal. Chim. Acta 110, 123-128.

Berthelot, M.P.E. (1859). Report de Chimie Applie-que, 284.

Celanese, Box 32414 Charlotte, North Carolina 28232, USA.

Haagensen, P. (1985). A general optiveelectronic detector and amplifier for flow analysis. Paper submitted to Flow Injection Analysis III, Birmingham, Great Britain.

Kümmel, M. and M.K. Nielsen (1982). Computer control of an activated sludge process. Proc. Int. Symp. on Process Engineering, 190-198, Kyoto, Japan.

Ruzicka, J. and E.H. Hansen (1981). Flow Injection Analysis. John Wiley & Sons Inc., New York.

Ruzicka, J. and E.H. Hansen (1984). Anal.Chim.Acta 161, 1-25.

Sirai, S.r.i. Via Masaccio, 6-20096 Piotello MI, Italy.

Slanina, J., F. Bakker, A. Bruyn-Hes and J.J. Möls (1980). Anal.Chim.Acta 113, 331-342.

Stewart et al. (1976). Anal.Chim.Acta 82, 137-144.

Tecator, A.B. Box 70, S-26301 Höganäs, Sweden.

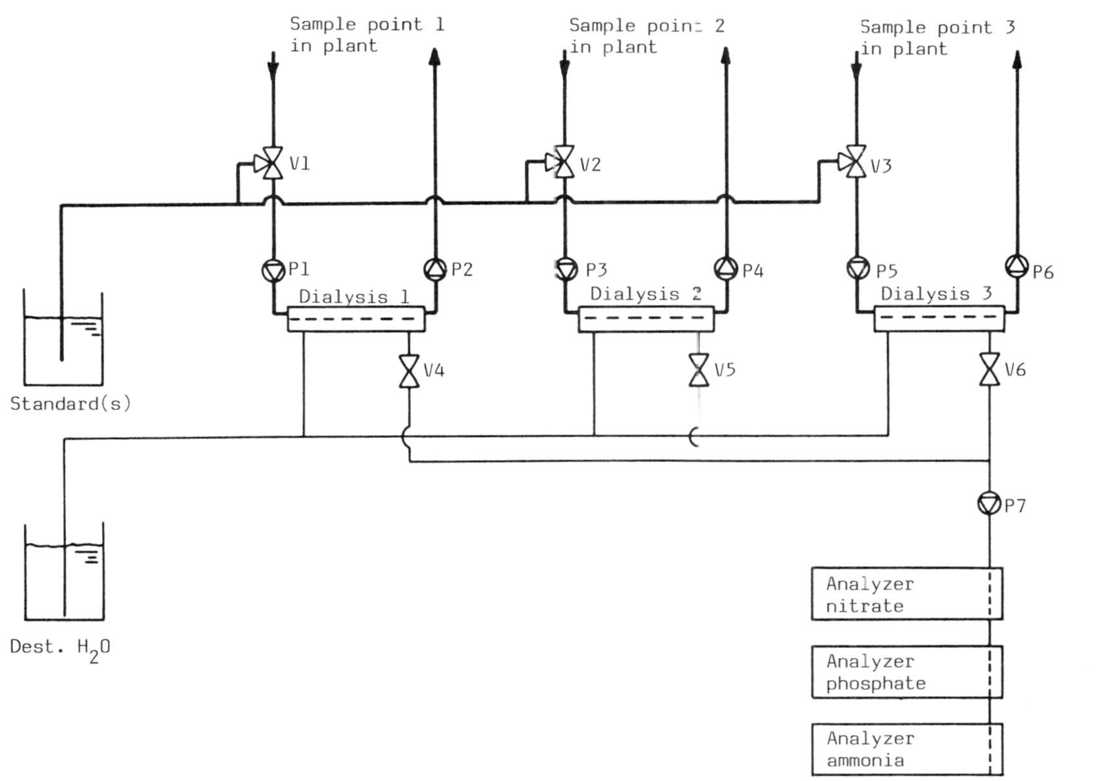

Figure 1:

The complete system for monitoring phosphate, nitrate and ammonia in sewage plant. Samples from the plant are transported to the dialysis-unit by peristaltic pumps. Dialysate from one of the three dialysis-units is analyzed in the three analyzers.

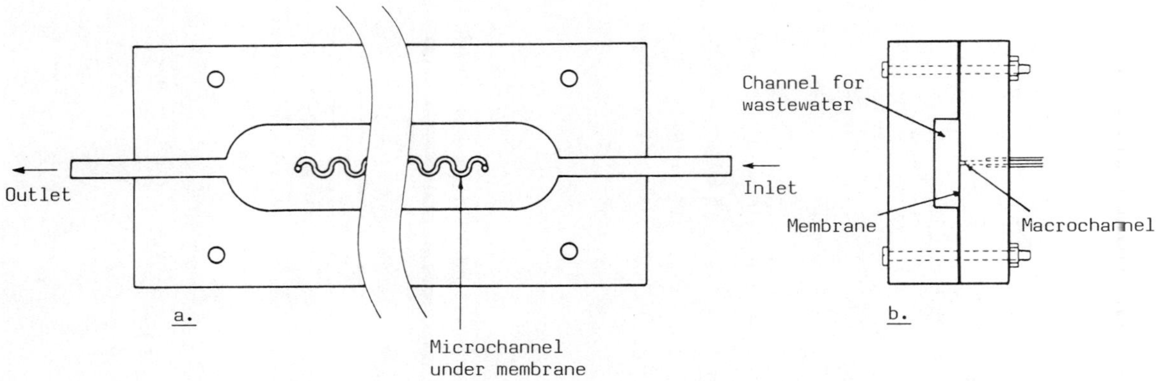

Figure 2a:

Dialysis-unit with inlet and outlet of water from the sewage plant. A sinusoidal microchannel covered with a polyprolylene membrane is situated at the bottom of the unit.

Figure 2b:

Cross-section of the dialysis-unit showing the membrane over the microchannel. PVC tube is connected to the microchannel.

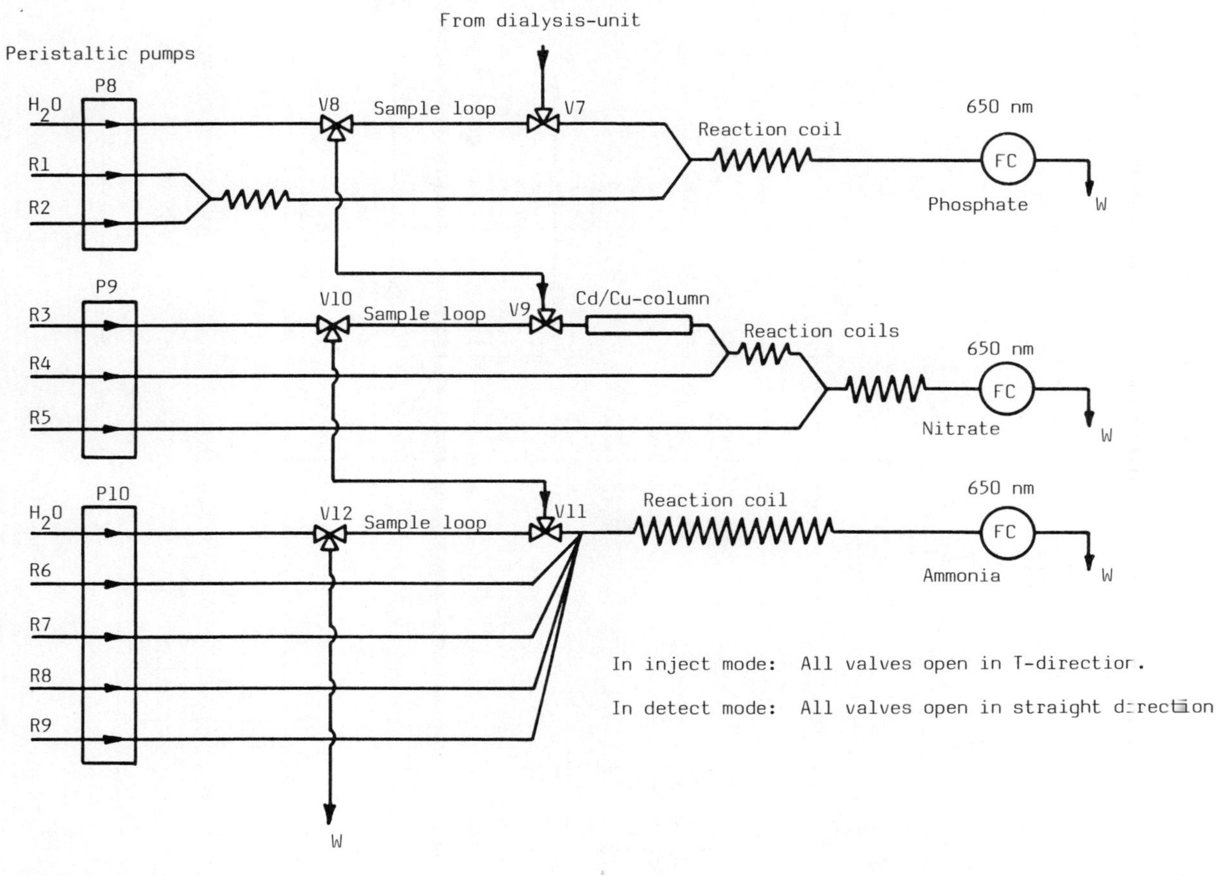

Figure 3:

Flowsheet for the three analyzers. Dialysate from the dialysis-unit is injected in the sample loops in the first period of the analysis-cycle. In the second period, reagents R1-R9 are mixed with the samples by diffusion during passage through the tubes. The colours formed by chemical reactions are detected in the flow cells (FC).

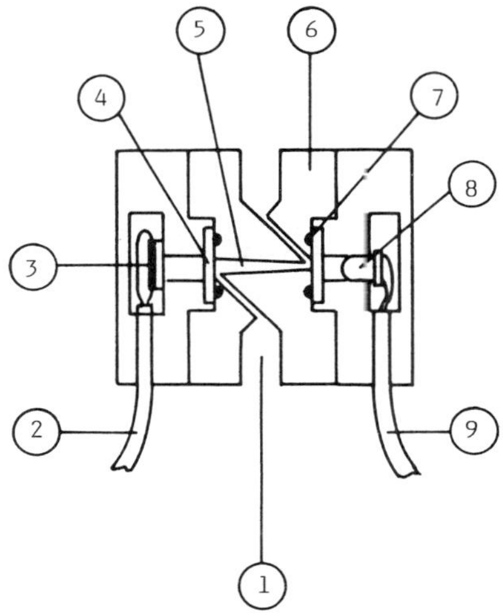

a.

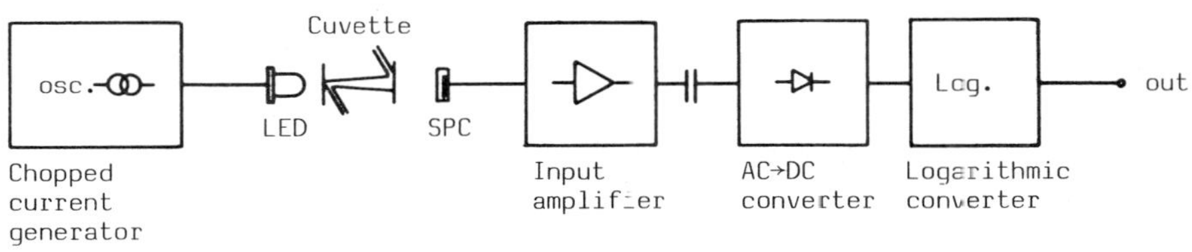

b.

Figure 4a: Flowcell

1) Hole for mounting of PVC tube with fitting
2) Cable for silicon photo diode (SPD)
3) Silicon photo diode (SPD) in solid state body
4) Glass window
5) Conic light path
6) Teflon/graphite body
7) o-ring
8) Light emitting diode (LED) in solid state body
9) Cable for light emitting diode (LED).

Figure 4b: Circuit for amplifier

A current generator generates an ac-current for the light emitting diode (LED). A silicon photo diode detects the light through the liquid-filled light path, and the current is amplified in a highly sensitive ac-amplifier. In the next step the signal is rectified and fed to the logarithmic amplifier. The concentration of the detected compound is proportional with the output of the logarithmic amplifier.

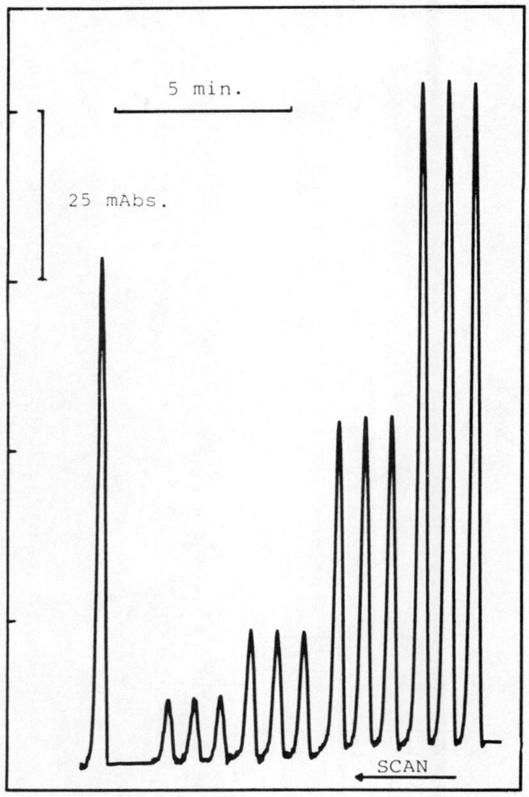

Figure 5:

Standards and dialysate from the dialysis-uni analyzed with the phosphate analyzer. Standards 500 µg-P/L, 250 µg-P/L, 100 µg-P/L, 50 µg-P/L Concentration of phosphate in dialysate: 0.4 mg-P/L.

OPTIMIZATION AND MODELLING OF THE MICROBIAL EPOXIDATION OF PROPENE IN AN ORGANIC-LIQUID-PHASE/IMMOBILIZED-CELL SYSTEM

L. E. S. Brink and J. Tramper

Department of Food Science, Food and Bioengineering Group, Agricultural University Wageningen, De Dreyen 12, 6703 BC Wageningen, The Netherlands

Abstract. Several fundamental aspects of an organic-liquid-phase/immobilized-cell system are studied using the epoxidation of propene as the model reaction. An automated experimental system, which facilitates the multiphase experiments, is described. Entrapment in calcium alginate gel provides high activity retention of the immobilized mycobacteria used. This gel is stable in the organic solvents tested and prevents aggregation and clotting of cells in the presence of these solvents, but does not seem to provide other protection against the solvents. High activity retention of the immobilized cells is favoured by low polarity and high molecular weight of the solvent. A bisubstrate kinetic model is used to describe the experimental consumption rates of oxygen and propene. Intraparticle substrate diffusion is found to limit the inherent kinetic reaction rate of the immobilized biocatalyst. The influence of particle diameter, catalytic activity and substrate level on epoxidation rates of pore-diffusion limited gel-entrapped cells is established under pseudo one-substrate conditions. Theoretical reaction rates, predicted by using an approximative pore diffusion model, correlate well with the experimental rates.

Keywords. Optimization; modelling; multiphase system; microbial epoxidation; mathematical analysis; sampled data systems.

INTRODUCTION

The number of applications in which enzymes, cell organelles or whole cells are used as catalysts has increased steadily over the last decade. The aqueous reaction medium is a common aspect of these bioconversions. For a long time it was thought that an aqueous medium is essential for biocatalysis, and this has restricted the number of practical applications. These restrictions can be related to factors like low solubility of (gaseous) substrates and/or products in water, substrate and/or product hydrolysis and laborious product recovery from dilute, aqueous solutions. Substrate or product inhibition and low operational stability of the biocatalyst are other well-known, fundamental problems of biocatalysis.

Immobilization of biocatalysts has become a widely applied technique to improve the product/biocatalyst separation and/or to enhance the operational stability. It has not changed the situation that bioconversions are still mostly performed in aqueous environments. Replacement of part of this water phase by a water-immiscible organic phase, however, may offer certain important advantages (Antonini, Carrea and Cremonesi, 1981; Lilly, 1982). High concentrations of poorly water-soluble substrates and/or products are possible in organic solvents. Reaction equilibria may be shifted favourably, and substrate and/or product hydrolysis can be largely prevented. Furthermore, substrate or product inhibition may be reduced as a consequence of a lower inhibitor concentration in the aqueous environment of the biocatalyst. Finally, recovery of product and biocatalyst may be facilitated. An obvious, major drawback of introducing an organic solvent is denaturation of the enzymes responsible for the desired bioconversion. Immobilization of the biocatalyst may then especially become interesting in view of a possible, protective effect by the support material.

In order to study the potential of the introduction of an additional organic phase, the stereospecific production of propene oxide from propene and oxygen by immobilized cells in a two-liquid-phase system has been investigated as a model (Habets-Crützen and co-workers, 1984, 1985; Tramper and co-workers, 1984). The system consists of a solid, immobilized-cell phase, an aqueous phase, an organic solvent and a gas phase containing the two gaseous substrates. The above mentioned advantages of immobilization and of an additional organic phase are highly relevant to this model system: decreased product inhibition by in situ-extraction of the toxic propene oxide into the organic phase, high solubility of oxygen and, especially, propene in a suitable organic solvent and, finally, facilitated recovery of biocatalyst and epoxide. Furthermore, this model is suited for fundamental studies in view of the relative simple molecules involved in the epoxidation. This facilitates quantitative gas analysis (Brink and co-workers, 1984) and the estimation of diffusion coefficients.

Many different factors and their possible mutual interactions have to be examined before a rational design of a multiphase bioreactor can take place. An optimal combination of immobilization technique and type of organic solvent must be found. Furthermore, modelling of epoxidation kinetics (bisubstrate, product inhibition), inactivation kinetics and of mass transfer effects (substrates, product, co-substrate) should precede the choice and design of a suitable bioreactor. An automated experimental system has been employed to facilitate the optimization and modelling. In this paper the automated system and the above mentioned aspects (immobilization, organic solvent, kinetics, mass transfer) are discussed.

EXPERIMENTAL SET-UP

The various aspects of the multiphase system are being studied in a laboratory scale batch-reactor system with gas circuit (Fig. 1). Full details of this experimental system are given elsewhere (Brink and co-workers, 1984). A small bubble-column bioreactor was used for contacting the immobilized cells and the liquid phase(s) with the circulating gas. The total volume of the biocatalyst and liquid phase(s) was always 250 cm^3. Ten cm^3 of the water phase was replaced by an equal volume of organic solvent when the solvent influence on biocatalyst activity was to be examined.

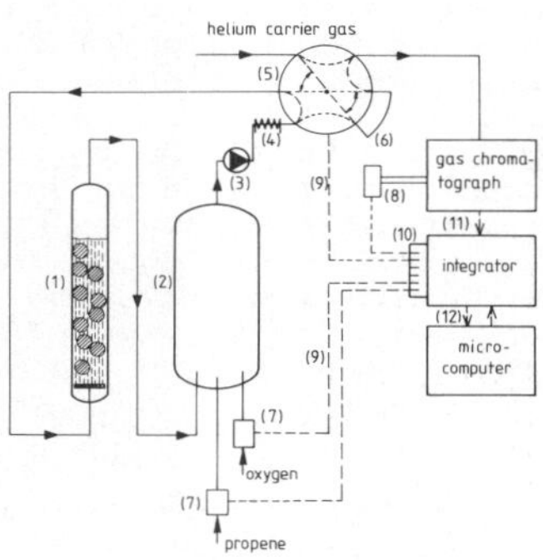

Fig. 1. Schematic representation of circulation batch reactor and control system: (1) bubble-column bioreactor; (2) gas-supply vessel; (3) diaphragm pump; (4) heating; (5) two-position gas-sampling valve showing the normal (-----) and inject (—→) mode; (6) sample loop; (7) flow controllers; (8) column-switching valve; (9) control signals; (10) external-control option; (11) analog g.c. signal; (12) digital communication.

The reaction course in the biocatalyst phase is followed by on-line gas chromatographic analysis of the gaseous substrates and volatile product in the circulation gas. This technique, so-called head-space gas chromatography, provides several important advantages in comparison to direct g.c. of the aqueous and/or organic liquid phase(s). Overloading the g.c.-column and the detector by water and interference and contamination of the column by the various solvents are much less severe in case of gas sampling. Furthermore, the automation of a gas-sampling valve is relatively easy as compared to that of liquid-sample withdrawal.

The on-line gas chromatograph is coupled to a programmable integrator and a micro-computer. This enabled automatic operation of the gas-sampling valve in conjunction with a column-switching valve. A Porapak Q column is automatically connected to the sampling valve, when propene and propene oxide are to be analyzed. A column packed with molecular sieve 13 X is selected for determination of oxygen. Besides data acquisition and reduction, the coupling to the computer has also made it possible to control the concentration levels of the two substrates in the experimental system. When the measured level of propene or oxygen falls below a user-defined setpoint, an appropriate amount of gaseous substrate can be added by computer-controlled opening of flow controllers. Level control of only one substrate is mostly used for establishing the kinetic and mass transfer effects of the other (limitating) substrate. Controlling both substrate levels is suited for measurement of the operational stability of the biocatalyst.

OPTIMIZATION OF THE MULTIPHASE SYSTEM

Two characteristics of immobilized-cell/two-liquid-phase systems are of paramount importance with respect to process optimization and modelling, that is immobilization technique and type of organic solvent. The method of immobilization not only influences the activity retention and stability of the cells, but also the amount of water surrounding the biocatalyst. When the latter quantity is very small, cell inactivation by the solvent is likely to occur. A large amount of protective water, however, may cause diffusion limitations as a consequence of long diffusion paths through water, small interfacial specific surface area and/or low solubility of substrates in water.

Entrapment of the cells in calcium alginate appeared to be a suitable immobilization technique with respect to retention of catalytic activity. The immobilized cells proved not to be inferior to the free cells (Fig. 2). This is not unexpected in view of the high water content of this support material. Another hydrophilic gel, κ-carrageenan, yielded similar results (Brink and Tramper, 1985a).
As can be seen in Fig. 2, an increasing difference is found between experimental epoxide concentrations and calculated values from propene consumption rates. This discrepancy can be attributed to microbial consumption of the formed epoxide (Brink and Tramper, 1985a). Immobilization in alginate (Fig. 2) or κ-carrageenan lowers the epoxide consumption rate, indicating an additional,

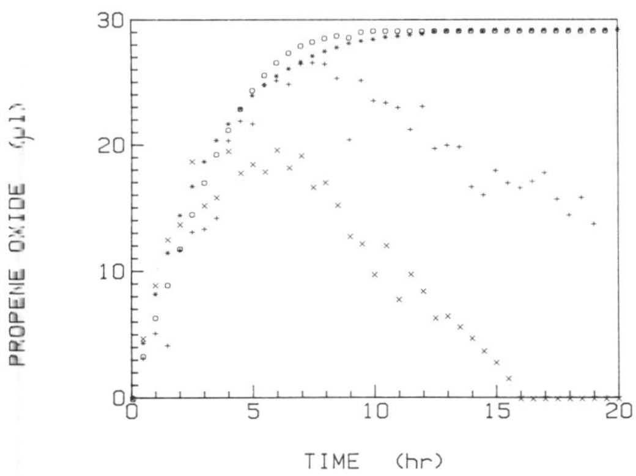

Fig. 2. Amount of produced propene oxide versus time: calculated from propene consumption: (*) free cells, (o) cells immobilized in calcium alginate; experimental: (x) free cells, (+) cells immobilized in calcium alginate.

at the moment unexplained advantage of immobilization. The problem of low rates of substrate diffusion in the hydrophilic calcium alginate gel will be quantified below.

Reaction kinetics and stability of the biocatalyst, chemical and mechanical stability of the carrier material and the partition of propene, oxygen and epoxide between the different phases are all likely to be influenced by the nature of the organic solvent. Over thirty, water--immiscible organic solvents from fifteen solvent classes were considered for use in the multiphase system (Brink and Tramper, 1985a). The structure of the calcium alginate gel appeared to be stable in all the solvents tested. Rapid loss of catalytic activity was often observed when water-suspended cells were exposed to an organic-liquid phase. It was not always clear to what extent this activity loss related to also observed, physical phenomena like cell aggregation at the water/organic interface and the formation of biomass clots. The latter phenomena could, however, be eliminated by using immobilized cells.

The type of organic solvent also appeared very critical when calcium alginate--entrapped cells were brought in contact with an organic phase (Brink and Tramper, 1985a). Little loss of activity was measured when the following solvents were applied: n-hexadecane, di-iso-pentyl ether, di-phenyl ether, di-n-alkyl phthalates, except the di-methyl derivative, and, finally, two perfluoro chemicals. To avoid time-consuming and expensive testing of large numbers of solvents, a relation was sought between the measured activity retentions of immobilized cells and readily available solvent properties. The combined use of two characteristics of organic solvents made it possible to explain the experimentally found activities (Fig. 3). The polarity, as expressed here by the Hildebrand solubility parameter δ, has already been mentioned in a qualitative manner (Antonini and co-workers, 1981; Omata, Tanaka and Fukui, 1980). The second characteristic, dimension of the solvent molecule (as expressed by the molecular weight or the molar volume), may relate to the degree of hindrance a solvent molecule would experience on his way from the organic bulk phase to the catalyzing enzymes in the immobilized cells. Most of the low activities can be found in an area of high polarity ($\delta > 8$) and small molecular weight ($M < 200$). Outside this area the higher activity retentions are found. The experiments with the larger di-n-alkyl phthalates and perfluoro chemicals provide good examples of this finding.

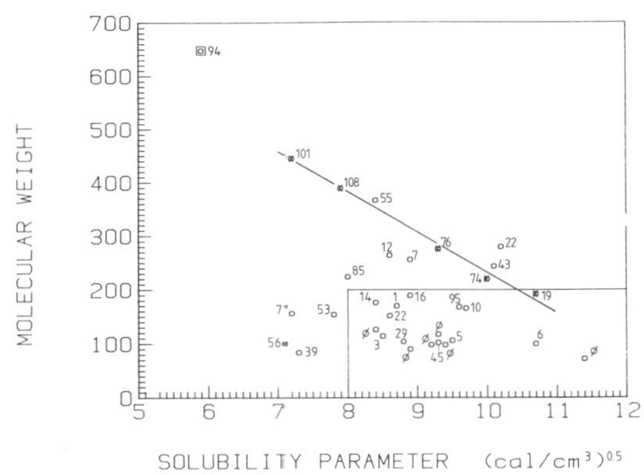

Fig. 3. Measured activity retentions (%) of immobilized cells exposed to organic solvents as a function of molecular weight and solubility parameter of the solvents; line and ⊗ : di-n-alkyl phthalates, ▣ : perfluoro chemical.

From the likewise response of free and immobilized cells to organic solvents, it could furthermore be deduced that immobilization in calcium alginate provides no additional cell protection against organic media (Brink and Tramper! 1985a).

The solubility parameter also proved useful for describing the solubilities of oxygen and propene oxide in various organic solvents. Values for the propene solubility in solvents (expressed in mM) did not seem to correlate with this parameter. The capacity of the solvents for propene is however always about two orders of magnitude higher than that of water. A higher solubility parameter (polarity) corresponds with a lower oxygen solubility, but with an increased extraction capacity for propene oxide. Thus, an optimum in the solvent polarity is likely, especially when the influence of polarity on cell activity retention is also considered. In choosing an organic solvent one must also take into account solvent properties, which will affect the epoxide recovery from the solvent (boiling point, water capacity), and, furthermore, general properties like viscosity, toxicity, flammability and price.

MODELLING THE MULTIPHASE SYSTEM

The apparent kinetic rate of the gel-entrapped cells may be limited due to diffusion of substrates and product to, in and from the biocatalyst particles. Modelling of the biocatalyst performance must therefore be based on a combination of epoxidation kinetics and mass transfer rates. Ultimately, this biocatalyst model should be conjoined with a reactor flow model to take into account factors like hydrodynamics, concentration gradients and non-idealities in the bioreactor.

Kinetic Parameter Estimation

The used *Mycobacteria* possess a mono-oxygenase, which can produce epoxides from gaseous alkenes and oxygen (Habets-Crützen and co-workers, 1984). The overall reaction equation of the enzyme-catalyzed, NADH-dependent epoxidation of propene is as follows:

$$\text{propene} + O_2 \xrightarrow{NADH + H^+ \rightarrow NAD^+} \text{propene oxide} + H_2O$$

It was found experimentally, that about one mole of oxygen was needed for the epoxidation of one mole of propene, and that the reaction was nearly irreversible.

A well-known kinetic model for bisubstrate enzyme-catalyzed reactions, the ping-pong mechanisme, was applied to describe the experimental propene consumption rates, v_1, of the gel-entrapped non-growing cells:

$$v_1(t) = -\frac{dS_1(t)}{dt} = \frac{V\, S_1(t) S_2(t)}{K_2 S_1(t) + K_1 S_2(t) + S_1(t) S_2(t)} \quad (1)$$

(V is the maximal epoxidation rate; S_1 and S_2 the propene and oxygen concentration, respectively)
The immobilized cells also appeared to consume oxygen in the absence of propene. This cell respiration was assumed to obey irreversible Michaelis-Menten kinetics with a maximal respiration rate of βV. Therefore, the total oxygen consumption rate, v_2, can be described as follows:

$$v_2(t) = -\frac{dS_2(t)}{dt} = -\frac{dS_1(t)}{dt} + \frac{\beta V\, S_2(t)}{K_R + S_2(t)} \quad (2)$$

The estimation of the five kinetic parameters (V, β, K_1, K_2 and K_R) was performed by fitting the theoretical model, as predicted by equations 1 and 2, to three types of experimental time course data (aqueous environment). It was verified that both the produced levels of the toxic propene oxide and the small inactivation of the immobilized cells did not have significant effects on the time-dependent consumption rates of propene and oxygen (Brink and Tramper, 1985b). Furthermore, diffusional influences (see below) were eliminated by proper selection of particle radius (~ 0.35 mm) and biocatalytic activity (~ 5 10^4 μmol min^{-1} (m^3 particle)$^{-1}$). The following three-step procedure was used to estimate the parameters:

i./ At constant S_2^* (239 μM) equation 1 can be rearranged as:

$$-\frac{dS_1(t)}{dt} = \frac{V_{M,1}\, S_1(t)}{K_{M,1} + S_1(t)} \quad (3)$$

with

$$V_{M,1} = \frac{S_2^*}{K_2 + S_2^*} V \quad (4)$$

$$K_{M,1} = \frac{S_2^*}{K_2 + S_2^*} K_1 \quad (5)$$

Fitting the integrated form of equation 3 to experimental $S_1(t)$ data, using a non-linear regression technique, gives then values for $V_{M,1}$ and $K_{M,1}$.

ii./ In the absence of propene, equation 2 becomes:

$$-\frac{dS_2(t)}{dt} = \frac{\beta V\, S_2(t)}{K_R + S_2(t)} \quad (6)$$

Values for βV and K_R are then determined from experimental $S_2(t)$ by applying the same fitting technique as above. An illustration of the resulting fit in a concentration versus time plot is given in Fig. 4.

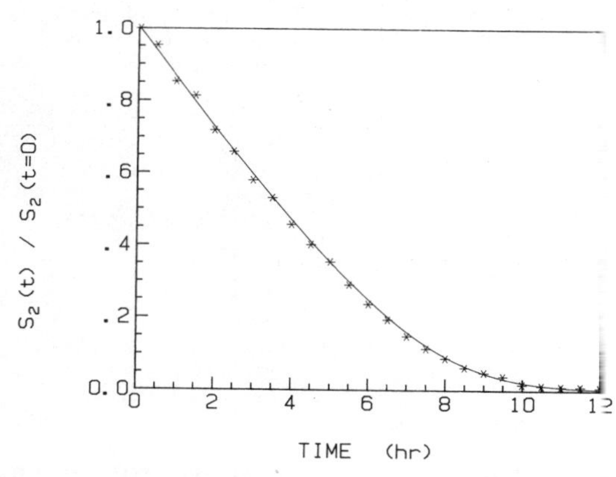

Fig. 4. Model description (βV = 4.6 μmol min^{-1} g dry weight^{-1}, K_R = 15 μM, equation 6) of experimental, dimensionless oxygen concentrations (*).

iii./ At constant $S_1 \gg K_1$ (160 μM) equations 2 and 4 can be rearranged to give:

$$-\frac{dS_2(t)}{dt} = \frac{V_{M,1}(K_2/S_2^* + 1) S_2(t)}{K_2 + S_2(t)} + \frac{\beta V\, S_2(t)}{K_R + S_2(t)} \quad (7)$$

Conversion of the experimental $S_2(t)$ to oxygen consumption rates, as described by Brink and Tramper (1985b), makes it possible to determine K_2, and subsequently V and K_1 (from equations 4 and 5) and β (values have already been assigned to i.e. βV and K_R in the above two steps). As can be seen in Fig. 5, equation 7 can be successfully applied, by adjustment of only K_2, to describe the experimental oxygen consumption rates.

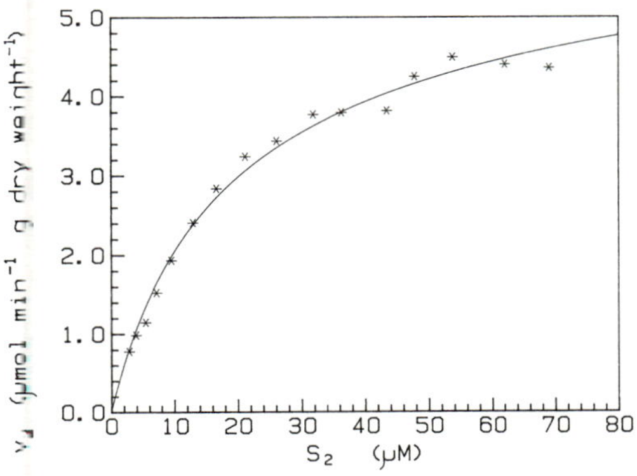

Fig. 5. Model description (K_2 = 108 µM, equation 7) of experimental oxygen consumption rates (*).

The results in Table 1 were obtained by following above described procedure. As different cell batches were used for the two measured series, it is not unexpected that the kinetic parameters V and β are also different. However, the (pseudo)-equilibrium constants (K_1, K_2 and K_R) of the two series do not vary to a large extent, which underlines the value of the model.

TABLE 1 Obtained Values for the Kinetic Parameters

parameter	V (1)	β (−)	K_1 (µM)	K_2 (µM)	K_R (µM)
first series	4.7	1.4	10	85	18
second series	2.1	2.2	7.0	108	15

(1) µmol min^{-1} g dry $weight^{-1}$

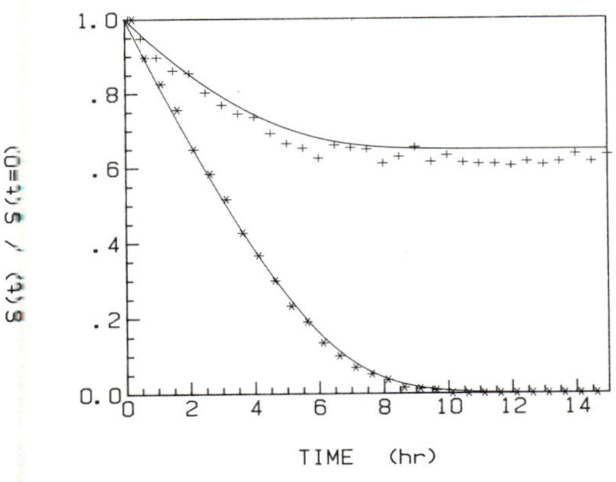

Fig. 6. Model predictions (equations 1 and 2) and experimental, dimensionless propene (+) and oxygen (*) concentrations.

To test the predictive capacity of this kinetic model, experiments were performed with both substrate concentrations decreasing with time. For this case, predictions were obtained by numerically solving equations 1 and 2 using the Runge Kutta procedure. The kinetic parameters of the second series (table 1) were utilized. A good agreement between predicted and measured propene and oxygen concentrations is obtained (Fig. 6).

Modelling Mass Transfer Effects

Several transport steps in the multiphase system may limit the epoxidation rate. First, the gaseous substrates have to be transfered to a liquid bulk phase. In Fig. 7 it is assumed that this is the organic phase. Subsequently, the substrate molecules must diffuse through organic and/or aqueous stagnant films around the biocatalyst particle. Finally, to reach all the entrapped cells, the substrates should diffuse through the gel matrix. In addition, though not shown in Fig. 7, the formed propene oxide has to diffuse out of the biocatalyst and into the bulk liquid. Some of these external and internal diffusion steps may limit the overall epoxidation rate, thereby lowering the biocatalyst efficiency. In our system the most serious limitaion was found to be the intraparticle diffusion of the two substrates in the calcium alginate gel (Brink and Tramper, 1985b).

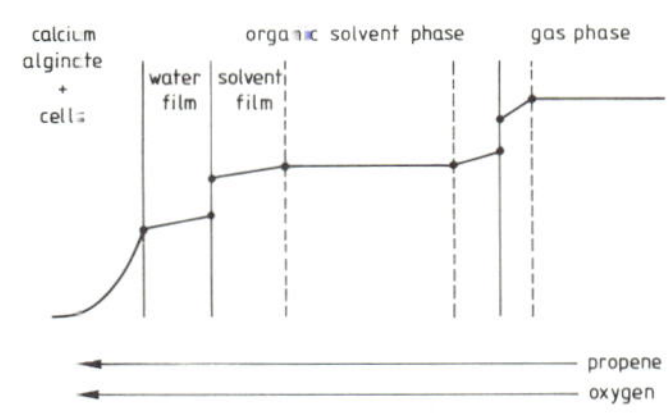

Fig. 7. Mass transfer steps in the multiphase system.

The theoretical description of coupled reaction and internal pore diffusion in permeable biocatalysts is well-established (Kasche, 1983; Radovich, 1985). Results of these theoretical analyses are often presented as relationships between the internal effectiveness factor of the immobilized enzymes or cells and a Thiele-type modulus. The effectiveness factor, defined as the ratio of observed reaction rate to the rate if no diffusion limitation existed, can be derived from the solution of the following, steady--state differential equation:

$$De\left(\frac{1}{r^2}\frac{d}{dr}\left(r^2\frac{dS(r)}{dr}\right)\right) = b\ v(S) \qquad (8)$$

describing the coupled diffusion and

(one-substrate) reaction in a spherical particle (De is the effective diffusion constant, r is the radial distance in the sphere and b is the cell load).

The mass transfer effects of propene and oxygen were investigated using the above described effectiveness factor concept, in combination with a pseuso-one substrate model. At present, the case of two limiting substrates is not considered. The high solvent capacity for propene, as compared to that of oxygen, makes it likely that only oxygen will limit the reaction rate in the multiphase bio-reactor. Therefore, two types of experiments were used to test the pseuso-one--substrate model:

i./ At a constant, high propene concentration ($S_1 >> K_1$) oxygen was considered the limiting substrate. The kinetic rate, to be inserted in equation 8, can then be derived from equation 2:

$$v(S)=v(S_2)=\frac{V\ S_2}{K_2+S_2}+\frac{\beta V\ S_2}{K_R+S_2}\approx\frac{V_{M,2}\ S_2}{K_{M,2}+S_2} \qquad (9)$$

ii./ At a constant, high oxygen concentration (239 μM) propene diffusion rates were studied. Equation 1 can then be re-written as:

$$v(S) = v(S_1) = \frac{V_{M,1}\ S_1}{K_{M,1}+S_1} \qquad (10)$$

($V_{M,1}$ and $K_{M,1}$ as defined by equations 4 and 5)

The error caused by the approximation in the right hand side of equation 9 can be shown to be negligible, when optimal values for $V_{M,2}$ and $K_{M,2}$ are chosen. The same diffusion model can now be applied for case i and ii.

Even for the relatively simple case of Michaelis-Menten kinetics, equation 8 can not be solved analytically. In this work, a simple, approximative solution of Kobayashi and Laidler (1973) was used. This approach makes use of the analytical solutions to equation 8 of two limiting cases, zero and first order reaction:

$$v^{PRED} = \eta_i\left(\frac{bV_M\ Ss}{K_M + Ss}\right) \qquad (11)$$

with the internal effectiveness factor:

$$\eta_i = \frac{\eta_{i,0} + (K_M/Ss)\ \eta_{i,1}}{1 + (K_M/Ss)} \qquad (12)$$

The zero and first order internal effectiveness factors, $\eta_{i,0}$ and $\eta_{i,1}$, are functions of a Thiele-type modulus, which can be calculated from known parameters (radius R, De, b, V_M and K_M) and from the substrate concentration at the particle surface, Ss. Considering the low molecular dimensions of the substrates and the high water content of the support material, the effective diffusion constant of propene and oxygen in the alginate gel was approximated by the corresponding diffusion constants in water. The surface concentration can be quantified by assuming, that, under steady-state conditions, the substrate consumption rate equals the substrate diffusion rate through a stagnant water film around the particle:

$$v^{PRED} = k_L\ a\ (\ Sb - Ss\ ) \qquad (13)$$

(Sb is the bulk concentration in the water phase, as organic solvents were not used in the diffusion experiments; a is the specific surface area)
The mass transfer coefficient in the water film k_L was estimated from the Sherwood number. In our experiments, the effect of external film diffusion was small as compared to the effect of internal pore diffusion. From combining equations 11 and 13, Ss, η_i and v^{PRED} can be found for different values for R, bV_M and Sb.

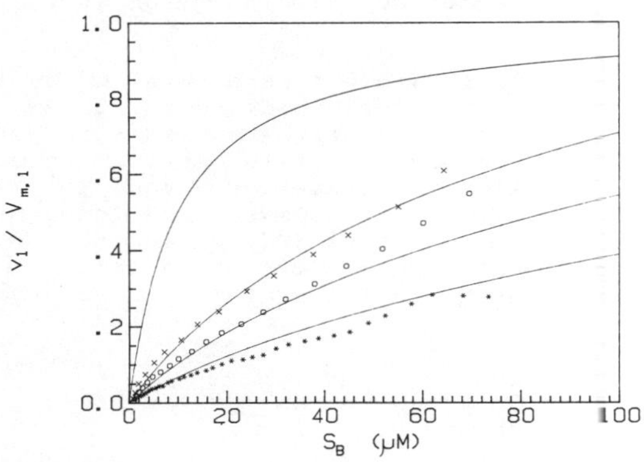

Fig. 8. Model predictions and experimental, dimensionless propene consumption rates of pore-diffusion limited, gel-entrapped cells (R = 1.0 mm, $K_{M,1}$ = 10 μM) at various levels of the propene bulk concentration; $bV_{M,1}$ (10^4 μmol min^{-1} (m^3 particle)$^{-1}$) 12 (x), 24 (o), 48 (*). Upper line represents the reaction rate if no diffusion limitation exists.

The predicted consumption rates of propene and oxygen were correlated to experimental rates from time-course experiments. The influence of catalytic activity (bV_M) on observed and predicted propene consumption rates is shown in Fig. 8. Figure 9 shows the effect of particle radius on observed and calculated oxygen consumption. From these and other results (Brink and Tramper, 1985b), it can be concluded, that the approximative pore diffusion model, used in this study, can be employed for predicting the effectiveness of the immobilized-biocatalyst preparation. This enables a proper selection of important bioreactor design parameters, like substrate concentration levels, particle radius and biocatalytic activity. Furthermore, under pseudo-one substrate conditions, this one-substrate model can describe the mass transfer effects of propene as well as of oxygen.

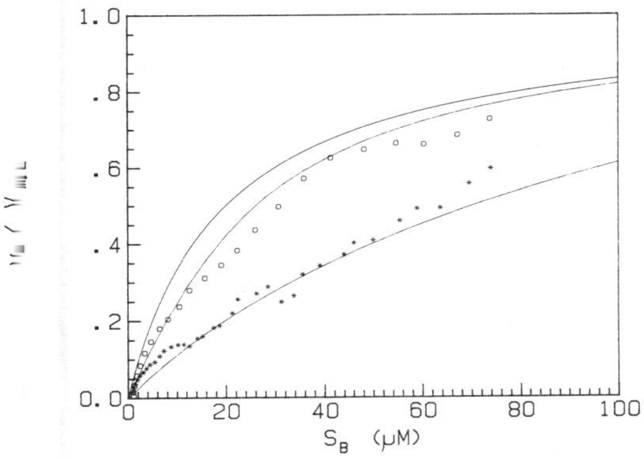

Fig. 9. Model predictions and experimental, dimensionless oxygen consumption rates of pore-diffusion limited, gel-entrapped cells ($bV_{M,2}$ = 15 10^4 μmol min^{-1} (m^3 particle)$^{-1}$, $K_{M,2}$ = 20 μM) at various levels of the oxygen bulk concentration; R (mm): 0.4 (o), 1.1 (*). Upper line represents the reaction rate if no diffusion limitation exists.

CONCLUSIONS

Many organic solvents cause rapid inactivation of the free or calcium alginate-entrapped cells. By using only two solvent properties, polarity (as expressed by the Hildebrand solubility parameter) and molecular weight, this loss of activity can be predicted. Solvents with low polarity and high molecular weight are preferable with respect to activity retention. The solubility parameter (polarity) is also useful for describing the capacity of organic solvents for substrates and product. A high propene oxide extraction rate requires a high solvent polarity. An optimum in polarity is therefore likely, as polar solvents may give rise to unacceptable low cell activities.

A kinetic model, based on the ping-pong mechanism fits experimental substrate consumption rates well. The bisubstrate model can also be used for predicting reaction rates. In case of a constant level of propene or oxygen, the consumption rate of the other substrate can be described by Michaelis-Menten kinetics. This facilitates modelling of internal-diffusion limitation. An approximative solution to the problem of coupled reaction and internal pore diffusion gives reliable predictions of substrate consumption rates. Intraparticle diffusion can decrease the effectiveness of the immobilized cells considerably. Small particle diameters and high substrate concentration levels are therefore important when using active immobilized-cell preparations. These parameters strongly affect the design of an organic-liquid-phase/immobilized cell bioreactor, the present aim of our investigations.

Acknowledgements. The authors are grateful to Prof. Ir. K. Ch. A. M. Luyben and Prof. Dr. Ir. K. Van 't Riet for helpful discussions. These investigations were supported (in part) by the Netherlands Technology Foundation (STW).

REFERENCES

Antonini, E., G. Carrea and P. Cremonesi (1981). Enzyme catalysed reactions in water-organic solvent two-phase systems. *Enzyme Microb. Technol.*, 3, 291-296

Brink, L. E. S., J. Tramper, K. Van 't Riet and K. Ch. A. M. Luyben (1984). Automation of an experimental system for the microbial epoxidation of propene and 1-butene. *Analytica Chimica Acta*, 163, 207-217.

Brink, L. E. S. and J. Tramper (1985a). Optimization of the organic solvent in multiphase biocatalysis. *Biotechnol. Bioeng.*, in press.

Brink, L. E. S. and J. Tramper (1985b). Modelling the oxygen and propene mass transfer effects on propene epoxidation kinetics of pore-diffusion limited gel-entrapped cells, submitted.

Habets-Crützen, A. Q. H., L. E. S. Brink, C. G. van Ginkel, J. A. M. de Bont and J. Tramper (1984). Production of epoxides from gaseous alkenes by resting-cell suspensions and immobilized cells of alkene-utilizing bacteria. *Appl. Microbiol. Biotechnol.*, 20, 245-250.

Habets-Crützen, A. Q. H., S. J. N. Carlier, J. A. M. de Bont, D. Wistuba, V. Schurig, S. Hartmans and J. Tramper (1985). Stereospecific formation of 1,2-epoxypropane, 1,2-epoxybutane and 1-chloro-2,3-epoxypropane by alkene-utilizing bacteria. *Enzyme Microb. Technol.*, 7, 17-21.

Kasche, V. (1983). Correlation of experimental and theoretical data for artificial and natural systems with immobilized biocatalysts. *Enzyme Microb. Technol.*, 5, 2-13.

Kobayashi, T. and K. J. Laidler (1973). Kinetic analysis for solid-supported enzymes. *Biochim. Biophys. Acta.*, 302, 1-12.

Lilly, M. D. (1982). Two-liquid-phase biocatalytic reactions. *J. Chem. Technol. Biotechnol.*, 32, 162-169.

Omata, T., A. Tanaka and S. Fukui (1980). Bioconversions under hydrophobic conditions: effect of solvent polarity on steroid transformations by gel-entrapped Nocardia rhodocrous cells. *J. Ferment. Tecnnol.*, 58, 339-343.

Radovich, J. M. (1985). Mass transfer effects in fermentations using immobilized whole cells. *Enzyme Microb. Technol.*, 7, 2-10.

Tramper, J., L. E. S. Brink, R. S. Hamstra, J. A. M. de Bont, A. Q. H. Habets-Crützen and C. G. van Ginkel (1984). Production of (chiral) epoxy alkanes in second-generation bioreactors. In: Third European Congress on Biotechnology, Vol. 2. Verlag Chemie, Weinheim. p. 269-276.

INFLUENCE OF SUBSTRATE DISTRIBUTION ON PRODUCTIVITIES IN COMPUTER CONTROLLED BAKER'S YEAST PRODUCTION

M. Reuss and U. Brammer

Fachgebiet Biotechnologie, Technical University of Berlin, Seestr. 13, D-1000 Berlin 65, FRG

Abstract. In recent years considerable emphasis has been placed on the application of computer control to substrate feeding in baker's yeast production. Many investigators have studied the dynamics of different control loops and also first steps have already been taken to optimize the control algorithms. As far as the experimental verification of different strategies is concerned, the published work is, however, restricted to laboratory scale only. From current practice in baker's yeast industry which is most often based on open loop control with fixed schedule for carbon substrate addition, it is well known that mixing of substrate is playing an important role. The effects of the dynamics of mixing process in mechanicaly stirred tanks upon the performance of baker's yeast fermentation with fixed schedule have been investigated earlier by using a circulation model (Bajpai and Reuss, 1982). In the present contribution this framework is used in conjunction with a closed computer control of carbon substrate feeding based on exhaust air analysis. From experimental results for computer controlled fermentation in 10 and 2 000 l scale a decrease of productivity with increasing circulation times of liquid is observed. These results can be confirmed with the model calculations.

Keywords. Baker's yeast. Computer coupled fermentor. Mixing. Recirculation distribution model.

INTRODUCTION

Several investigators have proposed the application of computer process control to the fed-batch baker's yeast fermentation (Aiba, Nagai and Nishisawa, 1976; Dairaku, Yamasaki, Kuki, Shioya and Takamatsu, 1981; Dairaku, Isumoto, Morikawa, Shioya and Takamatsu, 1983; Cooney and Swartz, 1982; Pons, Bordet, Engasser, 1982; Wang, Cooney and Wang, 1977 and 1979). The rational behind these strategies is the maximization of biomass yield and productivities. This goal requires a control of the growth rate below a critical value in order to prevent ethanol formation which occurs in response to the Crabtree effect. To meet the second objective, maximal productivity, growth rate should be controlled as close as possible to this critical value. The fed-batch operation is then achieved by controlling the supply of carbon and energy source applying open or closed loop strategies. The open loop control is based upon a precalculated time depending feeding rate. For realization of a closed loop strategy the required feed back information from the process can be either ethanol concentration in the fluid bulk or head space of the fermentor or alternatively the respiration coefficient of the culture obtained from exhaust air analysis.

A considerable piece of work has been devoted to the optimization of the different strategies. As far as the experimental verification is concerned, however, most of the published literature is restricted to only laboratory scale of operation. From the excessive experience in open loop control (Zulauf) in the established industrial fermentation process it is well known that due to the difference in mixing quality the expectations of the laboratory optima can not be duplicated when scale of operation is in the order of several hundred cubicmeters. In a previous paper (Bajpai and Reuss, 1982) we have suggested a mathematical model for evaluating the performance of open loop feeding strategies with a fixed schedule at different mixing qualities. These calculations were based on a circulation-model to account for flow of the biosuspension and a two-environment model to account for the extent micro- and macromixing in the vessel.

The present contribution is aimed to apply this model for coupling of mixing and microbial kinetics in a slightly modified version to the problem of the feed back control of sugar feeding in the baker's yeast fermentation. The results of theoretical calculations will be supported from experiments with computer controlled sugar feeding in 10 and 2 000 l scale of operation. Also some experimental results from measurements of the circulation time distribution will be presented because these informations are essential prerequisites for the application of the proposed model.

COMPUTER SIMULATIONS OF THE OPEN LOOP FEEDING STRATEGIES

Since the proposed approach employs the mathematical modelling of the coupling of microbial reaction kinetics with mixing of substrate in the baker's yeast fermentation, the ideas and results for the system with fixed feeding schedule are briefly repeated below. For a more detailed description of the proposed discrete simulation approach for the micro-macromixer model the original paper should be consulted (Bajpai and Reuss, 1982).

For the mechanically stirred bioreactor equipped with a single impeller the physical system was divided into two regions; a micromixer in the immediate vicinity of the impeller and a macromixer consisting of the rest of the reactor volume (Fig. 1). Flow pattern through the macromixer is defined by the residence time distribution of the various fluid elements in it. Thus, the frequency with which the elements reenter the micromixer with the circulation stream has a distribution. All the fluid elements passing through the micromixer lose their identity due to micromixing therein and enter the macromixer as fresh elements loaded with the energy and carbon source which is fed as a concentrated solution into the impeller stream region. In the discrete simulation procedure used to predict the performance of the microbial reaction taking place in the bioreactor the macromixer was divided into a number of discrete elements. In each of these elements, the reaction process was simulated for a short period of time. For the reaction kinetics of baker's yeast in the vicinity of the critical growth rate an empirical model of the following type was developed from the experimental data of van Meyenberg (1969):

<u>growth of biomass:</u>

$$1/X(dX/dt) = \mu = \mu_{max}(S/(K_S + S)) \qquad (1)$$

<u>production of ethanol:</u>

$$1/X(dP/dt) = \nu = 0 \text{ for } S < S_{crit} \qquad (2a)$$

$$= \nu_{max}((S-S_{crit})/(K_p+(S-S_{crit}))) \text{ for } S \geq S_{crit} \qquad (2b)$$

At the end of the short time interval all the elements present in the macromixer contribute towards the recycle stream to the micromixer in order to provide space for the next fresh element which will enter the micromixer in the next time interval. The individual fractions of contributions are calculated from the circulation time distribution which dictates the probability that a fluid element which has spent a certain time in the micromixer will recycle to the micromixer.

In context with the present paper only the results from simulations of the fed-batch process will be shortly discussed. These results are shown in Fig. 2, where the predicted integral productivities as well as yields of biomass are plotted against growth rate. The case of complete micro-mixing was simulated by solving the kinetics without circulation time distribution and represents the upper limits of the system. Due to the very high inlet sugar concentration in the feeding stream (S_0=350kg/m³), the effect of circulation is significant and results in a decrease of critical growth rate even for small circulation times. Due to the fact that maximal yield is economically more important then productivity in this process these results can be interpreted in the following manner: In order to prevent ethanol formation lower growth rates and thus, reduced sugar feeding rates are required with increasing circulation times. Productivity of the process is then automatically reduced.

The first task to be solved in order to apply these simulation results to a practical situation is the experimental determination of circulation time distribution for the bioreactor of interest.

MEASURMENTS OF THE CIRCULATION TIME DISTRIBUTION

In the investigations on measuring circulation time distribution for single and multiple impellers up to 3 000 l volume we have applied the magneto flow-follower technique. In this method (Fig. 3) a small magnet is adjusted to the fluid density by placement in a hollow plastic sphere as a flow-follower. This particle induces an electromotive force in a coil which has the form of a loop around the impeller. Every time the sphere passes the coil in a set distance, the signal is send to an amplifier and for further data acquisition to a process computer. Fig. 4 shows an example of the experimental results of the probability density function of circulation times. We have also performed statistical tests on different distribution functions from which the log-normal distribution was found to be the best for correlating the experimental results. This is in accordance to what has been reported in literature for agitated vessels. The measurements were performed in different vessels with changing geometrical properties (impeller and vessel diameter, liquid height). Fig. 5 summarizes the results of the measurements resulting in a dimensionless correlation for the mean circulation time Θ which has the form

$$N\Theta = 0.76(H/D_T)^{0.56}(D_T/d_i)^{2.7} \quad (3)$$

where N is the agitation speed, H liquid height, D_T and d_i tank and impeller diameter, respectively.

EXPERIMENTAL SETUP AND STRATEGY FOR COMPUTER CONTROLLED BAKER'S YEAST FERMENTATION

The experiments for the process computer controlled baker's yeast fermentation were performed in stirred bioreactors with working volumes of 10 and 2 000 l, respectively. The compact fermentation units (Bioengineering AG, Wald, Switzerland) are equipped with standard measurement and control units. The computer coupling is illustrated in Fig. 6. Data acquisition of gas flow rate $\dot{V}_G$ (mass flow meter), reaction volume V_L (load cells), exhaust air analysis (paramagnetic and infrared analyzers) and substrate feeding (load cells) was performed with a Data Acquisition Unit (HP 3497 A) coupled to a computer (HP 9816).

For control of sugar feeding we applied the strategy suggested by Wang, Cooney and Wang (1979) in a slightly simplified manner. Instead of on-line estimation of biomass concentrations and specific growth rates from measured oxygen consumption and CO_2-evolution rates the feed rate F_0 in the proportional controller for the feeding pump

$$F = F_0(1-K_p(Q_{CO_2}-RQ_{soll}Q_{O_2})) \quad (4)$$

(Q_{CO_2} = Carbon dioxide evolutionrate, Q_{O_2} = Oxygen consumption rate, RQ_{soll} = set point for respiration coefficient = 1.05, K_p = controller gain)

was estimated from

$$F_0 = \frac{V_0}{S_0} X_C \mu e^{\mu t}/Y_{X/S} f_{cor} \quad (5)$$

(V_0 = initial volume, S_0 = sugar concentration in the feeding, $Y_{X/S}$ = biomass yield = 0.5)

using a fixed value for the specific growth rate at the beginning of the process (μ = 0.2 h^{-1}). As suggested by Wang, Cooney and Wang (1979) a correction factor f_{cor} with an initial value of 1.0 was incorporated in eq.(5). The actual value of f_{cor} was calculated in the following manner:

if $\overline{RQ} < (RQ_{soll}-0.05)$ then $f_{cor}=f_{cor}1.1$

if $\overline{RQ} > (RQ_{soll}+0.05)$ then $f_{cor}=f_{cor}0.9$

$\overline{RQ}$ is the average respiration coefficient during a time period of 10min.

Although we found by experimentation in different scales that this strategy is not the best one for the purpose of scale up, it was possible to maintain fairly stable operation conditions after readjustment of the control parameter in the different scale of operation. In Fig. 7 and 8 are shown the RQ-values and total biomass vs. process time for two experiments in the 10 and 1 000 l scale. It must be emphasized that optimization of the control strategy was not the primary objective at this stage of the investigations. Our main interest was rather to compare the results obtained in the different scales. The outcome of this comparison is shown in Fig. 9 where biomass concentration obtained in the 10 l scale is plotted along with the results from 1 000l and 2 000l volume. As expected from the model simulations discussed above the productivity in the larger scale is smaller, because of the larger circulation time of the liquid. The increase of growth rate after a shift in agitation speed in the 1 000 l scale is a further proof for this hypothesis. Because oxygen was not limiting in these experiments the increase of productivity is caused by a better distribution of the sugar in the systems volume.

COMPUTER SIMULATION OF THE CLOSED LOOP FEEDING STRATEGY

For the special purpose of applying the framework of coupling the intensity of substrate distribution and microbial reaction to the experimental conditions of the computer controlled feeding using a constant respiration coefficient a simplified model version was selected. The idea behind this model reduction is the longterm objective of these investigations. One of the problems of interest is the simulation of the system dynamics which requires the incorporation of gas liquid mass transfer. Obviously the basic model structure should be more simple for this complex task.

The simplified model is shown in Fig.10 where the entire vessel is divided into three CSTR's in series. The first CSTR (V_1) which account for a very small fraction of the total volume is used to model the conditions in the vicinity of the injection ring of the concentrated sugar feed. It is assumed that the volume of this vessel is equal to the cross section of the reactor multiplied with the height of the impeller blade. The two other vessels account for the conditions in the largest part of the reaction volume. The vessel No. 2 is an averaged value estimated from measured recirculation time distribution according to

$$i = \Theta^2/\sigma^2$$

where σ is the variance of the distribution.

Assuming constant biomass concentration in the three vessels the substrate balance equations can be written in the following form:

$$dS_1/dt = FS_0/V_1 + (Q/V_1)S_3 - ((Q+F)/V_1)S_1 - (F/V_{total})S_1 - \mu_1 X/Y_{X/S} - \nu_1 X/Y_{P/S} \quad (6)$$

$$dS_2/dt = ((Q+F)/V_2)(S_1 - S_2) - (F/V_{total})S_2 - \mu_2 X/Y_{X/S} - \nu_2 X/Y_{P/S} \quad (7)$$

$$dS_3/dt = ((Q+F)/V_2)(S_2 - S_3) - (F/V_{total})S_3 - \mu_3 X/Y_{X/S} - \nu_3 X/Y_{P/S} \quad (8)$$

with the recirculation stream

$$Q = V_{total}/\Theta$$

and μ and ν the kinetic expression for growth and ethanol formation (Eq.(1) and (2)). The average growth and ethanol production rate are then calculated from

$$\mu_{total} = (V_1\mu_1 + V_2\mu_2 + V_3\mu_3)/V_{total} \quad (9)$$

$$\nu_{total} = (V_1\nu_1 + V_2\nu_2 + V_3\nu_3)/V_{total} \quad (10)$$

The overall oxygen consumption rate is predicted with

$$Q_{O_2} = \mu_{total} X/Y_{X/O_2} \quad (11)$$

and the CO_2-evolution rate with

$$Q_{CO_2} = RQ_{soll} Q_{O_2} + \nu_{total} X/Y_{CO_2/P} \quad (12)$$

where Y_{X/O_2} and $Y_{CO_2/P}$ are the yield coefficients in kg biomass/mol O_2 and mol CO_2/kg ethanol, respectively.

For simulation of the entire growth curve the system is considered in a pseudo steady state for Δt = 5 min. process time, respectively. Thus, Eqs. (6) to (8) are iteratively solved with $dS/dt = 0$. The three substrate concentrations S_1, S_2 and S_3 are then used to predict the average growth and ethanol production rate (Eq. (9) and (10)). Next oxygen consumption rate (Eq. (11)) carbon dioxide evolution rate (Eq. (12)) and the respiration coefficient

$$RQ = Q_{CO_2}/Q_{O_2}$$

are predicted. With aid of a search algorithm the feeding rate F_j is then calculated which minimizes the deviation between actual RQ and set point RQ_{soll} = 1.05. Finally, from

$$dX/dt = \mu_{total} X - XF_j/V_{total}$$

and

$$V_{j+1} = V_j + F_j \Delta t$$

biomass concentrations and reaction volumes for the next time period of 5 minutes are predicted.

The results of the simulations using the circulation times calculated from Eq.(3) and the following kinetic parameters:

Y_{X/O_2} = 0.042 Kg DWT/mol O_2
$Y_{CO_2/P}$ = 21.7 mol CO_2/kg ethanol
μ_{max} = 0.42 h^{-1}, K_S = 0.25 kg Gluc/m^3
ν_{max} = 0.89 kg ethanol/kg DWT h
K_P = 0.15 kg Gluc/m^3,
S_{crit} = 0.03 kg Gluc/m^3,
$Y_{P/S}$ = 0.51 kg ethanol/kg Gluc

are presented in Fig. 11. As observed in the experiments the increase in circulation times results in a corresponding reduction of growth rates. Due to the reduced recirculation stream the concentrated inlet stream to the reactor is not diluted as much as for lower circulation times and local spots of high concentrations would develop in the first vessel. The control strategy therefore reduces the feeding rate as a result of which sugar concentration in the larger part of the vessel is below the critical value to match the critical growth rate.

REFERENCES

Aiba,S., S.Nagai and Y.Nishizawa(1976). Fed-batch culture of Saccharomyces cerevisiae: A perspective of computer control to enhance the productivity in baker's yeast cultivation. Biotechnol.Bioeng., 18, 1001 - 1016.

Bajpai,R.K. and M.Reuss (1982). Coupling of mixing and microbial kinetics for evaluating the performance of bioreactors. Can.J.Chem.Eng., 60, 384 - 392.
Cooney,C.L. and J.R.Swartz (1982).Application of computer control to yeast fermentation. In: Modelling and Control of Biotechnical Processes (Ed.: A. Halme). IFAC-Publications, Pergamon Press, 243 - 251.
Dairaku,K., Y.Yamasaki, K.Kuki, S. Shioya and T.Takamatsu (1981). Maximum production in a baker's yeast fed-batch culture by a tubing method. Biotechnol.Bioeng., 23, 2069 - 2081.
Dairaku,K., E.Izumoto, H.Morikawa, S. Shioya and T.Takamatsu (1983). An advanced microcomputer coupled control system in a baker's yeast fed-batch culture using a tubing method. J.Ferment.Technol., 61, 189 - 196.
Meyenberg van,K. (1969). Katabolit-Repression und der Sporungszyklus von Saccharomyces cerevisiae. Diss. No. 4279, ETH Zürich.
Pons,M.N., J.Bordet and J.M.Engasser (1982). A micro-mini computer hierarchical control system for a laboratory scale fermentor. In: Modelling and Control of Biotechnical Processes (Ed.: A. Halme). IFAC-Publications, Pergamon Press, 259 - 265.
Wang,H.Y., C.L.Cooney and D.I.C.Wang (1977). Computer-aided baker's yeast fermentation. Biotechnol. Bioeng., 19, 67 - 86.
Wang,H.Y., C.L.Cooney and D.I.C.Wang (1979). Computer control of baker's yeast production. Biotechnol. Bioeng., 21, 975 - 995.

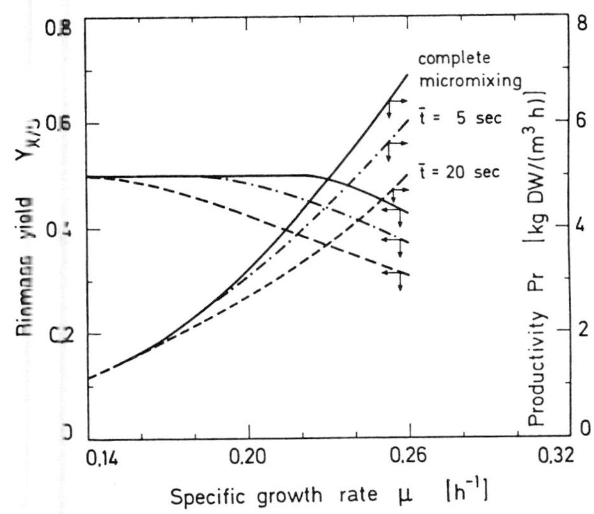

Fig. 2. Results of simulation of the open loop feeding strategie in baker's yeast production.

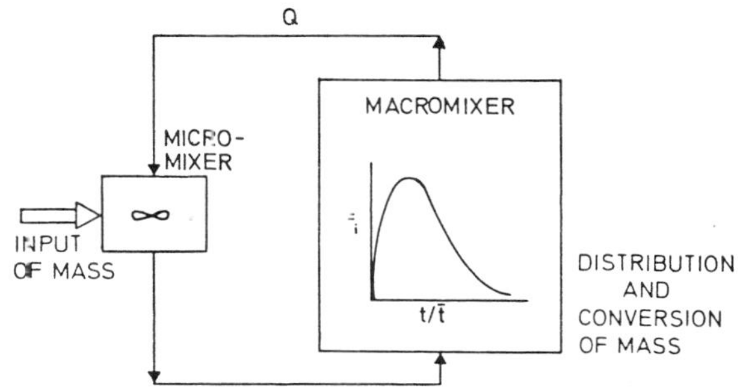

Fig. 1. Micro-Macromixer Model for mass distribution in an agitated vessel.

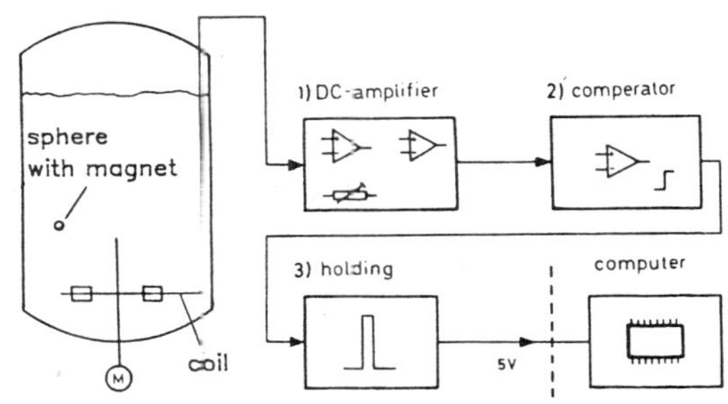

Fig. 3. Experimental setup for determination of circulation time distributions.

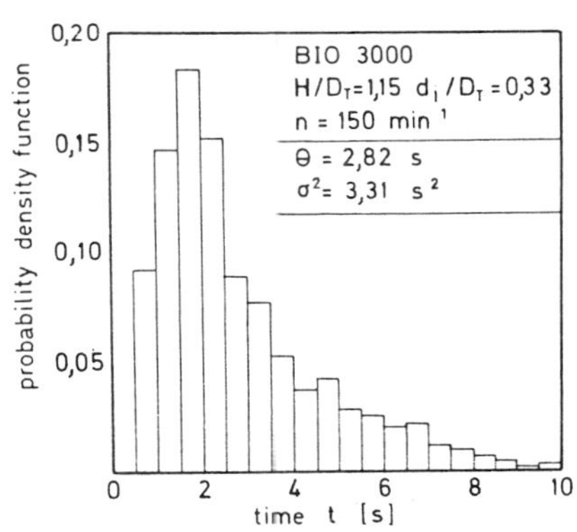

Fig. 4. Recirculation time distribution in the 3 000 l scale.

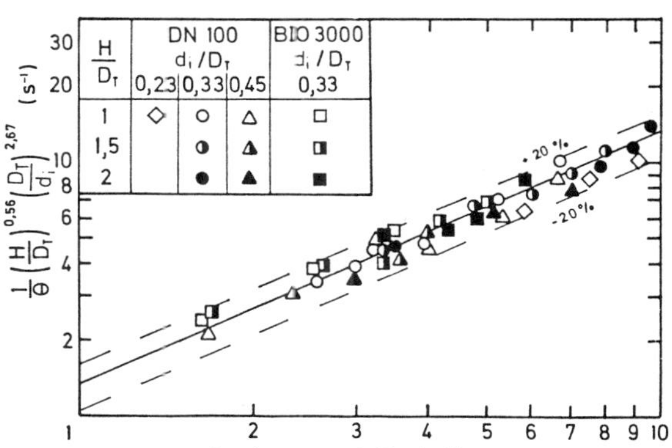

Fig. 5. Dimensionless correlation for mean circulation times.

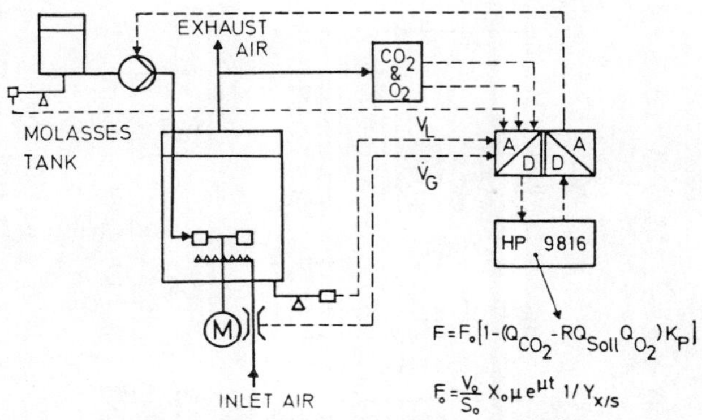

Fig. 6. Experimental setup for computer controlled baker's yeast production.

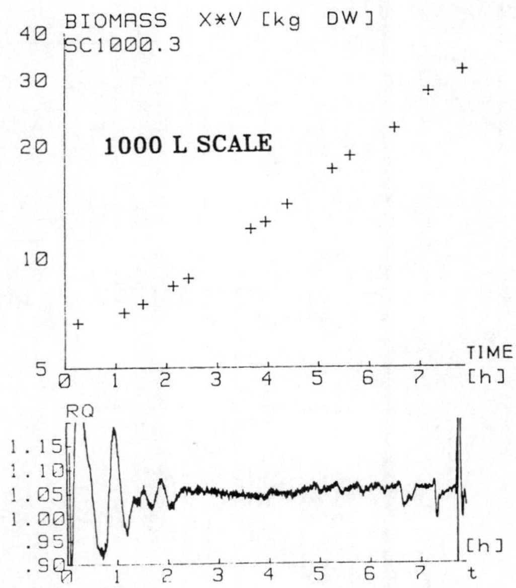

Fig. 8. Actual RQ and total biomass in the 1 000 l scale.

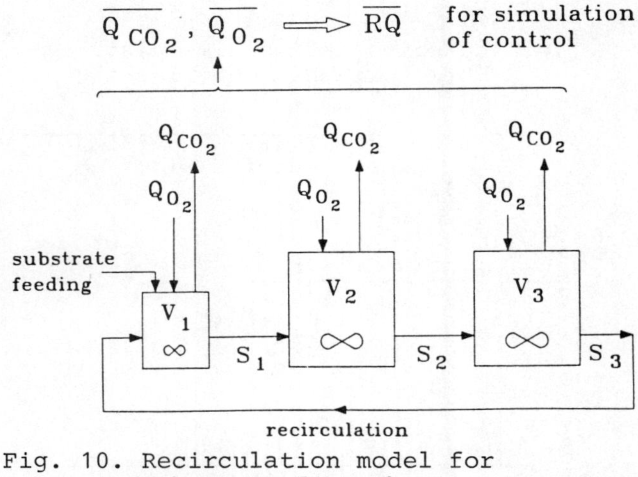

Fig. 10. Recirculation model for substrate distribution.

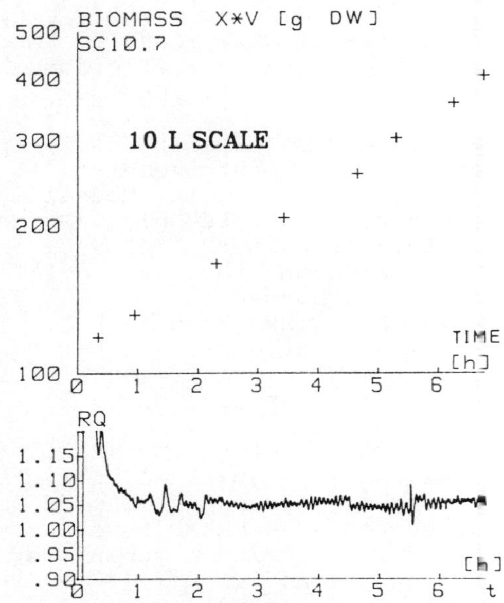

Fig. 7. Actual RQ and total biomass in the 10 l scale.

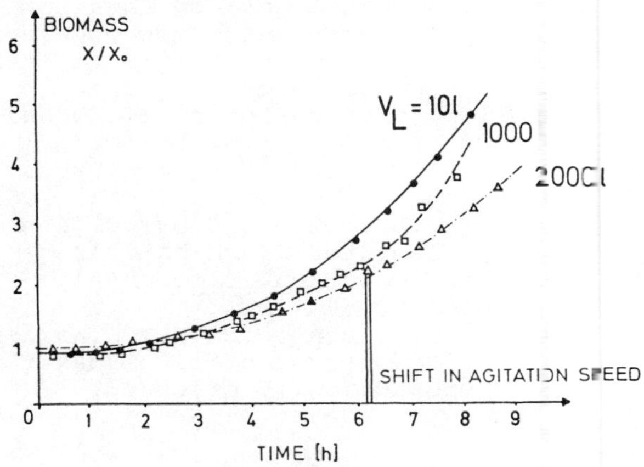

Fig. 9. Comparison of the growth curves in different scale of operation.

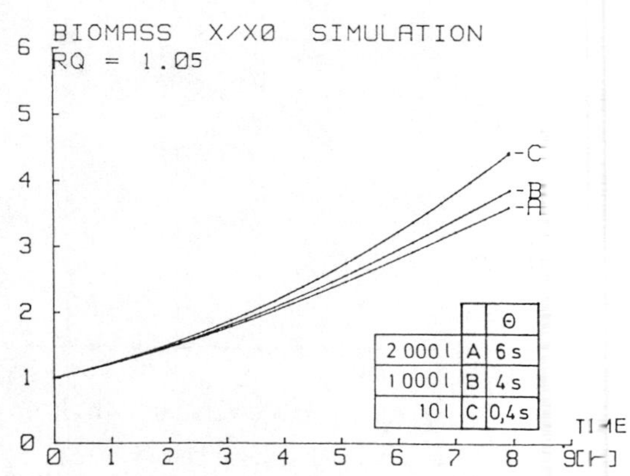

Fig. 11. Model predictions for growth curves in different scale of operation.

MATHEMATICAL MODELLING OF ALCOHOLIC FERMENTATION IN A GAS/SOLID BIOREACTOR — COMBINED EFFECTS OF SOLIDS MIXING AND NON-STEADY-STATE KINETICS

L. Röttenbacher, M. Schößler and W. Bauer

Technical University of Hamburg-Harburg, Eissendorfer Strasse 38, 2100 Hamburg 90, FRG

Abstract. A model is proposed to describe ethanol production and growth of yeasts in a gas/solid fluidized bed reactor. The model is based on the assumption that in the fluid bed each particle passes the spraying zone with a mean frequency, depending on solids mixing intensity, and gets a certain amount of substrate. Non steady state mass balances are established, considering substrate diffusion and product formation in the pellet. The balance equations are solved by numerical methods. From these results yield coefficients for ethanol and cell mass are derived and compared to the results of fermentation experiments performed in laboratory and technical scale fluid beds. The model is in good agreement with the experimental findings. The scale dependent mixing parameters, i.e. circulation time and specific substrate supply, are shown to influence reactor productivity significantly.

Key-words. Modelling, gas/solid fluidized bed fermentor; alcohol production; yeast growth; yield coefficients; reactor productivity

INTRODUCTION

Up to now fluidized bed technology has been applied to many operations including chemical, pharmaceutical and food processing. An entirely new way to use gas/solid fluidization is the application to biocatalysed reactions, e.g. production of ethanol, glutathione and growth of the yeast S. cerevisiae (Bauer 1984a, Bauer 1984b, Moebus 1982, Mishra 1982).

In the gas/solid fluid bed fermenter small pellets of microorganisms are fluidized with air or inert gas, the water activity of the microorganisms being in equilibrium with the fluidizing gas to assure high metabolic activity. Liquid substrates are sprayed directly upon the pellets by two phase nozzles.

For ethanol production the gas/solid fluidized bed offers some advantages compared to conventional fermentation techniques. As the cell density in the reactor is high (about 160 g dw/l) high volumetric productivities can be achieved. Product inhibition is very small, as the produced alcohol is continuously carried off by the fluidizing gas. As only gaseous substrates leave the reactor, there is no loss of sugar, which is the main cost factor in alcohol production. The biological heat can be directly used for the evaporation of water, so no external cooling is necessary.

Bauer (1984b) showed, that under anaerobic conditions the sugar can be completely converted to ethanol, while under aerobic conditions part of the substrate is used for cell growth.

The ratio between ethanol production and cell growth varied for different substrate supply and temperatures. A mathematical model to describe these effects is proposed in this paper.

REACTOR MODEL

The model to describe the performance of the fluidized bed reactor is based on the assumption that for each particle in the reactor there is a mean time interval τ between two spraying events, i.e. feed supplies and that for given reaction conditions the quantity of glucose $\Delta m_{G,p}$ fed to a particle passing the spraying zone A_s is constant (Fig. 1). During this period the particle is considered to be at nonstationary reaction conditions. Figure 2 shows the various possibilities of substrate utilization within a single pellet, depending to a high degree on circulation time τ and specific substrate supply $\Delta m_{G,p}$.

As known from aerobic cultivation of S.cerevisiae in submerged culture the glucose concentration has a large influence on cell yield and ethanol production. At very low glucose concentration no alcohol is found outside the cells. For high concentrations the main product is ethanol. Woehrer (1981) pointed out that the values for critical glucose concentration (the concentration, where the substrate is partly metabolized to ethanol) varies between 92 mg/l and 391 mg/l, depending on strain and physiological conditions used.

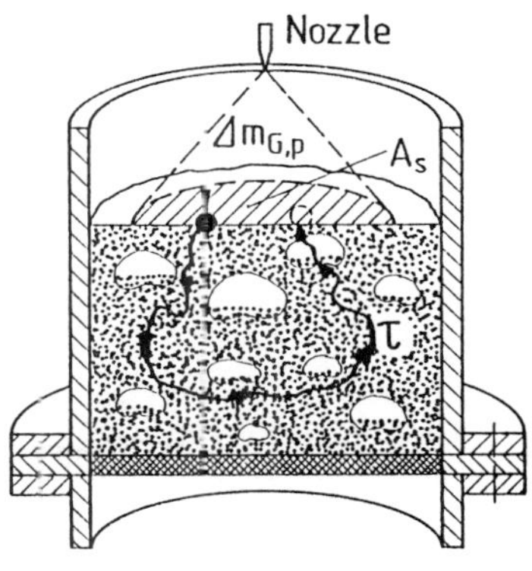

Fig. 1 Particle circulation and substrate supply in the fluidized bed gas/solid fermenter

Batch experiments (v. Meyenburg, 1969) showed that the alcohol produced is reconsumed for respiration and growth, when glucose is nearly completely consumed.

Thus inside the yeast pellets the following reactions have to be considered for mass balances.

1. Alcohol production from glucose
   $C_6H_{12}O_6 \rightarrow 2\ C_2H_5OH + 2\ CO_2$
2. Growth on glucose
   $\alpha_1\ C_6H_{12}O_6 + \alpha_2\ NH_3 + \ldots \rightarrow \beta_1\ C_6H_7O_3\ NH_2 + \beta_2\ H_2O + \ldots$
3. Growth on ethanol
   $\alpha_3\ C_2H_5OH + \alpha_4\ NH_3 + \ldots \rightarrow \beta_3\ C_6H_7O_3\ NH_2 + \beta_4\ H_2O + \ldots$
4. Oxidation of ethanol
   $C_2H_5OH + 3\ O_2 \rightarrow 2\ CO_2 + 3\ H_2O$

Oxidation of glucose is considered by combinating reactions 1. and 4. These four reactions take place simultaneously inside the pellet. If a pellet for a given substrate supply returns to the spraying zone after a short time (τ_1 in Fig. 2) the particle glucose concentration will be high and the main product will be ethanol. If the circulation time is large ($\tau > t_2$ in Fig. 2) glucose concentration will fall below a critical concentration, with ethanol production and consumption being equal. For very large circulation times, (e.g. τ_3 in Fig. 2), the produced ethanol will partly be reconsumed for growth and respiration. The ethanol yield will be low then, while cell yield increases.

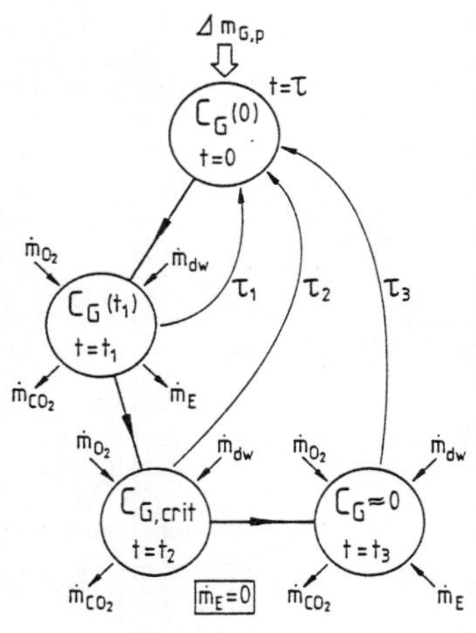

Fig. 2 Metabolism of a yeast pellet in the fluidized bed fermenter

MATHEMATICAL ANALYSIS

The following chapter deals with the mathematical description of fluidized bed behaviour combined with reaction kinetics. Particle circulation time and specific glucose supply are derived from fluidized bed hydrodynamics. Material balances for oxygen, glucose, ethanol, carbon dioxide and cell mass are established and theoretical yield coefficients for the reaction products are derived as functions of reaction and mixing parameters.

Circulation time and specific glucose supply

The calculation of circulation time τ bases on the fact that in fluidized beds for particle diameters larger than 100 µm the upward mass flow of solids $\dot{m}_{s,u}$ is solely caused by the action of bubbles rising in the bed, carrying particles with them (Rowe, 1965). This upward transport is counterbalanced by a downward stream of particles $\dot{m}_{s,d}$ (Fig. 3). The solids mass balance for the whole reactor is given by

$$\frac{d\,m_s}{dt} = \dot{m}_{s,u} - \dot{m}_{s,d} \overset{!}{=} 0 \qquad (1)$$

with

$$\dot{m}_{s,u} = \bar{\varepsilon}_b \cdot V \cdot \bar{u}_b \cdot f(1 - \varepsilon_{mf}) \cdot \rho_s \qquad (2)$$

$$(\bar{u}_b = \bar{u}_{s,u})$$

$$\dot{m}_{s,d} = \bar{u}_{s,d}(1-\bar{\varepsilon}_b) \cdot V \cdot (1-\varepsilon_{mf}) \cdot \rho_s \qquad (3)$$

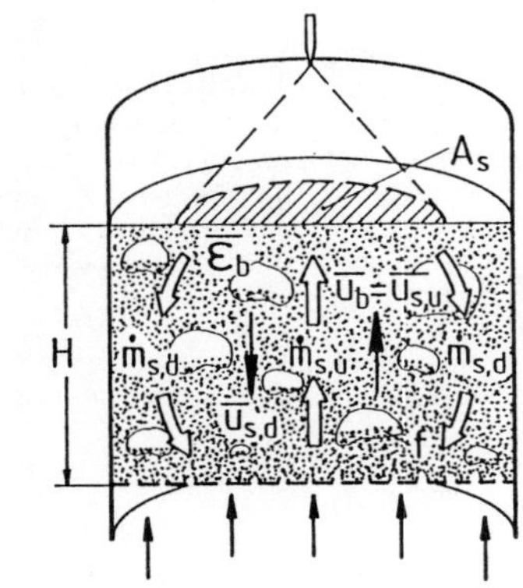

Fig. 3 Solids movement in a gas/solid fluidized bed

The mean values $\bar{\varepsilon}_b$ and $\bar{u}_b$ can be calculated by integration of the local values. For Geldart group B powders i.e. for the yeast pellets used here, these can be calculated according to Hilligardt (19[illegible]5). In Eq. (2) f is a factor describing the amount of solids carried with the bubbles compared to total bubble volume. According to Rowe (1973), f is about 0.6. From Equs. (1), (2) and (3) $\bar{u}_{s,d}$ can be calculated

$$\bar{u}_{s,d} = \frac{\bar{\varepsilon}_b}{(1-\bar{\varepsilon}_b)} \cdot f \cdot \bar{u}_b \qquad (4)$$

The time a particle needs to return to the bed surface is

$$t_z = H\left(\frac{1}{\bar{u}_{s,d}} + \frac{1}{\bar{u}_b}\right) = \frac{H}{\bar{u}_b}\left(\frac{1-\bar{\varepsilon}_b}{\bar{\varepsilon}_b \cdot f} + 1\right) \qquad (5)$$

where H is the expanded bed height. Taking into account that the spraying zone covers part of the bed surface only, the mean circulation time is given by

$$\tau = t_z \cdot \frac{A}{A_S} \qquad (6)$$

The specific substrate supply to a single particle $\Delta m_{G,p}$ for given feed $\dot{m}_G$, solid volume V_s and circulation time is given by

$$\Delta m_{G,p} = \frac{\dot{m}_G \cdot \tau}{V_S} \qquad (7)$$

Material balances

Oxygen. As Oxygen is continuously supplied with the gas stream and the specific oxygen consumption rate σ by the yeast can be considered a zero order reaction, the material balance for a spherical particle can be written in the stationary from

$$D_{O_2}\left(\frac{d^2c_{O_2}}{dr^2}+\frac{2}{r}\frac{dc_{O_2}}{dr}\right)-X\cdot\sigma=0 \tag{8}$$

with the boundary conditions

$$c_{O_2}=c_{O_2,0} \qquad \text{for } r = R$$

$$\left.\frac{dc_{O_2}}{dr}\right|_{r=0}=0$$

where R is the particle radius and $c_{O_2,0}$ the concentration of dissolved oxygen in water at saturation. Analytical solution of Eq. (8) yields,

$$c_{O_2}/c_{O_2,0}=\Phi^2/6\ \ r^2/R^2+1-\Phi^2/6 \tag{9}$$

where Φ is the Thiele module defined by Eq. (10)

$$\Phi=R\sqrt{\frac{X\cdot\sigma}{D_{O_2}\cdot c_{O_2,0}}} \tag{10}$$

Eq. (9) yields the distance Δr from the pellet surface where oxygen concentration becomes zero:

$$\Delta r=R(1-\sqrt{1-6/\Phi^2}) \tag{11}$$

A set of data for the calculation of Φ and Δr is presented in Table 1.

TABLE 1 Data for calculation of the Thiele module

$c_{O_2,0}$	$= 8,3 \cdot 10^{-3}$	g/l (Oxygen saturation for T = 25°C)
X	= 300 g/l	
D_{O_2}	$= 1,98\ \ 10^{-5}$	cm^2/s
R	$= 4\ \ 10^{-4}$	m
σ(25°C)	= 0,056	h^{-1}

For these data the Thiele module is Φ = 6,84 and Δr = 26,5 µm. This means that only for a part V_a of the pellet oxygen is available.

$$V_a=4/3\ \pi\ (R^3-(R-\Delta r)^3) \tag{12}$$

Therefore respiration and growth on ethanol, which require oxygen have to be considered in that part of the pellet only.

Glucose. The material balance is established under the following assumptions: i) the pellet is considered to be a sphere; ii) the substrate feed is distributed homogeneously on the pellet surface; iii) substrate diffusion inside the pellet can be described by Fick's law; iiii) product inhibition can be neglected. The balance equation is then given by

$$\frac{\partial c_G(r,t)}{\partial t}=D_G\left(\frac{\partial^2 c_G(r,t)}{\partial r^2}+\frac{2}{r}\cdot\frac{\partial c_G(r,t)}{\partial r}\right)-\frac{X}{\varepsilon_p}\cdot\left(\frac{M_G}{2\ M_E}\cdot\nu\,(c_{G(r,t)})+\frac{M_G}{M_{dm}}\cdot\mu_G(c_{G(r,t)})+k_m\right) \tag{13}$$

with the initial values

$$c_G(r,0)=\begin{cases}c_{G,0} & \text{for } r = R\\ 0 & \text{for } r \neq R\end{cases}$$

and the boundary conditions

$$r=0:\ \frac{\partial c_G}{\partial r}=0 \tag{13a}$$

$$r=R:\ \frac{\partial c_G}{\partial t}=-\frac{4\pi R^2\cdot c_{G,0}\cdot D_G}{\Delta m_{G,p}}\frac{\partial c_G}{\partial r}\quad \text{for } t\neq n\cdot\tau,\ \ n=0,1,2,\ldots \tag{13b}$$

$$c_G(R,t)=c_{G,0}\qquad t=n\cdot\tau \tag{13c}$$

Eq. (13c) holds, as the sugar concentration on the pellet surface at the end of one circulation is small compared to the feed concentration $c_{G,0}$.

As product inhibition can be neglected the specific ethanol production rate ν and the glucose bound growth rate μ_G can be described by Michaelis-Menten type kinetics.

$$\nu=\nu_{max}(T)\ \frac{c_G}{K_{S,\nu}+c_G} \tag{14}$$

$$\mu_G=\mu_{max_G}(T)\ \frac{c_G}{K_{S,\mu}+c_G} \tag{15}$$

The temperature dependence of ν_{max} and μ_{max_G} up to the temperature optimum of T = 36°C is given by Arrhenius equations

$$\nu_{max}(T)=\nu_0\ \ \exp(-E_\nu/R\ T) \tag{16}$$

$$\mu_{max_G}(T)=\mu_{0,G}\ \ \exp(-E_{\mu_G}/R\ T) \tag{17}$$

The values of the kinetic parameters derived from batch experiments in a stirred tank reactor are sumarized in TABLE 2.

TABLE 2. List of kinetic parameters

ν_0	= 6,15	10^9	h^{-1}
E_ν	= 57,6		kJ/mol
$K_{S,\nu}$	= 0,59		g/l
$\mu_{0,G}$	= 1,18	10^{10}	h^{-1}
$E_{\mu,G}$	= 61,5		kJ/mol
$K_{S,\mu}$	= 0,4		g/l
$\sigma_{0,E}$	= 1,2	10^{11}	h^{-1}
E_σ	= 70		kJ/mol
$\mu_{0,E}$	= 2,48	10^3	h^{-1}
$E_{\mu,E}$	= 24,6		h^{-1}
k_m	= 0,012		h^{-1}

Eq. (13) is solved numerically by an inplicit difference method with a regulation of time steps and small radial steps near the pellet surface, where the steepest gradients occur.
For each t = n τ the surface concentration is corrected according Eq. (13c) simulating a new substrate supply. The procedure is finished, when the concentration profiles at the end of two following circulation times are equal, i.e.

$$c_G(r,t)=c_G(r,t+\tau) \tag{18}$$

As according to the model a mean circulation time is used for each particle, steady state for the whole reactor is reached, when Eq. (18) is fullfilled.

Ethanol. The ethanol balance is established for the assumption that the amount of ethanol inside the pellet is in equilibrium with the ethanol content in the gas stream according to Henry's law for ethanol/water mixtures.

$$p_E=H(T)\ \ x_E \tag{19}$$

$$H(T) = p_E^o\ (T)\quad \gamma_E(T) \tag{19a}$$

The validity of this assumption was proved by measuring the adsorption of ethanol by yeast pellets (Röttenbacher, 1985). Figure 4 shows values for the Henry constant H derived from these experiments at different temperatures. These values are compared to the correlation of Williams (1983) with γ_E calculated according to Hamer (1965). The good agreement between measured and correlated values of H demonstrates that the moist pellets behave like a two phase system of water and ethanol.

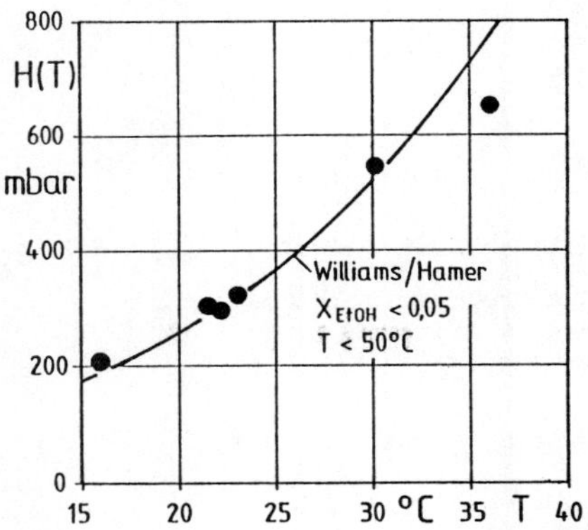

Fig. 4 Phase equilibrium ethanol /water for yeast pellets

Sorption experiments in a small fluidized bed showed that even for shallow beds sorption equilibrium is achieved. So at steady state of the reactor the ethanol concentration inside the pellets can be considered constant with time, yielding the following balance equation

$$\frac{dm_E}{dt} = 0 = X\int_{V_p} \nu(c_G(r,t)dV_p - X\quad V_a\ (\frac{3M_E}{M_{dw}}\ \mu_E + \frac{M_E}{3M_{O_2}}\ \sigma_E) - \dot{m}_{E,g} \tag{20}$$

where ν, μ_E, σ_E are the specific rates of ethanol production, growth on ethanol and oxidation of ethanol, V_a is the part of the pellet, where oxygen is available, $\dot{m}_{E,g}$ is the mass flow of ethanol to or from the gas stream for a single pellet.

The ethanol mass produced by one pellet during one circulation time is then given by

$$\Delta m_E = \int_\tau \dot{m}_{E,g}dt = X\int_\tau\int_{V_p} \nu(c_G(r,t))dV_p dt - \tag{21}$$
$$- X\ V_a(\frac{3M_E}{M_{dw}}\ \mu_E + \frac{M_E}{3M_{O_2}}\ \sigma_E\ )\ \tau$$

Eq. (21) was solved numerically from the values of $c_G(r,t)$ at steady state conditions, calculated from Eq. (13).

Cell mass. The increase of cell dry matter $\dot{m}_{dw}$ for one pellet with time is given by

$$\dot{m}_{dw}(t) = X\ (\int_{V_p} \mu_G(C_G(r,t)\ dV_p + \mu_E\ V_a) \tag{22}$$

The dry mass produced in one circulation time is

$$\Delta m_{dw} = X\ (\int_\tau \int_{V_p} (\mu_G(r,t))dV_p + \mu_E\ V_a\ \tau) \tag{23}$$

Eq. (23) was solved numerically using results for $C_G(r,t)$ from solving Eq. (13).

Carbon dioxide. As the solubility of CO_2 in water is very small the assumption is made that the CO_2 produced by a particle is quantitatively transferred to the gas phase, yielding:

$$\frac{dm_c}{dt} = 0 = X\ \frac{M_C}{M_E}\ \int_{V_p} \nu(C_G(r,t))\ dV_p\ +$$
$$+ X\ V_a\ \frac{2M_C}{3M_{O_2}}\ \sigma_E - \dot{m}_{C,g} \quad (2\text{-})$$

The produced mass of CO_2 by one pellet during one circle is given by

$$\Delta m_C = \int_\tau \dot{m}_{C,g}\ dt = \quad (25$$
$$= X\ \frac{M_C}{M_E}\int_\tau\int_{V_p} \nu(C_G(r,t)dV_p\ dt + X\ V_a\ \frac{2M_C}{3M_{O_2}}\ \sigma_E\ \tau$$

Eq. (25) can be solved numerically for a given glucose concentrationprofile at stationary conditions.

Yield coefficients

The yield coefficients for ethanol, cell mass and CO_2 can be calculated as the ratio of the amount produced during one circulation time and Δm_G, the sugar supplied

$$\Delta m_G = \Delta m_{G,p}\quad V_p \tag{26}$$

Thus the yield coefficients are:

$$\text{Ethanol}\qquad :\ Y_{P/S} = \frac{\Delta m_E}{\Delta m_G} \tag{27}$$

$$\text{Cell mass}\qquad :\ Y_{X/S} = \frac{\Delta m_{dw}}{\Delta m_G} \tag{28}$$

$$\text{Carbon dioxide}\qquad :\ Y_{C/S} = \frac{\Delta m_C}{\Delta m_G} \tag{29}$$

with Δm_E, Δm_{dw}, Δm_C calculated from Equs. (21), (23), (25).

FERMENTATION EXPERIMENTS

Experimental Equipment and Procedure

Two reactors of different size are used for the experiments (Fig. 5 and 6).

The yeast mass in the laboratory scale fermenter was varied from 1.5 to 3 kg having a mean pellet diameter of 800 µm while in the demonstration scale reactor the bed mass was 20 kg, the mean diameter being 500 µm. The calculated circulation times ranged from 5.6 s to 10.8 s in the laboratory scale and 13.5 s in the technical reactor.

Inlet air was humidified to avoid drying of the pellets. The fluidization behaviour was controlled by an optoelectronical system. The substrate containing glucose concentrations ranging from 50 - 200 g/l, ammonium sulphate, yeast extract and KH_2PO_4 was continuously fed to the reactor by a two phase nozzle. The ethanol, CO_2 and water content in the fluidizing gas was continously monitored by IR-analyzers. The fermentation temperature was varied from 20 - 34°C. The specific glucose flow rate ranged from 19 - 190 g/l_p h.

Figure 7 shows the time course of ethanol, CO_2 and water content in the exit gas stream. The initial maxima for ethanol and CO_2 are typical for these experiments and may be caused by a short lag phase for growth, while the ethanol production starts immediately. Steady state conditions for the reactor with flow rate of ethanol, CO_2 and water staying constant was reached between 15 and 45 minutes after start up. The total time for the experiments ranged from 1 to 11 hours. At the end of the experimental runs the increase in yeast dry matter $m_{dw,e}$ was estimated by weighing and measuring the moisture content of the pellets.

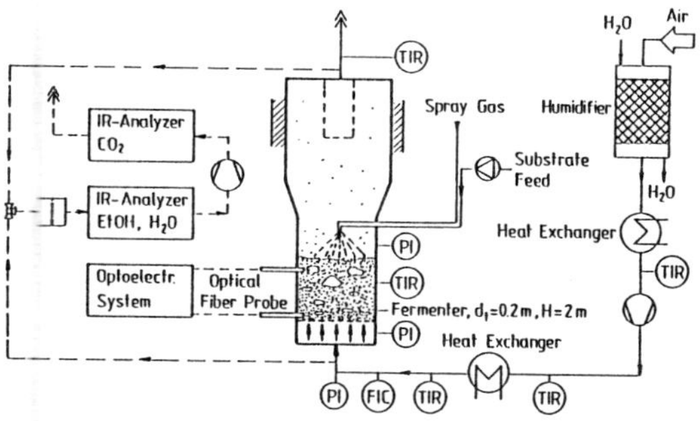

Fig. 5 Laboratory scale fluidized bed gas/solid fermenter

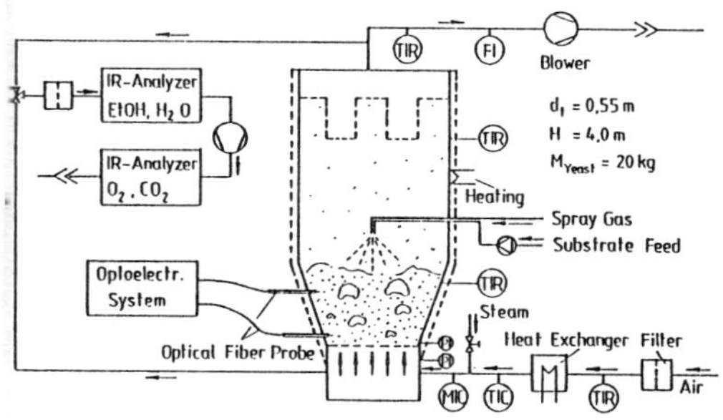

Fig. 6 Demonstration scale fluidized bed gas/solid fermenter

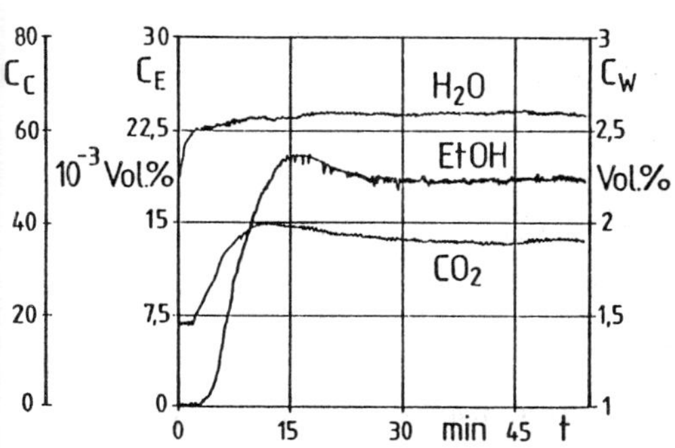

Fig. 7 Continuous gas analysis for determination of steady state in fluid bed fermentation

Determination of yield coefficients

The cell yield for each experiment was calculated as the ratio of increase in cell dry matter $m_{dw,e}$ to the total amount of sugar, supplied to the reactor $m_{G,e}$.

$$Y_{X/S_e} = \frac{m_{dw,e}}{m_{G,e}} \tag{30}$$

The ethanol and CO_2 yield is given by the ratio of mass flow out of the system and the glucose flow rate $\dot{m}_G$ to the reactor at stationary conditions.

$$Y_{P/S} = \frac{\dot{m}_{E,st}}{\dot{m}_G} \tag{31}$$

$$Y_{C/S} = \frac{\dot{m}_{C,st}}{\dot{m}_G} \tag{32}$$

RESULTS AND DISCUSSION

Yield coefficients with respect to substrate supply

Figure 8 shows the results of fermentation experiments in the laboratory scale reactor at a temperature of T = 21 - 22°C and various specific feed rates $\dot{m}_{G,V}$.

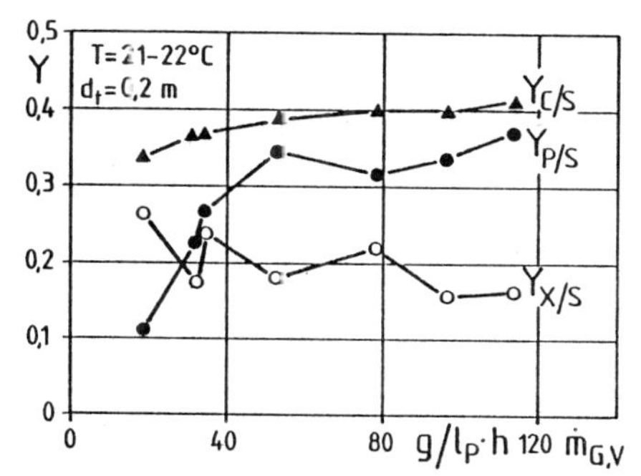

Fig. 8 Yield coefficients for experiments in the laboratory scale reactor for different substrate feed rates

There is an immediate decrease for ethanol yield, when the feed rate is below a critical value. The carbon dioxide yield instead decreases only slightly while the cell yield starts to increase.

Comparison of experimental and theoretical derived yield coefficients

Figures 9 - 11 show the comparison between theoretical and the measured values of the yield coefficients of ethanol, cell mass and carbon dioxide.

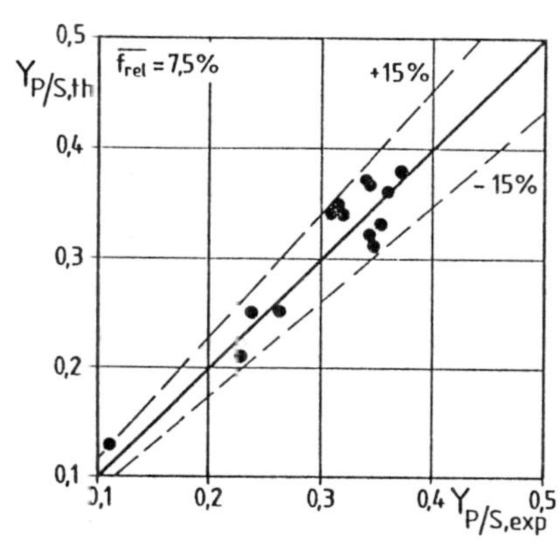

Fig. 9 Comparison of theoretical and experimental ethanol yield

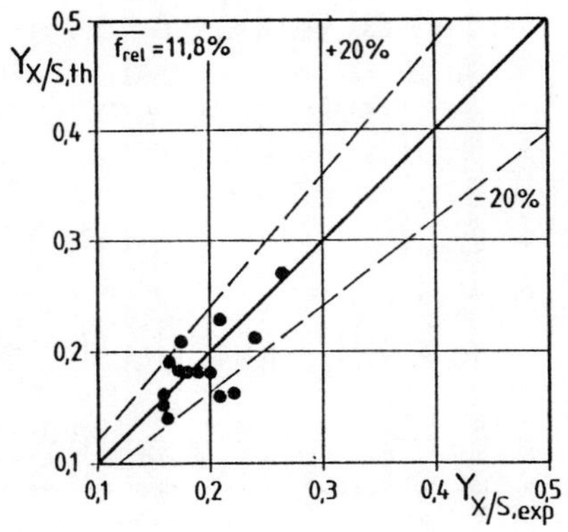

Fig. 10 Comparison of theoretical and experimental cell yield

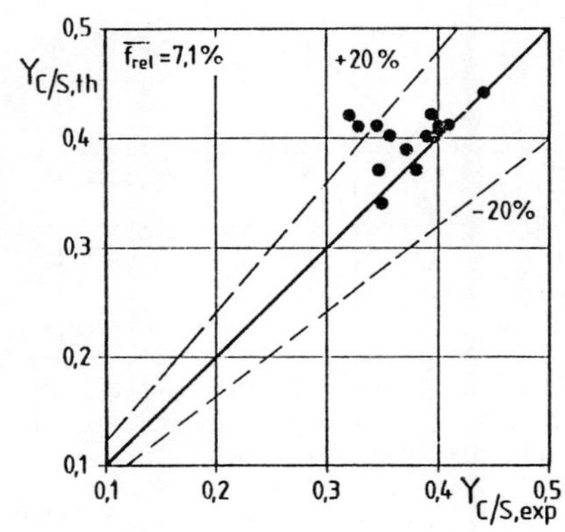

Fig. 11 Comparison of theoretical and experimental CO_2-yield

The difference between the experimental values for $Y_{P/S}$ and the yield coefficient predicted by the model does not exceed 15 %, with a mean deviation of 7.5 %. For carbon dioxide yield the mean deviation is smallest. In the relatively short time of the experimental runs only small amounts of biomass are produced compared to the cell mass, fed to the reactor . Therefore little errors in weighing and estimation of moisture content will be of great influence on the measured values. Taking this into consideration it can be stated that the model gives a good description of the experimental data.

Reactor efficiency for ethanol production with respect to fluid bed hydrodynamics and reaction kinetics

The ethanol production rate $\dot{m}_{E/dw}$ based on total dry mass in the reactor for reaction conditions being constant is given by

$$\dot{m}_{E/dw} = \frac{Y_{P/S}}{X\ V_s}\ \dot{m}_G \tag{35}$$

The yield coefficient $Y_{P/S}$, depending on the parameters $\Delta m_{G,p}$ and τ derived from solids mixing as well as on the kinetics at given temperature is calculated from Eq. (27). Dividing Eq. (35) by the maximum specific ethanol production rate ν_{max} for a given reactor temperature yields the efficiency factor η_{ws} for the fluidized bed reactor.

$$\eta_{ws} = \frac{\dot{m}_{E/dw}}{\nu_{max}} = f(\tau,\ \Delta m_{G,p},\ T) \tag{36}$$

Figure 12 shows the calculated reactor efficiency as a function of the specific glucose supply $\Delta m_{G,p}$ for different circulation times and temperatures together with the results of fermentation experiments in the laboratory scale reactor at T = 21°C and 22°C respectively and three different circulation times and the technical reactor with τ = 13,5s and a temperature of 29°C.

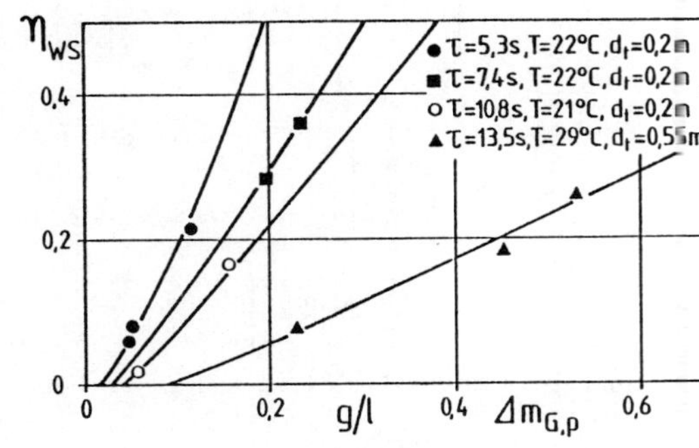

Fig. 12 Efficiency of the fluidized bed gas/solid fermenter as a function of mixing parameters

Even though the volumetric productivity of the system is high (in the demonstration scale reactor more than 40 g/l h) the efficiency of the fermenter can still be raised. As the specific substrate supply by one nozzle is limited by agglomeration effects, an optimization of the fermenter productivity can be achieved by increasing the number of spraying nozzles as well as by increasing solids mixing. Short circulation times will result, and high reactor efficiencies will be achieved as Fig. 12 clearly demonstrates.

Sugar concentration profiles inside a single pellet

Figures 13 and 14 show calculated sugar concentration profiles inside a pellet with a radius of 400 µm for varying specific substrate feed and a circulation time of τ = 5 s. It can be seen that only in the case of high substrate supply (Fig. 14) the whole pellet is supplied with glucose. For low sugar supply the glucose is distributed only over part of the pellet volume resulting in a small specific productivity of the pellet. This clearly demonstrates that high efficiency of the reactor can only be achieved by optimizing substrate supply and mixing behaviour of the system.

Conclusion

The presented model which combines the mixing behaviour of a gas/solid fluidized bed fermenter for ethanol production and the non stationary reaction kinetics inside a single yeast particle is appropriate to describe the reactor performance. From this model the ethanol- and cell yield for given reaction conditions can be predicted. The efficiency of the reactor was shown to be strongly dependent on mixing intensity and feed distribution.

Up to now the model is restricted to the description of experiments of relatively short times where changes in pellet-volume and diameter due to growth can be neglected. So further investigations will be made to expand the model to situations where growth effects must be taken into consideration.

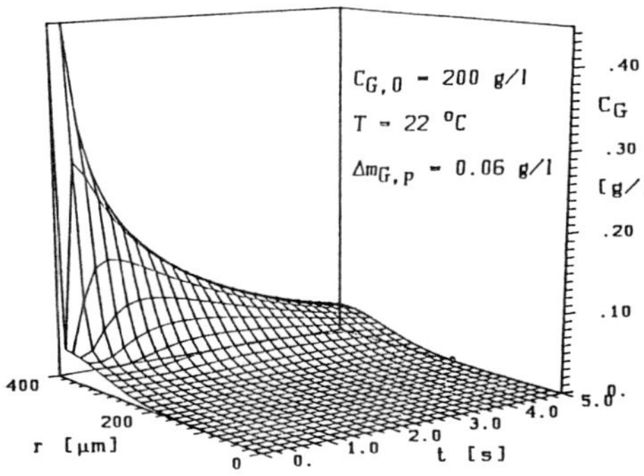

Fig. 13 Calculated sugar concentration profile inside a pellet at low glucose feed

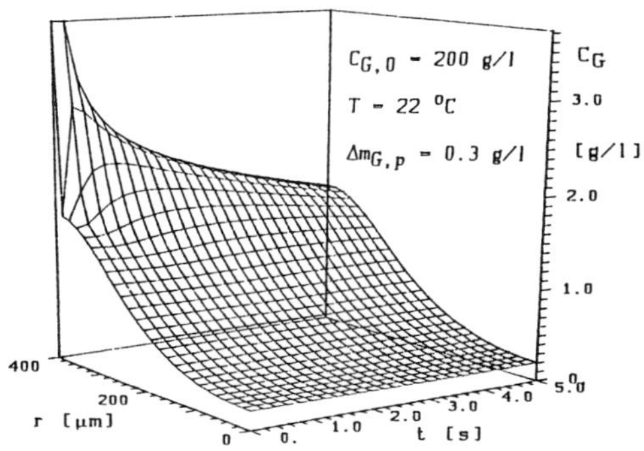

Fig. 14 Calculated sugar concentration profile inside a pellet at high glucose feed

Symbols

A bed surface, m^2
A_s bed surface to which substrate is supplied, m^2
c_i concentration of species i, g/l
D_i diffusion coefficient of species i, m^2/s
E activation energy, J/mol
H expanded bed height, m
m maintenance coefficient, h^{-1}
$K_{S\nu}$ Michaelis-Menten constant for ethanol production, g/l
$K_{S\mu}$ Michaelis-Menten constant for growth on glucose, g/l
M_i molecular weight of species i, g/mol
m_i mass of species i, g
$\dot{m}_i$ mass flow of species i, g/s
p_i partial pressure of species i, mbar
p_i^o vapour pressure of pure species i, mbar
r radial distance, m
R pellet radius, m
R gas constant, J/mol K
t time, s
T temperature, K
u velocity, m/s
V expanded bed volume, m^3
V_a pellet volume, where oxygen is available, m^3
V_p pellet volume, m^3
X density of dry mass inside the pellet, g/l
x_i mole fraction of species i, -
$Y_{i/j}$ yield coefficient between species i and j, -
γ_E activity coefficient for ethanol, -
$\bar{\varepsilon}_b$ mean bubble gas holdup, -
ε_{mf} bed porosity at minimum fluidization, -
ε_p particle porosity, -
η_{ws} efficiency factor for fluid bed fermenter, -
μ_G spec. growth rate on glucose, h^{-1}
μ_E spec. growth rate on ethanol, h^{-1}
ν spec. ethanol production rate, h^{-1}
ρ density, kg/m^3
σ spec. oxygen consumption rate, h^{-1}
τ circulation time, s
Φ Thiele module, -

Subscripts

b bubble
C carbon dioxide
dw dry weight
E ethanol
G glucose
g gas
p particle
s solid
st stationary
W water

References

Bauer, W., L. Röttenbacher (1984a). Regelung der Substratzufuhr eines Gas/Feststoff-Wirbelschichtfermenters über die Bestimmung des Fluidisationsverhaltens. Zeitschrift f. Lebensmitteltech., 1, 18 - 23

Bauer, W. (1984b). Gas/Solid Fermentation Systems-Development and Application. Prepr.34th CschE Conference, Quebec, Kanada, 29.09. - 03.10.84

Hamer, G. (1965). Stripping in sparged fermentation systems. Biotech. a. Bioeng., 12, 199-214

Hilligardt, K., J. Werther (1985). Chem. Ing. Tech. in press

Mishra, I.M., S.A. El-Temtamy and K. Schügerl (1982). Growth of Saccharomyces cerevisiae in Gaseous Fluidized Beds. Europ. J. Appl. Microb. Biotechnol., 16, 197 - 203

Moebus, O., M. Teuber, H. Reuter (1981). Growth of Saccharomyces cerevisiae in form of solid particles in a gaseous fluidized bed. Kieler Milchwirtschaftliche Forschungsberichte, 33

Röttenbacher, L. (1985), Thesis in preparation

Rowe, P.N., B.A. Partridge, A.G. Cheney, G.A. Henwood, E. Lyall (1965). Trans. Inst. Chem. Eng., 43, T 271

Rowe, P.N. (1973). Estimation of solids circulation rate in bubbling fluidized beds. Chem. Eng. Science, 28, 979 - 980

Von Meyenburg, K. (1969). Katabolit-Repression und der Sprossungszyklus von Saccharomyces cerevisiae. Vierteljahresschr. Naturforsch. Ges. Zürich, 114, 113 - 222

Williams, L.A. (1983). Theory and modelling of ethanol evaporative losses during batch alcoholic fermentation. Biotech. a. Bioeng., 25, 1597 - 1612

Woehrer, W., M. Roehr (1981). Regulatory aspects of baker's yeast metabolism in aerobic fed-batch cultures. Biotech. a. Bioeng., 23, 567 - 581

ON PARAMETER IDENTIFICATION FOR COMPLEX BIOTECHNICAL SYSTEMS

A. Munack

Institut für Regelungstechnik, Universität Hannover, Hannover, FRG

Abstract. The parameter identification problem is treated for an SCP process in a tower loop bioreactor (fed batch process). The emphasis is laid on the sensors needed for proper identification of the unknown four parameters in the given model. In particular the following three questions are solved: a)Which types of measurements must be used (dissolved oxygen in liquid phase of the reactor - cell mass - outlet gas analysis), b) how many sensors for dissolved oxygen must be installed, c) where should these sensors be posed along the reactor column? The calculations are based on evaluations of the Fisher information matrix. Identification results are given, based on measurements taken at a bench-scale plant.

Keywords. Parameter identification; tower loop reactors; SCP process; distributed-parameter models; optimal sensor allocation.

INTRODUCTION

A fundamental task in design of biotechnical processes and their control is system modelling. This task consists of a series of steps, starting with structural modelling of the reactor and the biological reactions. Various degrees of complexity may be attained in this step, depending on the degree of depth involved in modelling, which means for the biological equations the decision whether an unstructured, a structured, or even a genetically structured model is used to describe the behaviour of the microorganisms.

In this paper, another case is treated, where an unstructured model is used to describe biomass formation, whereas the complexity is introduced by the reactor model. An SCP process is considered which is performed in a tower loop bioreactor. This results in a set of partial differential equations describing the time- and space-dependent concentration profiles along the reactor column and the external loop or bypass.

A further step in modelling of these systems is the identification of unknown system parameters, which may be physical, chemical or biological parameters. However, it turns out that some of these parameters may not be fixed by identification procedures in advance, but have to be identified on-line in order to yield meaningful results for control of these processes. In particular, predictive control algorithms, like the OLFO-algorithm - Schubert(1970) - need actual parameter estimates in order to predict the system behaviour sufficiently precisely for the near future and to allow for computation of control inputs in order to optimize a given economical performance index.

However, as already stated above, for biotechnical systems the model equations are quite complex, and therefore several problems do arise with on-line computations. The main questions in this context are

- can the model equations be simplified or - in case of distributed parameter models - may they be approximated by lumped parameter systems,
- may decomposition/coordination methods be applied to the process equations in order to accelerate computation of system parameters and input control functions,
- can these measurements be determined, which give most information about the process, such that only a small number of sensors is needed (which also results in considerable savings of computation time, as will be demonstrated).

While the second problem was addressed in former publications, cf. Munack(1984), Munack(1985), the first will be considered briefly at the end of this paper, and the third will be discussed in detail in the following.

DESCRIPTION OF THE PROCESS AND THE MODEL EQUATIONS

A schematic diagram of the used double-reactor system (column and loop) is shown in Fig. 1. Since only the column is gassed from the bottom, gas phase (G) balances have only to be formulated in this part, while liquid phase balances appear in the column (F) and in the bypass (B). In modelling the growth process, only mass balances are taken into account since isothermal conditions are assured by control, and momentum is negligible. The task of the control is to guarantee maximal SCP production with minimal cost, which means low substrate (S) and oxygen (O) consumption.

A general model, based only on the nutrients O_2 and S, the products X (cell concentration) and CO_2, and the spatially varying gas velocity u_G leads to a nonlinear system of six parabolic partial differential equations (PDEs) and five plug flow equations, cf. Luttmann(1980), Luttmann/Munack/Thoma(1985). For control purposes in a limited area of operating conditions, several simplifications may be introduced in order to reduce this large number of system equations; these are:

1) biomass and substrate are well mixed in the liquid phase,
2) the substrate concentration is nowhere growth-limiting (extended culture),

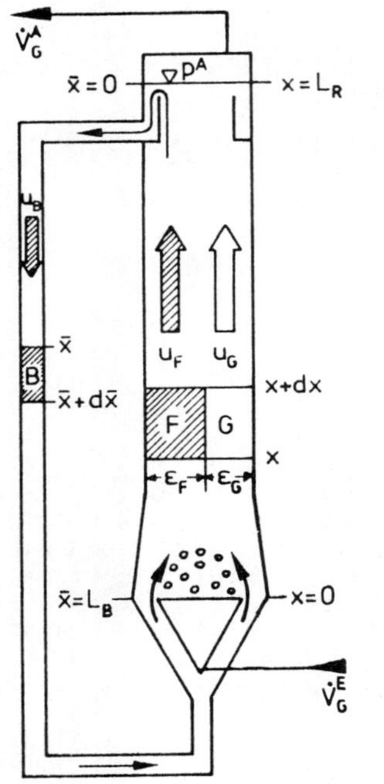

Fig.1: The tower loop bioreactor

3) the oxygen mole fraction falls quadratically in the gas phase of the reactor,
4) the respiration quotient is rq=0, (RQ=1); therefore the velocity of the gas phase is reciprocal to the pressure profile in the reactor,
5) residence time in the loop is small in comparison with that one in the column. Conditions in the loop are assumed to be quasistationary.

By these means the model is reduced drastically to a quasilinear PDE of parabolic type, describing the dissolved oxygen concentration in liquid phase of the column, an algebraic equation for the gas phase in the reactor, an implicit algebraic equation for the dissolved oxygen concentration in the loop, and an ordinary differential equation for the biomass. These simplified model equations are given in eqs. (A1,a-d) in the appendix. There also the most important symbols are summarized.
The system equations are coupled via the states and the boundary conditions; this means that c_X enters into the dissolved oxygen equations, and on the other hand both c_{OF} and c_{OB} enter into the cells' equation. c_{OF} and c_{OG} are also coupled, and the physical connection of reactor column and loop is modelled via coupling terms in the boundary conditions.

PARAMETER IDENTIFICATION

Four parameters have turned out to be unknown; these are two fluiddynamical parameters, k_La and K_{St}, both describing oxygen transfer from gas phase into liquid phase along the reactor column, and two biological parameters, the metabolic quotient $q_{O/Xm}$ and the yield coefficient $y_{X/O}$. Note that the first two parameters only enter into the oxygen equations and the last one only into the cells' equation. The metabolic quotient is to be found in both balances.

Measurements can be taken of the dissolved oxygen concentration c_{OF} at distinct points in the reactor column, of the cells' concentration, and of the outlet gas mole fractions, yielding the overall oxygen transfer rate $\overline{OTR}$. Since an adaptive optimal control law (predictive control) was considered for this fed-batch process, a fast parameter identification routine was needed. Therefore, and due to the easy decomposition of the system equations into a fluiddynamical (eqs. A1a,b c) and a biological (eq. A1d) part, and the almost perfect grouping of the unknown parameters, decomposition/coordination methods were used for resolution of the parameter identification problem. The treatment of the metabolic quotient, which enters into both subsystems, requires some additional considerations from the theoretical point of view, cf. Munack/Thoma(1983).

The parameter identification problem was treated by using the method proposed by Chavent (1974). It consists in reformulating the problem as optimization problem with output least squares error criterion and solving this by optimal control theory, cf. Lions (1971).

For parameter identification a model of the system is used with estimated parameters, and corresponding measurements to those at the system are calculated with this model.This enables to define measurement errors which are weighted in a quadratic criterion. To be concrete, the following functional is used (^ denotes modelled measurements with estimated parameter set $\left[\widehat{k_La^E}, \widehat{K_{St}}, \widehat{q_{O/Xm}}, \widehat{y_{X/O}}\right]^T = \hat{\underline{P}}$)

$$J(\hat{\underline{P}}) = \int_0^T \Big\{ W(t) \Big[\sum_{i=1}^{N} \left(\widehat{c_{OF}}(x_i,t) - c_{OF}^M(x_i,t) \right)^2 - + \sigma \cdot \left(\widehat{\overline{OTR}}(t) - \overline{OTR}^M(t) \right)^2 + \kappa \cdot \left(\widehat{c_X}(t) - c_X^M(t) \right)^2 \Big] \Big\} dt.$$

Due to the decomposition/coordination method, the functional is divided into subfunctionals, which means that the first two terms are used to identify k_La, K_{St}, and $q_{O/Xm}^{(O)}$ together with equations (A1,a-c), and the last term is used to identify $q_{O/Xm}^{(X)}$ and $y_{X/O}$ using the cells' equation (A1d).

The equality $q_{O/Xm}^{(O)} = q_{O/Xm}^{(X)}$ has to be satisfied at the end of the iteration process. The functional modifications which are necessary due to the decomposition/coordination theory (zero-sum modifications) do not play a major role in this case of a very weakly coupled system and may be neglected for the computations.

Since parameter identification in the cells' equation is very easy, in the following only the first identification problem will be discussed further. In order to apply highly efficient numerical optimization algorithms, the gradient of the functional J w.r.t. the unknown system parameters is needed. It can be evaluated by means of the so-called adjoint state ζ , which is given by eq. (A2).
Then the gradient components are computed by the calculations given by eqs. (A3,a-c) and (A4).

In formulas (A3) and (A4) some modifications are introduced in comparison to the expressions published in Luttmann/Munack (1983). These additions take into account the quadratic decrease of the gas phase and the coupling of the oxygen concentrations in gas phase and liquid phase of the reactor. Particularly in case of high oxygen consumption in the reactor and low gas phase velocities, these additional terms result in definitely higher accuracy.
The quite lengthy formulas give an insight into the problems involved in resolution of parameter identification problems for such processes with several balances.

OPTIMAL CHOICE OF SENSORS AND THEIR ALLOCATION

General considerations

As demonstrated, for parameter identification the computation of the optimality system - consisting of the system equations and the adjoint equations - has to be performed on-line. Using the discrete space/discrete time method for the solution of the PDEs, the computation time needed is more or less directly proportional to the number of grid points involved. It is very useful and time-saving, to take non equidistant spatial discretizations, with smaller step-size in these areas where the norm of the second spatial derivative of the solution profile is large. For the system equations, this is the case in the area from the bottom up to 15 to 20 % of height of the reactor column. For the adjoint system, every pointwise measurement introduces jumps of the first spatial derivative; thus it is advisable to pose some more node points around the measurement locations. This leads to the fact that taking more and more sensors, the computation time for the adjoint equation becomes longer and longer. Together with the increasing costs for sensors and measuring devices if a larger number of sensors is installed this leads to the necessity to solve the following questions:

- **How many pick-ups** for measuring the dissolved oxygen concentration along the reactor column should be used?
- **Which are the best positions** along the reactor column to install these sensors?

Furthermore, a more general consideration should also give some hints with respect to a quantification of the information coming from the different types of measurements. This leads to the question

- **which types of measurements** should be used for identification of the system parameters: dissolved oxygen measurements - outlet gas analysis, yielding the overall oxygen transfer rate $\overline{OTR}$ - cell mass measurements (quite problematic to take on-line)?

A solution to these questions can be given by means of application of optimal sensor allocation methods. A great variety of such methods was discussed in the literature, cf. eg. the survey by Kubrusly/Malebranche(1983). A relatively simple approach is used here, which is based on the Fisher information matrix - cf. Qureshi/Ng/Goodwin(1980).

Description of the used method

To apply this method, the measurements are assumed to be corrupted by white noise, yielding

$$s_1^j(t) = c_{OF}^M(x_j,t) = c_{OF}(x_j,t) + w_1^j(t),$$

$$s_2(t) = \overline{OTR}^M(t) = \overline{OTR}(t) + w_2(t),$$

$$s_3(t) = c_X^M(t) = c_X(t) + w_3(t).$$

w_1^j denotes the measurement noise of the j-th dissolved oxygen sensor, $w_2(t)$ the noise of the $\overline{OTR}$ measurements, and $w_3(t)$ of the cells' concentration measurements, respectively. Spatially uncorrelated white noise is considered with covariance

$$E\left\{w_1^j(t)\, w_1^k(s)\right\} = (\sigma_1^j)^2\, \delta_{jk}\, \delta(t-s) \quad .$$

Similar conditions are assumed to hold for w_2 and w_3 with covariances $(\sigma_2)^2$ and $(\sigma_3)^2$, resp.. The measurements may be combined in a measurement vector $\underline{s}$, where

$$\underline{s} = (s_1^1,\, s_1^2, \ldots, s_1^N,\, s_2,\, s_3)^T \;,$$

and the noise covariances form a matrix

$$\underline{C} = \mathrm{diag}_{N+2}\left[(\sigma_1^1)^2, \ldots, (\sigma_1^N)^2, (\sigma_2)^2, (\sigma_3)^2\right] .$$

Here it is assumed that N sensors are installed for measuring the dissolved oxygen concentration in the reactor and that both $\overline{OTR}$ and c_X measurements are used. In cases where some of these measurements are not taken, the corresponding vector and matrix components are omitted.

The Fisher information matrix is then given by

$$\underline{F} = \int_0^T \left(\frac{\partial \underline{s}}{\partial \underline{P}}\right)^T \underline{C}^{-1} \left(\frac{\partial \underline{s}}{\partial \underline{P}}\right) dt \;,$$

where

$$\frac{\partial \underline{s}}{\partial \underline{P}} = \begin{bmatrix} \frac{\partial s_1^1}{\partial P_1} & \cdots & \frac{\partial s_1^1}{\partial P_4} \\ \vdots & \ddots & \vdots \\ \frac{\partial s_3}{\partial P_1} & \cdots & \frac{\partial s_3}{\partial P_4} \end{bmatrix} .$$

The measurement sensitivities can be computed by sensitivity analyses of the system equations using either the sensitivity equations or difference formulae as approximations. For fixed cultivation conditions, which means fixed input functions and fixed courses of the system parameters, $\underline{F}$ is only a function of the sensor locations and the sensor types used. It can be shown that the Fisher information matrix is the inverse of the covariance matrix of the best linear estimator for the unknown parameters. Thus the aim is to maximize the performance of this estimator, which means to maximize a scalar function of $\underline{F}$, and usually the determinant is taken for this purpose. The objective for sensor allocation thus is to find

$$\max_{x_i} \det \underline{F} \;.$$

RESULTS

One dissolved oxygen measurement. Figure 2a shows the results for allocation of a single dissolved oxygen sensor in the reactor column. The best position for this sensor is always near the bottom of the reactor, regardless whether c_X, $\overline{OTR}$, or c_X + $\overline{OTR}$ - measurements are used in addition. However, as can be seen from this figure, the problem of using the adequate **type** of measurements is much more critical than an optimal **allocation** of the c_{OF} sensor. While det $\underline{F}$ varies about 2.5 decades by sensor allocation, it may be increased by more than 8 decades when measuring c_X and $\overline{OTR}$ in addition to c_{OF}. It can also be stated, that $\overline{OTR}$ is the most valuable measurement of all; while c_X results in an increase by 3 decades, $\overline{OTR}$ gives an increase by 7 decades, and further addition of c_X to c_{OF} and $\overline{OTR}$ only adds 1.5 decades.

Several dissolved oxygen measurements. Figure 2a also contains incomplete results for the allocation problem in case of two and three sensors, where c_X and $\overline{OTR}$ measurements are used in addition. There the first sensor is fixed at the bottom and -in case of three sensors- the second is fixed at the top of the reactor. Complete results may also be drawn, cf. Fig. 3, but there the situation is complicated to visualize. Fig. 2a demonstrates already, that taking more than two dissolved oxygen sensors does not add very much information to the identification process.

Of course further computations may be carried out for increasing numbers of c_{OF} sensors. This leads to the result, that even taking ten measurements of the dissolved oxygen concentration does not gi-

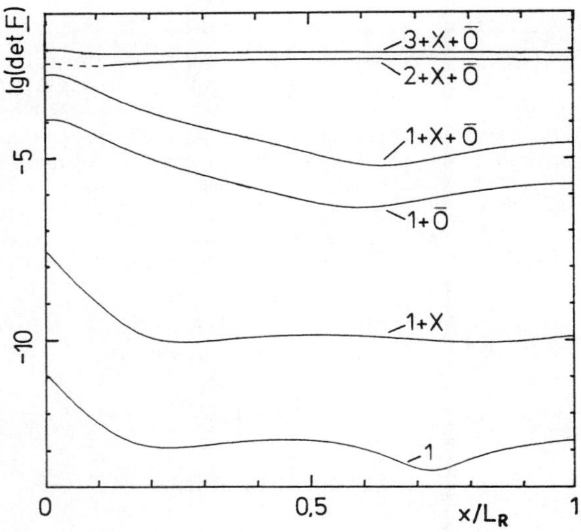

Fig.2a: Evaluation of det $\underline{F}$ for 1,2, and 3 sensors; identification interval = complete cultivation time.

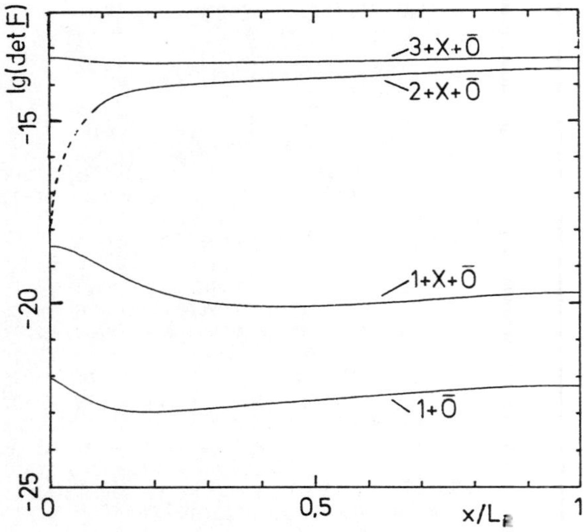

Fig.2b: Evaluation of det $\underline{F}$ for 1,2, and 3 sensors; identification intervals = 3 hours each.

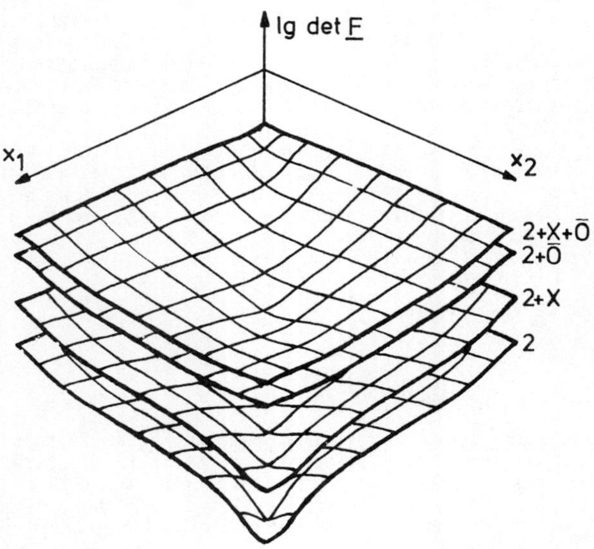

Fig.3: Evaluation of det $\underline{F}$; complete figure for two sensors.

ve as much information than only one c_{OF} sensor near the bottom of the reactor combined with the $\overline{OTR}$ measurement.

When the number of oxygen sensors is increased, the optimal locations do not spread uniformly over the whole spatial area of the reactor; taking more and more sensors partly leads to the same positions (bottom or top) as already occupied by other sensors. This is due to the assumption of spatially uncorrelated noise, and one would have to be careful with the installation of the sensors in order to fulfil this condition in practice.

Cyclic identifications. If the parameter identification routine is applied on-line, one has to guarantee that the parameter estimates are of good quality in each of the time intervals which are the basis for the identification procedure. In case of the tower loop reactor producing SCP, this time interval was set to three hours. In this time, parameter identification and optimization of input control functions can be calculated, both together forming a so-called OLFO (Open Loop Feedback Optimal) control strategy. Then the objective for sensor allocation and choice of measurements is to find

$$\max_{x_i} \min_{j=2,..,T/\tau} \det \underline{F}_j ;$$

where τ denotes the length of the time interval. The corresponding results to those of Fig. 2a are given in Fig. 2b. They show a similar behaviour in case of sufficient measurement equipment. The cases of pure c_{OF} or $c_{OF} + c_X$ measurements, however are totally insufficient for identification purposes.

The identified parameters. Figs. 4a-d show the identified parameters. The results are in general meaningful and correspond to former identification runs with a simpler (linear) model for the gas phase, cf. Munack(1980). To check the sensor allocation results, two different cases were considered: First, the identification was carried out with the whole measured data available, which were ten c_{OF} measurements as well as c_X and $\overline{OTR}$ data. To compare this with a reduced information set, only three c_{OF} measurements were used in the second case, together with c_X and $\overline{OTR}$ data. The results clearly show, that the important biological parameters are in fact identified very reliably. The fluiddynamical parameters vary about 10%, which however does not have great influence onto the profiles, since the state of the system is not very sensitive to variations of these parameters.

CONCLUSIONS

Parameter identification for biotechnical systems may be a complex problem, either due to sophisticated models for the biological process, or due to reactor models incorporating various balances for the important substances and involving partial differential equations for non well mixed reactors. Especially in on-line calculations (e.g. for adaptive strategies) there is a need to speed up the computational calculations. This may be achieved by application of decomposition/coordination methods to decouple some of the balances. These methods may lead to significant reductions of computation time needed for parameter identification, as already discussed in an earlier paper, cf. Munack/Thoma(1983).

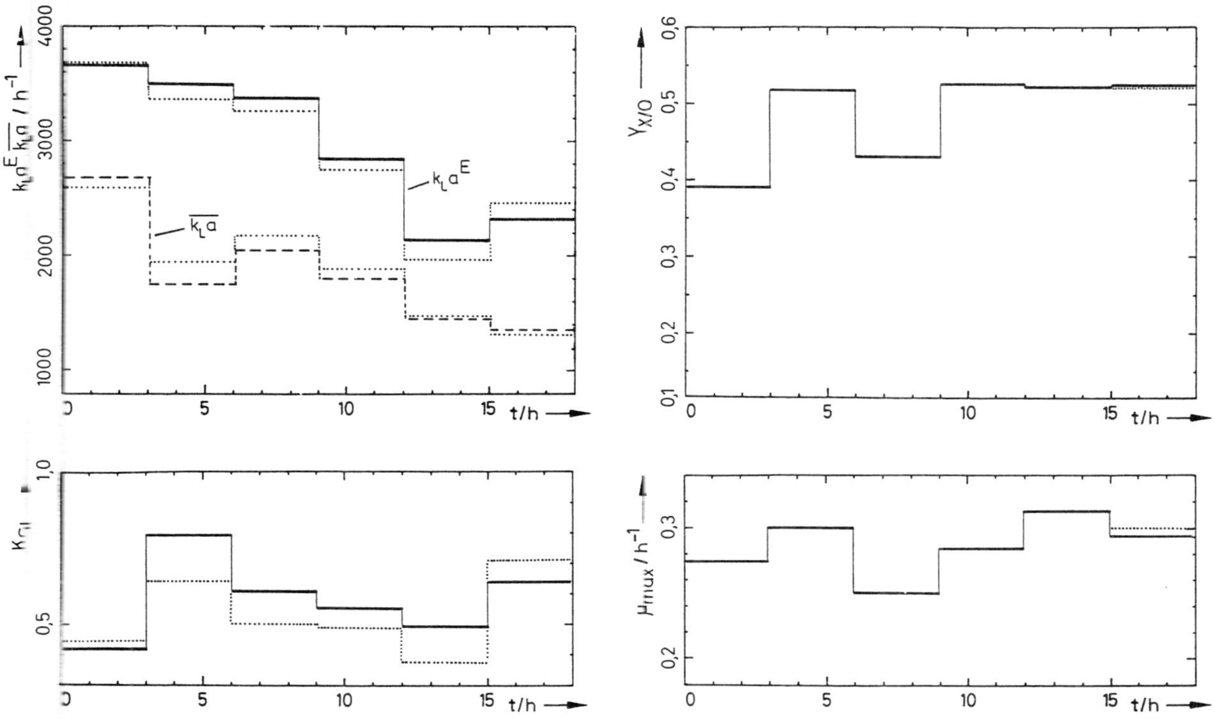

Fig.4: Course of the identified parameters; —— 10 c_{OF} sensors, ······· 2 c_{OF} sensors.

Another possibility would be to approximate the distributed parameter system equations by a low-order lumped system. It was found in recent work - Ott(1985) - that an approximation by a cascade of 7 CSTRs gives quite satisfactory results for the process under consideration. Computation times, however, were longer than expected, compared with the distributed system calculations using 101 node points. This was partly due to the fact that the distributed system simulation package used is very fast; and on the other hand some of the calculations involved are only needed once per time step. These take nearly as much time for the lumped system as for the distributed system. - At the same time it could be shown for the system under consideration, that early lumping and late lumping lead to the same formulas for the adjoint state and the gradients.

In this paper, optimization of type, number, and location of sensors was discussed for an SCP fed-batch process in a tower loop bioreactor. The calculations demonstrated the importance of the outlet gas analysis. Two or three sensors for the dissolved oxygen concentration in the liquid phase of the reactor were shown to be sufficient, where the positions should be near the bottom and near the top of the reactor. As the adjoint system then only shows two or three δ-functions on the right-hand side of the equations, the number of node points may be reduced for the calculations which results in faster computation. The measurements of the cells' concentration also add some information to the identification procedure; however, this information is by far not as significant as the outlet gas analysis.

The comparison of the identification results gained with the originally installed sensors and these with only two c_{OF} sensors indicated that the important biological parameters are identified with excellent agreement. The fluiddynamical parameters differ up to ten percent. However, due to the low sensitivity of the cultivation to variations in these parameters, these deviations may be tolerated.

The latter statement should be justified by calculations of the loss in optimality of the control performance index due to imprecise parameter estimates. Research in this direction will be carried out in the near future.

REFERENCES

Chavent, G. (1974). Identification of Functional Parameters in Partial Differential Equations. In R.E. Goodson and M. Polis (Eds.), Identification of Parameters in Distributed Systems. ASME, New York, pp. 31-48.

Kubrusly, C.S., and H. Malebranche (1983). A Survey on Optimal Sensors and Controllers Location in DPS. 3rd IFAC Symp. on Control of Distributed Parameter Systems. Pergamon, Oxford, New York, pp. 59-73.

Lions, J.L. (1971). Optimal Control of Systems Governed by Partial Differential Equations. Springer, Berlin, Heidelberg, New York.

Luttmann, R., and A. Munack (1983). Economic Optimization of Distributed Parameter Fermentation Processes (a Case Study). 1st IFAC Workshop on Modelling and Control of Biotechnical Processes. Pergamon, Oxford, New York, pp. 145-159.

Luttmann, R., A. Munack, and M. Thoma (1985). Mathematical Modelling, Parameter Identification and Adaptive Control of Single Cell Protein Processes in Tower Loop Bioreactors. Advances in Biochemical Engineering, to appear. Springer, Berlin, Heidelberg, New York.

Munack, A. (1980). Zur Theorie und Anwendung adaptiver Steuerungsverfahren für eine Klasse von Systemen mit verteilten Parametern. Dissertation, Universität Hannover.

Munack, A. (1984). Optimal Sensor Allocation for Identification of Unknown System Parameters in a Bubble-Column Loop Reactor. Proc. 6th Int. Conf. on Analysis and Optimization of Systems. Lecture Notes in Control and Information Sciences, Vol. 63. Springer, Berlin, Heidelberg, New York, pp. 415-433.

Munack, A. (1985). Parameter Identification Problems for Interconnected Distributed Parameter Systems and Applications to a Biotechnological Plant. Proc. 2nd Int. Conf. on Control Theory for Distributed Parameter Systems and Applications. Lecture Notes in Control and Information Sciences, to appear. Springer, Berlin, Heidelberg, New York.

Munack, A., and M. Thoma (1983). Application of Decomposition/Coordination Methods to Control and Identification of Distributed Parameter Systems. Proc. 3rd IFAC Symp. on Control of Distributed Parameter Systems. Pergamon, Oxford, New York, pp. 539-544.

Ott, H. (1985). Modellierung, Identifikation und Regelung von Kultivierungen in Blasensäulenschlaufenfermentern unter Einbeziehung der Substratbilanz. Diplomarbeit, Universität Hannover, Inst. f. Regelungstechnik (unpublished).

Qureshi, Z.H., T.S. Ng, and G.C. Goodwin (1980). Optimum Experimental Design for Identification of Interconnected, Distributed Parameter Systems. Int. J. Control, 31, pp. 21-29.

Schubert, H. (1976). Ein Beitag zur adaptiven Regelung von linearen Einfachsystemen. Dissertation, Universität Hannover.

APPENDIX

List of the most important symbols

x	: longitudinal coordinate in the column
$z=x/L_R$	: dimensionless longitudinal coordinate in the column
$\bar{z}=\bar{x}/L_B$	: dimensionless longitudinal coordinate in the loop
$\tau=t/T_N$	: normalized time
Bo	: normalized Bodenstein number
k_0	: normalized Monod saturation constant
$k_L a$	: volumetric oxygen mass transfer coefficient
p	: normalized pressure
q_{02}^E	: normalized oxygen supply rate
$q_{0/Xm}=\mu_m/y_{X/0}$	: normalized maximum metabolic quotient
μ_t	: normalized death rate
v	: normalized velocity
$y_{X/0}$	: normalized yield coefficient
Ψ	: coalescence function

For reasons of a concise presentation, arguments of the dependent variables are omitted whenever possible as well as dependencies of parameters from the gas phase velocity.

Normalized dissolved oxygen concentration in the liquid phase of the reactor

$$\text{PDE:}\quad \frac{\partial c_{OF}}{\partial t} = \frac{1}{Bo_F}\cdot\frac{\partial^2 c_{OF}}{\partial z^2} - v_F\cdot\frac{\partial c_{OF}}{\partial z} + k_L a^E\cdot\Psi(K_{St})\cdot\left(p\cdot c_{OG} - c_{OF}\right) - q_{0/Xm}\cdot\frac{c_{OF}}{k_0 + c_{OF}}\cdot c_X$$

$$\text{IC:}\quad c_{OF}(0) = p(0)$$

$$\text{BC:}\quad \left.\frac{\partial c_{OF}}{\partial z}\right|_{z=0} = Bo_F\cdot v_F\cdot\left(c_{OF}(0) - c_{OB}(1)\right) \;;\qquad \left.\frac{\partial c_{OF}}{\partial z}\right|_{z=1} = 0 \tag{A1 a}$$

where

$$\Psi(K_{St}) = \begin{cases} e^{-K_{St}\cdot\frac{z}{a}} & 0 \le z < a \\ e^{-K_{St} - a\cdot K_{St}\cdot(z-a)} & a \le z \le 1 \end{cases} \qquad (a = 0.1)$$

Normalized oxygen concentration in the gas phase of the reactor

$$c_{OG} = 1 - \frac{U_{02}}{2 + Bo_G v_G^E}\cdot\left(2 + Bo_G v_G^E\cdot(2z-z^2)\right)$$

where

$$U_{02} = \frac{\overline{OTR}}{q_{02}^E} = \frac{1}{q_{02}^E}\cdot\left[\frac{\mu_t}{y_{X/0}}\cdot c_X + v_F\cdot(c_{OF}(1)-c_{OB}(1)) + q_{0/Xm}\cdot c_X\cdot\int_0^1 \frac{c_{OF}}{k_0 + c_{OF}}\,dz\right] \tag{A1 b}$$

Normalized dissolved oxygen concentration in the loop

$$c_{OB}(\bar{z}) - c_{OB}(0) + k_0\cdot\ln\frac{c_{OB}(\bar{z})}{c_{OB}(0)} = -q_{0/Xm}\cdot\frac{\bar{z}}{v_B}\cdot c_X \;;\qquad \text{BC:}\quad c_{OB}(0) = c_{OF}(1) \tag{A1 c}$$

Normalized concentration of cells

$$\frac{dc_X}{d\tau} = y_{X/0}\cdot\left[\frac{v_B}{v_B+v_F}\cdot q_{0/Xm}\cdot c_X\cdot\int_0^1 \frac{c_{OF}}{k_0+c_{OF}}\,dz + \frac{v_B\cdot v_F}{v_B+v_F}\cdot\left(c_{OB}(0)-c_{OB}(1)\right) - \frac{\mu_t}{y_{X/0}}\cdot c_X\right] \qquad \text{(A1 d)}$$

IC: $c_X(0) = 1$

Adjoint state for identification of parameters k_La^E, K_{St}, and $q_{0/Xm}$

PDE:
$$\frac{\partial\zeta}{\partial\tau} = -\frac{1}{Bo_F}\cdot\frac{\partial^2\zeta}{\partial z^2} - v_F\cdot\frac{\partial\zeta}{\partial z} + \left(k_La^E\cdot\Psi(K_{St}) + q_{0/Xm}\cdot\frac{k_0\cdot c_X}{(k_0+c_{OF})^2}\right)\cdot\zeta -$$

$$- k_La^E\cdot\Psi^2(K_{St})\cdot p\cdot\frac{2+Bo_Gv_G^E\cdot(2z-z^2)}{2+Bo_Gv_G^E}\cdot\frac{OTR_m^E}{q_{02}^E+OTR_m^E\cdot I_2}\cdot\int_0^1\zeta\,dz - \qquad \text{(A2)}$$

$$- 2W\sum_{i=1}^{N}\left(\hat{c}_{OF}(z_i) - c_{OF}^M(z_i)\right) + 2\sigma W\cdot\left(\widehat{\overline{OTR}} - \overline{OTR}^M\right)\cdot\frac{q_{02}^E\cdot CTR_m^E\cdot\Psi(K_{St})}{q_{02}^E+OTR_m^E\cdot I_2}$$

FC: $\zeta(T) = 0$

BC:
$$\left.\frac{\partial\zeta}{\partial z}\right|_{z=0} = 0 \quad ; \quad \left.\frac{1}{Bo_F}\cdot\frac{\partial\zeta}{\partial z}\right|_{z=1} + \left.v_F\cdot\zeta\right|_{z=1} = \zeta(0)\cdot v_F\cdot\frac{c_{OB}(1)\cdot(c_{OF}(1)+k_0)}{c_{OF}(1)\cdot(c_{OB}(1)+k_0)}$$

Gradient formulae

$$\frac{\partial J}{\partial k_La^E} = \int_0^T\int_0^1 \zeta\cdot\Psi(K_{St},z)\cdot\left(p\cdot c_{OG} - c_{OF}\right)dz\,d\tau + \int_0^T \frac{OTR_m^E\cdot I_1\cdot q_{02}^E}{(q_{02}^E+OTR_m^E\cdot I_2)^2\cdot k_La^E}\cdot\left[2\sigma W\cdot\left(\widehat{\overline{OTR}} - \overline{OTR}^M\right)\cdot q_{02}^E - I_2^\zeta\right]d\tau$$

(A3 a)

$$\frac{\partial J}{\partial K_{St}} = k_La^E\cdot\int_0^T\int_0^1 \zeta\cdot\frac{\partial\Psi(K_{St},z)}{\partial K_{St}}\left(p\cdot c_{OG} - c_{OF}\right)dz\,d\tau +$$

$$+ \int_0^T\left[\frac{OTR_m^E\cdot I_3}{q_{02}^E+OTR_m^E\cdot I_2} - \frac{OTR_m^{E^2}\cdot I_4}{(q_{02}^E+OTR_m^E\cdot I_2)^2}\right]\cdot\left[2\sigma W\cdot\left(\widehat{\overline{OTR}} - \overline{OTR}^M\right)\cdot q_{02}^E - \frac{k_La^E\cdot OTR_m^E}{q_{02}^E}\cdot I_2^\zeta\right]d\tau$$

(A3 b)

$$\frac{\partial J}{\partial q_{0/Xm}} = -\frac{c_X(0)\cdot k_La^E}{OTR_m^E}\cdot\left[\int_0^T\int_0^1 \zeta\cdot\frac{c_{OF}\cdot c_X}{k_0+c_{OF}}\,dz\,d\tau + \int_0^T\frac{v_F\cdot c_X\cdot c_{OB}(1)\cdot\zeta(0)}{v_B\cdot(c_{OB}(1)+k_0)}\,d\tau\right] \qquad \text{(A3 c)}$$

where
$$I_1 = \int_0^1 \Psi(K_{St})\cdot(p - c_{OF})\,dz\ ,$$

(A4)

$$I_2 = \int_0^1 \Psi(K_{St})\cdot p\cdot\frac{2+Bo_Gv_G^E\cdot(2z-z^2)}{2+Bo_Gv_G^E}\,dz\ , \qquad I_3 = \int_0^1\frac{\partial\Psi(K_{St})}{\partial K_{St}}\cdot(p - c_{OF})\,dz\ ,$$

$$I_2^\zeta = \int_0^1 \Psi(K_{St})\cdot p\cdot\frac{2+Bo_Gv_G^E\cdot(2z-z^2)}{2+Bo_Gv_G^E}\cdot\zeta\,dz\ , \qquad I_4 = \int_0^1\frac{\partial\Psi(K_{St})}{\partial K_{St}}\cdot p\cdot\frac{2+Bo_Gv_G^E\cdot(2z-z^2)}{2+Bo_Gv_G^E}\,dz\ .$$

MODEL REDUCTION OF AN OXYGEN ENRICHED INDUSTRIAL WASTE-WATER TREATMENT PROCESS

A. Johnson and A. R. Tiedemann

Department of Applied Physics, Delft University of Technology, Prins Bernhardlaan 6, 2628 BW Delft, The Netherlands

Abstract. This paper describes the reduction of a 10th order nonlinear model of an oxygen enriched industrial waste-water process to a 3rd order linear model. With the aim of providing good control of the settling tank sludge volume a model has been developed which describes the dynamical behaviour of the bioreactor and settling tank. Two crucial properties of the reduced model are: (i) that it variables may still be associated with the physical variables of the waste-water treatment process, (ii) that all the unknown parameters of the reduced model may be estimated from steady-state measurements.

Keywords. Waste-water treatment; activated sludge process; modelling; model reduction; parameter estimation.

INTRODUCTION

The waste-water treatment process is a large (ca. 1200 m^3/hr of effluent with an average of 1500 gCOD/m^3) chemical effluent processing plant at SHELL's Pernis refinery in Holland. The waste-water is first neutralised (pH = 6) and any solids or oil is removed. The temperature is also regulated at 35°C. Two equalisation tanks and a spill tank (each with a volume of 8540 m^3) are provided to damp out flow and substrate concentration variations in the waste-water before this is pumped to the bioreactors. Three parallel trains of UNOX bioreactors, each of four compartments, and containing activated sludge, form the heart of the process. Pure oxygen is supplied to the first compartment of each bioreactor to assist the degradation of the substrate. The sludge from each bioreactor is concentrated in one of the three settling tanks before being partially recycled.

Firstly, in this paper, a process description is given of the bioreactor and settler and a nonlinear 10th order model is developed which describes the dynamical behaviour of the bioreactor and settling tank. The model of each bioreactor is based upon the assumption that the organisms and substrate can each be treated as one homogeneous mass; using Monod kinetics this leads to an 8th order model with four states describing biomass concentrations and four the substrate concentrations. The settling tank model has been developed from the limiting flux theory and contains two states, one modelling the sludge volume in the settling tank and one the recycled biomass concentration. Sedimentation is expressed by two parameters in the model.

Next, the model is reduced in order. To reduce the complete model, three assumptions are made. Firstly, that the dynamics of the decomposition of the substrate are negligible when compared to the dynamics of the growth of the biomass. Secondly, that the dynamics of the recycled biomass concentration can be neglected, and lastly that it is sufficient to use a two stage representation of each bioreactor. Simulations have confirmed that there is excellent agreement between the original and the reduced order model responses. Our studies have predicted an almost maximum possible decomposition of the substrate, so that the control of the sludge volume in the settling tank becomes of paramount importance due to the stringent penalties incurred in the loss of the sludge in the overflow of the settler.

Finally, simulations of both the high and low order model are presented.

PROCESS DESCRIPTION

Figure 1 shows one of the three (identical) UNOX bioreactors and its associated settling tank. Some of the important operating parameters are given in Table 1.

TABLE 1. Operating Parameters and Nomenclature used in Fig. 1

V^c:	bioreactor compartment volume	1000 m^3
A^s:	settling tank area	875 m^2
V^s:	settling tank volume	2820 m^3
V^b:	settling tank sludge volume	2000 m^3
C :	substrate concentration	g COD/m^3
F :	flowrate	m^3/hr
X :	active biomass concentration	g VSS/m^3
F_r :	sludge recycle flowrate	0-600 m^3/hr
F_w:	waste sludge flowrate	0-30 m^3/hr
F_f:	waste-water feed flowrate (average for one reactor)	400 m^3/hr
C_f:	waste-water feed substrate concentration (average)	1500 g COD/m^3

Subscripts used:

o : entering bioreactor		f: waste-water feed
1-4: number of compartment		r: sludge recycle
e : effluent		w: sludge waste

Each of the four closed, continuously stirred, compartments of the bioreactor contains active biomass. Pure oxygen is supplied to the first compartment and surface aerators ensure a high percentage of the oxygen is dissolved with only a relatively low stirred power consumption. It is possible to control the sludge recycle, waste sludge flowrate, the oxygen supply and the aerator submension depth in each compartment. It is also possible to measure the active biomass concentration in the outflow of the bioreactor and the settler and the sludge volume in the settler. These measurements are recorded every 20 minutes on a computer log. Other measurements are taken less frequently by the operators.

BIOREACTOR MODEL

The model of each bioreactor is based upon the assumption that the organisms and substrate can each be treated as one homogeneous mass. The composition of the waste-water flow to the process is characterized by its Chemical Oxygen Demand (COD), modified to account for nonremovable substances. The active biomass is characterized by its Volatile Suspended Solids (VSS). Using linearized Monod kinetics this leads to an 8th order model with four states describing active biomass concentrations and four the substrate concentrations. The equations describing the model are given by

$$\dot{X}_n(t) = D(t)X_{n-1}(t) + [GC_n^+(t)-\mu_s-D(t)]X_n(t) \quad (1a)$$

$$\dot{C}_n^+(t) = D(t)C_{n-1}^+(t) - [GX_n(t)/Y+D(t)]C_n^+(t) \quad (1b)$$

$$G = \mu_m/K_s,\ D(t) = [F_f(t)+F_r(t)]/V^c \quad (1c,d)$$

$$C_o^+(t) = [C_f^+(t)F_f(t)+C_r^+(t)F_r(t)]/[F_f(t)+F_r(t)] \quad (1e)$$

$$X_o(t) = X_r(t)F_r(t)/[F_f(t)+F_r(t)] \quad (1f)$$

$$C_n^+(t) = C_n(t) - C^-,\ n = 1\text{-}4 \quad (1g)$$

where

$D(t)$	: dilution rate	hr^{-1}
G	: μ_m/K_s	m^3/g COD hr
K_s	: Michaelis-Menten constant	g COD/m^3
Y	: yield coefficient	g VSS/g COD
μ_m	: maximum growth rate	hr^{-1}
μ_s	: term for endogene respiration and cell decay	hr^{-1}

The estimation of the unknown parameters of the bioreactor model is done with the help of steady-state measurements of biomass and substrate concentrations in each compartment of the reactor, the recycle flow and the influent flow. Under steady-state conditions equations (1a) and (1b) can be written as

$$X_n = [D/(D+\mu_s-GC_n^+)]X_{n-1},\ C_n^+ = [YD/(YD+GX_n)]C_{n-1}^+,\quad n = 1\text{-}4 \quad (2a,b)$$

We can write equation (2) in matrix form resulting in

$$\begin{bmatrix} X_1/D & C_1^+-C_o^+ \\ X_2/D & C_2^+-C_1^+ \\ X_3/D & C_3^+-C_2^+ \\ X_4/D & C_4^+-C_3^+ \end{bmatrix} \begin{bmatrix} \mu_s \\ Y \end{bmatrix} = \begin{bmatrix} X_o-X_1 \\ X_1-X_2 \\ X_2-X_3 \\ X_3-X_4 \end{bmatrix} \quad (3a)$$

$$\begin{bmatrix} X_1/D \\ X_2/D \\ X_3/D \\ X_4/D \end{bmatrix} [G/Y] = \begin{bmatrix} C_o^+/C_1^+-1 \\ C_1^+/C_2^+-1 \\ C_2^+/C_3^+-1 \\ C_3^+/C_4^+-1 \end{bmatrix} \quad (3b)$$

That is

$$Ax = b \quad (4a)$$

where x is the unknown parameter vector.
The least-squares estimate, $\hat{x}$, of the parameter vector is given by

$$\hat{x} = (A^TA)^{-1}A^Tb \quad (4b)$$

Various sets of steady-state measurements of D, C_{0-4}, C^- and X_{0-4} have been obtained by taking samples in each compartment of the bioreactor.

Using this data in equation (4b) we find the matrix A^TA is ill conditioned with respect to inversion. The least-squares estimations of μ_s, Y and G with the corresponding estimation faults are as given in Table 2, where comparable values from the literature are also shown. Note however that the quoted values do not pertain to chemical plant waste-water treatment, which explains our low value of G.

TABLE 2. Estimations of the Parameters μ_s, Y and G

Parameter	Estim.	Estim. Fault	Grutsch (1980)	Takamatsu and others (1981)
$\mu_s(hr^{-1})$	0.004	±0.002	0.009	0.0017
Y(gVSS/gCOD)	0.45	±0.1	0.35	0.55
G(m^3/gCODhr)	0.0003	±0.0001	0.0007	0.0033

An estimate of C^- is C^- = 500 gCOD/m^3 ± 100 gCOD/m^3.

SETTLING TANK MODEL

Limiting flux theory (Tracy and Keinath, 1974; Attir, Denn and Petty, 1977; Attir and Denn, 1978; Heynen, 1984) can be used to derive a satisfactory settling tank model. Through any chosen horizontal cross-sectional area of the settling tank the biomass flux, W(t), is determined by the gravitational forces and the flow out of the settling tank, and may be written as

$$W(t) = X(t)[\alpha X(t)\exp(-X(t)/\beta)+(F_r(t)+F_w(t))/A^S] \quad (5)$$

Sedimentation is expressed by the parameters α and β. How W(t) in equation (5) depends on the settling tank outflow $F_r(t) + F_w(t)$ and on the biomass concentration X(t) is shown in Fig. 2. From this figure it is apparent that a limiting biomass concentration $X_l(t)$ exists, coupled to a maximal transport capacity, $W_l(t)$; these are shown in Fig. 3 as a function of the tank outflow.

For the purpose of the model, the settling tank is divided into three horizontal layers, the clarification, compacting and thickening zone (Heynen, 1984), where the biomass concentration is $X_e(t)$, $X_l(t)$ and $X_r(t)$ respectively (see Fig. 4). We assume that there is no biomass in the clarification zone ($X_e(t) = 0$) and that the biomass from the bioreactor (concentration $X_4(t)$) is immediately at the limiting concentration $X_1(t)$. In practice $X_e(t)$ is not exactly zero. The value of $X_e(t)$ however is low, but unknown. Tracey and Keinath (1974) modelled the compacting zone as a series of variable volume completely mixed reactors. This leads to a complicated dynamic model for the behaviour of the compacting zone. For our simple model we modelled the compacting zone as one variable volume completely mixed reactor with biomass concentration $X_1(t)$ and variable volume $V^1(t)$. The sludge volume in the settler can be expected to increase if the biomass flux entering the settling tank is greater than the limiting flux between the compacting and thickening zone. This latter zone has a fixed volume V^2 and serves to concentrate the biomass from $X_1(t)$ to $X_r(t)$. From Fig. 3 it can be seen that for given values of α, β and A^s, $X_1(t)$ is only a function of $F_r(t)$ and $F_w(t)$. In practice $F_r(t) \gg F_w(t)$, so that a good representation of $X_1(t)$ for small changes of $F_r(t)$ around $F_r = 400\ m^3/hr$ can be given by a linear equation of the form:

$$X_1(t) = \gamma - \varepsilon F_r(t) \qquad (6)$$

Later on we will prove that small changes in $X_1(t)$ due to small changes in $F_r(t)$ are insignificant for the static- and dynamic behaviour of the settling tank.

It is now possible to derive a simple dynamic model for the settling tank that is valid for small changes in $F_r(t)$ and $F_w(t)$

$$\dot{V}^b(t) = \dot{V}^1(t) = [(F_f(t)+F_r(t))X_4(t) - X_1(t)(F_r(t)+F_w(t) + A^s\alpha X_1(t)\exp(-X_1(t)/\beta))]/X_1(t) \qquad (7a)$$

$$\dot{X}_r(t) = [X_1(t)(F_r(t)+F_w(t)+A^s\alpha X_1(t)\exp(-X_1(t)/\beta)) - (F_r(t)+F_w(t))X_r(t)]/V^2 \qquad (7b)$$

$$V^b(t) = V^1(t) + V^2,\ X_1(t) = \gamma - \varepsilon F_r(t) \qquad (7c,d)$$

$$C_e(t) = C_4(t),\ C_r(t) = C_4(t) \qquad (7e,f)$$

We assume that no further decomposition of the substrate takes place in the settling tank.

The unknown parameters α and β in this model (γ and ε can be derived if α and β are known) can be estimated from available stationary measurements of F_r, F_w, X_1 and X_r.

In the steady-state

$$X_r = X_1 + [A^s\alpha X_1\exp(-X_1/\beta)]/(F_r+F_w) \qquad (8a)$$

$$W_1 = X_4(F_f+F_r)/A^s = X_r(F_r+F_w)/A^s \qquad (8b)$$

must hold. Unfortunately, equation (8) is not linear in the parameters α and β. Approximate values of α and β are therefore obtained by pointwise fitting the minimum of the biomass flux curve W (see Fig. 2) to the point X_1, W_1. In Table 3 estimations are given of α and β with the corresponding estimation faults. From the values of α and β and with the help of Fig. 3 we find $\gamma = 13250$ g VSS/m^3 and ε = 4.5 g VSS hr/m^6. Steady-state values of V^1, V^2 and V^b are $V^1 = 1825\ m^3$, $V^2 = 175\ m^3$ and thus $V^b = 2000\ m^3$.

TABLE 3. Estimations of the Parameters α and β

Parameter	Estim.	Estim.Fault
α (m^4/gVSShr)	0.01	± 0.002
β (gVSS/m^3)	1600	± 300

MODEL REDUCTION

Together, equations (1) and (7) form the 10th order nonlinear model of the biotreater, which because of its high order is not suitable for controller design. Since our studies have predicted an almost maximum possible substrate decomposition, the control of the settling tank sludge volume ($V^b(t)$ in our model) becomes of paramount importance due to the stringent penalties incurred in the loss of sludge in the overflow of the settler. In practice $X_e(t)$ is approximately zero providing there is good sludge volume control. Thus any reduced model must faithfully reproduce the dynamic behaviour of the sludge volume.

An extensive literature exists concerning model reduction techniques (see, for example, Bultheel (1983) for a bibliography with about 200 items). A big disadvantage of most of them is that the reduced model state variables are normally no longer associated with real physical variables. Two crucial properties of the reduced model which is introduced next are that its variables may still be recognized as physical variables and that the unknown parameters of the reduced model may be estimated from steady-state measurements.

In order to study the dynamic behaviour of the 10th order model, linearisation is performed around an equilibrium point. Such an equilibrium point is given in Table 4. The values of the parameters C^-, μ_s, Y, G, α and β are given in Table 2 and 3 (estim.). The equilibrium point is the steady-state solution of the 10th order nonlinear model. Table 4 also contains the value of the F/M ratio (F/M), the sludge recycle ratio (REC) and the sludge age (SA) defined respectively by

$$F/M = (C_fF_f24)/[(X_1+X_2+X_3+X_4)V^c] \qquad (9a)$$

$$REC = F_r/F_f \qquad (9b)$$

$$SA = [(X_1+X_2+X_3+X_4)V^c+X_1V^1+X_rV^2]/(X_rF_w24)\ (days) \qquad (9c)$$

TABLE 4. Normal Steady-State Operating Conditions

C_f = 1500 gCOD/m^3	X_r = 13660 gVSS/m^3
C_0 = 1000 "	X_0 = 6830 "
C_1 = 573.3 "	X_1 = 6987 "
C_2 = 510.8 "	X_2 = 6980 "
C_3 = 501.6 "	X_3 = 6949 "
C_4 = 500.2 "	X_4 = 6915 "
C_r = 500.2 "	X_1 = 11430 "
C_e = 500.2 "	X_e = 0 "
F_f = 400 m^3/hr	V^b = 2000 m^3
F_r = 400 m^3/hr	F/M = 0.52 gCOD/gVSS
F_w = 5 m^3/hr	REC = 1
	SA = 31 days

It has been noted that the substrate is almost completely decomposed to its nondegradable point C^-, so that $C_e^+ = 0$ and $C_r^+ = 0$. In practice the purification efficiency is a bit lower then our model predicts. In the last section we stated that changes in $X_1(t)$ due to changes in $F_r(t)$ (see equation (6)) are insignificant for the static- and dynamic behaviour of the settling tank. In Fig. 5 simulations are given for the response of the sludge volume $V^b(t)$ to step changes in the recycle flow rate $F_r(t)$ (see also Table 5) for the case that $X_1(t)$ is a linear function of $F_r(t)$ (see equation (6)) and for the case X_1 has the constant value given in Table 4. The simulations have been done with the 10th order nonlinear model. In the sequel we have chosen X_1 constant, which makes the linearization much easier.

TABLE 5. Step changes in the Recycle flow rate $F_r(t)$ used in the Simulation of Fig. 5

time (hr)	$F_r(t)(m^3/hr)$
0 - 25	400
25 - 50	500
50 - 75	400
75 -100	300
100 -125	400

Defining now the following state, control and disturbance vectors $\Delta x(t)$, $\Delta u(t)$ and $\Delta v(t)$ as

$$\Delta x^T(t) = [\Delta X_1(t)\Delta C_1^+(t)\Delta X_2(t)\Delta C_2^+(t)\Delta X_3(t)\Delta C_3^+(t) \Delta X_4(t)\Delta C_4^+(t)\Delta V^b(t)\Delta X_r(t)] \quad (10a)$$

$$\Delta u^T(t) = [\Delta F_r(t)\Delta F_w(t)], \ \Delta v^T(t) = [\Delta F_f(t)\Delta C_f(t)] \quad (10b,c)$$

the linear 10th order model may be written as

$$\Delta\dot{x}(t) = A\Delta x(t) + B\Delta u(t) + E\Delta v(t) \quad (11)$$

where the matrices A, B and E are defined as

$$A = \begin{bmatrix} A_1 & & 0 & 0 & 0 & 0 & 0 & 0 & 0 & a_1 \\ & & 0 & 0 & 0 & 0 & 0 & 0 & 0 & 0 \\ D & 0 & A_2 & & 0 & 0 & 0 & 0 & 0 & 0 \\ 0 & D & & & 0 & 0 & 0 & 0 & 0 & 0 \\ 0 & 0 & D & 0 & A_3 & & 0 & 0 & 0 & 0 \\ 0 & 0 & 0 & D & & & 0 & 0 & 0 & 0 \\ 0 & 0 & 0 & 0 & D & 0 & A_4 & & 0 & 0 \\ 0 & 0 & 0 & 0 & 0 & D & & & 0 & 0 \\ 0 & 0 & 0 & 0 & 0 & 0 & a_2 & 0 & 0 & 0 \\ 0 & 0 & 0 & 0 & 0 & 0 & 0 & 0 & 0 & a_3 \end{bmatrix} \quad (12a)$$

$$A_i = \begin{bmatrix} GC_n^+ - \mu_s - D & GX_n \\ -GC_n^+/Y & -D-GX_n/Y \end{bmatrix}, \ n = 1\text{-}4 \quad (12b)$$

$$a_1 = F_r/V^c, \ a_2 = (F_f+F_r)/X_1, \ a_3 = -(F_r+F_w)/V^2 \quad (12c)$$

$$B = \begin{bmatrix} (X_r-X_1)/V^c & 0 \\ -C_1^+/V^c & 0 \\ (X_1-X_2)/V^c & 0 \\ (C_1^+-C_2^+)/V^c & 0 \\ (X_2-X_3)/V^c & 0 \\ (C_2^+-C_3^+)/V^c & 0 \\ (X_3-X_4)/V^c & 0 \\ (C_3^+-C_4^+)/V^c & 0 \\ (X_4-X_1)/X_1 & -1 \\ (X_1-X_r)/V^2 & (X_1-X_r)/V^2 \end{bmatrix} \quad (12d) \qquad E = \begin{bmatrix} -X_1/V^c & 0 \\ (C_f^+-C_1^+)/V^c & F_f/V^c \\ (X_1-X_2)/V^c & 0 \\ (C_1^+-C_2^+)/V^c & 0 \\ (X_2-X_3)/V^c & 0 \\ (C_2^+-C_3^+)/V^c & 0 \\ (X_3-X_4)/V^c & 0 \\ (C_3^+-C_4^+)/V^c & 0 \\ X_4/X_1 & 0 \\ 0 & 0 \end{bmatrix} \quad (12e)$$

Table 6 gives the eigenvalues of the A matrix calculated from the values given in Table 4.

TABLE 6. Eigenvalues of the A matrix (hr^{-1})

$\Delta C_1^+(t)$-$\Delta C_4^+(t)$	: -5.7 ± 0.2 j and -5.2 ± 0.2 j
$\Delta X_1(t)$-$\Delta X_4(t)$	: -0.83 ± 0.03 j and -0.77 ± 0.03
$\Delta X_r(t)$	: -2.3
$\Delta V^b(t)$	: 0

It is evident that substrate decomposition occurs a factor 7 times as fast as the biomass growth in the reactor, so that compared to the dynamics of the biomass growth the dynamics of the substrate decomposition can be neglected. Moreover, Table 6 indicates that the dynamics of the biomass recycle concentration is 3 times as fast as the dynamics of the biomass growth in the reactor. If a step change occurs in the recycle flow rate $F_r(t)$ simulations have proved that the influence of the dynamics of the recycle biomass concentration $X_r(t)$ on the dynamics of the biomass growth in the reactor is negligible. This is due to the small changes in the recycle biomass concentration $X_r(t)$ when the recycle flow $F_r(t)$ changes. So far five of the state variables of the model ($\Delta C_n^+(t)$, n = 1-4; $\Delta X_r(t)$) have been disgarded.

From Tabel 4 it is apparent that the largest biological activity occurs in the first reactor compartment. There, 85% of the (biologically decomposable) substrate is decomposed, with an increase in biomass concentration of 157 $gVSS/m^3$. In compartments two to four, 15% of the substrate is decomposed and there is a decrease in biomass concentration of 72 $gVSS/m^3$. Therefore it is possible to represent the bioreactor by a two-tank model. The first state $X_1'(t)$ represents the biomass concentration in the first compartment, while the second state $X_2'(t)$ represents the biomass concentration in the other three compartments.

The model reduction just described results in a nonlinear 3rd order model given by

$$\dot{X}_1'(t) = D(t)X_0(t)+[GC_1'^+(t)-\mu_s-D(t)]X_1'(t) \quad (13a)$$

$$\dot{X}_2'(t) = D(t)X_1'(t)/3+[GC_2'^+(t)-\mu_s-D(t)/3]X_2'(t) \quad (13b)$$

$$C_1'^+(t) = \frac{D(t)C_o'^+(t)}{GX_1'(t)/Y+D(t)}, \quad C_2'^+(t) = \frac{D(t)C_1'^+(t)}{3GX_2'(t)/Y+D(t)} \qquad (13c,d)$$

$$\dot{V}^{b}{}'(t) = [(F_f(t)+F_r(t))X_2'(t)-(F_r(t)+F_w(t))X_r(t)]/X_1 \qquad (13e)$$

$$C_o'^+(t) = C_f^+(t)F_f(t)/[F_f(t)+F_r(t)], C_r' = C_e' = C^- \qquad (13f,g)$$

$$X_r(t) = X_1+[A^S\alpha X_1 \exp(-X_1/\beta)]/[F_r(t)+F_w(t)] \qquad (13h,i)$$

The equilibrium point of the 3rd order nonlinear model is given in Table 7. Note that the same parameters have been used as in the 10th order nonlinear model, resulting in practical the same steady-state operating conditions.

TABLE 7. Normal Steady-State Operating Conditions

F_f=400 m^3/hr	C_f= 1500 gCOD/m^3	X_r=13660 gVSS/m^3
F_r=400 "	C_o'= 1000 "	X_o= 6830 "
F_w= 5 "	C_1'= 573.3 "	X_1'= 6988 "
	C_2'= 504.1 "	X_2'= 6913 "
V^b = 2000 m^3	C_r'= 500 "	X_1=11430 "
	C_e'= 500 "	X_e= 0 "

Defining now the following state vector $\Delta x'(t)$

$$\Delta \bar{x}'(t) = [\Delta X_1'(t) \Delta X_2'(t) \Delta V^b{}'(t)] \qquad (14)$$

with the control and disturbance vector defined in eqns. (10b,c) the linear 3rd order model may be written as

$$\Delta \dot{x}'(t) = A'\Delta x'(t) + B'\Delta u(t) + E'\Delta v(t) \qquad (15)$$

where the matrices A', B' and E' are defined as

$$A' = \begin{bmatrix} a_{11} & 0 & 0 \\ a_{21} & a_{22} & 0 \\ 0 & a_{32} & 0 \end{bmatrix}, \quad B' = \begin{bmatrix} b_{11} & b_{12} \\ b_{21} & 0 \\ b_{31} & b_{32} \end{bmatrix}, \quad E' = \begin{bmatrix} e_{11} & e_{12} \\ e_{21} & e_{22} \\ e_{31} & 0 \end{bmatrix} \qquad (16a,b,c)$$

with the elements of A', B' and E' given by

$$a_{11} = GD^2C_o'^+/\lambda_1^2-\mu_s-D, \quad a_{32} = (F_f+F_r)/X_1 \qquad (17a,b)$$

$$a_{21} = D/3-G^2D^2C_o'^+X_2'/(Y\lambda_1^2\lambda_2) \qquad (17c)$$

$$a_{22} = GD^3C_o'^+/(\lambda_1\lambda_2^2)-\mu_s-D/3 \qquad (17d)$$

$$b_{11} = (X_r-X_1')/V^c+G^2X_1'^2C_o'^+/(YV^c\lambda_1^2)-GX_1'C_f^+F_f/(V^{c^2}D\lambda_1) + b_{12} \qquad (17e)$$

$$b_{12} = -A^S\alpha X_1^2 \exp(-X_1/\beta)F_r/[V^c(F_r+F_w)^2] \qquad (17f)$$

$$b_{21} = (X_1'-X_2')/3V^c+G^2DX_2'C_o'^+[D(X_1'+3X_2')+6GX_1'X_2'/Y]/(YV^c\lambda_1^2\lambda_2^2)-GX_2'C_f'^+F_f/(V^{c^2}\lambda_1\lambda_2) \qquad (17g)$$

$$b_{31} = X_2'/X_1-1, \quad b_{32} = -1 \qquad (17h,i)$$

$$e_{11} = -X_1'/V^c+G^2X_1'^2C_o^+/(YV^c\lambda_1^2) + GX_1'C_f^+F_r/(V^{c^2}D\lambda_1) \qquad (17j)$$

$$e_{21} = (X_1'-X_2')/3V^c+G^2DX_2'C_o'^+[D(X_1'+3X_2')+6GX_1'X_2'/Y]/(YV^c\lambda_1\lambda_2) + GX_2'C_f^+F_r/(V^{c^2}\lambda_1\lambda_2) \qquad (17k)$$

$$e_{12} = GX_1'F_f/(V^c\lambda_1), \quad e_{22} = GDX_2'F_f/(V^c\lambda_1\lambda_2) \qquad (17l,m)$$

$$e_{31} = X_2'/X_1, \quad \lambda_1 = GX_1'/Y+D, \quad \lambda_2 = 3GX_2'/Y+D \qquad (17n,o,p)$$

In eqns. (18a-c) the numerical values of A', B' and E' are given which are calculated using the data given in Table 7.

$$A' = \begin{bmatrix} -0.801 & 0 & 0 \\ 0.266 & -0.271 & 0 \\ 0 & 0.070 & 0 \end{bmatrix}, \quad B' = \begin{bmatrix} 4.446 & -2.198 \\ 0.033 & 0 \\ -0.395 & -1 \end{bmatrix} \qquad (18a,b)$$

$$E' = \begin{bmatrix} -6.632 & 0.154 \\ 0.054 & 0.008 \\ 0.605 & 0 \end{bmatrix} \qquad (18c)$$

From Eqn. (18a) it can be seen that the elements a_{11}, a_{21} and a_{22} of the A' matrix are dominated by respectively the flow elements -D, D/3 and -D/3.

Table 8 gives the eigenvalues of the A' matrix calculated from eqn. (18a).

TABLE 8. Eigenvalues of the A' matrix (hr^{-1})

$\Delta X_1'(t)$: -0.801;	$\Delta X_2'(t)$: -0.271;	$\Delta V^b{}'(t)$: 0

SIMULATIONS

Simulations have been used to prove how closely the dynamic behaviour of the reduced 3rd order linear model (eqns. (14) to (17)) matches that of the original 10th order nonlinear model (eqns. (1) and (7)). We choose to present here only simulations (Figs. 6 and 7) showing responses to the sludge volume, the first compartment biomass concentration and the biomass concentration leaving the reactor to step changes in the controls $F_r(t)$ and $F_w(t)$ (see also Table 9) and simulations (Figs. 8 and 9) showing how the same variables react to step changes in the disturbances $F_f(t)$ and $C_f(t)$ (see also Table 10).

TABLE 9. Step changes in $F_r(t)$ and $F_w(t)$ used in the simulations of Figs. 6 and 7

time (hr)	$F_r(t)$ (m^3/hr)	$F_w(t)$ (m^3/hr)
0-25	400	5
25-50	450	5
50-75	400	5
75-100	350	5
100-125	400	5
125-150	400	10
150-175	400	5
175-200	400	0
200-225	400	5
225-250	450	10

TABLE 10. Step changes in $F_f(t)$ and $C_f(t)$ used in the simulations of Figs. 8 and 9

time (hr)	$F_f(t)$ (m^3/hr)	$C_f(t)$ ($gCOD/m^3$)
0-25	400	1500
25-50	450	1500
50-75	400	1500
75-100	350	1500
100-125	400	1500
125-150	400	1700
150-175	400	1500
175-200	400	1300
200-225	400	1500
225-250	450	1700

CONCLUSIONS

In this paper we have derived a simple dynamic model, eqns. (13) to (17), for an oxygen-enriched industrial waste-water process. Not only may the model parameters be estimated from steady-state measurements, but the model variables may still be associated with the physical variables of the waste-water treatment process.
Further research will be directed towards establishing how robust the model is and towards the development of control systems using the linearized model of the process.

ACKNOWLEDGEMENTS

The authors acknowledge the cooperation of SHELL Nederland B.V. Pernis in this research, and permission to publish the results contained here in.

REFERENCES

Attir, U., M.M. Denn and C.A. Petty (1977). Dynamic simulation of continuous sedimentation. AIChE Sympos. Ser., no. 167, 73, 49-54.

Attir, U. and M.M. Denn (1978). Dynamics and control of activated sludge waste-water process. AIChE Journal, 24, 693-698.

Bultheel, A. (1983). Report TW61, Department of Computer Science, K.U. Leuven, Heverlee, Belgium.

Grutsch, J.F. (1980). The α-β-γ's of waste-water treatment. Environmental Science & Technology, 14, 276-281.

Heynen, J.J. (1984). Biological industrial waste-water treatment minimizing biomass production and maximizing biomass concentration. Ph.D Thesis, Delft University Press, Delft, The Netherlands.

Takamatsu, T., S. Shioya, K. Yokoyama, Y. Kurome and K. Morisaki (1981). On line monitoring and control of biochemical reaction processes. IFAC 8th Triennial World Congress, Kyoto, 3679-3684.

Tracy, K.D. and T.M. Keinath (1974). Dynamic model for thickening of activated sludge. AIChE Sympos. Ser. , no. 136, 70, 291-308.

CAPTIONS

Fig. 1. Schematic diagram of the waste-water treatment process

Fig. 2. Biomass flux W as a function of biomass concentration X for different values of the tank outflow $F_r + F_w$

Fig. 3. Limiting biomass flux W_l and limiting biomass concentration X_l as a function of the tank outflow $F_r + F_w$

Fig. 4. Schematic diagram of the settling tank

Fig. 5. Responses of the sludge volume $V^b(t)$ to step changes in the recycle flow $F_r(t)$ (see Table 5) in the case X_l = 11430 $gVSS/m^3$ (line 1) and $X_l(t) = 13250 - 4.5\ F_r(t)$ ($gVSS/m^3$) (line 2)

Fig. 6. Responses of the sludge volume (line 1 non linear 10th order model (NL 10th M); line 2: linear 3rd order model (L 3rd M)) to step changes in the recycle sludge flow $F_r(t)$ and sludge waste flow $F_w(t)$ (see Table 9)

Fig. 7. Responses of the biomass concentration in the first compartment (line 1: NL 10th M; line 2: L 3rd M) and the biomass concentration leaving the reactor (line 3: NL 10th M; line 4: L 3rd M) to step changes in the recycle sludge flow $F_r(t)$ and sludge waste flow $F_w(t)$ (see Table 9)

Fig. 8. Responses of the sludge volume (line 1: NL 10th M; line 2: L 3rd M) to step changes in the waste-water feed flow $F_f(t)$ and substrate concentration $C_f(t)$ (see Table 10)

Fig. 9. Responses of the biomass concentration in the first compartment (line 1: NL 10th M; line 2: L 3rd M) and the biomass concentration leaving the reactor (line 3: NL 10th M; line 4: L 3rd M) to step changes in the waste-water feed flow $F_f(t)$ and substrate concentration $C_f(t)$ (see Table 10)

Fig.1

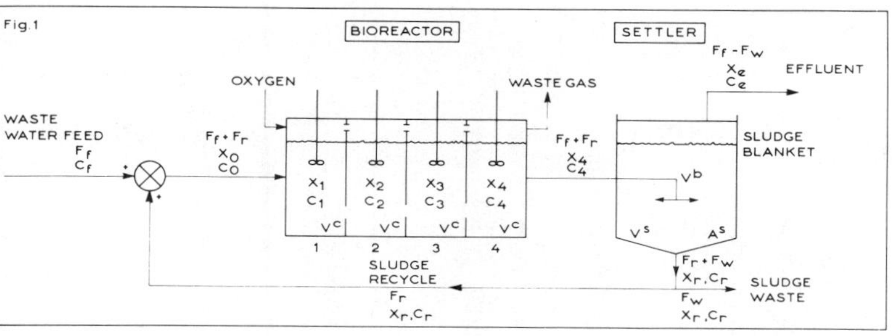

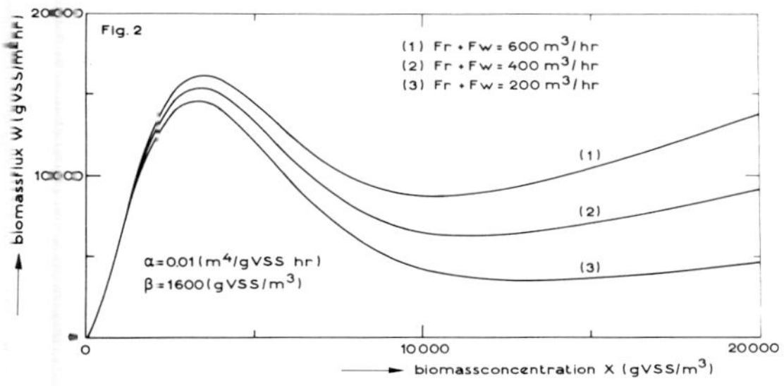
Fig. 2
(1) Fr + Fw = 600 m³/hr
(2) Fr + Fw = 400 m³/hr
(3) Fr + Fw = 200 m³/hr
α = 0.01 (m⁴/gVSS hr)
β = 1600 (gVSS/m³)
biomassflux W (gVSS/m² hr)
biomassconcentration X (gVSS/m³)
0
10000
20000

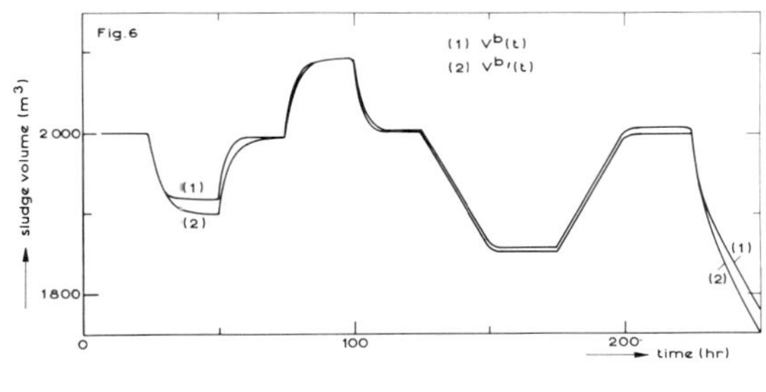
Fig. 6
(1) Vᵇ(t)
(2) Vᵇ'(t)
sludge volume (m³)
2000
1800
0
100
200
time (hr)

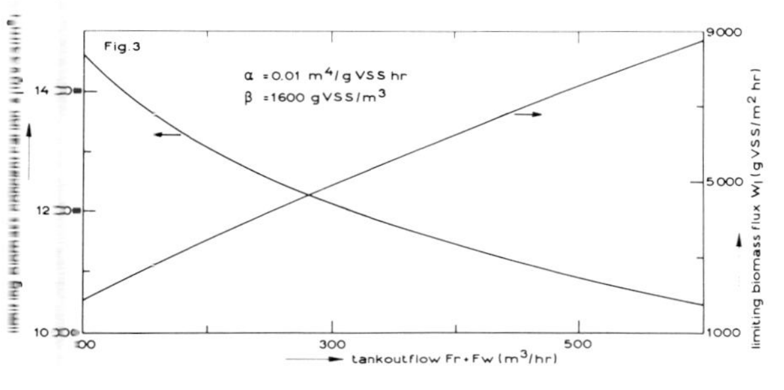
Fig. 3
α = 0.01 m⁴/gVSS hr
β = 1600 gVSS/m³
tankoutflow Fr + Fw (m³/hr)
limiting biomass flux W (gVSS/m² hr)
300
500
9000
5000
1000

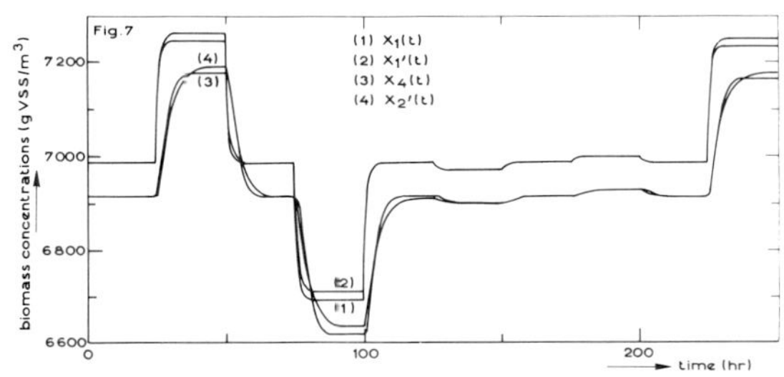
Fig. 7
(1) X₁(t)
(2) X₁'(t)
(3) X₄(t)
(4) X₂'(t)
biomass concentrations (gVSS/m³)
7200
7000
6800
6600
0
100
200
time (hr)

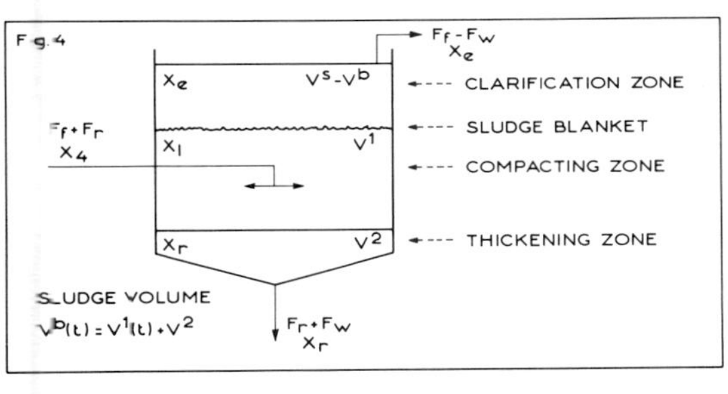
Fig. 4
Fr - Fw
Xe
Vˢ - Vᵇ
CLARIFICATION ZONE
SLUDGE BLANKET
COMPACTING ZONE
THICKENING ZONE
Ff + Fr
X₄
X₁
V¹
Xr
V²
SLUDGE VOLUME
Vᵇ(t) = V¹(t) + V²
Fr + Fw
Xr

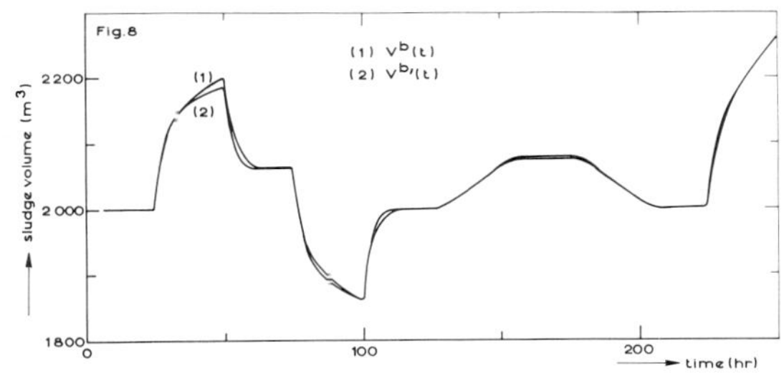
Fig. 8
(1) Vᵇ(t)
(2) Vᵇ'(t)
sludge volume (m³)
2200
2000
1800
0
100
200
time (hr)

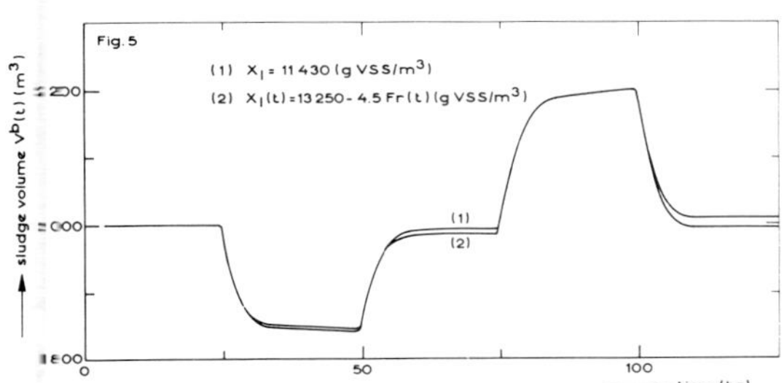
Fig. 5
(1) X₁ = 11430 (gVSS/m³)
(2) X₁(t) = 13250 - 4.5 Fr(t) (gVSS/m³)
sludge volume Vᵇ(t) (m³)
0
50
100
time (hr)

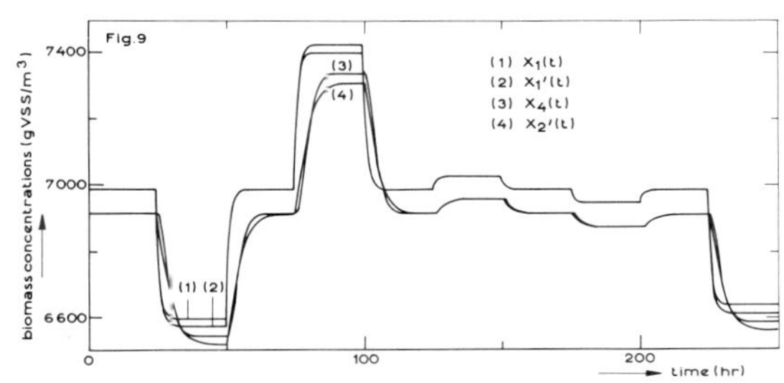
Fig. 9
(1) X₁(t)
(2) X₁'(t)
(3) X₄(t)
(4) X₂'(t)
biomass concentrations (gVSS/m³)
7400
7000
6600
0
100
200
time (hr)

AN EXPERIMENTAL APPROACH TO IMPROVE THE MONOD MODEL IDENTIFICATION

C. Vialas*, A. Cheruy and S. Gentil****

**MIT, Departement of Nutrition and Food Science, Cambridge, Mass. 02139, USA*
***Laboratoire d'Automatique, CNRS UA 228, ENSIEG, National Polytechnic Institute of Grenoble, BP 46, 38042 St-Martin-D'Hères, France*

Abstract. Although the theoretical identifiability of models including Michaelis-Menten type non-linearities has been proved, the practical identification of those models remains difficult.
This paper deals with a simple and pragmatic approach to improve the estimation of the k_s-parameter and to minimize the amount of required experimental work, by choosing adequately the acquisition times of experimental data.

Keywords. Identification, identifiability, parameter estimation, Monod model, biotechnological process, experimental design.

INTRODUCTION

The Michaelis Menten equation, which was originally proposed to describe enzymatic reaction rate as a function of substrate concentration, is widely used in biotechnological modelling to describe microorganism growth. This result in the Monod model (Monod 1942) where the microorganism growth rate μ is expressed as a function of limiting substrate concentrations by the following equation :

$$\mu(S) = \frac{\mu_{max}\, S}{K_s + S} \qquad (1)$$

where μ_{max} is the maximum growth rate and K_s the Michaelis Menten constant.

Indeed, in a continuous culture, microorganism growth can be described by a Monod model as following :

$$\frac{dX}{dt} = \mu(S)\, X - DX \qquad (2)$$

$$\frac{dS}{dt} = -\frac{1}{Y}\, \mu(S)\, X - D(S - S_o) \qquad (3)$$

where X is the concentration of the biomass, S_o the concentration of limiting substrate in the feed flow, D the flowrate and Y the yield coefficient.

Holmberg and Ranta (1982), Nihtilä and Virkkunen (1977) have proved the identifiability of the three model parameters, μ_{max}, K_s and Y when the transient states of X and S during the growth phase are known (perfect measurements data). For that purpose they used the Pohjanpalo test which is based on a Taylor series expansion of the output vector (Pohjanpalo, 1978).
However, they also concluded that the practical identifiability of those parameters remained a problem. That was also noted by others authors working on several kinds of models (Brown, 1978 ; Carson 1979).

Indeed, measurements of chemical or biochemical variables often provide incomplete and noisy data since there is a lack of accurate and automated on line instrumentation in those fields. Besides, microorganism cultures are alive and are sensitive, and hence, in general, the inputs of such systems cannot be excited freely. Thus, it is very important to investigate the practical identifiability of the model (Young, 1982), which leads to the following question : is it possible to estimate correctly the model parameters from noisy and incomplete data ?

This problem will be investigated only for the parameters K_s, which can be justified in practice by the fact that, in general μ_{max} and Y are easily deduced from batch experiments. Besides, in most of the cases, K_s has a very small order of magnitude relative to μ_{max} and Y, which makes it more difficult to estimate. Holmberg (1982) showed that noisy data generally give a quite incorrect value for K_s, although the experimental curves fit is very good. The reason that the K_s-estimation has to be improved is that it represents the affinity of the organisms for the substrate. Thus, it is important to know the K_s-value for different substrates and for various microorganisms in order to compare them and to evaluate their performances in a process.

NUMERICAL METHODS

In what follows (Vialas, 1984) a modified Gauss-Marquardt method is used to minimize a cost function which is non linear in the parameters. This method is well suited for parameter estimation.

The cost function is :

$$J = \frac{1}{2} \sum_{i=1} \left[(X_m^i - X_f^i)^2 - (S_m^i - S_f^i)^2 \right] \qquad (4)$$

where N is the number of measurement points, X_m^i and X_f^i respectively represent the ith measurement and fit points of the biomass, and S_m^i and S_f^i represent the measured and fitted values of the substrate.

The Gauss-Marquardt method needs the computation of the objective function (4) and its gradient. So, it uses the sensitivity functions of the model, that is, first order derivatives of the variables X and S with respect of the parameters. The gradient of the cost function with respect to K_s can thus be expressed as :

$$\frac{\partial J}{\partial K_s} = \sum_{i=1}^{N} \left[\frac{\partial X_f^i}{\partial K_s} (X_f^i - X_m^i) + \frac{\partial S_f^i}{\partial K_s} (S_f^i - S_m^i) \right] \quad (5)$$

The variance of the estimate $\hat{K}_s$ is approached by :

$$\sigma^2_{\hat{K}_s} = \frac{1}{\sum_i \left(\frac{\partial X}{\partial K_s}\, i\right)^2 + \left(\frac{\partial S}{\partial K_s}\, i\right)^2} \cdot \frac{J(\hat{K}_s)}{2N-1} \quad (6)$$

The confidence interval is computed according to :

$$K_s = \hat{K}_s \pm t(N-1\ ,\ 1-\frac{\alpha}{2})\ \sigma_{\hat{K}_s} \quad (7)$$

where $t(N-1,\ 1-\frac{\alpha}{2})$ is the t-distribution value for 2N-1 degrees of freedom (2N measurement points, one parameter) and for a probability of $\alpha/2$ to be on the interval.

INFLUENCE OF THE REPARTITION OF DATA ON THE K_s ESTIMATE

The sensitivity curves of the variables X and S with respect to the parameter K_s emphasize a very high peak value just before the steady state occurs.

Fig. 1a gives with the same time scale, the simulation of the state variables (X, S) and the corresponding sensitivity curves $(\partial X, \partial K_s)$; $\partial S/\partial K_s)$. The K_s value is 0.001 mM/1. When the K_s value is higher the transient state is longer and the sensitivity peak is smoother (Fig. 1.b ; K_s = 0.1 mM/1).

Figure 2 illustrates the local effect on simulation curves of large K_s variations (0.1 mM/1,0.001 mM/1).

As the sensitivity function represents the information contained about a parameter in a measurement, choosing regular sampling times during the full experimental run involves the implicit assumption that this information is uniformely distributed over time. Obviously, it is not well suited to a correct identification of K_s. So, in order to study the influence of the sampling, a comparison of the identification results follows in three cases :

a - the sampling times are regular and numerous on the full experiment (one sampling time each about 15 minutes i.e. 60 points on the experiment).

b - the sampling times are regular and not very numerous (15 points instead of the previous 60).

c - only those 15 sampling points for which the sensitivity with respect to K_s is the highest are chosen from the previous 60 (case a).

In order to perform the identification on noisy data, two types of noise were simulated :

* an additive white gaussian with different variances

$$Z_N^k = Z^k + W \qquad (k=1,2) \qquad (8)$$

where W is uniformly distributed over an interval $|-\varepsilon\ ;\ \varepsilon|$.

Fig. 3 shows a typical run, with perfect and noisy simulations. The identification results are reported on table 1. They are always much better in case (c) than in case (b), although the number of measurements is the same.

When the noise is an additive gaussian one (8), experiment (c) provides also better values for K_s than experiment (a) where the number of samples is four times greater (runs 3, 4, 5).

When the noise is a proportional one (9), the values of K_s, with the three experiments (a) (b) (c) are roughly the same, but the standard deviations of the K_s values are always better in case (c) than in case (a) and (b) (runs 7, 8, 9).

INTERPRETATION OF THE RESULTS

Obviously, these results prove that the very small influence of K_s is generally hidden by the noise when the sampling is regular. To correctly estimate K_s, the sampling times have to be choosen when this parameter is influent, i.e. in the high sensitivity zone ; all the points which are not in this zone bring no information, but perturbation to the estimation problem.

During the optimization of cost function J (eq. 4), the progression in the parameter space is computed according to the gradient, eq(5). In this expression appear the sensitivity functions $\partial/\partial K_s$, which are near zero in most of the measurements points, unless in case (c). The differences between measured and fitted variables due to a wrong parameter value are also near zero except in the sensitivity zone, so the deviation is for the major part due to the noise. The first half of the table 1 is easily explained in view of these comments.

In case (b), the gradient value results essentially from the noise, and the optimization hardly converges ; as the gradient is low, the parameter standard-deviation is high (eq. 6).

In case (a), the gradient is influenced by the noise and the optimum is not as well located as in case (c) nevertheless the standard durations are near the same.

In the case of proportional noise, it must be noted that as X or S are zero, respectively at the beginning and at the end of the experiment, the noise is negligible at these times and the gradient is much less influenced than in the gaussian case. So the convergence can occur in all experiments but only in case (c) the gradient is high, and thus the standard deviation is low.

From a theoretical point of view, it is interesting to test in what extent these results are in agreement with the statistical approach used in optimal design of experiments (Fedorov 1972, Endrenyi 1982). In this case, the purpose is to choose the experimental conditions leading to the globaly most precise estimated parameters. This precision can be measured by various criteria ; the most usual ones are the so-called D-optimization and A-optimization criteria which select experimental conditions minimizing respectively the determinant (D-optimization) or the trace (A-optimization) of the variance-covariance matrix of the estimates. These criteria yield optimal parameters in the sense of minimizing the volume of their joint confidence regions.

Consider the following expression of the parameters

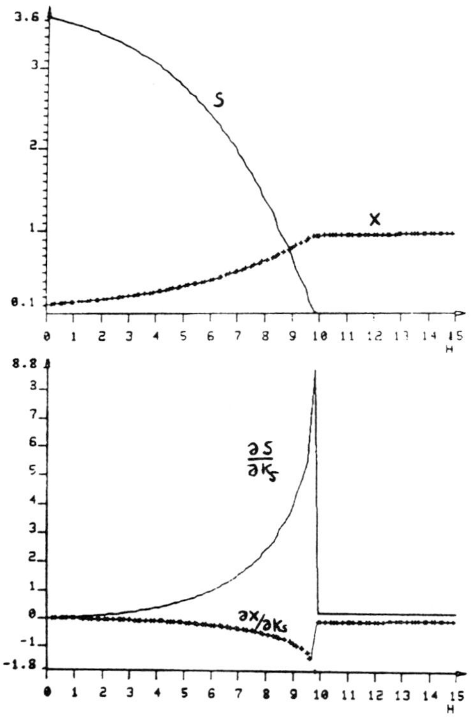

Figure 1a

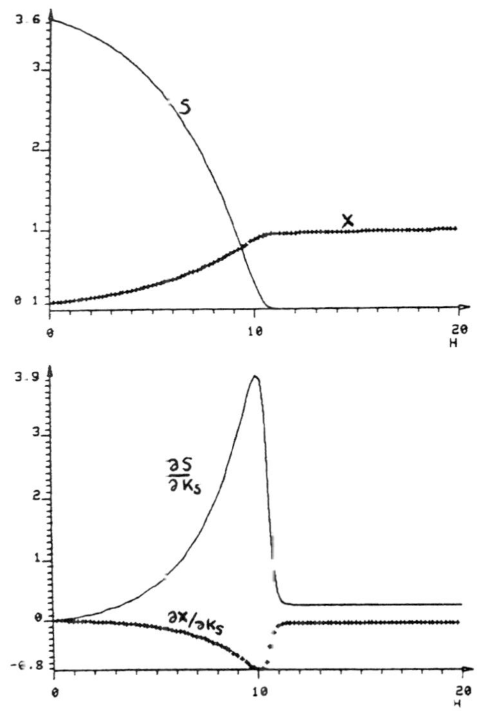

Figure 1b

Simulation curves of the state variables and of the corresponding sensitivity functions

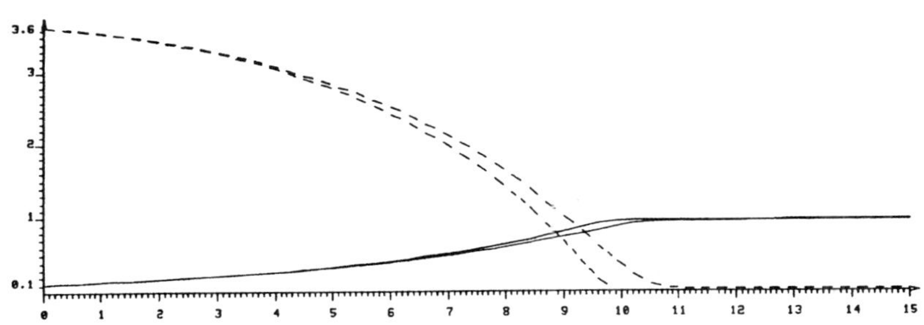

Figure 2

Simulations of X and S with 2 different K_s values

-------- S : Substrate ——— X : biomass

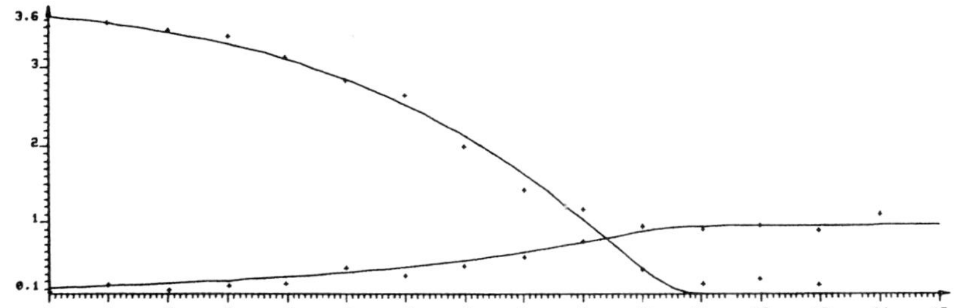

Figure 3

A typical run of simulation

+++ with noise

—— without noise

variance-covariance matrix V (approximated by a limited expansion).

$$V = \left| \left(\frac{\partial Z}{\partial \theta}\right)^T \Gamma_b \left(\frac{\partial Z}{\partial \theta}\right) \right|^{-1}$$

where $(\partial Z/\partial \theta)$ is the vector of sensitivity functions (θ parameter, Z state variable).

Γ_b = variance-co variance matrix of the noises added to the observations.

It appears that, if Γ_b is a diagonal matrix (which holds if the noises are uncorrelated), to minimize the determinant on the trace of V is the same as to minimize the determinant or the trace of $(\partial Z/\partial \theta)^T (\partial Z/\partial \theta)$. And that amounts to choose the conditions for which sensitivity functions are maximum. Therefore, that is in agreement with the above results.

CONCLUSION

Optimal experiment design is an important and difficult problem for biotechnological processes : on one hand, the number of experiments has to be minimized because biological analysis are generally expensive, are difficult to perform and the available sensors number is limited ; on the other hand, the model parameters have to be identified with the best possible accuracy because it is not just a matter of curve fitting but they are further used as typical biological characteristics of the process under study. Up to now, this problem is not satisfactorily solved because the theory of optimal design (Fedorov, 1972) is only valid for models linear with respect to the parameters and the models used in biotechnology are usually non linear.

The Monod model is a typical example of these difficulties. This paper clearly illustrates that, although theoretical identifiability study of a model is a useful tool, it is not sufficient to insure that in practise the parameters are identifiable from the available measurements. Clearly that depends on the noise level, the amount of data and their sampling times. These have to be chosen adequately with respect to the process dynamics. This study provides a guide to choose a minimal number of measurement points which give an improved estimation with respect to a regular and high frequency sampling strategy.

REFERENCES

[1] BROWN, R.F. and GODFREY, K.R. (1978)
Problems of determinacy in compartmental modelling with applications to bilirubin kinetics. Math. Biosci, 40, 205-224.

[2] CARSON, E.R., COBELLI, C. FINKELSTEIN, L.(1979)
The identification of metabolic systems, IFAC Symposium on identification and system parameter estimation, DARMSTADT, (24-28 sept.)

[3] ENDRENYI, L., (1981)
"Kinetic data analysis" Plenum Press.

[4] FEDOROV, V. (1972)
"Theory of optimal experiments" Academic Press.

[5] HOLMBERG, A., RANTA, J. (1982)
Procedures for parameter and state estimation of microbial growth processes. AUTOMATICA, 18, 181-193.

|6| HOLMBERG, A. (1982)
On the practical identifiability of microbial growth models incorporating Michaelis Menten type nonlinearities. Math. Biosci. 62, 23-43.

|7| MONOD, J. (1942)
The growth of bacterial cultures. Ann. Rev. Microb. 3, 371-394.

|8| NIHTILÄ, M., VIRKKUNEN. (1977)
Practical identifiability of growth and substrate consumption Models. Biotech. Bioengng 19, 1831-1850.

|9| POHJAMPALO, H., (1978)
Systems Identifiability based on the power series expansion of the solution Math. Biosci., 41, 21.

|10| VIALAS, C., (1984)
Ph. D Thesis, INPG, Grenoble (1984).

|11| WALTER, E., (1982)
Identifiability of state space models Lecture Notes in biomathematics, 46, Springer Verlag.

|12| YOUNG, P., (1982)
General theory of modeling for badly defined systems. IFIP on modeling, Identification and control in Environmental Systems, Vansteen kiste ed.

Run	Noise characteristics		K_s initial	K_s true	$\hat{K}_s$: estimated value and standard deviation		
	σ	ε			a	b	c
1	0	0	10^{-5}	10^{-3}	$1.004\ 10^{-3}$ $\pm\ 0.007\ 10^{-3}$	$1.004\ 10^{-3}$ -	$1.003\ 10^{-3}$ $\pm\ 0.006\ 10^{-3}$
2	10^{-3}	0	10^{-5}	10^{-3}	$1.04\ 10^{-3}$ $\pm\ 0.07\ 10^{-3}$	$1.3\ 10^{-3}$ -	$1.03\ 10^{-3}$ $\pm\ 0.06\ 10^{-3}$
3	10^{-2}	0	10^{-5}	10^{-3}	$1.43\ 10^{-3}$ $\pm\ 0.7\ 10^{-3}$	$4.05\ 10^{-3}$ -	$1.26\ 10^{-3}$ $\pm\ 0.6\ 10^{-3}$
4	0.1	0	10^{-5}	10^{-2}	0.013 $\pm$ 0.006	0.00112 $\pm$ 0.01	0.012 $\pm$ 0.006
5	0.5	0	10^{-5}	10^{-1}	0.14 $\pm$ 0.04	0.23 $\pm$ 0.76	0.12 $\pm$ 0.04
6	10^{-1}	0	10^{-4}	10^{-1}	$1.078\ 10^{-1}$ $\pm\ 0.08\ 10^{-1}$	$1.36\ 10^{-1}$ $\pm\ 0.15\ 10^{-1}$	$1.078\ 10^{-1}$ $\pm\ 0.08\ 10^{-1}$
7	0	0.1	10^{-4}	10^{-1}	$9.82\ 10^{-2}$ $\pm\ 0.8\ 10^{-2}$	$9.65\ 10^{-2}$ $\pm\ 1.6\ 10^{-2}$	$9.78\ 10^{-2}$ $\pm\ 0.45\ 10^{-2}$
8	0	0.5	10^{-4}	10^{-1}	$9.07\ 10^{-2}$ $\pm\ 4\ 10^{-2}$	$8.24\ 10^{-2}$ $\pm\ 8.0\ 10^{-2}$	$8.9\ 10^{-2}$ $\pm\ 2.2\ 10^{-2}$
9	0	10^{-3}	10^{-3}	10^{-3}	$0.74\ 10^{-3}$ $\pm\ 0.73\ 10^{-3}$	$0.69\ 10^{-3}$ $\pm\ 1.2\ 10^{-3}$	$0.71\ 10^{-3}$ $\pm\ 0.36\ 10^{-3}$

Table 1

Results of the identification with three different experimental procedures -a, b, c, design

MODELLING AND PARAMETER IDENTIFICATION OF BATCH FERMENTATION PROCESS

A. Nakrachi and R. Laurent

Centre d'Automatique, Université des Sciences et Techniques de Lille, 59655 Villeneuve d'Ascq, France

Abstract. We have studied the first step of methane fermentation, in our case the transformation of saccharose into lactic acid by an anaerobic microorganisms population. The first part of the study is based on a graphical method analysis of the various functions and their first derivatives obtained by using cubic smoothing splines. It shows that the first derivatives must be not linearly dependant. We propose a model derived from Teissier'model where the term of inhibition appears in exponential form. If the inhibitory substance is substract itself, the Andrew's model can be found again by approximation. The last part is the determination of the parameters' numerical values for some experimental data.

Keywords. Methanefermentation; modelling; smoothing splines; identification; iterative method; parameter estimation.

INTRODUCTION

The advantage of the methane fermentation is that it can at the same time suppress undesirable organic wastes and produce methane gas.
The modelling and identification of such a process was considered as a necessary step toward the implementation of regulation schemes for industrial-scale methanizers.
One of the major problems of this process, considered in the aggregate, is the variety of intermediate products (Volatile fatty acid).
In order to make a model mathematically tractable, some authors propose the following scheme :

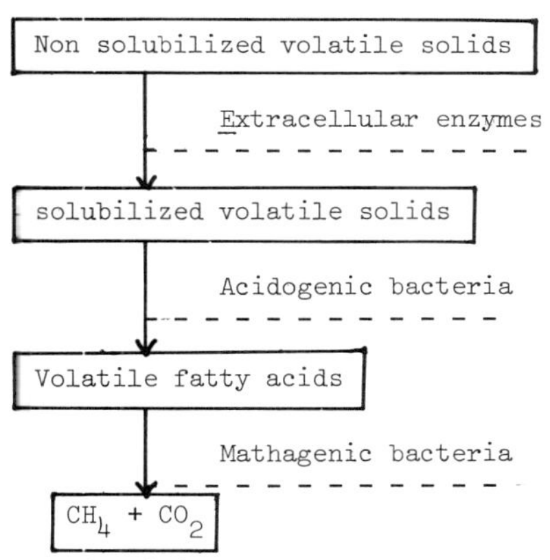

Our work is limited to the study of the first step.

MODELLING

Generally continuous fermentation is less common than batch fermentation, under the circumstances a lot of models were developed in the batch case (Monod; Teissier; Andrews; Laurent), models which especially stress the fact of the growth itself (with or without inhibition).
Smoothing experimental curves by cubic smoothing allowed us to have access to the values of the first derivatives, which are more exact than those obtained by the direct reckonings of the slopes at the different experimental points.

It is proved that the different rates are not proportional between themselves (Fig.2); on the other hand the growth rate $\mu = 1/X(dX/dt)$ and that proposed by Andrews are different (Fig.3).
If we trace the function

$$FCT = 1/X . dX/dt(\exp(+.24\ P))$$

(X, S and P are respectively the biomass, substract and product concentrations)
in terms of S we obtain a curve of the type proposed by Monod and Teissier (Fig.4).
Since Monod's model could be considered as Teissier's model approximation, we opted for the latter.
Consequently, the model which represents the growth could be written :

$$dX/dt = kX\ EXP(-\alpha P)(1-EXP(-\beta S))$$

And in the case where inhibitory substance is the substract itself the model expression became

$$dX/dt = kX\ EXP(-\alpha S)(1-EXP(-\beta S))$$

and that of Andrew's model

$$dX/dt = \mu_{max}\ X \frac{1}{1+Ks/S+S/Ki}$$

If we substitute: $EXP(x) = 1+x$ we find the Andrews's model expression with

$\mu_{max} = k\beta/(\alpha+\beta)$; $Ks = 1/(\alpha+\beta)$; $Ki = \alpha\beta/(\alpha+\beta)$.
And if $\alpha=0$ which means, without inhibition, we find

$$dX/dt = kX \frac{S}{S + 1/\beta}$$

which means the Monod's model expression; when it was not obvious to find again Monod's model expression from that of Andrew's model expression when there is not inhibition.
A study of same kind has been carried out for substract elimination rate.
The expression of aggregate model becomes then :

$$dX/dt = kX\ EXP(-\alpha P)(1-EXP(-\beta S))$$

EXTENDED KALMAN FILTERING TECHNIQUE FOR THE ON-LINE CONTROL OF *CANDIDA UTILIS* PRODUCTION

M. Ghoul*, M. -N. Pons, J. -M. Engasser** and J. Bordet****

**Laboratoire de Génie Alimentaire et Biochimique, 1, rue Grandville, F-54042 Nancy Cedex, France*
***Laboratoire des Sciences du Génie Chimique CNRS-ENSIC, 1, rue Grandville, F–54042 Nancy Cedex, France, Institut National Polytechnique de Lorraine*

Abstract. Extended Kalman Filter (EKF) is used for the on-line estimation of glucose, ethanol and biomass concentrations during the batch and fedbatch growth of Candida utilis on glucose. A physiological dynamic model describes the behavior of the yeast on glucose. The experimental systen is presented. The performances of EKF are evaluated under both cultivation modes using classical observation variables: carbon dioxide production, oxygen and ammonia consumption. Although care has to be taken in the numerical procedures to obtain EKF algorithm convergence, successful on-line estimation of biomass and glucose concentrations are performed.

Keywords. Kalman filters - Fermentation - Control theory - Biocontrol

INTRODUCTION

One of the challenges of the biotechnological process engineer is the development of optimal control strategy for microbial cultures. This task is made difficult by the scarceness of on-line direct sensors especially for such important state variables as biomass. Indirect estimation techniques are therefore required. One of them is the Extended Kalman Filter (EKF) method: using a kinetic model describing the dynamic evolution of the state variables and linear relations between observation and state variables, EKF provides on-line estimation of the state variables.
Some applications of EKF in bioprocesses have been described. Swiniarski and coworkers tested EKF on the microbial degradation of cellulose. Bellgardt used this method to follow the on-line evolution of a batch fermentation of Sacharomyces cerevisiae on glucose. In our opinion EKF methods should be still more useful in fedbatch processes when the estimation of states variables could be included in the control strategy (Dekkers).
We described in this paper the setting up of such a control structure. The process is the fedbatch aerobic production of Candida utilis on glucose.

DYNAMIC PHYSIOLOGICAL MODEL

To simulate the growth of Candida utilis in a wide range of conditions a physiological model has been developped. It is based on three different physiological states of the yeast. Each state has its particular activity, and during the fermentation process, the yeast can shift from one state to the other. The three states are:
State X_1, the fermentation of glucose: the growth of the yeast is accompagnied by an important transformation of glucose into ethanol. With Candida utilis the produced ethanol remains essentially intracellular.
State X_2, the respiration of glucose: glucose is metabolized through the respiratory pathway to sustain cellular growth.
State X_3, the respiration of ethanol: in the presence of oxygen ethanol is reconsumed for cellular growth.
Each state has its own kinetics: Monod-type growth kinetics with inhibition by intracellular ethanol, constant yields for substrate conversion into biomass and product. The different rate equations for cellular growth, substrate uptake and product formation are:

$$r_{1Gl} = \frac{1}{Y_{1XGl}} \; \frac{\mu_{1max} Gl}{K_{1Gl}+Gl} \; \frac{X_1}{1+E/K_{1E}} \quad [1]$$

$$r_{2Gl} = \frac{1}{Y_{2XGl}} \; \frac{\mu_{2max} Gl}{K_{2Gl}+Gl} \; \frac{X_2}{1+E/K_{2E}} \quad [2]$$

$$r_{1E} = \frac{Y_{1EGl}}{Y_{1XGl}} \; \frac{\mu_{1max} Gl}{K_{1Gl}+Gl} \; \frac{X_1}{1+E/K_{1E}} \quad [3]$$

$$r_{3E} = \frac{1}{Y_{3XE}} \; \frac{\mu_{3max} E}{K_E+E} \; X_3 \quad [4]$$

$$r_{1X} = \frac{\mu_{1max} Gl}{K_{1Gl}+Gl} \; \frac{X_1}{1+E/K_{1E}} \quad [5]$$

$$r_{2X} = \frac{\mu_{2max} Gl}{K_{2Gl}+Gl} \; \frac{X_2}{1+E/K_{2E}} \quad [6]$$

$$r_{3X} = \frac{\mu_{3max} E}{K_E+E} \; X_3 \quad [7]$$

The model also allows for reversible transitions between the three states, at rates determined by the glucose level in the broth and the intracellular ethanol concentration, as shown in Figure 1. State X_1 is favored by high glucose concentration, state X_2 by low glucose and ethanol levels and state X_3 by high ethanol concentration. The transition rates are expressed as:

$$r_{12} = \frac{K_{12}}{1+Gl/G_{12}} \; \frac{X_1}{1+E/E_{12}} \quad [8]$$

$$r_{21} = \frac{K_{21}Gl\ X_2}{C_{21}+Gl} \quad [9]$$

$$r_{23} = \frac{K_{23}E\ X_2}{C_{23}+E} \quad [10]$$

$$r_{32} = \frac{K_{32}}{1+Gl/G_{32}} \cdot \frac{X_3}{1+E/E_{32}} \quad [11]$$

$$r_{31} = \frac{K_{31}Gl\ X_3}{C_{31}+Gl} \quad [12]$$

$$r_{13} = \frac{K_{13}E\ X_1}{C_{13}+E} \quad [13]$$

Mass balances over the batch and fedbatch fermentor yield to the following set of differential equations:

$$dGl/dt = -r_{1Gl}-r_{2Gl}+(dV/dt)(S_0-Gl)/V \quad [14]$$

$$dE/dt = r_{1E}-r_{3E}-(dV/dt)E/V \quad [15]$$

$$dX_1/dt = r_{1X}-r_{12}-r_{13}+r_{21}+r_{31}-(dV/dt)X_1/V \quad [16]$$

$$dX_2/dt = r_{2X}-r_{21}-r_{23}+r_{12}+r_{32}-(dV/dt)X_2/V \quad [17]$$

$$dX_3/dt = r_{3X}-r_{31}-r_{32}+r_{13}+r_{23}-(dV/dt)X_3/V \quad [18]$$

$$dV/dt = q + q_{NH3} - q_{sample} \quad [19]$$

with $q = 0$ in batch and $q = q(t)$ in fedbatch

We verified that in our cultivation conditions $dV/dt \approx 0$ in batch and is small in fedbatch. The terms $X_i(dV/dt)/V$, $E(dV/dt)/V$ and $Gl(dV/dt)/V$ are negligible and will not be further considered. However precise volume integration is performed.

The parameters of the model were determined from several batch and continuous cultures of Candida utilis at different initial or feed concentration of glucose and ethanol.

Figures 2 and 3 show that, with the same set of parameters, the model is able to correctly simulate the batch and fedbatch aerobic growth of the yeast on glucose. The experimental data of Figure 3 were taken from a fedbatch with S_0 = 250 g/l run at residual glucose concentration constantly maintained below 0.1 g/l and the experimental feedrate was used for the simulation. It has been verified that the initial biomass repartition between the three states is not a critical parameter. These results encouraged us to pursue our work with the implementation of an EKF estimation policy using this model.

EQUIPMENT DESCRIPTION

The experimental set up is described in Figure 4. A microprocessor-based dissolved oxygen control system has been added to the fully instrumented 20-liter Chemapec fermentor. On-line glucose determination through an automated continuous flow device is possible. The real-time data -acquisition, -treatment and -logging as well as the control of the substrate pump is performed by a KONTRON-PSI80 microcomputer. Two digital load-cells (SARTORIUS) are connected via RS-232C lines. The PSI80 is connected by RS232C to a SEMS-MITRA 115 in charge with the EKF algorithm.

ESTIMATION PROCEDURE

The complete EKF equations can be easily found in litterature (Radix) and will not be explicited here. The chosen state variables are glucose, ethanol and the three biomass concentrations (X_1, X_2 and X_3). None of them are observed directly. The observation variables are classical ones: the carbon dioxide production (CPR, g/l), the oxygen consumption (OUR, g/l) and the ammonia solution consumption (NUR, g/l) (added for pH control) per liter of broth (Figure 5).

The evolution of the state vector $\underline{x}(t)$ is described by:

$$\underline{x}(t) = F[\underline{x}(t),\underline{u}(t)] + \underline{w}(t)$$

$$\text{with } \underline{x}(t) = \begin{bmatrix} Gl \\ E \\ X_1 \\ X_2 \\ X_3 \end{bmatrix} \qquad \underline{u}(t) = q\,S_0/V$$

F = model functional relations explicited previously [eqs 1 to 18]

The observation equation is:

$$\underline{y}(t) = H\ \underline{x}(t) + \underline{v}(t)$$

$$\text{with } \underline{y}(t) = \begin{bmatrix} OUR \\ CPR \\ NUR \end{bmatrix} \text{ and } H = \begin{bmatrix} 0.46 & 0. & 0. & 1. & 1.5 \\ 0.35 & 0. & 2.2 & 1.65 & 1.3 \\ 0. & 0. & 4.2 & 4.2 & 4. \end{bmatrix}$$

The non-linear model equations are linearized to fit into the discrete EKF algorithm. The optimal estimate of the state vector at time t = k T_e (T_e = Kalman estimation period), $\hat{x}_{opt}(k)$ is given by the following relation:

$$\hat{x}_{opt}(k) = \hat{x} + K\ [\underline{y} - H\hat{x}]$$

where $\hat{x}$ is the state vector estimate obtained from integration of the linearized discrete model equations (initial point = $\hat{x}_{opt}(k-1)$) and K the Kalman gain matrix.

The state and observation noises (respectively $\underline{w}(t)$ and $\underline{v}(t)$ are supposed to be uncorrelated zero mean white noises. The choice of the different covariance matrices (positive definite state (Q), observation (R), noises and error filtering (P) covariances matrices) is one of the most difficult task of the implementation of EKF. They can be approximately set from examination of previous runs data but they have to be considered as important tuning parameters affecting the filter convergence. After tests on simulated processes it appeared that precautions have to be taken to deal with numerical information losses. It is especially important to verify that the P matrix is keeping its symmetry.

APPLICATION TO BATCH PROCESSES

The first on-line test of EKF was performed on batch fermentations. Figure 6 shows the results of such an experiment with only CPR and OUR as observation variables. The experiment of Figure 7 was done with the complete set of observation variables. The agreement between the estimated state variables and the off-line measurements (glucose and total biomass) is quite good. The estimation period T_e was 10 mn. No convergence problem was encountered on the batch fermentations. The choice of estimate of covariance matrices was not critical.

APPLICATION TO A FEDBATCH PROCESS

The next step was the application to a fedbatch fermentation at constant residual glucose concentration for the estimation of process state variables. The substrate concentration in the feeding solution was 250 g/L. CPR and OUR were the observation variables during the experiment described in Figure 8. The agreement between estimated and measured total biomass is good. The glucose concentration in the broth was below detectable limit (<0.1 g/l) during the experiment.

The estimation period was 10 mn. Although the choice of estimate of covariance matrices was not critical, fedbatch EKF estimation seems more sensitive than batch EKF estimation to numerical uncertainties. The integration of the model equations with Euler method which produces state vector estimate requires a small time step (0.005 hr). The estimation period should not be greater than 15 mn. This seems to be related to the facts that:

- a fedbatch process is highly nonlinear, with a short response time of microorganisms to control action (overfeeding);
- two of the state variables, glucose and ethanol, have very small values with respect to the biomass concentrations.

CONCLUSIONS

Extended Kalman Filter have been applied to the on-line monitoring of *Candida utilis* growth on glucose. A sophisticated model takes into account the metabolic behavior of the yeast and correctly simulates both batch and fedbatch cultures. Although the implementation of EKF requires to take precautions in the numerical procedures, the evaluation of the performances of the estimator leads to encouraging results. Easily available observation variables have been used: the carbon dioxide production, the oxygen and ammonia consumptions. The next step to this work is the uses of EKF estimations in the adaptative control strategy of optimal fedbatch, i.e. the online determination of the substrate feedrate.

REFERENCES

Swiniarski R., Lesniewski A., Dewski N.A.M., Leigh J.R. (1982) *1st IFAC Workshop on Mod. and Control of Biotech. Proc.* 231

Bellgardt K.H. Dissertation (1984) *Modellbildung des Wachstums von Saccharomyces cerevisae in Rührkesselreaktoren*

Dekkers R.M. *1st IFAC Workshop on Mod. and Control of Biotech. Proc.* (1982)

Radix J.C. (1970), *Introduction au filtrage numérique*, Eyrolles Paris

NOTATIONS

C_{ij} limitation constant on transition reaction state i -> state j (g/l)
CPR carbon dioxide production (g/l)
E intracellular ethanol concentration (g/l of broth)
E_{ij} inhibition constant for ethanol on transition reaction state i -> state j (g/l)
F model fonctional relations
Gl glucose concentration (g/l)
G_{ij} inhibition constant for glucose on transition reaction state i -> state j (g/l)
H observation matrix
k number of estimation period
K Kalman gain matrix
K_{iGl} Monod's constant on glucose for X_i (g/l)
K_{iE} Monod's constant on ethanol for X_i (g/l)
K_E Monod's constant on ethanol
K_{ij} state i -> state j transition rate constant (hr^{-1})
NUR ammonia consumption (g/l)
OUR oxygen consumption (g/l)
P error filtering covariance matrix
q substrate feedrate (l/hr)
q_{NH3} ammonia flowrate (l/hr)
q_{sample} sample flowrate (l/hr)
Q state noise covariance matrix
r_{iGl} glucose uptake rate of X_i (g/lhr)
r_{iE} ethanol uptake rate of X_i (g/lhr)
r_{iX} growth rate of X_i (g/lhr)
r_{ij} state i -> state j transition reaction rate (g/lhr)
R observation noise covariance matrix
S_0 glucose concentration in the feed (g/l)
t time
T_e estimation period
$\underline{u}$ control vector
$\underline{v}$ observation noise vector
V volume (l)
$\underline{w}$ state noise vector
X_i state i biomass concentration (g/l)
$\underline{x}$ state vector
$\hat{\underline{x}}$ state vector estimate
$\hat{\underline{x}}_{opt}$ optimal state vector estimate
$\underline{y}$ observation vector
Y_{iXGl} yield biomass X_i/glucose
Y_{1EGl} yield ethanol/glucose

μ_{imax} maximum specific growth rate on X_i (hr^{-1})

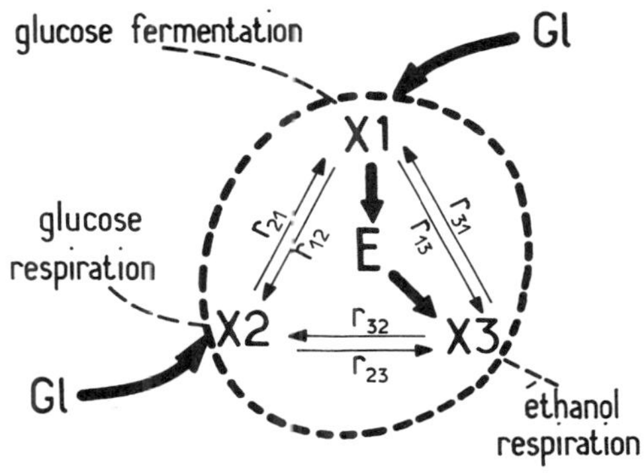

Fig. 1 - *Candida utilis* metabolic reactions

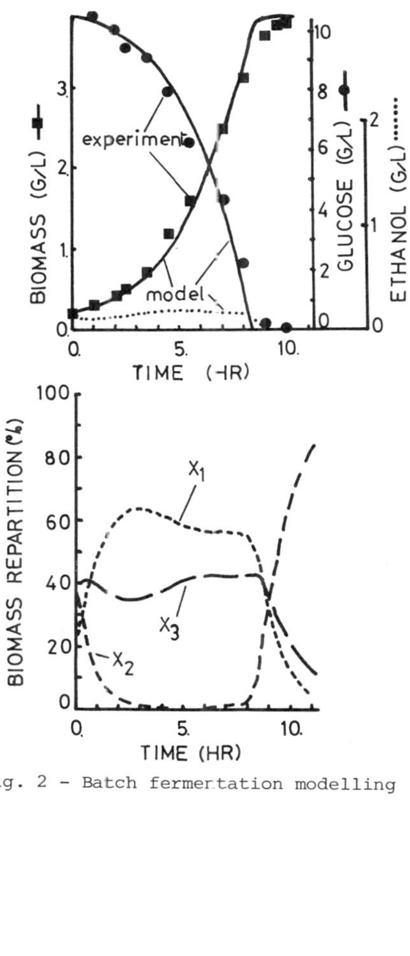

Fig. 2 - Batch fermentation modelling

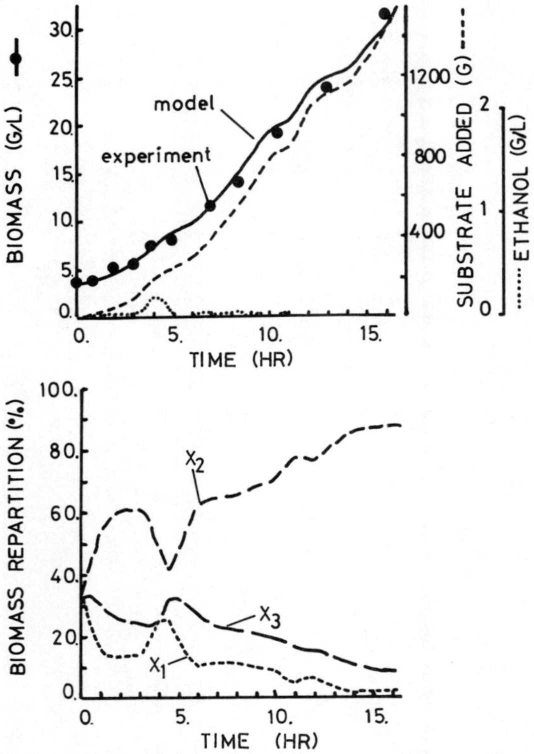

Fig. 3 - Fedbatch fermentation modelling

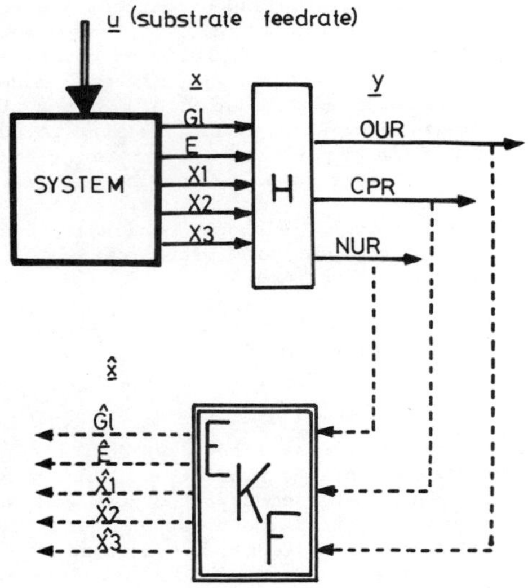

Fig. 5 - EKF block diagram

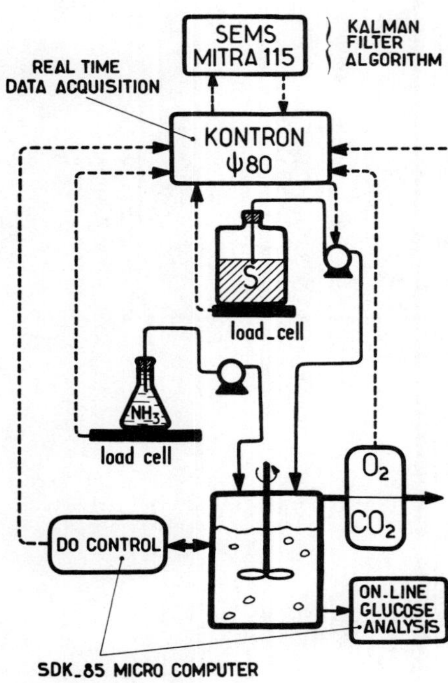

Fig. 4- Experimental setup

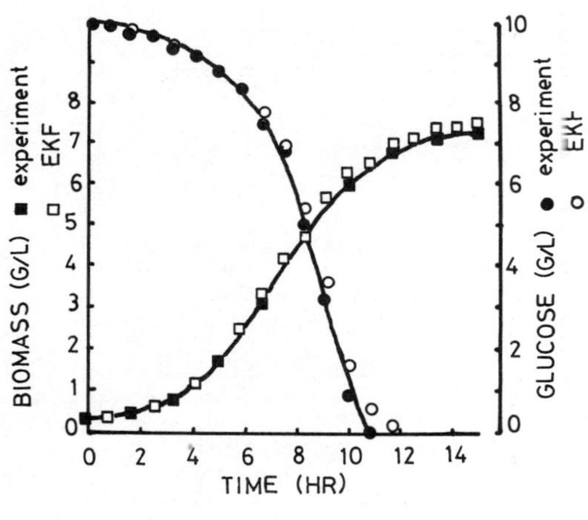

Fig. 6 - Batch EKF estimation with CPR and CUR as observation variables

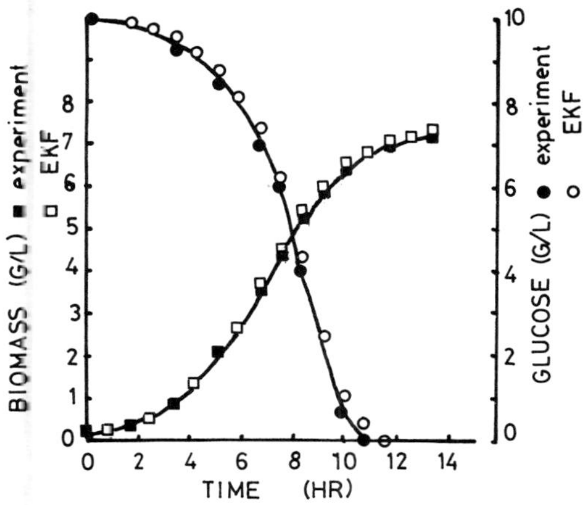

Fig. 7 - Batch EKF estimation with CPR, OUR and NUR as observation variables

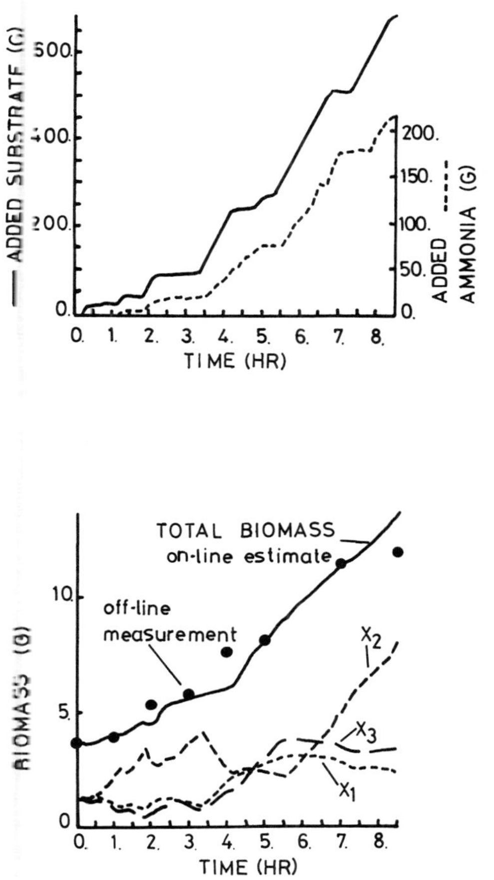

Fig. 8 - Fedbatch EKF estimation with CPR and OUR as observation variables

ESTIMATION OF THE PROGRESS OF STREPTOMYCES CLAVULIGERUS FERMENTATIONS FOR IMPROVED ON-LINE CONTROL OF ANTIBIOTIC PRODUCTION

L. A. Tarbuck, M. H. Ng, J. R. Leigh and J. Tampion

School of Biotechnology, Faculty of Engineering and Science, The Polytechnic of Central London, 115 New Cavendish Street, London W1M 8JS, UK

Abstract. Progress in the optimisation of the production of an antibiotic in aerobic batch fermentations is described. The application of a variant of the Kalman filter to the data indicate that model fit is improved using a time-varying parameter to predict biomass development during the fermentations. Production of the antibiotic is shown to be improved under conditions of limited growth, and the progress in the description of the controlling parameters and factors is discussed.

Keywords. Biocontrol; fermentation process; modelling; non-linear estimation; Kalman filter; antibiotics; Streptomyces.

INTRODUCTION

The aims of the project are to provide reliable on-line estimates of biomass and secondary product formation in stirred batch reactors that are suitable for integration into control strategies for industrial fermentation. Previous papers have described progress in the investigation of variants of the extended Kalman filter applied to the estimation of biomass and product formation in sorbose and cellulose fermentations (Swiniarski et al., 1982; Leigh and Ng, 1984). The major problem of unmodelled batch-to-batch deviations in nominally identical batches was treated in these particular fermentations by identifying by sensitivity analysis those parameters that vary from batch to batch and that most affect model fit. From these, one particular parameter, Yxc, was then redesigned as a time varying parameter, Yxc(t), to be estimated on-line along with process states (Leigh and Ng, 1984).

The present paper describes progress in the application of the Kalman filter, including the adaptive feature described above, to a typical antibiotic-producing system. The product, clavulanic acid, is typical of the secondary metabolite produced by the Streptomyces species of current pharmaceutical interest. Specific bio-chemical pathways leading to product formation have not yet been elucidated (Butterworth, 1984) and optimisation of production is limited to manipulation of the fermentation process, using wild-type strains of the bacteria (ie. non-genetically improved or engineered strains).

The work described is supported by the Science and Engineering Research Council at the School of Biotechnology and the School of Electrical, Electronic and Control Engineering.

EQUIPMENT

The fermentations were carried out using an upgraded LH Engineering Series 1000 10 litre fermenter, instrumented as shown in Fig. 1. Dissolved oxygen concentrations were measured using a LH Engineering % oxygen meter and autoclavable electrode E1-01-18. Carbon dioxide in the exit gases was measured using an infra-red detector (Analytical Development Co. Ltd), and exit gas oxygen concentrations were determined using a paramagnetic analyser (Sybron, model 540A). These data were logged on a LSI 11 microprocessor together with pH, temperature, stirring speed and volume of acid and alkali added to maintain pH control. Data is logged at 30 minute intervals, and is communicated in bursts to the VAX computer for storage and off-line data handling and model development. On-line control of stirring speed, dissolved oxygen tension and air flow rate will be established with the installation of the Motorolla 68000 hardware under development (Fig. 2). Software under development will allow the real-time continuous graphical display of the progress of the fermentations, and application of the models developed.

FERMENTATION DETAILS

ORGANISM. The bacterium Streptomyces clavuligerus was first described by Higgens and Kastner (1971), and the strain used in all fermentations was obtained from the American Type Culture Collection (ATCC) No. 27064.

MEDIA. No manufacturing process information has yet been published and most information available is based on laboratory scale production of the type used in this project. Fermentation media suitable for the production of clavulanic acid have been described in various patents issued (British patent No. 1,563,103; Japanese patents 53-104796 and 55-162993). All patents show soya-bean flour to be the prefered industrial substrate. For the early stages of the project this was unsuitable, as it produces an opaque, particulate medium that make biomass estimations difficult. Such estimations would of course be available once the Kalman filter was fully developed, and indicate the potential of the development of this method. A relatively clear medium was selected for the project, containing glycerol, malt extract and bacteriological peptone described in the British patent. (Soyabean flour medium will be used to ensure comparability with industrial processes when modelling has progressed sufficiently). A defined medium described by Aharonowitz and Demain (1977) containing glycerol, asparagine, phosphate and trace salts was also used for a limited number of experiments. All fermentations were carried out at 26°C.

INOCULA PREPARATION

Standardisation and control of the preparation of the inocula has been shown to be important in reducing the batch-to-batch deviations. To prepare inocula 10ml agar slants were grown for 7 days at 26°C. The resulting aerial mycelia and spores were suspended in 10ml fermentation broth and transferred to 250 ml liquid medium in a vegetative stage flask. The flasks were incubated at 26°C on a rotary shaker (220 rpm) for 72 hours. A total of 1 litre of inoculum was then used to inoculate 9 litres of medium in the fermenter.

BIOMASS AND CLAVULANIC ACID ASSAYS

Biomass was measured using optical density of the cultures. The bacterial flocs were disrupted by agitation with 3.5mm diameter glass beads before dilution and measurements were made, typically 5 g beads to 10ml culture. Optical density measurements were made on a Perkin-Elmer 402 double beam spectrophotometer at 650nm. The wavelength was chosen as that at which the culture medium absorbance is minimal. All OD_{650} measurements were converted to dry cell mass (DCM) as g L^{-1} using a standard calibration graph prepared for S. clavuligerus.

Clavulanic acid concentrations in the broths were determined by high performance liquid chromatography (HPLC) using a modification of the method described by Foulstone and Reading (1982). Broth filtrates were derivatised at room temperature with imidazole, and separated using a Spherisorb ODS2 5u column with 0.05M KH_2PO_4 buffer + 9% acetonitrile (pH 3.2). Standards were prepared from a laboratory reference preparation of potassium clavulanate provided by Beechams Pharmaceuticals.

MODELLING AND STATE ESTIMATION

The model used has already been described in a previous publication (Leigh and Ng, 1984). The combination of Leudeking and Pirt and Koga's models has been used to describe the type of process considered here. The substrate consumption consists of two parts. The first is used in bacterial growth and the second is consumed in the product formation. The model is as follows:

Dynamic equations:

$$\frac{dx}{dt} = \frac{\mu_m sx}{K_s + s} - K_d x$$

$$\frac{ds}{dt} = -\frac{1}{Y_{xs}} \cdot \frac{\mu_m sx}{K_s + s} - \frac{1}{Y'_{xs}} \frac{\mu'_m sx}{K'_s + s}$$

$$\frac{dp}{dt} = \frac{\mu'_m sx}{K'_s + s} + m_p x$$

Measurement equation

$$CER = \frac{1}{Y_{xc}} \frac{\mu_m sx}{K_s + s} + m_c x$$

The extended Kalman filter with time-varying parameter developed previously is applied to estimate the biomass and product formation.

RESULTS AND DISCUSSION

The results presented represent the initial studies using the fermenter under manual control to provide data on the characteristics of S. clavuligerus fermentations. These studies allowed the main controlling parameters to be deduced and provided the data base for model development and off-line simulation. The approach to the initial control strategy was determined by the decision to use a complex undefined medium, where no single limiting substrate is discernable and substrate utilisation rates are not easily determined. Growth rates and clavulanic acid production were followed in nominally identical batches with air flow rates of 1.0, 0.5 and 0.25 v/v min, and stirring speeds between 250 and 500rpm. The conditions and results for seven batches are given in Table 1.

Clavulanic acid production was found to vary inversely with the stationary phase biomass density, Fig. 3. There was little correlation between the actual growth rates and the biomass density in the stationary phase. The actual doubling time during the growth phases did, however, show a trend associated with clavulanic acid production, indicating an optimum doubling time of approximately 6 hours (Fig. 4). These results indicate two separate controlling factors that can be used to enhance antibiotic production. The conditions that control these factors were identified as the oxygen transfer rate, controlled by stirring speed and air flow rate, and shear forces in the liquid, also a factor of the stirring speeds. The results given in Table 1 show that in batches 1 and 2, where all conditions were the same except air flow rate, the higher aeration produced a faster doubling time during the growth phase, higher stationary phase biomass density but lower clavulanic acid production. If these batches are compared with batch 5, where stirring speed was increased, it can be seen tht the increased oxygen transfer resulted in relatively high stationary phase biomass densities, but the growth rate was reduced, expressed as a lengthening of the doubling times. Such a reduction in the growth rate would appear to be a function of the shear forces at the higher speeds.

In initial batches pH was not controlled, and during the course of the fermentations it varied between pH 6 and pH 8. Clavulanic acid is relatively stable between pH 6.2 and 7.5. Control of the pH during the experiments at two points within this range was introduced to reduce the batch-to-batch deviations. Antibiotic production at pH 7.2 was poor. At pH 6.5 the results support the trend shown in those without pH control of increased antibiotic production with reduced biomass densities. The doubling times were affected by the control of pH, being longer at pH 6.5 than in any other experiments, and indicate the sensitivity of growth rates to changes over a relatively narrow pH range. The optimum pH for antibiotic production was not determined by these experiments.

Carbon dioxide evolution rates (CER) show a rapid utilisation of nutrients during the growth phase in those batches with the higher growth rates, and also later falling CER indicative of nutrient depletion (figs. 5 and 6). Batches with lower growth rates and higher antibiotic production produced distinctive CER patterns suggsting a more ordered sequential removal of substrates. (Figs. 7 and 8). Relatively stable CER were given in the antibiotic phases of these batches. These results show that the more controlled uptake of nutrients during the growth phase with lower doubling times and restricted stationary phase biomass densities avoids nutrient depletion before the antibiotic production phase. In other bacterial antibiotic

Table 1. Summary of Batch Conditions and Results for S. clavuligerus

Batch	Air Flow ($v/v\ min^{-1}$)	Speed (rpm)	Doubling time (h)	Biomass max ($g\ l^{-1}$)	[a]CA max $mg\ L^{-1}$	[b]Specific CA	pH Set point
1	0.5	250	4.7	5.5	5.2	1.01	-
2	0.25	250	7.9	4.5	2.0	2.68	-
3	0.25	250	5.5	3.9	17.1	4.79	-
4	0.25	375	6.0	3.4	36.0	10.70	-
5	0.25	500	9.5	6.6	<0.05	<0.05	7.2
6	0.25	400	17.8	2.9	6.7	2.31	6.5
7	1.00	375	12.0	10.1	5.5	0.53	6.5

a - CA = clavulanic acid
b* amount of clavulanic acid produced per unit of biomass (mg. g biomass . L^{-1}).

Table 2. Summary of Batch Conditions and Results for S. clavuligerus Grown on Defined Medium

Batch	Air ($v/v\ min^{-1}$)	Speed (rpm)	Doubling time (h)	Max Biomass ($g\ L^{-1}$)	CA max ($mg\ L^{-1}$)
8	0.40	500	7.4	4.25	trace
9	0.25	500	13.6	2.81	trace
[a]10	0.40	500	9.9	3.67	1.4
[b]11	0.25	500	6.34	5.50	2.7

a - with 0.1 g L^{-1} yeast extract
b - with 0.3 g L^{-1} yeast extract

producing systems idiolite production has been shown to be highly sensitive to nutrient uptake rates over a very narrow range. The rate of nutrient uptake may be controlled by incorporating a fed-batch system into the optimisation strategy. The use of complex medium precludes the identification of which substrates were used in the sequence of removal, but the RQ values in most of the batches suggest more oxidised substrates are used in preference to glycerol during the growth phase (RQ >0.75), and this may not be required to produce adequate growth prior to antibiotic production (Fig. 9).

Several batches were run with a defined medium containing glycerol and arginine to determine if the dominant controlling factor in antibiotic production was the growth rate or the variety of carbon sources in the undefined medium. These experiments are summarised in Table 2. The first two batches showed only traces of clavulanic acid produced, but the doubling time and maximum biomass densities were controlled by the aeration rate, as in the undefined medium. The second pair of batches were supplemented with yeast extract (a complex nitrogen source) and clavulanic acid production improved. This indicated the nitrogen nutrition was a determining factor in the antibiotic production. The amounts of clavulanic acid produced, although low, were in the range found for fermentations with similar conditions in the defined medium, indicating the growth rate rather than varieties of carbon source was the main factor. The results also supported the trend noted in the complex medium that clavulanic acid production was improved through a reduction in the doubling times.

CER, RQ and Kla wer all tested in the development of the biomass model. Of these CER provided the best fit, as shown in Fig. 10. Application of the Kalman filter has produced good fit of the predicted biomass to the actual values (Fig. 11). Yield coefficient of carbon dioxide Yxc is used as the time varying parameter. The apparent underestimation of product shown in Fig. 11 is through the use of the model based on proportional biomass/production, rather than the inverse relationship described in the results above. The development of the idiolite production model is at present continuing, using Kla, RQ, biomass and dissolved oxygen tension as the controlling parameters. The main difficulty encountered has been the non-Monod biomass development in the higher yielding batches, so that growth and stationary phases are not distinct. The system being modelled is more complex than those considered during the development of the variants of the Kalman filter, as a multi-step reaction where the biosynthetic pathway is not yet known.

SUMMARY

The results for the optimisation of clavulanic acid show the factors related to the production are the doubling time during the growth phase and the stationary phase biomass density. Control of these factors may be achieved by a combination of control of oxygen transfer, via air flow rates and stirring speed, and by shear of turbulence control, also a factor of the stirring speed. Model development and application of the Kalman filter to provide reliable estimates of biomass have been proven, although the development of the idiolite model is continuing.

REFERENCES

Aharonowitz, Y. and Demain, A.L. (1977). Influence of inorganic phosphate and organic buffers on cephalosporin production by Streptomyces clavuligerus. Arch. Microbiol., 115, 169-173.

Butterworth, D. (1984). Clavulanic acid: properties, biosynthesis and fermentation. In E.J. Vandamme (Ed.), Biotechnology of Industrial Antibiotics. Marcel Dekker, N.Y. and Basle.

Foulstone, M. and Reading, C. (1982). Assay of amoxicillin and clavulanic acid in biological fluids with high performace liquid chromatography. Antimicrobial Agents and Chemotherapy, 22 (5), 753-762.

Higgins, C.E. and Kastner, R.E. (1971). Streptomyces clavuligerus Sp. Nov., a β-lactam antibiotic producer. Int. J. Syst. Bacteriol., 21, 326-332.
Leigh, R.J. and Ng, M.H. (1984). Estimation of biomass and secondary product in batch fermentations. Presented at the 6th International Conference on Analysis and Optimisation of Systems.
Swiniarski, R. et al. (1982). Progress towards estimation of biomass in a batch fermentation process. Proceedings of IFAC Workshop, Modelling and Control of Biotechnical Processes, Helsinki, Finland.

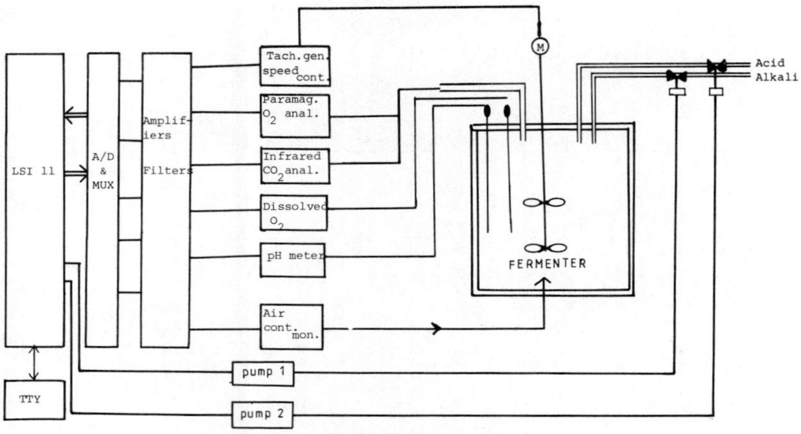

Figure 1. Configuration of Data Logging System Hardware

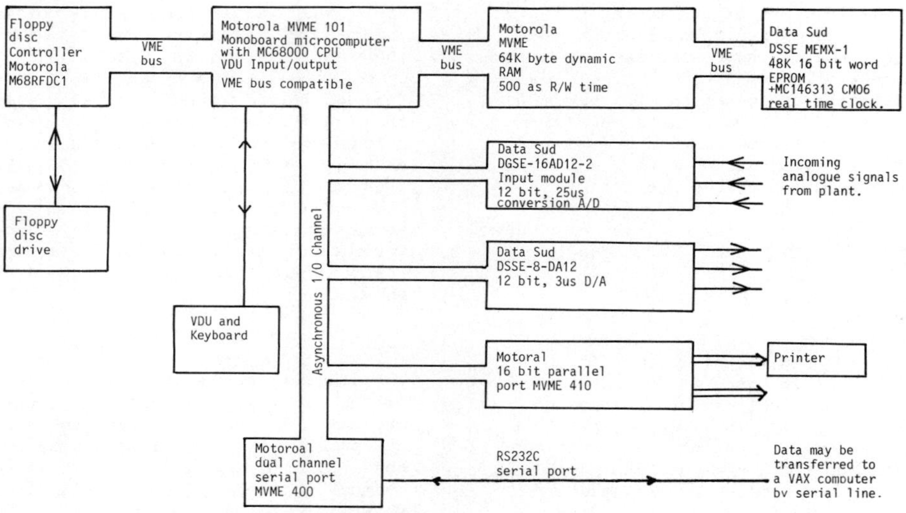

Fig 2. Configuration of M68000 System.

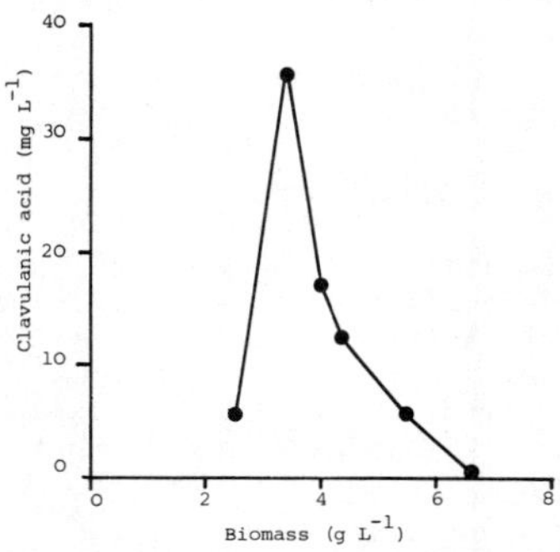

Figure 3. Relationship between clavulanic acid produced and stationary phase biomass density.

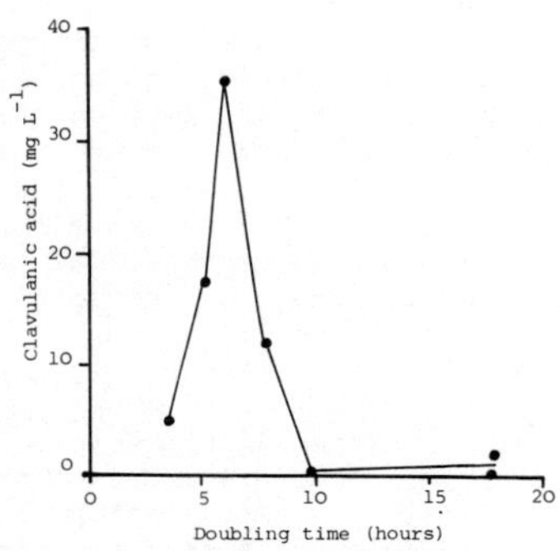

Figure 4. Relationship between doubling time of Streptomyces clavuligerus in batch culture and clavulanic acid produced

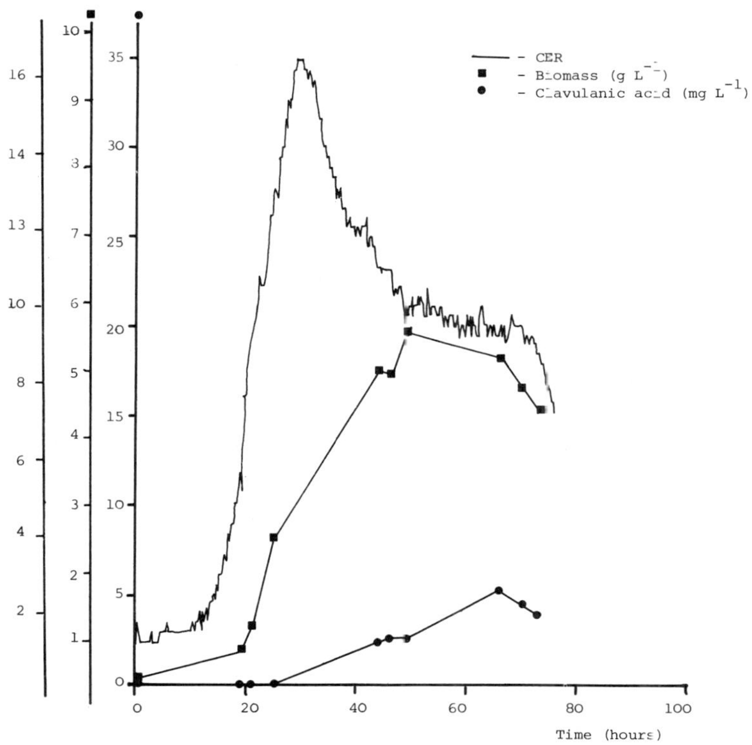

Figure 5. Carbon dioxide evolution rate. Batch 1.

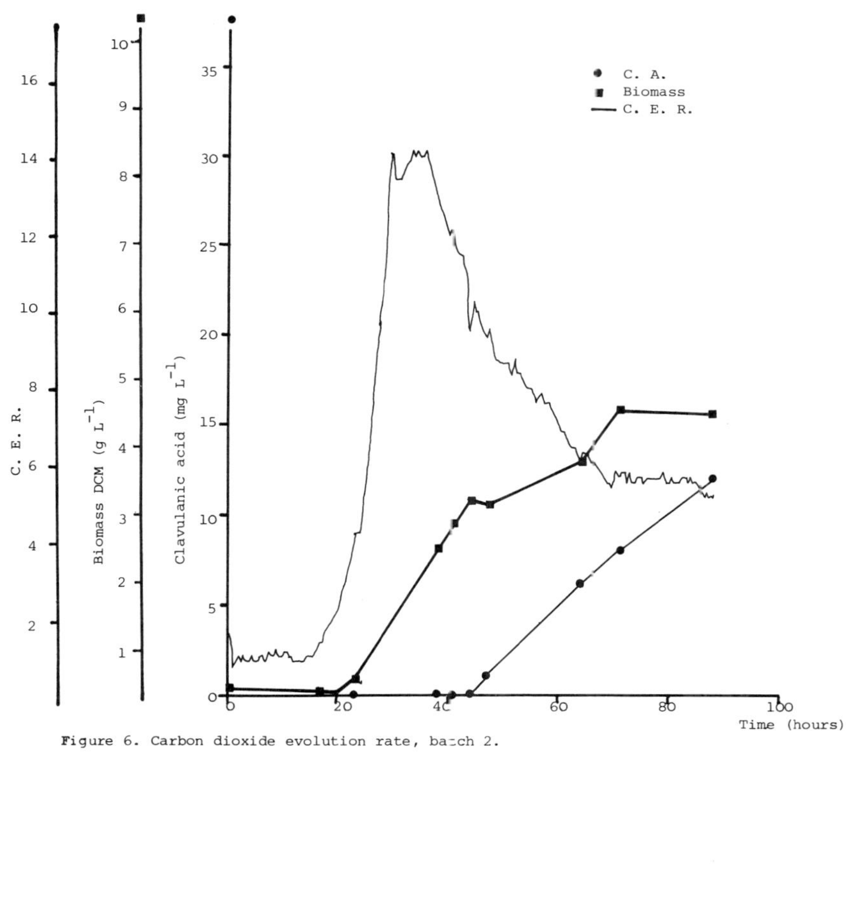

Figure 6. Carbon dioxide evolution rate, batch 2.

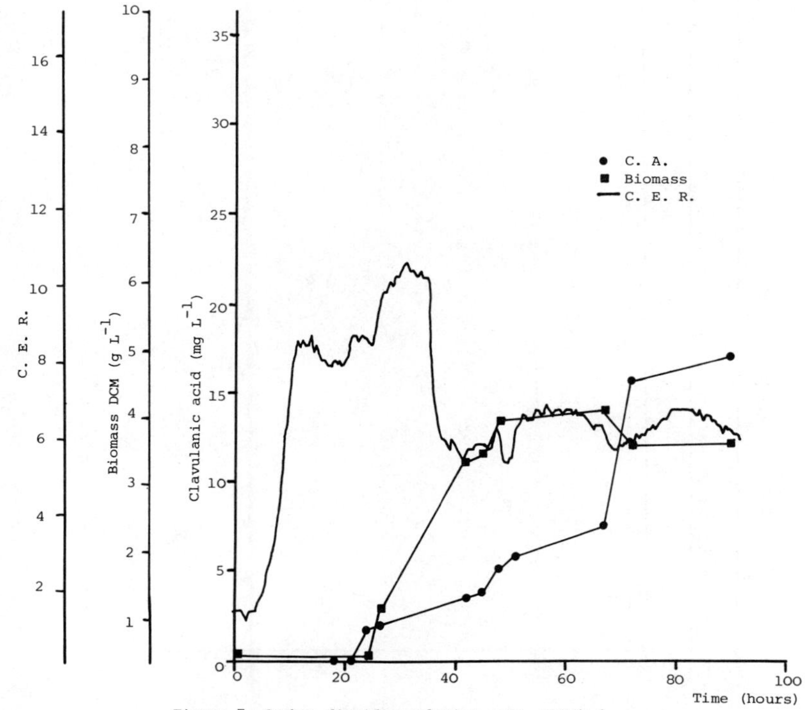

Figure 7. Carbon dioxide evolution rate. Batch 3.

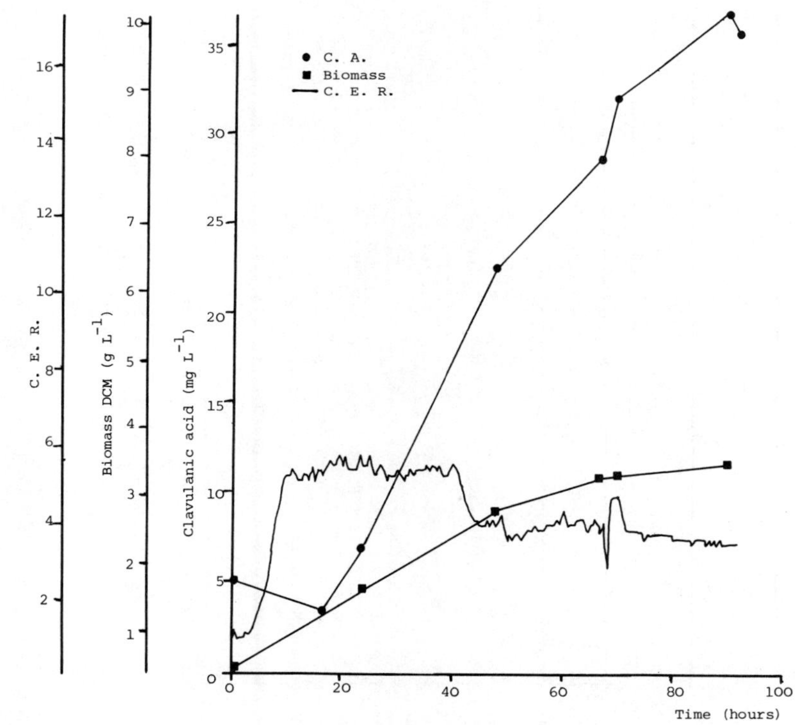

Figure 8. Carbon dioxide evolution rate. Batch 4.

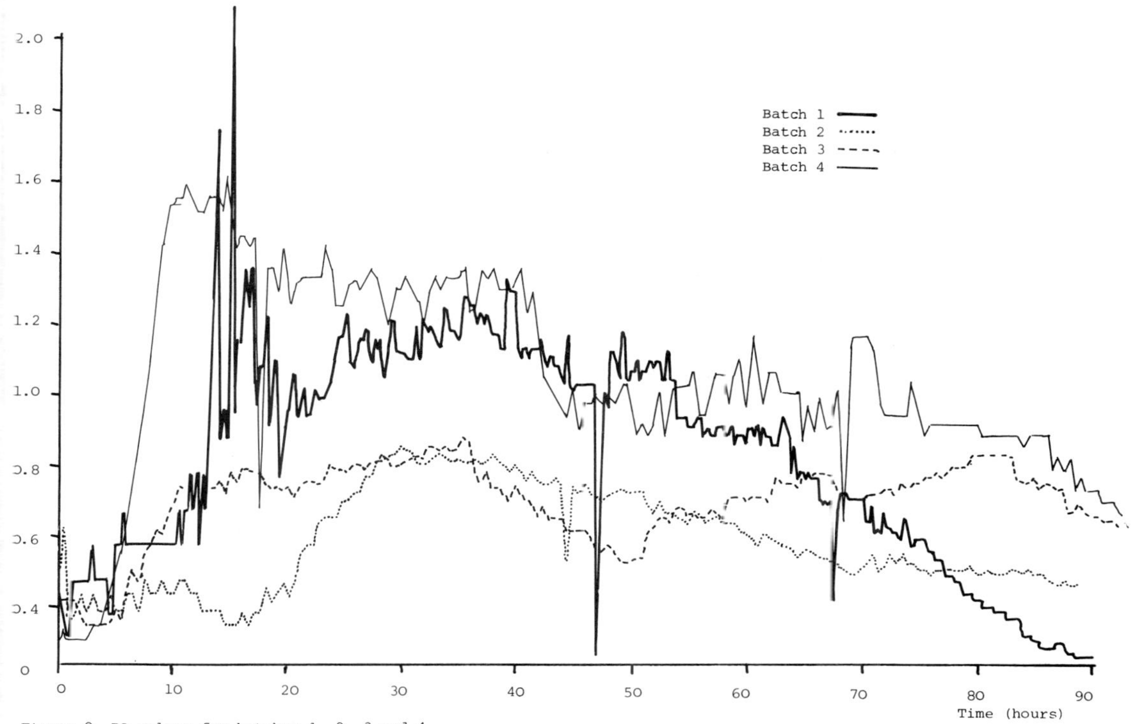

Figure 9. RQ values for batches 1, 2, 3 and 4.

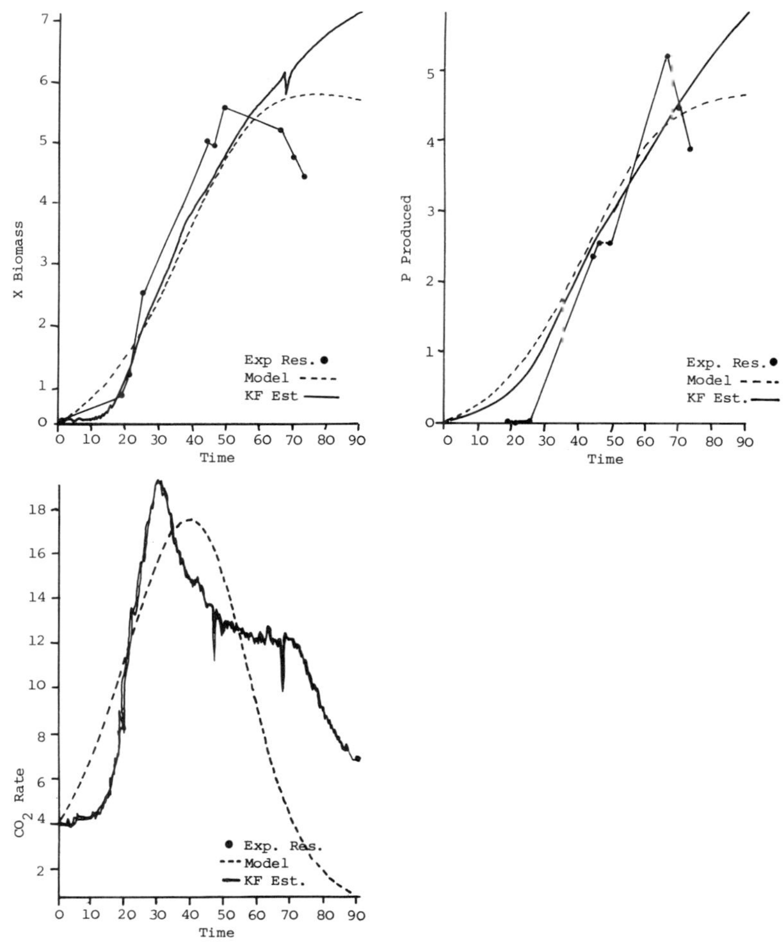

Figure 10. State estimation on Clavulanic acid fermentation (I).

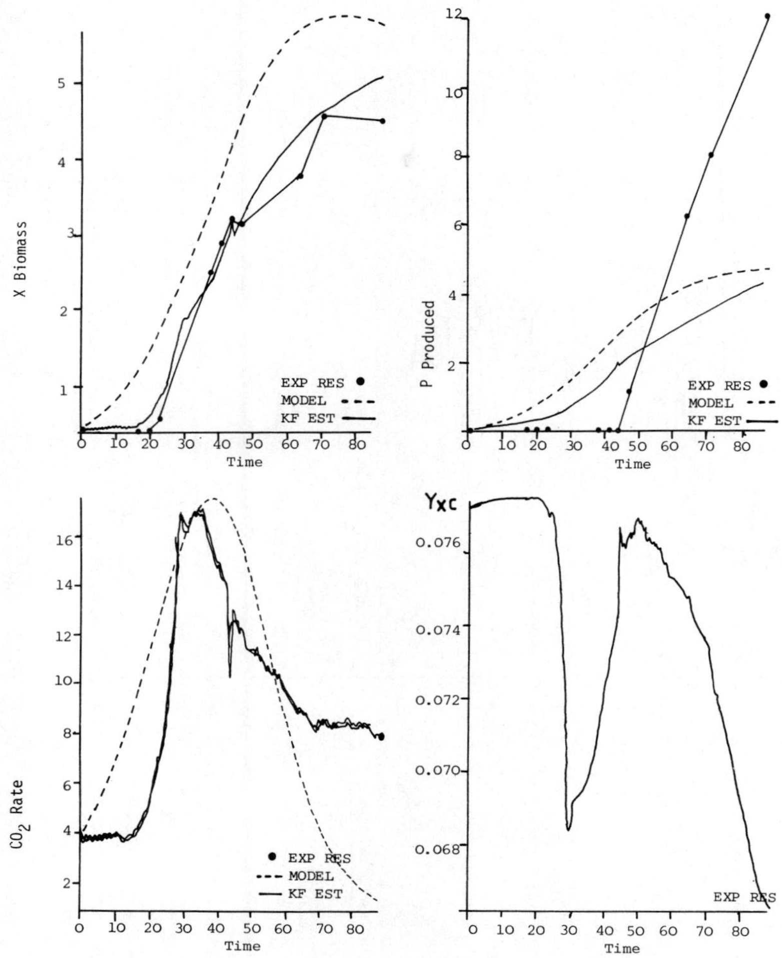

Figure 11. State estimation on clavulanic acid fermentation (II).

A METHOD TO CONSIDER DELAYED LABORATORY ANALYSIS IN STATE AND PARAMETER ESTIMATION OF BIOREACTORS

A. Halme*, R. Kuismin and M. Korteniemi****

**Department of Electrical Engineering, Helsinki University of Technology, Espoo, Finland*

***Department of Process Engineering, University of Oulu, Oulu, Finland*

Abstract. The experimental data obtained from a bioreactor usually includes analysis of laboratory samples. These are necessary either because no on-line instrument exists for certain variables or just to check the proper functioning of an instrument. The samples may be taken with a constant rate, but usually the time intervals between samples vary more or less due to practical reasons. Another practical fact is a delay in analysis which can vary from a tens of minutes to several hours. Because of these difficulties the information included in the laboratory samples is not usually utilized in on-line estimation procedures developed for bioreactors. The paper presents a method of how such information can be included in estimation in a practical way. The method itself is not bound to any specific estimation technique. The results are, however, presented when using a nonlinear extended Kalman-type filtering algorithm. Both state variables, like biomass or substrates, and parameters like yield coefficients can be estimated simultaneously. An illustrative example is presented when the method is applied to a semibatch baker's yeast fermentation.

Keywords. Biotechnical processes; state and parameter estimation; filtering; delays

INTRODUCTION

It is a well known fact that sensors available today for bioreactor control do not cover all necessary measurements. In spite of recent development in optical, ion selective and enzyme sensors most of the concentration variables in the fermentation liquid phase cannot be measured reliably on-line. Therefore a tedious laboratory sample analysis procedure is usually needed to support fermentation runs. In addition to be expensive, sampling produces delayed values which causes further problems when utilizing them in control. Due to the complex nature of biotechnical processes it is, however, unrealistic to think that bioreactors could be run entirely without sampling even there will be considerable progress in sensor development.

In the other hand, computers are nowadays more and more common in fermentation installations and on-line signal processing and estimation methods can be easily utilized to improve the information available from the existing sensors and laboratory analysis. Recently successful application have been reported by several authors (Swiniarski and co-workers, 1982; Stephanopoulos and San, 1984; Pihtilä and co-workers, 1984). It is interesting to notice that all of them use filtering and state space representation as the basic approach to the problem, probably because it helps to preserve the original biochemical model of a bioreactor (opposite to the ARMA representation and RLS method widely used otherwise in control engineering). None of them has, however, considered the case where irregular and unfrequent sample analysis information is included to improve the information obtained regularly and frequently from continuous measurements. The purpose of the present paper is at first hand to introduce a method to include also the sample analysis information at the time instant when it is available in the laboratory. At second hand we also introduce a not new but less known nonlinear filtering method, which have been successfully applied in this context for simultaneous parameter and state estimation of bioreactors. The method has certain algorithmic advantages if compared e.g. to the wellknown EKF-method. An illustrative example is presented related to a fed-batch baker's yeast fermentation.

A FILTER BASED ESTIMATION METHOD

Suppose that the state of a bioreactor can be represented by a discrete-time stochastic model

$$x(k+1) = f(x(k),u(k);a(k))+w_x(k), \qquad (1)$$

where x is the state, u possible control action and a a time-dependent parameter vector. The disturbance vector w_x represents, as usual, the modelling errors and process disturbances and is supposed to have white gaussian character. As seen later on, the state may include besides biomass, substrate and other similar concentration variables also the growth of the micro-organisms. The parameter vector include those of the biochemical (or other) parameters, like the specific growth rate, yield coefficients etc., which are defined as the estimated parameters. The dynamics of the parameters can be modelled in several different way. Here we suppose the simplest possible way

$$a(k+1) = a(k)+w_a(k), \qquad (2)$$

which means that slow variations are permitted in practice. The disturbance w_a describes the general uncertainty related to the parameters and is necessary in the method. All the measurements available, whether they are continuous type or samples, are modelled via the observation equation

$$y(k) = h(x(k),k; a(k))+v(k), \qquad (3)$$

where v represents the measurement errors and disturbances. It is explained later on how the sample analysis values are included in practice into the observation equation. Like w_x both w_a and v are supposed to have a white gaussian character.

Further, like in Kalman filtering, we suppose to know the covariance matrices Q_x, Q_a and R related to the disturbances w_x, w_a and v respectively.

The nonlinear filter used is developed by Halme and Selkäinaho (1982) and gives basically a bayesian maximum likelihood estimate for the state variable x when it's dynamics are covered by the state equation (1) and observations from it are obtained via the equation (3). Both the dynamics and observation equations are supposed to have linear parts of forms

$$x(k+1) = A(k)x(k)+f_1(x(k),k), \tag{4}$$

$$y(k) = C(k)x(k)+h_1(x(k)), \tag{5}$$

where the pair (A(k),C(k)) is supposed to be complete observable in mean. It can be shown (a complete proof can be found in Selkäinaho, 1984) that under these circumstances the state estimate can be given in the form

$$\hat{x}(k+1|k) = \hat{x}(k+1|k-1)+K_x(\hat{x}(k|k-1))\cdot (y(k)-h(\hat{x}(k|k-1),k)), \tag{6}$$

$$\hat{x}(k+1|k-1) = f(\hat{x}(k|k-1),u(k),k), \tag{7}$$

where the gain K_x is calculated from the equation

$$K_x(\hat{x}(k|k-1) = A(k)P(k)C(k)^T[C(k)P(k)C(k)^T+R]^{-1} + A(k)[P(k)^{-1}+C(k)^T R^{-1}C(k)]^{-1}\cdot [\frac{\partial}{\partial x} h_1(x,k)_{x=x(k|k-1)} - C(k)]R^{-1} \tag{8}$$

where in turn the covariance matrix P is calculated off-line from the Riccati equation

$$P(k) = A(k)[P(k-1)^{-1}+C(k)R^{-1}C(k)]^{-1}A(k)^T+Q \qquad P(0) = P_o. \tag{9}$$

If compared with the EKF algorithm the most noticeable difference is in the gain calculation, where the equation (9) can be solved independently from the prediction-correction equations (6)-(7). In the case where A and C are time invariant the limit gain

$$K_x(x) = K_1+AP_+[\frac{\partial}{\partial x} h_1(x)-C]^T R^{-1} \tag{10}$$

where

$$K_1 = AP_-C^T[CP_-C^T+R]^{-1}, \tag{11}$$

$$P_- = AP_+A^T+Q_x, \quad \text{(covariance before measurement)} \tag{12}$$

$$P_+ = (P_-^{-1}+C^T R^{-1} C)^{-1} \quad \text{(covariance after measurement)} \tag{13}$$

can be used. When applying the filter to the simultaneous parameter and state estimation we do not use the state extention method which may lead extensive calculations due to increased dimensions but separate the state and parameter estimation tasks for different filters. The state filter is in the form given above in the equations (6)-(9). The parameter filter uses the dynamic equation (2) but the observation equation is written in the form

$$y(k) = h(f(\hat{x}(k|k-1),u(k-1);a(k-1)), \tag{14}$$

where the parameter vector a is in an explicite position. The gain K_a for the parameter estimator

$$\hat{a}(k|k) = \hat{a}(k|k-1)+K_a(y(k)-h(\hat{x}(k|k-1))) \tag{15}$$

$$\hat{a}(k|k-1) = \hat{a}(k-1|k-1) \tag{16}$$

can be calculated now following the general equations (8)-(13).

In many processes, also in the case of bioreactors, the original continuous state dynamics are nonlinear. It is difficult or even impossible to obtain the corresponding exact discrete-time version analytically. In practice the most convenient way is to work out a linearized model of the form

$$x(k+1) = A(a)x(k)+B(a)u(k) \tag{17}$$

$$y(k) = C(a)x(k), \tag{18}$$

where the matrices A, B and C have been formed via derivation so that an explicite dependence on the parameter vector a is known. This is important because we usually like to estimate directly the biotechnical parameters such as yields, the specific growth rate etc., and not the parameters of a linear black-box model. The general filter equations can be applied to the model (17)-(18) in a straightforward manner. The result is a state filter with a constant, off-line calculated gain and a parameter filter with a variable gain, where the variable part is an explicite function on the state estimate and control vector.

CONSIDERING THE INFORMATION CARRYING BY THE SAMPLE VALUES

The observation vector y may contain both continuously measured values like oxygen content in the exhaust gases and values obtained from sample analysis. In the previous case an instrument measurement value is obtained without delay after every basic sampling time. In the latter case an analysis result is obtained with a variable delay from the laboratory. The sample time instant can be usually supposed to coincide with a good accuracy with the basic sample time running typically with 10-1[illegible] minutes interval. Instead, the time interval between samples cannot suppose to be a constant or at least doing so is a clear restriction. Further, the average sampling time may also vary between variables, for example the substrate and biomass concentrations may be analyzed more frequently than the nutrient concentrations. Handling both measurement information types together in the same filter requires special arrangements. There are several possibilities available. All of them should utilize, however, the fact that close to a sample time instant we know that the sampled variable has about the same value than the analysis result because of the slow dynamics of the process. The further the running time is from the nearest sample point the broader is the band within which the value can be located. A natural way to consider this is to define a window arround each sample point where the value of the sample variable is supposed to be equal to or calculated in a known way from the analysis result. Outside the window there is no knowledge about the value of the variable. This can be accomplished either by using two different sets of observation equations or by using just one but defining the gain components corresponding the sampled variables equal to zero outside the windows and non-zero within them. The fact that the uncertainty of the observation increases with the time distance from the sample point (supposed to be at the center of the window) can be considered by defining a time variable component for the corresponding measurement in the covariance matrix R. We have applied the following gaussian distribution type function with a good success

$$r(k) = r_1-(r_1-r_o)\exp(-\delta(\frac{k-k_o}{h})^2), \qquad (19)$$

where h is the width of the window as a multiple of the sampling time and δ is a positive constant. The sample point is at time k_o, r_o and r_1 are the variance of the laboratory analysis and the variance measuring the uncertainty of the measurement at the both ends of the window respectively. If r_1 is made very large the gain approaches zero at the ends of the window. The gain maximum is attained at the time $k=k_o$ i.e. at the sample time. It should be emphasized, that this is only one possible choise. The simplest choise where r(k) is a constant works, but the filter is more difficult to tune because of switching the correction term on and off is not so smooth than in the previous case. When considering the variables like biomass, substrate and nutrient concentrations, whose dynamics are slow, a suitable width of the window is 1-3 hours in practice.

The window technique described above makes it possible to include the laboratory analysis in the observation vector and run the estimators with the basic discretion interval. Still there is a problem how to consider the delays always present in the laboratory analysis. Again there are several possibilities. One would be to estimate predicted values. Although sound in theory this principle has two serious drawbacks in practice. Firstly, using a prediction model presupposes the knowledge of the delay, which is not the case in practice. Secondly, the non-delayed information cannot be utilized as effectively as possible. Another possibility is to use a non-predictive ordinary model but to rerun it in history through the time instant where the sample was taken as soon as it's analysis value is available. The additional information available usually generates different estimates for those variables whose estimation error depends on the sample value in question. The principle is illustrated in the Fig. 1. At the real time the estimates will jump after rerunning the estimator, which may cause a problem if the estimates are used e.g. in closed loop control. This problem is, however, easily overcome by smoothing out the jump using a properly tuned ordinary RC-filter. Another problem related to the method is a variable need to store the history data. However, a maximum delay after which the laboratory analysis data does not carry any more relevant information for the real time estimates can be usually easily judged in practice. Thus the maximum storage capacity needed can be also estimated. When several variables are under sampling the oldest sample determines at each time the actual need to store the on-line measured signals. When the sample value is fed to the computer it at the same time releases some of the storage capacity needed. This property may be used to decrease the actual storage capacity in practice.

AN ILLUSTRATIVE EXAMPLE

A fed-batch or continuous culture of baker's yeast, or in fact also many other organisms, can be described in fully aerobic conditions by the following simple model

$$\dot{X} = (\mu-D)X \qquad (20)$$

$$\dot{S} = D(S_o-S) - \frac{1}{Y_{X/S}}\mu X, \qquad (21)$$

where X represents the biomass and S the substrate concentration. The parameters μ and $Y_{X/S}$ are the usual specific growth rate and substrate yield respectively. We do not suppose any further model for the specific growth rate. It is usually easy to measure continuously the total oxygen utilization rate, denoted in the following by OUR. A reasonable model for OUR is

$$OUR = \frac{1}{Y_{X/O_2}} G + M_{O_2/X} X, \qquad (22)$$

where G is the growth rate, $Y_{X/O2}$ the oxygen yield and $M_{O2/X}$ the maintenance coefficient. The biomass and substrate concentrations may be measured in certain cases by optical and glucose sensors, but we suppose here that only laboratory samples are available. In the experiments done to verify the estimation method samples were taken every 2-3 hours. The analysis delay was approximately 1 h.

A commercial strain of saccharomyces cerevisae was used in the experiments. Molasses was used as a sole carbon and energy source. The initial volume of the fermenter was 5 l, in which the growth medium contained: $(NH_4)_2SO_4$, 8.7 g; $(NH_4)_2HPO_4$, 1.5 g; $MgSO_4 \cdot 7H_2O$, 0.4 g. The feed contained sugar expressed as a glucose 198 g/l. Temperature and pH was adjusted to 30°C and 4.5, respectively. Substrate feeding schedule was arranged so, that operator could choose the percentage, by which the feed rate was raised every ten minute. Fig. 2 shows the basic data of a typical run. This data is used later on in all estimation tests illustrated in Figs. 3-5.

Let us suppose that we like to estimate on-line the biomass and substrate concentrations together with some of the parameters by using all the information available in the OUR-measurement and samples. It is a wellknown fact (Holmberg, 1981) that the model of a bioreactor includes easily too many parameters to be well identifiable. In the present case it is easy to conclude by intuition that four parameters are too much as to the measurement information available. So we suppose one, the oxygen yield $Y_{X/O2}$, to be a constant. Suppose we are also interested on the growth rate and like to estimate it on-line. A possibility would be to calculate an estimate for it from the estimates of μ and X. However, a more accurate and smoother result is obtained if the growth rate is estimated directly by adding dynamics for it. Supposing that the specific growth rate varies relatively slowly the following simplified model can be easily justified

$$\dot{G} = (\mu-D)G . \qquad (23)$$

This equation together with the equations (20) and (21) form now the state equations. Discretizing them according to the general linear form (17) gives

$$\begin{bmatrix} G(k+1) \\ S(k+1) \\ X(k+1) \end{bmatrix} = \begin{bmatrix} 1+(\mu-D)T & 0 & 0 \\ -\frac{1}{Y_{X/S}}T & 1 & 0 \\ 0 & 0 & 1+(\mu-D)T \end{bmatrix} \begin{bmatrix} G(k) \\ S(k) \\ X(k) \end{bmatrix} + \begin{bmatrix} 0 \\ S_o T \\ 0 \end{bmatrix} D(k). \qquad (24)$$

where T is the basic sampling time. The observation equation (18) gets the form

$$\begin{bmatrix} OUR \\ \hline S_s \\ X_s \end{bmatrix} = \left[\begin{array}{ccc} \frac{1}{Y_{X/O_2}} & 0 & M_{O_2/X} \\ \hline 0 & 1 & 0 \\ 0 & 0 & 1 \end{array}\right] \begin{bmatrix} G(k) \\ S(k) \\ X(k) \end{bmatrix} \tag{25}$$

where S_s and X_s denote the sampled values of the substrate and biomass concentrations. The full observation equation is valid only during the windows arround the sample points. The upper part is valid all the time. The gains for the state and parameter filters can be formed through a straightforward procedure. The details are not shown here, but in the final form the gains have the following structures

$$K_x = \left[\begin{array}{c|cc} h_{11} & 0 & 0 \\ h_{21} & h_{22} & h_{23} \\ h_{31} & h_{32} & h_{33} \end{array}\right] \tag{26}$$

$$K_a = \left[\begin{array}{c|cc} k_{11} & 0 & k_{13} \\ 0 & k_{22} & 0 \\ k_{31} & 0 & 0 \end{array}\right] + \left[\begin{array}{c|cc} l_{11}(G,X,M) & 0 & l_{13}(G,X,M) \\ 0 & l_{22}(G,X,M) & 0 \\ l_{31}(G,X,M) & 0 & 0 \end{array}\right] \tag{27}$$

where the non-zero elements indicate which part of the estimation error vector is contributing. The second part in the parameter filter gain K_a is time-variable through it's dependence on the state estimates. The matrices can be divided again to two parts of which the right hand one is active only during the windows. In the figures shown in the following the width of the window at each sample point is 90 min and the basic sampling time T=10 min. Fig. 3 shows first the case where the maintenance coefficient $M_{O2/X}$ was a constant ($M_{O2/X}$= 0.12 $gO_2/gX/hr$). This is the simplest case because the estimated parameters, i.e. μ and $Y_{X/S}$, are both in the dynamic equation. Three different initial values was given to the specific growth rate μ. As is seen all of them converges to the same values after about 6.5 hours. Each time when the values of the laboratory samples are fed into the computer the convergence is clearly improved stepwise. Note, that speaking about convergence is unexact because the specific growth rate changes all the time. Fig. 4 shows the case where also the maintenance coefficient has been taken under estimation. Three different initial values has been given for this parameter. Although the biomass, substrate and growth estimates behave now approximately in the same way than in the previous case there is a clear difference in the convergence properties of the parameter estimates. There is now an indication that the estimate of $M_{O2/X}$ does not converge during the run. In a biochemical sense some of the values obtained at the end of the run are also somewhat too high because of quite a young and active cell population should have a clearly lower value. This is a clear indication that the parameters of the observation equation are more difficult to estimate than the process parameters. Intuitively, it is quite clear that if there would be more relevant information available the situation may be improved. In this case we supposed that new measurements cannot be added. In the practice such variables as ammonia consumption in pH-control could be used. It is, however, interesting to note that by reorganizing the oxygen utilization rate equation to the form

$$OUR - \frac{1}{Y_{X/O_2}} \mu X_s = M_{O_2/X} X, \tag{28}$$

which is supposed to hold only during the windows we can add a new artificial measurement variable which turns out to improve the estimation of the maintenance coefficient. Adding a new measurement equation requires recalculation of the filter gains, which is again a straightforward procedure. Fig. 5 illustrates the situation after this change. As we can now see estimates from two different initial values converges to approximately same values after 6,5 hours. The values obtained at the end of the fermentation run are also closer to biologically sensible values.

CONCLUSIONS

The presented method gives a flexible possibility to consider delayed laboratory analysis in on-line estimation of the state and parameters in bioreactors. All the experiences the authors have this far indicate that the method works in practice and is relatively easy to apply. The idea how to consider the sample values in the observation information is not related to the filter based estimation algorithm used in the study but any other online estimation algorithm can be used. The windowing technique may be modified in several ways to fit the specific problem at hand. The filter approach is, however, quite natural when estimation is based on the kinetic models of the bioreactor. The nonlinear filter algorithm used is close to the constant covariance extended Kalman filter and has similar properties. The illustrative example indicates that although the estimation of state components, like biomass and substrates, is done with a good accuracy there may be still problems in the estimation of parameters especially when they are in the observation equation. This is basically due to the identifiability problem related to the bioreactor models. The problem may be solved by adding new information to the observation equation as was done in the example. The method used, where the equation of OUR was manipulated is, however, more or less of ad-hoc-type and may not apply in a more general connection.

REFERENCES

Halme. A., J. Selkäinaho (1982). Application of a non-linear filter to multivariable parameter adaptive control in a distributed micro-computer. 6th IFAC Symposium on Identification and System Parameter Estimation, Washington, USA.

Holmberg, A. (1981). A systems engineering approach to biotechnical processes. Experiences of modelling, estimation and control methods. Acta Polytechnica Scandinavica, Mathematics and computer science series No. 33.

Nihtilä, M., P. Harmo and M. Perttula (1984). Real-time growth estimation in batch fermentation. 9th World Congress of IFAC. Budapest, Hungary.

San, K-Y., G. Stephanopoulos (1984). Studies on on-line bioreactor identification. II. Numerical and experimental results. Biotechnol. & Bioeng. 26, 1189-1197.

Selkäinaho, J. (1984). A new nonlinear filter for real time estimation. University of Oulu, Report C3.

Stephanopoulos, G. and K-Y San (1984). Studies on on-line bioreactor identification. I. Theory. Biotechnol. & Bioeng. 26, 1176-1188.

Swiniarski, R., A. Lesniewski, M.A.M. Dewski, M.H. Ng and J.R. Leigh (1982). Progress towards estimation of biomass in a batch fermentation process. 1st IFAC Workshop on Modelling and Control of Biotechnical Processes, Espoo, Finland.

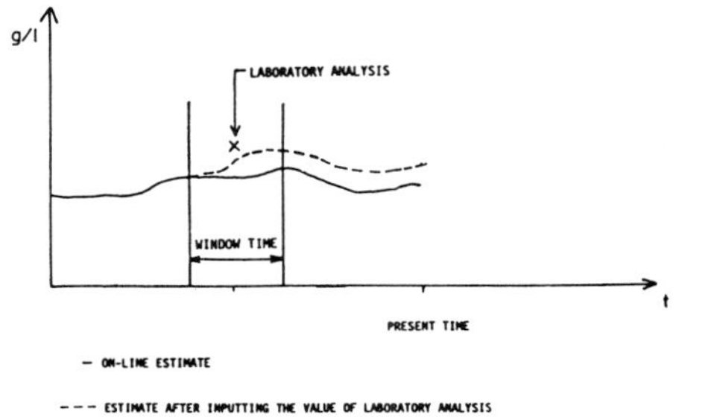

Fig. 1. Illustration of the window technique. When the value of a laboratory analysis is available the estimation algorithm is rerun through the time window arround the sample point. The analysis information contributes to the estimation result when the time index is within the window.

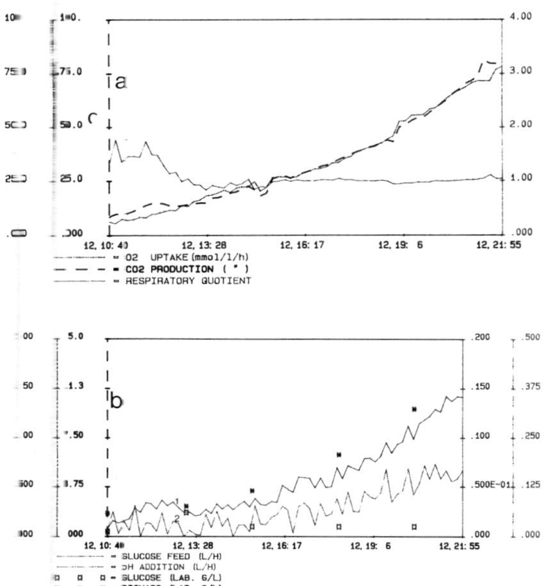

Fig. 2. Experimental data used in estimation tests; a) gas analysis; b) laboratory samples of biomass (*) and substrate (□), substrate feed rate (1) and ammonia addition rate (2).

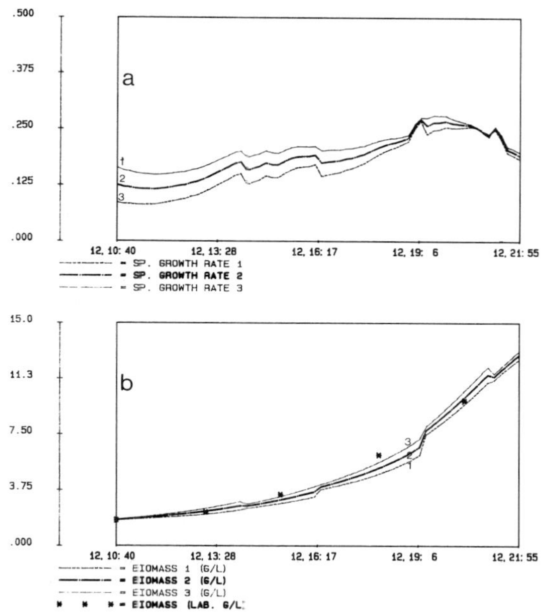

Fig. 3. Estimation of biomass concentration and the specific growth rate by varying the initial value of the specific growth rate (μ=0.09; 0.13; 0.17) and keeping the maintenance coefficient at a constant value (=0.12 $gO_2/gX/hr$); a) specific growth rate; b) biomass concentration.

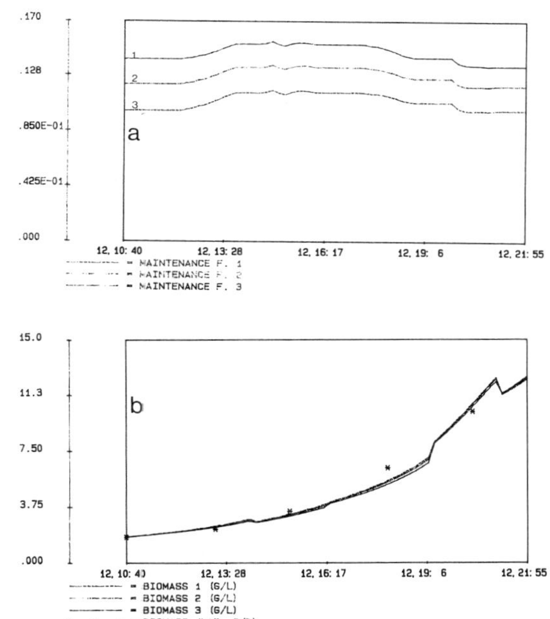

Fig. 4. The same estimation experiment as in Fig. 3 but also the maintenance coefficient is now estimated by starting it from different initial values ($M_{O2/X}$=0.10; 0.12; 0.14); a) maintenance coefficient; b) biomass concentration.

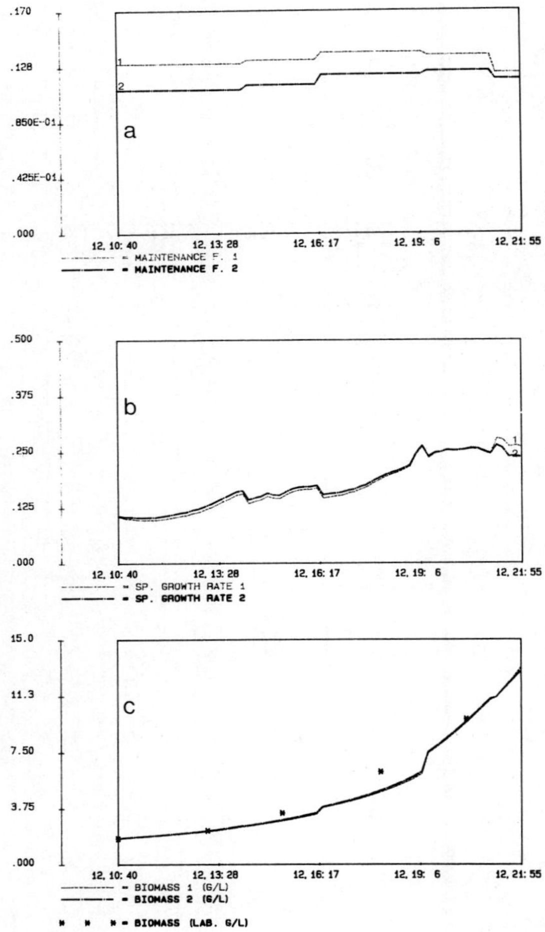

Fig. 5. The same estimation experiment as in Fig. 4 but now the observation equation has been augmented with the measurement equation obtained from the modified OUR model (see text); a) maintenance coefficient starting from two different initial values ($M_{O2/X}$=0.12; 0.14); b) specific growth rate; c) biomass concentration.

SIMULTANEOUS ON-LINE ESTIMATION OF OXYGEN TRANSFER RATE AND RESPIRATION RATE

Ulf Holmberg and Gustaf Olsson

Department of Automatic Control, Lund Institute of Technology, PO Box 118, S-22100 Lund, Sweden

1. INTRODUCTION

The oxygen uptake rate or respiration rate (R) is an essential variable in both fermentation and activated sludge systems. This paper presents an estimation method to find R directly from measurements of dissolved oxygen concentration and air flow rate in an open aerator system. This idea was proposed by Brouzes (1968), among others. The difficulty in calculating R directly from the DO concentration is that the oxygen transfer rate k_La also has to be known. To estimate k_La is not trivial, since it is varies slowly along with R (see Olsson (1984)). Therefore k_La and R have to be estimated simultaneously.

Estimation of k_La and R has been tried elsewhere. Since information about the oxygen transfer rate is useful in several applications it has been estimated separately, while R has been assumed to be constant but unknown, (see Olsson-Hansson (1976), Holmberg (1981)). The value of k_La may indicate both changes in e.g. the water quality and clogging of the air diffusers. Even a membrane failure in the DO sensor may be detected in this way. In an adaptive control simulation study (Ko et al (1982)) R has been assumed constant, which is unrealistic. The servo problem, varying DO setpoint, has been considered instead of the real regulator problem, varying load. In another approach only deviations of R from an (unknown) steady state value could be estimated (see Cook et al (1981)).

The main contribution of this paper is to show that the values of the two quantities can be estimated on-line simultaneously without any steady-state off-set even if both quantities are time varying.

The R and the k_La cannot be estimated simultaneously unless a good use of the system structure is made. Here the nonlinear structure has been retained. Still the model only includes the DO mass balance. Two approaches are presented in the paper. The first is a Kalman filter approach, where the two parameters are tracked simultaneously. It is demonstrated that unbiased estimates are obtained. The second approach uses large singular changes in the air flow signal to rapidly update the oxygen transfer rate, a so-called deadbeat estimator. The two methods can be combined.

The estimator design begins with an appropriate process model, described in the following section. Then the estimator structure is developed first to the Kalman approach and then to the deadbeat estimator. Both simulations and open loop plant experiments are used to evaluate the two methods.

2. PROCESS MODEL

The oxygen mass balance equation for the system under consideration is:

$$\frac{dy}{dt} = -\beta\cdot y + \alpha\cdot u\cdot[c - y] - R \tag{2.1}$$

where

$y(t)$	dissolved oxygen concentration [mg/l]
$u(t)$	air flow rate [m^3/min]
$R(t)$	respiration rate [mg/(l·h)]
$\alpha(t)$, $\beta(t)$, $c(t)$	time variant coefficients
$\alpha\cdot u = K_La$	oxygen transfer rate [1/h]

Our goal is to identify the parameters α and R. The coefficient α is an unknown function of aerator type, water depth, basin shape and even air flow rate. Aerator clogging accounts for much of the variation in α. Therefore we will instead let α be time dependent and estimate it on-line. The respiration R is the total oxygen utilization rate of the microorganisms in both the primary respiration, used in synthesis, and the endogenous respiration, during decay. Respiration can vary unpredictably due to different loads on the plant and also will be estimated on-line.

In practice β and c are known. The parameter β corresponds to a flow term in the mass balance and flows can be measured rather easily. The saturation concentration c is a function of oxygen partial pressure, temperature, salinity and surfactant concentration. All of these quantities are assumed to be known and are almost constant. It ought to be sufficient to update c now and then without risking accuracy.

3. ESTIMATOR STRUCTURE

Descriptions of time-varying dynamics can be achieved with different mixes of formality and intuition. The respiration R will change significantly in a number of hours while α differs just slightly from day to day, so that in terms of parameter variation α and R have their own estimation requirements.

Earlier attempts have been made to base the estimator on the linearized form of (2.1). This approach has several disadvantages. Since the system seldom is at steady-state there is no obvious point to linearize about. Moreover, the parameters of the sampled system will be complex functions of the parameters of the continuous system (α and R). This will make it difficult for the estimator to take advantage of the knowledge of different time scales of α and R.

Instead of linearizing the right-hand side of (2.1), which yields an unwieldy estimator structure, we prefer to estimate the derivative term, allowing use of the continuous time system directly as the estimator model set. Then the estimator will have a parameter vector $\theta=(\alpha\ R)^T$, consisting of the parameters we are interested in.

An appropriate estimator structure may look like:

$$\hat{\varsigma}(t) = \hat{\theta}^T(t)\varphi(t)$$

where

$$\hat{\theta}(t) = \begin{bmatrix} \hat{\alpha}(t-1) \\ \hat{R}(t-1) \end{bmatrix} \quad \varphi(t) = \begin{bmatrix} u(t-1)[c(t-1)-y(t-1)] \\ -1 \end{bmatrix}$$

$$\hat{\varsigma}(t) = \left[\frac{dy(t-1)}{dt}\right]_{est} + \beta(t-1)\cdot y(t-1)$$

where the first term on the right-hand side is the approximated derivative.

An Euler Approximation of the Derivative

A difference approximation of the kind

$$\frac{dy(t)}{dt} \approx \frac{y(t+1)-y(t)}{h} \tag{3.1}$$

will work well when the sampling interval h is short enough. But in measuring there is always noise which makes it undesirable to use a too short sampling period. A simple way to make the estimator less noise sensitive is to lengthen the sampling interval. The difference approximation will then become poorer as an estimation of the derivative, causing a bias in the parameter estimates.

A Modified Approximation of the Derivative

Assume for the moment that $u(t)$, $\alpha(t)$, $\beta(t)$, $c(t)$ and $R(t)$ are constant during the sampling interval. Then the corresponding sampled mass balance equation from the computer's point of view is:

$$y(kh+h) = y(kh)$$
$$+ h^*\left[-\beta(kh)y(kh) + \alpha(kh)u(kh)[c(kh)-y(kh)] - R(kh)\right]$$

where $$h^* \equiv \frac{1-e^{-\gamma(kh)\cdot h}}{\gamma(kh)} = h - \gamma(kh)\frac{h^2}{2} + \cdots$$

$$\gamma(kh) = \beta(kh) + \alpha(kh)\cdot u(kh)$$

Then by identification of parts we can easily solve for a modified expression for the derivative:

$$\frac{dy(kh)}{dt} = \frac{y(kh+h)-y(kh)}{h^*} \tag{3.2}$$

The above expression for the derivative is exact at the sampling instants if the parameters are constant during the sampling period, which can reasonably be assumed for all the parameters except for R. We can cover this assumption in the case of R by selecting an appropriate sampling interval.

An important difference between the two approximations is that the accuracy of (3.1) is dependent on the total process dynamics, while (3.2) is crucially dependent on just a part of it, R(t). This means that when the derivative is of most importance, during an excitation with u(t), (3.1) will be inaccurate while (3.2) remains accurate so long as R(t) is relatively constant.

4. SIMULTANEOUS TRACKING

A straight-forward recursive least squares approach with a forgetting factor would not be able to take into account the different rates of change of the parameters. Of course that problem is easy to overcome by introducing one suitable forgetting factor for each parameter. However, a severe problem occurs when there is poor excitation; the covariance matrix P in the recursive algorithm may explode.

A better formal treatment is the Bayesian approach, see Ljung and Söderström (1983). Here the parameter vector θ is considered to be a random variable. This approach will not lead to any P matrix explosion risks. The description of the system is a linear regression model with stochastically varying dynamics.

Assume that the true value of the parameter vector θ varies according to

$$\theta(t+1) = \theta(t) + w(t)$$

where $\{w(t)\}$ is a sequence of independent Gaussian random vectors such that $w(t)$ has zero mean and covariance matrix $R1(t)$. This allows us to separate the time scales of α and R by making an educated guess as to their respective variances. The overall description of the system becomes:

$$\theta(t+1) = \theta(t) + w(t)$$
$$\varsigma(t) = \varphi^T(t)\theta(t) + e(t)$$

where

$$\theta(t) = \begin{bmatrix}\alpha(t-1) \\ R(t-1)\end{bmatrix}; \quad \varphi(t) = \begin{bmatrix}u(t-1)[c(t-1)-y(t-1)] \\ -1\end{bmatrix}$$

$$\varsigma(t) \equiv \frac{dy(t-1)}{dt} + \beta(t-1)\cdot y(t-1)$$

and $Ew(t)w^T(s) = R1(t)\delta_{ts}$; $Ee(t)e(s) = R2\delta_{ts}$
$Ew(t)e(s) = 0$; $Ew(t) = Ee(t) = 0$

Applying the Kalman filter gives the estimates

$$\hat{\theta}(t) = \hat{\theta}(t-1) + L(t)\varepsilon(t)$$
$$\varepsilon(t) = \varsigma(t) - \varphi^T(t)\theta(t-1)$$
$$L(t) = \frac{P(t-1)\varphi(t)}{R2(t) + \varphi^T(t)P(t-1)\varphi(t)}$$
$$P(t) = P(t-1) + R1(t) - \frac{P(t-1)\varphi(t)}{R2(t) + \varphi^T(t)P(t-1)\varphi(t}$$

Even though the Kalman filter can be shown to converge to the true parameter values, as will be demonstrated in Section 7, it may take a long time if there is no a priori knowledge of the estimates. A way to quicken the estimation is to take advantage of large input excitations to rapidly update the estimates. An approach where singular excitations are utilized is described below.

5. MAKING USE OF SINGULAR EXCITATIONS

Now consider a situation where the air flow rate is changed stepwise. Reconsider the system (2.1) which can be rewritten as:

$$\varsigma(t) = \alpha(t-1)\cdot\varphi(t)$$

where
$$\varsigma(t) \equiv \frac{dy(t-1)}{dt} + \beta(t-1)\cdot y(t-1) + R(t-1)$$

$$\varphi(t) \equiv u(t-1)\cdot[c(t-1) - y(t-1)]$$

Let the estimator structure be:

$$\hat{\varsigma}(t) = \hat{\alpha}(t-1)\varphi(t)$$

where
$$\hat{\varsigma}(t) = \frac{y(t)-y(t-1)}{h^*} + \beta(t-1)\cdot\hat{y}(t-1) + R(t-1)$$

For this estimator a new equation error ε can be defined.

$$\varepsilon(t) = \hat{\varsigma}(t) - \hat{\alpha}(t-1)\varphi(t)$$

The "Deadbeat" Estimator Idea

Consider the equation error ε at two different times t_1 and t_2. If ε is negligible and the derivative is estimated without error, the following two equations are satisfied:

$$\varepsilon(t_1) = 0 \Rightarrow \alpha_1 u_1[c-y_1] - R_1 = \hat{\alpha}_1 u_1[c-y_1] - \hat{R}_1$$

$$\varepsilon(t_2) = 0 \Rightarrow \alpha_2 u_2[c-y_2] - R_2 = \hat{\alpha}_2 u_2[c-y_2] - \hat{R}_2$$

To ensure the requirement that $\varepsilon = 0$, the estimate of R is updated with a large gain. This results in a linear system consisting of two equations with four unknown variables, $\alpha_1=\alpha(t_1)$, $R_1=R(t_1)$, $\alpha_2=\alpha(t_2)$ and $R_2=R(t_2)$. To be able to solve this system the number of unknowns has to be reduced. Continuity reasons give:

$$t_1 \to t_2 \Rightarrow \begin{cases} \alpha_1 \to \alpha_2 \\ R_1 \to R_2 \end{cases}$$

Thus if Δt (= $t_2 - t_1$) is chosen small enough to ensure that even the fastest parameter R varies negligibly during the time interval there are only two unknowns. Note that this linear system consists of two different equations only if the input u is significantly different at the time instants considered. Since the time interval must be small such an input change is a sudden singular excitation. If Δt is chosen to be the sampling period, α and R are updated in one step. Thus the name deadbeat.

In practice, during periods of low excitation when noise is a large part of the variation, the Kalman filter is more appropriate than the deadbeat strategy. The algorithm can be divided naturally into two sections. The first is the Kalman filter and the second is built on the special deadbeat corrector. The criterion for switching to the deadbeat corrector depends on the input signals, parameter bias and noise level, etc.. This is the subject of present research.

6. SIMULATION EXAMPLES

This section will demonstrate deadbeat and Kalman parameter estimation using simulation plots. The inputs to the estimator, u and y, are shown in Figs. 6.1 a and b respectively. For simplicity the singular excitation is made by abrupt square wave changes of the air flow rate u. Consider an excitation to be singular (large enough) if $\Delta u > 5$.

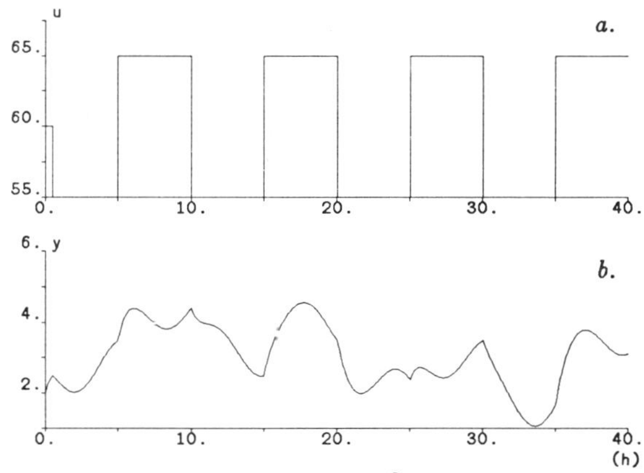

Fig. 6.1 Air flow rate (a) [m^3/min] and dissolved oxygen concentration (b) [mg/l] as input and output to the simulated model.

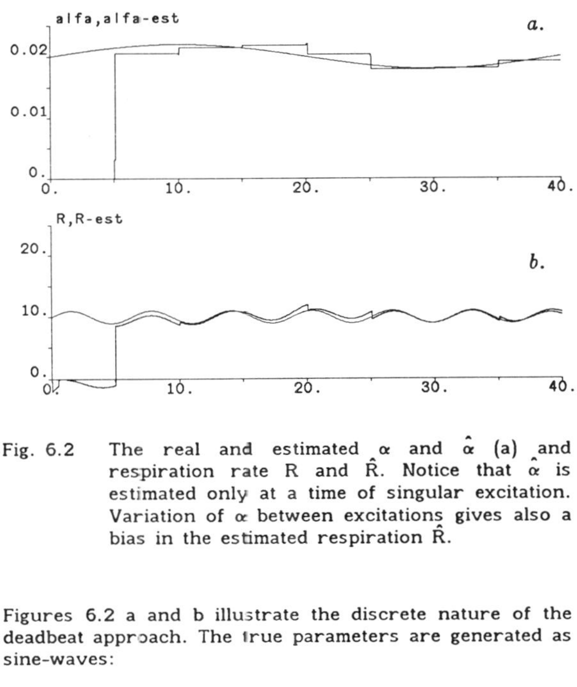

Fig. 6.2 The real and estimated α and $\hat{\alpha}$ (a) and respiration rate R and $\hat{R}$. Notice that $\hat{\alpha}$ is estimated only at a time of singular excitation. Variation of α between excitations gives also a bias in the estimated respiration $\hat{R}$.

Figures 6.2 a and b illustrate the discrete nature of the deadbeat approach. The true parameters are generated as sine-waves:

$$\begin{cases} \alpha(t) = 0.02\cdot[1 + 0\ 1\cdot\sin(2\pi \frac{t}{T_\alpha})] \quad , T_\alpha = 40 \\ R(t) = 10\cdot[1 + 0.1\ \sin(t)] \end{cases}$$

The inaccurate in the deadbeat estimates is due to the rapid variation of R during the excitation. The estimates give, however, a good picture of the parameter dynamics. The deadbeat estimator is designed to be used only during singular excitation periods since it is otherwise sensitive to plant and measurements noise. Therefore it has to be combined with a Kalman filter, and can give a quick start-up of the estimates. Also the deadbeat estimator can be executed intermittantly at input excitations to check the parameter tracking.

Figures 6.3 a and b illustrate the tracking capability of the Kalman filter when noise is present. Here, the output y has an additive component of white noise with the standard deviation 0.01. The period of α is $T_\alpha = 80$, i.e. a change of 10 % in 20 hours.

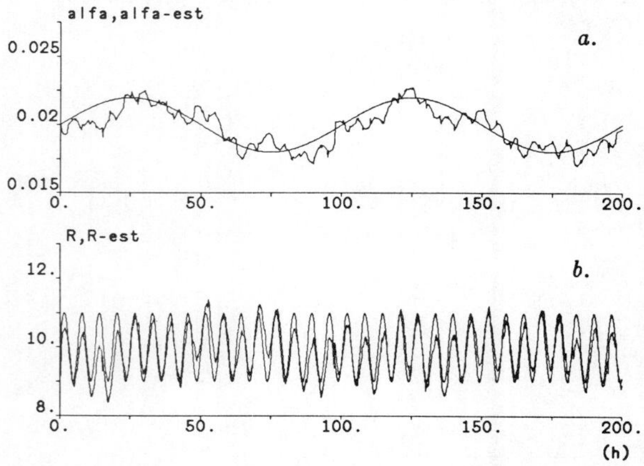

Fig. 6.3 The simulated values and Kalman estimations of α and $\hat{\alpha}$ (a) and respiration rate R and $\hat{R}$ (b) respectively. Despite signal noise and large variation in the parameters, the estimates follow the true values quite well. The dynamic model input is a square wave.

7. ESTIMATION ON PLANT DATA

Two sequences of sampled data of different length and characteristics will now be used. Both sequences come from Käppala Sewage Works, Lidingö, Sweden.

The first one has a length of 5 hours and a special excitation by the air flow rate, (see Fig. 7.1 a). The output (Fig. 7.1 b) responds as a first order system and thus we expect the parameters α and R to vary quite slowly. The main influences on the dissolved oxygen dynamics (y) come from the air flow rate (u). Different choices of initial values for the estimates were chosen to illustrate that the algorithm converges, (see Figs. 7.1 c and d). However, the deadbeat estimator would be helpful to make the convergence faster.

The second data sequence on the other hand, as shown in Figs. 7.2 a and b, is 50 hours. The excitation by the air flow rate (u) is poor with steps at only a few occasions. However, the possibility to estimate α is dependent on the excitation from $\varphi = u(c - y)$, which is fortunately affected by rapid changes in y. Thus we get relevant information for the α estimation when there is large variation in R even though u is constant. The figures show that the output y changes significantly even though u is constant, indicating that the main influences on the dissolved oxygen concentration (y) here is due to variations in the parameters R and α. The initial value of the estimates are not crucial as illustrated in Figs. 7.2 c and d.

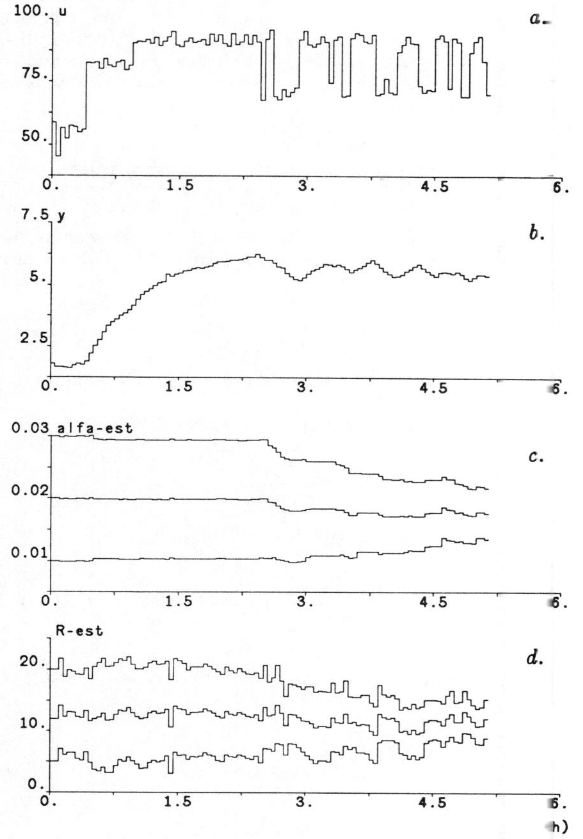

Fig. 7.1 Air flow rate (a) $[m^3/min]$ and dissolved oxygen concentration (b) [mg/l] from the Käppala Sewage Works, Lidingö, Sweden. The sampling interval is 3 minutes. Estimated $\hat{\alpha}$ (c) and $\hat{R}$ (d) started with different initialization of the Kalman filter to show convergence.

8. ESTIMATION AND CONTROL. CONCLUSIONS

The knowledge of K_La and R may be an end in itself. It has already been mentioned that K_La reflects the efficiency of the aeration system. Knowing R, the specific growth rate or the variations of substrate concentration can be calculated. Also, the knowledge of R is essential for good sludge inventory control in wastewater treatment, see Olsson (1984). The specific oxygen utilization rate in the aerator can be used to specify the target for return sludge control or step feed control in activated sludge systems.

The estimator can be part of an adaptive controller for dissolved oxygen. However, the need for extra air flow excitation should be noted. The use of self-tuning control for the DO has been demonstrated in a full scale application, see Olsson et al (1985). In this application, however, R and k_La were not estimated explicitly. Research is being conducted to learn more about the combination of this estimator with controllers.

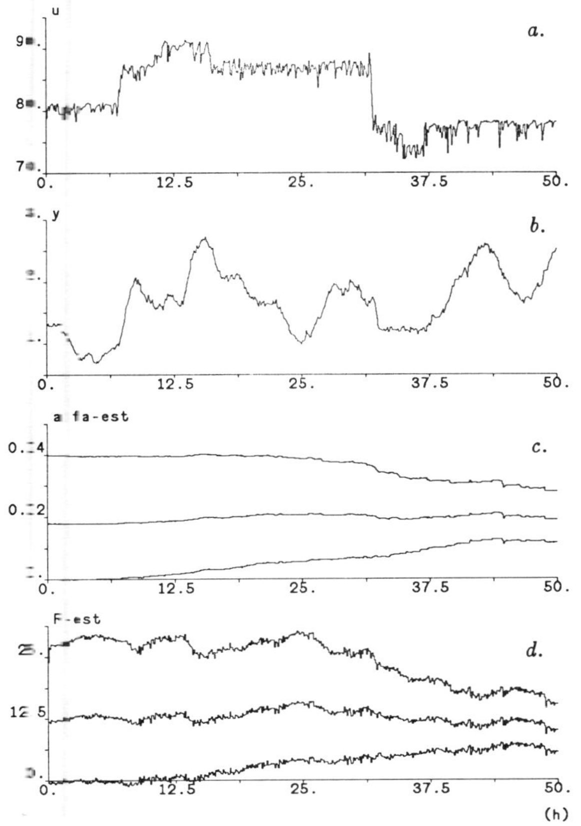

Fig. 7.2 Air flow rate (a) and dissolved oxygen concentration (b) from Käppala. The air flow rate excitations are less frequently than in Fig. 7.1. Crude initial guesses of the estimates $\hat{\alpha}$ (c) and $\hat{R}$ (d) give acceptable convergence. The sampling period is 6 minutes.

Acknowledgements

The work has been supported by the Swedish Board for Technical Development (STU, contract 82-3352). Data has been provided from the Käppala Sewage Works, Lidingö, which is gratefully acknowledged. The simulations have been performed with the interactive simulation program package Simnon, developed at the Department of Automatic Control, Lund Institute of Technology. Thanks to Craig Elevitch, Lund Institute of Technology, for constructive criticism.

REFERENCES

Brouzes, P. (1968). Automated Activated Sludge Plants with Respiratory Metabolism Control. Fourth Int. Conf. on Water Pollution Control Res., Prague.

Cook, S., and S. Marsili-Libelli (1981) Estimation and Control Problems in Activated Sludge Processes. Paper, IAWPRC Workshop on Practical Experiences of Control and Automation in Wastewater Treatment and Water Resources Management, June 20-26, Munich.

Holmberg, A. (1981). Microprocessor-based Estimation of Oxygen Utilization in the Activated Sludge Wastewater Treatment Process. Int. J. Systems Sci., 12, 6, 703-718.

Ko, K.Y., B. C. McInnis and G. C. Goodwin (1982). Adaptive Control and Identification of the Dissolved Oxygen Process. Automatica, 18, 727-730.

Ljung, L., and T. Söderström (1983). Theory and Practice of Recursive Identification. The MIT Press Series in Signal Processing, Optimization and Control, **4**, chap. 2.6, pp 54-56.

Olsson, G., and O. Hansson (1976). Stochastic Modeling and Computer Control of a Full Scale Wastewater Treatment Plant. Proc. Symp of Systems and Models in Air and Water Pollution, The Inst. of Measurement and Control, London, Sep.

Olsson, G. (1984). Control Strategies for the Activated Sludge Process. Comprehensive Biotechnology, 3, ch. 30.

Olsson, G., L. Rundqwist, L. Eriksson, and L. Hall (1985). Self Tuning Control of the Dissolved Oxygen Concentration in Activated Sludge Systems. Paper 4th IAWPRC Workshop on Instrumentation and Control of Water and Wastewater Treatment and Transport Systems, Houston, Texas.

ON-LINE RESPIROMETRY AND ESTIMATION OF AERATION EFFICIENCIES IN AN ACTIVATED SLUDGE AERATION BASIN FROM DISSOLVED OXYGEN MEASUREMENTS

J. A. Howell and B. O. Sodipo

Department of Chemical Engineering, University College of Swansea, Singleton Park, Swansea SA2 8PP, Wales

Respiration rates are a useful indication of the state of a microbial population and in the case of activated sludge plants with mixed populations and mixed nutrients where detailed knowledge of the growth rates of individual populations, or uptake rates of individual chemical species is extremely difficult to ascertain such a general measure as respiration rate is one of the few directly measurable parameters available. Mass transfer efficiency can be used to monitor the state of aeration equipment and can indicate whether sudden changes have occurred in the influent stream which will affect the overall performance of the population of the organisms in the sludge.

A method is described in which both respiration rates and aeration efficiencies can be determined on-line using only the measurement of the dissolved oxygen concentration in the aeration basin and processing these measurements using state observer theory with the aid of a computer. The dissolved oxygen concentration can then be controlled optimally using the results of the parameter determinations. For recursive parameter estimation however, it is necessary to use a sub-optimal tuning of the controller in order to generate sufficient dynamics in the dissolved oxygen concentration. In practice this only need vary from the control value by no more than about 0.1 mg/l.

The activated sludge aeration basin was simulated by a complex structured model and the dissolved oxygen concentration so obtained perturbed by a random amount to account for measurement noise. This value was then used as the input to the state observer algorithm which both devised the control scheme and estimated the parameters continuously. The method was able to cope with dissolved oxygen concentration probe noise of up to 45% of the correct reading. Fluctuations in respiration rate caused by either organic load, or mass transfer efficiency were followed well by the estimator which effectively tracked a sinusoidally varying respiration rate continuously. It was necessary that the initial estimate of the mass transfer coefficient be reasonably accurate, but several methods are known for obtaining this value.

Keywords State & Parameter Estimation, Kalman filter, Dissolved oxygen control.

INTRODUCTION

Apart from temperature and flow rates which are non-chemical measures, there is only the dissolved oxygen concentration which is regularly monitored on activated sludge treatment plants. There are known problems in such measurements but reliable probes are now available and if they are kept clean they show good reliability in practice. Stephenson et al. have described devices in which fouling build up on the probes can be minimised by installing the probe in an external loop. The measurement of solids concentration is also possible though it is hard to interpret exactly what such measurements imply in terms of the performance of the plant itself. There are on the market various respirometers which are supposed to give on-line measurements of the respiration rate of the sludge. These devices normally operate by withdrawing a sample of sludge into a sealed chamber, aerating it and then following the decline in the dissolved oxygen with time. Unfortunately, the dissolved organic concentration in the mixed liquor is normally so low in a well operated plant that by the time the reading is partially complete, all food in the liquor is exhausted and endogenous respiration begins to dominate. This is modified by the effect of luxury storage, but in any case the state within the respirometer differs substantially from that in the plant itself.

A continuous measure of the oxygen utilisation rate is an excellent indication of the load on the plant and can also indicate surges of organics both nutrient and toxic to the organisms. It can be used in conjunction with a solids measurement to indicate the specific oxygen uptake rate of the solids and thus their biological activity. Andrews et al. have shown that such information can be used to effect an excellent control of the activated sludge process.

Activated sludge plants are essentially oxygen/sludge contact devices and certainly in industrial plants the size of the plant is controlled by the maximum efficiency of the aerators. It is thus essential that good information is available regarding the performance of aeration devices in the field. This requires the determination of mass transfer coefficients on the plant which by traditional methods is time consuming and a difficult process. Generally the measurements must be taken by experts and it is necessary to interrupt the normal functioning of a plant for up to a day at a time. Recently Holmberg, and later Howell et al. have shown how by using a personal microcomputer it is possible to obtain such information quite easily with normal plant labour.

All that is required is the initiation of a change in the air flowrate and the computer records the measurements and interprets them. In the case of the method reported by Howell et al, the computer programme also reports the consistency of the data obtained and suggests whether the determination should be repeated or can be considered reliable. Even so, changes to the plant must be made manually and though moderate they are significant amounting to perhaps a 25-50% increase in the air flowrate. The information when obtained can be used to indicate the efficiency of the equipment installed and thus follow time dependent deterioration such as occurs when diffuser becomes blocked. They can also be useful if step feed changes are being used and perhaps an area of the basin has been inadequately equipped with diffuser domes. Ingress of surfactants to the plant can also be detected by this means.

Simultaneous measurement of respiration rate and mass transfer coefficient K_La is possible if they are calculated from the measurements of dissolved oxygen concentration that has been caused to vary slightly. The work of Holmberg calls for quite significant variation and the present work seeks to demonstrate that such large variations need not be necessary. It is necessary that the dissolved oxygen concentration be controllable and thus the technique will not be applicable to a majority of plants. Larger plants are now being equipped with dissolved oxygen (D.O.) controllers as the energy savings so achieved make the capital cost worth while. This situation is likely to be more prevalent in the future as frequency controllers which control the frequency of current fed to the blowers become much cheaper. They are the most effective method of controlling blower output within a reasonable range of the design output. Currently, large scale frequency controllers are a major cost in the control scheme. Other methods of D.O. control are less energy effective.

Many methods of controlling dissolved oxygen that have been discussed recently such as those of Olsson use self tuning controllers and have time series models whose parameters are recursively re-estimated. Unfortunately the parameters in these models bear no direct relationship to the parameters normally of physical interest. However, in the case of the D.O. control a first order moving average model of the Astrom type will yield respiration rate and D.O. following some manipulation. If the model turns out to be second order then the parameters have no directly interpretable physical meaning.

THEORY

The use of state observer theory was developed since it treats the current problem easily and can be extended to the more difficult problem of estimating the specific activity of the mixed liquor solids. In each case the parameters have directly interpretable physical meanings.

State observer theory requires that the plant be modelled as a function of the controllable inputs to produce estimates of the measurable outputs, and accepts that the process and the measurements are subject to stochastic variations. The difference between the estimates and the measurements is used to correct the estimates of the measured variable and also the parameters of the model (which can be non-measurable variables). In the case under consideration the process of oxygen transfer is modelled by equation 2.1

$$\frac{dC}{dt} = K_L a\,(C^* - C) - R + e \qquad \text{.........(2.1)}$$

where

C = D.O. concentration (mg/l)

C* = Saturated D.O. concentration (mg/l)

R = Oxygen utilisation rate (mg/l/hr)

t = Time (hr)

e = Noise, assumed to be white.

The value of the mass transfer coefficient (K_La) is assumed to be directly proportional to the air flowrate. Although in fermenters a power law relationship has been found between air flowrate and mass transfer coefficient it has been established that a linear law is more appropriate for activated sludge aerators (6.9). There are then two unknown parameters in the above equation - the constant of proportionality, k, between air flowrate and K_La, and the respiration rate, R. Both are assumed to be constant over the short term and the estimator is written accordingly. It can be shown that such an estimator is observable for C, k, and R although C* must be known previously. The equations are then integrated over the fixed time step interval to provide a transition matrix representation between the states at t and at t + h. The state variable C is supplemented by the parameters R and k which are also treated as state variables. Appendix B shows how the full equations for the estimator are set up and used. The plant is to be controlled optimally if possible, although a moderate degree of detuning may be necessary in order to generate sufficient dynamics in the process that the parameters be observable. The parameters are not observable in a steady-state process.

Various estimators have been used to solve such problems from the Kalman-Bucy filter to a series of more complex algorithms all based on the same premise. Given an existing system state (in this case values of k, C and R) the state after a time interval h can be estimated from the model. However, the model is probably in error since it does not accord exactly with the latest observation. Then again the observation is certainly subject to experimental error. If there is a great deal of error in the observation then it should hardly be taken into account at all, only after many readings should it be allowed to affect the best estimate of the state at the next time interval. On the other hand if the observations are excellent, one can almost discount the model for the directly observed variables. In the case of a single state variable x(t) which is also measured, the estimate at the next time interval is given by:-

$$x(t + h) = x(t) + K(t)(z(t) - y(t)) \qquad \text{.......(2.2)}$$

where

K(t) = Filter gain

z(t) = Observed value of the state

x(t) = State variables

y(t) = Expected measurement using filter if the estimate prediction were true.

If the filter gain is close to 1.0 the observation z is dominant and the model is insignificant. On the other hand, if the gain is close to 0 the observations are neglected in favour of the model. Clearly the estimation of the correct value of K is important and the algorithms used when there is more than one variable and K is a matrix are not so simple. In addition there are no clear ways of choosing some of the parameters and their probabilities used in the algorithm especially when the state equations are nonlinear. Considerable trial and error is required to establish for any given set of equations the best approach. These trials can be performed with simulated models and are essential before the method is applied in the field.

SIMULATION

The activated sludge plant was simulated using a mathematical model in which the biomass was represented in a structured manner in order that the dynamics of the plant could be faithfully represented. The full model is described in Appendix A. The model which comprises of a set of differential equations was integrated through time to give the simulated behaviour of the plant. One of the plant outputs so simulated was the dissolved oxygen concentration (D.O). A simulated observation was derived from the simulated concentration by taking this latter concentration and adding to it a random noise to account for the measurement error. This error term was stochastic with a normal distribution and generated from a random number algorithm. The simulated plant was under dissolved oxygen control using an optimal control algorithm that assumed knowledge of the mass transfer coefficient and the respiration rate. These latter were derived from the measured D.O. concentration using the algorithms described in the previous section.

The D.O. set point was 2.0mg/l which is considered the minimum desirable value to avoid any oxygen limitation occurring in the activated sludge flocs. The control was exercised on the air flow rate. It was assumed that the mass transfer coefficient was directly proportional to the applied air flow rate, and thus increased air flow resulted in increased D.O. levels in the plant. The air flow was adjusted so that the plant D.O. value was expected to approach closer to the set point in the next time step. An optimal controller would have achieved this within one time step. However, such control was attempted and it was found to be too good to allow adequate parameter estimation. The controller was then detuned so that the existing deviation at the beginning of a control time interval was to be reduced by 80% by the end of the interval rather than by the full 100%.

The estimation of the mass transfer coefficient, the D.O. and the respiration rate was achieved through the use of the UDU^T covariance filter due to Bierman (3) as the estimator. It uses a simple model based on equation 2.1, the observed D.O. measurement and information about the model error and measurement error in the observations to be assumed by the filter.

The response of the estimator to changes in the substrate loading, to the aerator medium oxygen transfer ability and the ability to cope with increased levels of error in the observed D.O. measurement was tested in order to determine the filter robustness.

ALGORITHM IMPLEMENTATION

For many biological waste treament plants, an important area for improving the water effluent quality is the aeration basin. In this study the secondary clarifier and the settler dynamics were neglected. The mathematical model used in simulating the plant behaviour was based on the Polysar pilot plant in Burlington, Canada with an aeration volume of $100m^3$ and the capacity to treat 800 m^3/day of waste. The influent waste concentration was 300 mg/l and 100mg/l for the soluble and particulate organics respectively. The kinetic parameters used in the simulation were typical values from the literature and are given in Table 1.

To design a Kalman filter a further requirement to the system model and the observations is the specification of the statistical knowledge of both process and measurement noise as well as of the initial values of the state and parameters to be estimated. For linear systems, the filter performance which involves the evolution of the variances and the gains, can be analysed independently of the model. However, for nonlinear systems a typical filter design would require linearisation about some nominal trajectory (in the case of the extended Kalman filter, the previous estimate) and because of this, the variances and the gains are coupled to the system model.

It was necessary to test different models of incorporating system disturbances in order to obtain the filter design that yielded the optimal filter performance. Initially the parameters k and R were treated as constants subject to initial conditions but the filter produced poor estimates. The model structure was then modified so that the parameters were driven by white noise in addition to the initial conditions. The disturbances in the K_La were taken to be the random fluctuations in the local hydrodynamic conditions as well as being due to the step changes in the air flow rate, while those in the respiration rate were due to local fluctuations in the amounts of D.O. and substrate reaching the microbes. The initial conditions were selected such that the dynamics were due in the main to the changing influent conditions and high initial covariances were chosen to reflect the degree of uncertainty in the choice of the initial state and parameter values.

The measurements were assumed to be those produced by a reliable D.O. probe and the disturbances were errors due to probe responses. To treat the case of fouled probes, the variances of the uncertainty only needed to be increased.

The filter was tuned so that the estimates from it matched the plant outputs as closely as possible. This meant that the design parameters were the variances of both the process and the measurement disturbances, Q and R'. This was such that the gains calculated were not too small and at the same time the variances being small as it resulted in the filter 'learning' its states too well and thereby not being capable of tracking any subsequent changes that occurred. The values used in the filter design are given in Table 2.

RESULTS AND DISCUSSION

The simulation of the plant is shown in Figs. 1 to 4. The feed flow and organic strength are shown in Fig. 1. They vary sinusoidally with a diurnal period. The efficiency of the suboptimally tuned controller is shown in Fig. 2 where the D.O. value from the simulator is shown. Although there is a slight variation with time, the control is held reasonably well and the variation from the set point of 2 mg/l is always within acceptable limits. Also shown on that figure is the estimated value of D.O. Not shown however is the measurement error added to the signal received from the probe which in this case was 10% of the signal. Figure 3 shows the respiration rate also varying sinusoidally with a lag of hours behind the feed rates. Again the result of the estimator is shown and it can be seen to follow the simulated value closely. Finally in Fig. 4 there is a plot of the value of K_La (factor k times the air flow rate). The air flow rate was assumed to be known exactly and so the estimate was of k alone. The estimate as before closely follows the value determined by the simulation.

During the simulation/estimation procedure it was found that after about 24 hours of simulated time the covariance matrix which is used in the calculation of the Kalman filter began to drift towards rather unsatisfactory values. Accordingly it was re-initialised on a daily basis to a value of 100I where I is the identity matrix. There was also an unsatisfactory result if noise in excess of 45% occurred in the D.O. measurement. The Kalman gain tended to become negligible and D.O. readings were being virtually ignored. Errors in the predictions of R and k could clearly not be corrected.

The performance of an estimator of this type in the field depends on its ability to cope with errors in the initial estimates of R, k, and C, and also in its ability to vary quickly if there is a real change in any of the states. This was tested and the results are shown in Figs. 5 to 9, as well as in Fig. 2 where both the controller and the estimator were tested simultaneously. The estimator started with a low value and the plant with a value that was too high. Very quickly the plant was brought under control and the estimate corrected. In Figures 5 and 6 a simulation of a sudden change in the influent organic load is shown. This causes a sharp change in the respiration rate which is immediately detected by the estimator. In Figure 7 a sharp change in the value of k is shown and again the estimator responds correctly. In Figure 8 the effect of an initially poor value of the respiration rate is shown. This results in a slow adaptation to the correct value but this is achieved within two hours. The effect of an incorrect value of the factor k is not easily coped with. Errors of around 10% are corrected but larger initial errors result in the correct value of k never being estimated. This cascades onto the respiration rate which is also incorrect by the same factor. The error is attenuated from the initial error in k but is still present. The comparative behaviour remains good and changes in the plant can still be followed although the exact values are not given. This can easily be dealt with however since the method of Howell _et al_ (7) for instance can be used initially to establish the first value of k and subsequently the estimator will keep track of any changes.

REFERENCES

1. Andrews, J. F. and Stenstrom, M. K. (1976). "Control systems for the reduction of effluent variability from the activated sludge process", Prog. water tech. vol.8, 1, 41-68.

2. Astrom, K. (1970). "Stochastic control theory", Academic Press, New York.

3. Bierman, G. J. (1976). "Factorization methods for discrete sequential estimation", Academic Press, New York.

4. Busby, J. B. and Andrews, J. F. (1975). "Dynamic modelling and control strategies for the activated sludge process". J.WPCF 47, 5, 1055-1080.

5. Holmberg, A. (1981). "A microprocessor based estimation of oxygen utilization in the activated sludge waste water treatment process", Int. Sys. Sci. vol. 12, 6, 703-718

6. Holmberg, A. (1982). "Modelling of the activated sludge process for microprocessor based state estimation and control", Water res. vol. 16, 1233-1246

7. Howell, J. A., Yust, L. J., and Reilly P. (1984). "On-line measurement of respiration and mass transfer rates in an activated sludge aeration tank", J.WPCF vol. 56, 4, 319-324

8. Olsson, G. and Hansson, O. (1976). "Modelling and identification of an activated sludge process", Proc. 4th IFAC symposium on Sys. Ident.

9. Stephenson, J. P., Luxon, E. G., Monaghan, B. A., and Gillespie, R. G. (1981). "Evaluation of instruments for continuous activated sludge monitoring", Water Sc. Tech. vol. 13, 11/12, 721-728.

APPENDIX A

Mathematical Model

Control strategies for water quality systems are often difficult to determine because of the size, the cost, and the knowledge required of the process. Mathematical models of the plant are used to determine the most suitable control strategy for the system, in this case an activated sludge plant.

To develop the mathematical model used in simulating the plant, the activated sludge process was assumed to consist of a single aeration basin with a recycle of the suspended solids from the secondary clarifier. Oxygen mass transfer to the mixed liquor medium was through well distributed spargers to ensure the reactor was well mixed. In the clarifier, the separation was assumed to be instantaneous and the amplitude of the diurnal variations in both the hydraulic and organic loadings was of the ratio 2.1/0.5 (peak/mean min).

The sludge mass itself is mathematically structured into three components:- the total mass structured into the active mass and inert solids, with the active mass being further structured to account for a variable specific activity which is dependent on the history and the environment. Changes in the specific activity ere due to the rates of formation and decay as well as the dilution within the cell due to cell growth of the enzymes responsible for cell activity respectively.

Taking mass balances over the plant results in a model of the form -

$$\frac{dS}{dt} = Q(S_0 - S) - \frac{ASX_a}{K_1 + S} \qquad \ldots\ldots (1)$$

$$\frac{dX_a}{dt} = -uX_a - K_dX_a + \frac{ASX_a}{(K_1 + S)} \qquad \ldots\ldots (2)$$

$$\frac{dX_I}{dt} = QX_I - uX_I + Y_dK_dX_a \qquad \ldots\ldots (3)$$

$$\frac{dA}{dt} = \frac{K_3S}{(K_2+S)} - K_4A - \frac{A^2S}{(K_1+S)} \qquad \ldots\ldots (4)$$

$$\frac{dC}{dt} = K_La(C^* - C) - \frac{Y_1ASX_a}{(K_1+S)} + Y_2K_dX_a \qquad \ldots\ldots (5)$$

$$u = Q((1+R) - (RX_a)/X_T) \qquad \ldots\ldots (6)$$

where

[illegible] = Net removal rate of solids from aerator (hr^{-1})

[illegible] = Specific flowrate through the plant (F/V) (hr^{-1})

[illegible] = Recycle ratio

[illegible]$_o$ = Influent substrate concentration, TOC (mg/l)

[illegible] = Effluent substrate concentration, TOC (mg/l)

[illegible]$_a$ = Active mass in the aerator, TOC (mg/l)

[illegible]$_l$ = Inert solids in the aerAtor, TOC (mg/l)

[illegible]$_r$ = Return suspended solids, TOC (mg/l)

[illegible]$_T$ = Total suspended solids, TOC (mg/l)

[illegible] = Specific activity (hr^{-1})

[illegible]$_i$ = Kinetic constants

[illegible]$_i$ = Stoichoimetric coefficients

Integrating the set of differential equations with a suitable integrator such as the Runge-Kutta Butcher technique will yield at designated times the dissolved oxygen concentration.

In practice measurements are not always accurate and the errors involved are stochastic and to obtain corrupted noise, random amounts are added to the obtained dissolved oxygen values. White noise which is Gaussian is modelled by using a numerical random number generator that is normally distributed with a zero mean and the variance that depends on the strength of the measurement uncertainty obtained in practice.

APPENDIX B

State Observer Theory

For many physical systems, some of the variables of interest are often not directly measurable or their determination through analytical methods are not suited to on-line monitoring and in such cases they could be obtained indirectly through estimation using variables that are measurable by combining the previous values of the unmeasured variables with a weighted fraction of the measured ones.

State observer theory is that approach where a simple mathematical model of the physical system is used in conjunction with a set of measurements to estimate the desired variables. The simple model such as a linear one of the system is used to predict values of the measurements and any deviation from the actual measurements is then due to (i) the model being inadequate (ii) errors in actual measurements (iii) incorrect initial values of the unmeasured variables. The use of estimation techniques such as the covariance filter allows the correction of the estimates after sufficient observations to eliminate the random errors of the model inadequacy and the measurements. For many measurements, the accuracy is sufficiently high that in the context of the problem little error need be assumed and for dissolved oxygen concentration in aerators this is the case.

The variables that will be estimated using the model are the overall respiration rate R_o, the oxygen mass transfer coefficient K_La as well as the dissolved oxygen concentration which is needed for feedback control purposes. For the aerator conditions, the respiration rate R_o as well as the transfer coefficient K_La are assumed to vary slowly and changes in the dissolved oxygen levels in the aerator are mainly due to the difference between the rate of oxygen transfer to the sludge and the respiration rate of the sludge. The filtered measurement is the dissolved oxygen concentration obtained from a probe located downstream from the aerator.

The state observer model becomes

$$\frac{dC}{dt} = kF(C^* - C) - R_o \qquad \text{.........(i)}$$

$$\frac{dR_o}{dt} = 0 \qquad \text{........(ii)}$$

$$\frac{dk}{dt} = 0 \qquad \text{.......(iii)}$$

where

C^* = Saturation constant of the dissolved oxygen ($mg.l^{-1}$)

From observations of the activated sludge process the relationship between the mass transfer coefficient K_La and the flowrate F is linear; using this and integrating the set of differential equations over the estimation interval yields the difference equations -

$$C(t+h) = \alpha C(t) + hC^*K_La(t) - hR_o(t) \qquad \text{......(iv)}$$

$$R_o(t+h) = R_o(t) \qquad \text{............(v)}$$

$$K_La(t+h) = (F_t/F_{t-h})K_La(t) \qquad \text{............(vi)}$$

where

$$\alpha = Exp(-kF_th)$$

For linear dynamic systems, the set of equations can be put into compact form through the use of the matrix - vector notation. The relationship between the observed measurements and the variables of interest (states) is also linear.

To account for errors in the filter model and the measurement equation noise terms are added linearly. In order to simplify the solution to the estimation problem, the state variables are assumed to be Gaussian random ones thus requiring only the mean and the covariance values. The noise terms are assumed to be white and mean values of zero.

If

$$X(t) = [C(t)\ R_o(t)\ K_La(t)]^T$$

$$Z(t) = C(t)$$

$$X(0) = X_o$$

then

$$X(t+h) = A\ X(t) + w(t) \qquad \text{...........(vii)}$$

$$Z(t) = H^TX(t) + v(t) \qquad \text{...........(viii)}$$

where

$E[w(t)] = 0$; $E[v(t)] = 0$;

$E[w(t)v(t)^T] = 0$; $E[v(t)w(t)^T] = 0$;

$E[X(0)X(0)^T] = P_o$; $E[w(t)w(t)^T] = Q$;
$E[v(t)v(t)^T] = R'$

P_o, Q, and R' are the covariances of the initial state, the model noise and the measurement noise respectively.

$$A = \begin{bmatrix} \alpha & -h & hC^* \\ 0 & 1 & 0 \\ 0 & 0 & F_t/F_t \end{bmatrix}$$

$$H^T = [1 \quad 0 \quad 0]$$

In order that the solution to the estimation problem converge or to track the plant behaviour, the system must be observable; i.e. the 3 x 9 matrix L_o should be of rank at least 3.

$$L_o = [H^T : A^TH^T : (A^T)^2H^T]$$

The algorithm used for the estimation problem is the UDU^T covariance filter(3). It is based on the conventional Kalman filter but the modification includes the structuring of the initial state and subsequent estimate covariances into a unit upper triangular and a diagonal matrix using the Cholesky decomposition and a rank one Agee-Turner update. This improved the numerical stability and offered computational savings. The time update was achieved using the modified Gram-Schmidt orthogonalization method.

In order to keep the energy costs down, the dissolved oxygen concentration was controlled to a setpoint value of 2mg/l

$$J = \sum_{i=1}^{N} [C_s - C(t+h)]^2 \qquad \text{...............(ix)}$$

from (iv)

$$F_t = F_{t-h}\frac{[f(C_s - C(t) + hR_o(t)]}{hK_La(t)[(C^* - (1-f)C_s - fC(t)]} \qquad \text{........(x)}$$

where

C_s = The D.O. setpoint value (mg/l)
f = Controller tuning parameter set at 0.8

TABLE 1. KINETIC CONSTANTS AND OPERATING CONDITIONS FOR THE PLANT SIMULATION

Parameter	Value	Parameter	Value
k_d	0.015 hr^{-1}	S_o	300.0 mg/l
K_m	50.5 mg/l	Q_o	0.3 hr^{-1}
K_p	0.015 hr^{-1}	C*	9.6 mg/l
K_A	15.00 mg/l	F_o	10.0 l/hr
V_m	0.75 hr^{-1}	X_o	100.0 l/hr
K_L	1.00 l/hr.	S_e	20.0 mg/l
R_c	0.5	A	0.4
Y_s	0.600	X_A	450.0 mg/l
Y_d	0.20	X_I	550.0 mg/l
Y_b	0.20	C_o	3.0 mg/l
Y	0.667	R_o	36.5 mg.O_2/hr
		K_La	10.0 $hr.^{-1}$

TABLE 2. KALMAN FILTER DESIGN PARAMETERS

C	= X (o)	=	1.0 mg/l
R	= X (o)	=	30.0 mg.O_2/hr
K_La	= X a(o)	=	5.0 hr^{-1}
P_{ij}	= 0.0		i = j, i, j = 1, 2, 3
P_{11}	= 100		
P_{22}	= 100		
P_{33}	= 100		
Q_{11}	= 0.0005		
Q_{22}	= 0.05		
Q_{33}	= 0.0004		
R	= 0.0033		

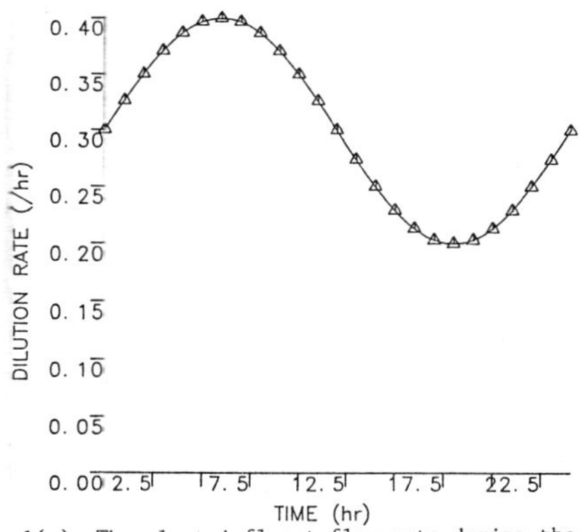

Fig 1(a) The plant influent flow rate during the 24. hr period. The mean value is $Q_o = 0.3\ hr^{-1}$.

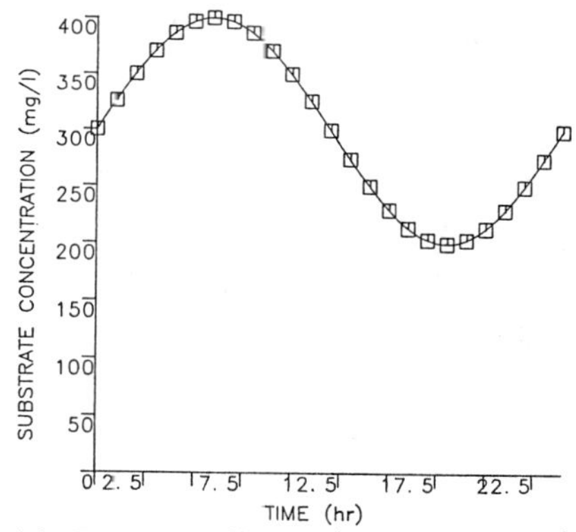

Fig 1(b) The plant influent substrate concentration during the 24 hr period and the mean value is S_o = 300 mg/l .

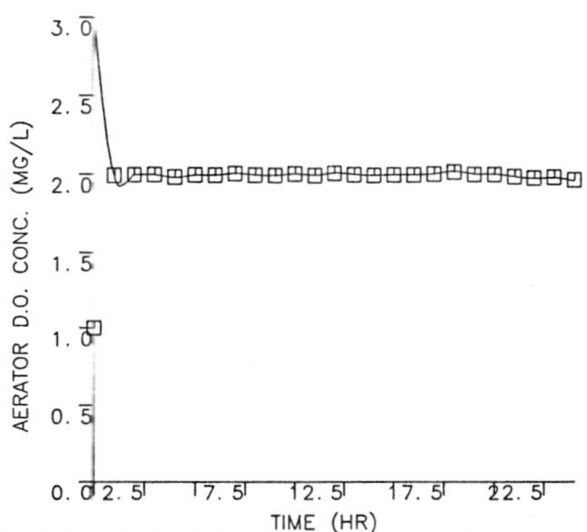

Fig 2 Estimation of the aerator dissolved oxygen concentration for the 24 hr. period. The continuous line shows the plant values and the points show the estimate values.

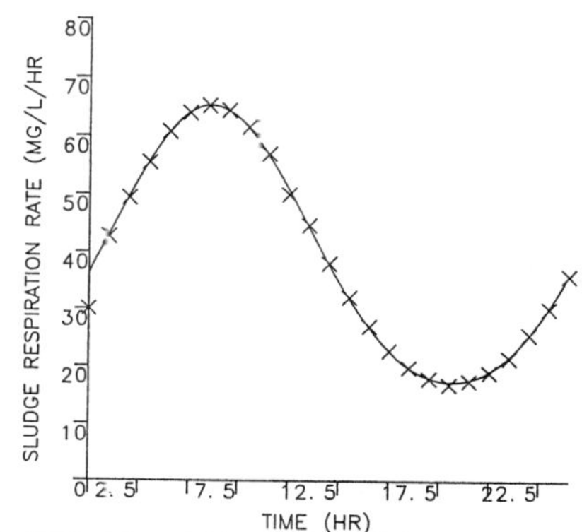

Fig 3 Estimation of the aerator sludge respiration rate for the 24 hr. period. The continuous line shows the plant values and the points show the estimate values.

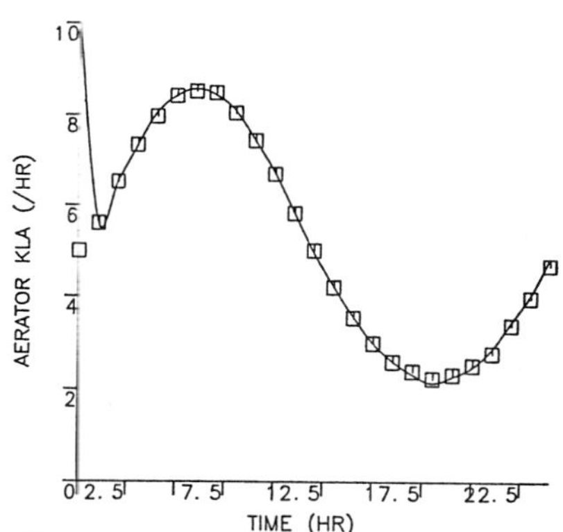

Fig 4 The aerator oxygen transfer coefficient K_La estimation for the 24 hr. period. The continuous line shows the plant values, the points the estimate values.

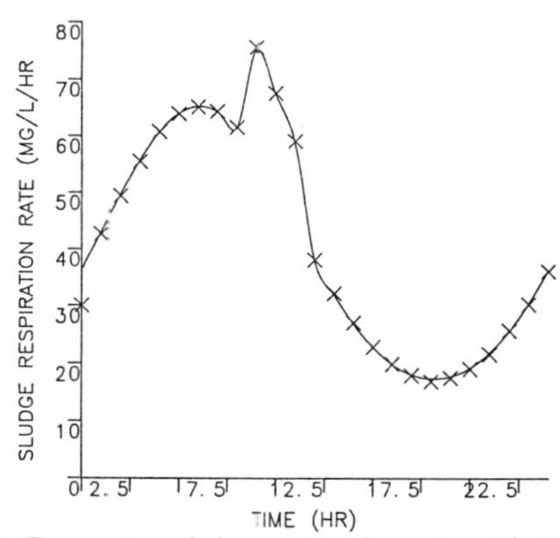

Fig 5 The aerator sludge respiration rate estimation when the influent substrate concentration is changed from S_o = 300 mg/l to S_o = 400 mg/l during the hours 8 - 11 .

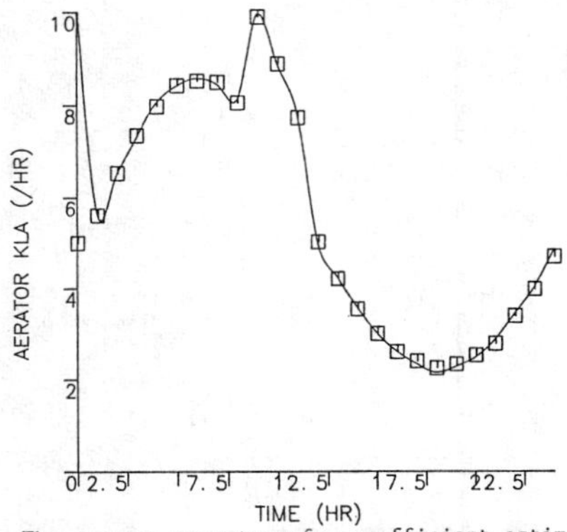

Fig 6 The aerator mass transfer coefficient estimation when the plant influent substrate concentration is changed from S_o = 300 mg/l to S_o = 400 mg/l during the hours 8 - 11 .

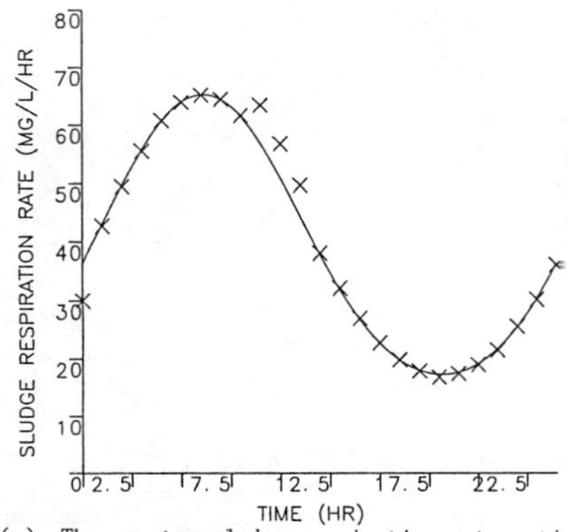

Fig 7(a) The aerator sludge respiration rate estimation when the plant liquid was such that the K_La changed during the hours 8 - 11. The change was due to the proportionality constant k. (k = 0.9).

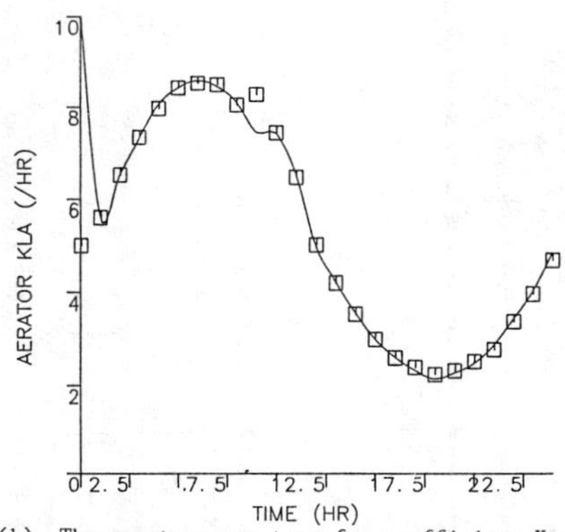

Fig 7(b) The aerator mass transfer coefficient K_La estimation when the plant aerator medium resistance to the oxygen transfer was reduced during the hours 8 - 11. (k = 0.9).

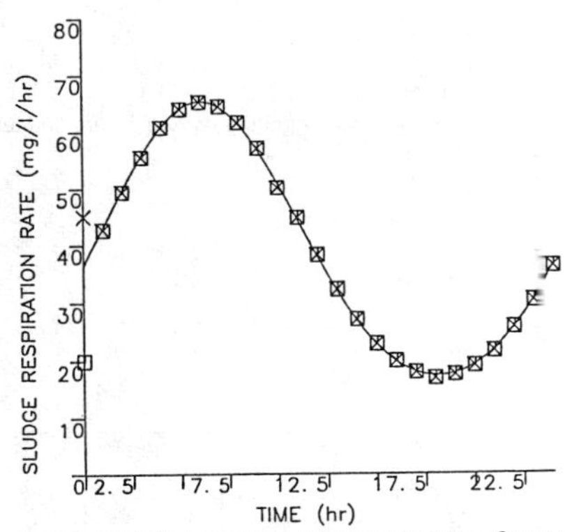

Fig 8 Aerator sludge respiration estimation for two different initial filter respiration rate values one R = 20 mg.O_2/hr., the other R = 45 mg.O_2/hr.

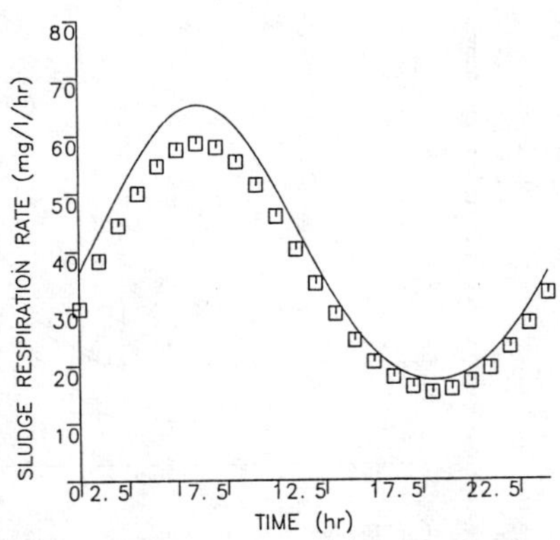

Fig 9(a) The aerator respiration rate estimation for a different initial value for the filter. K_La = 4.0 hr^{-1}.

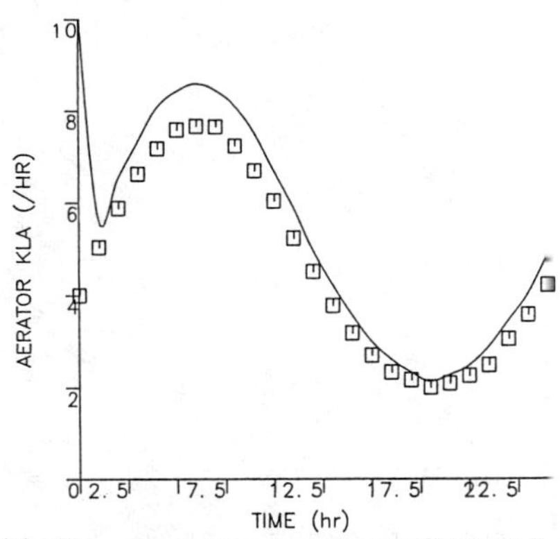

Fig 9(b) The aerator mass transfer coefficient K_La estimation for a different initial value for the filter. This corresponds to a change in k. K_La = 4.0 hr^{-1}.

DETERMINATION OF pH AND TEMPERATURE PROFILES FOR THE GROWTH OF *ESCHERICHIA COLI*

P. Reinikainen*, J. Olkku, P. Markkanen*, A. Holmberg*****

*Helsinki University of Technology, Department of Chemistry, Kemistintie 1, SF-02150 Espoo, Finland
**Orion Corporation Ltd., PO Box 65, SF-02101 Espoo, Finland
***Rintekno Oy, PO Box 146, SF-02101 Espoo, Finland

Abstract. Escherichia coli batch fermentations were studied in a 10 l laboratory fermenter. The first stage of the work consisted of 18 experiments with static environmental conditions. A logistic glucose-, pH- and temperature-dependant model for bacterial growth and static regression models for several fermentation parameters were derived from the results. The models were used to study the characteristics of the process. The possibilities of improving yields of biomass and acids using dynamic changes in pH were studied experimentally. Regression models could be used to construct control profiles. Using suitable profiles, the yields of biomass and acids were increased by 30-40 % compared with the corresponding yields in static conditions.

Keywords. Fermentation; modelling; optimal control;optimization

INTRODUCTION

The optimal growth of microbes and optimal product formation may require different conditions during the fermentation process. By varying the set-points of certain variables, i.e. by using suitable control profiles the yield may be increased or the duration of the process decreased considerably. A good knowledge of control profiles is particularly important in fermentations based on gene technology, because in such processes growth of the microbes, amplification of the recombinant gene and formation of the product seem to occur at different times in different chemical, physical and physiological conditions.

Computer-based control techniques both facilitate and demand the use of mathematical models of the process. Models can be used for studying the basic phenomena of the system and they can be employed in predicting process bahaviour. One interesting application is to use models for determining optimal control profiles (Cheruy and Durand, 1979).

When mathematical models are used it is important that the observations and measurements are reliable and that they represent the real state of the modelling object. Special attention should be paid to experimental planning and accuracy of the primary data. Because the state of living organisms changes continuously, storage of the microbial strain, revival of the culture and cultivation of the inoculum should be carried out using an exactly standardized method. When the state of the culture at the beginning of fermentation process is always as similar as possible, the effects of altered process conditions during the fermentation are not overshadowed by the variations of inoculum state.

The aim of this work was to use mathematical models for studying some characteristics of an Escherichia coli fermentation process and to develop a methodology for determining optimal conditions for bacterial growth.

MATERIALS AND METHODS

Strain and Media

Escherichia coli K-12 HB101 was grown in a 10 l laboratory fermenter on LB-broth (Bertani, 1951) supplemented with glucose. LB-broth was sterilized with the fermenter. The glucose solution was sterilized separately and added to the fermenter together with the inoculum.

Assays

For each experiment an inoculum was grown using a standardized method. The bacterial strain was stored at 4 ^oC on a nutrient agar slant. Preparation of the inoculum was started from 3...7 days old agar cultures. An inoculum consisting of E. coli cells in 500 ml LB-broth was grown at 37 ^oC in shaking flasks (240 RPM), always to the same optical density value representing the point at which the growth rate began to decrease.

During the fermentations, samples were collected at one-hour intervals. Growth was followed by measuring optical density with a spectrofotometer (OD_{660}) or with a Klett-Summerson colorimeter (filter no.66). and by determining the dry cell mass (g/l). Glucose was determined by the GOD-Perid-method (Boehringer Mannheim). Production of acids was monitored by the consumption of 6M NaOH used in pH control. In experiments with dynamic pH profiles the amount of base needed for the change of pH was considered when calculating the production of acids (Reinikainen et al., 1984).

Equipment

The experiments were carried out in a Biostat E fermenter (B Braun Melsungen) with a working volume of 10 liters. The fermenter was equipped with the laboratory fermenter control system LFCS (Rintekno Oy) (Oinas, 1982), which takes care of data acquisition, calculation, reporting, supervision and control tasks. Temperature, pH, agitation speed, agitation current, acid/base feed rates and dissolved oxygen were continuously recorded. Temperature, pH, agitation speed and dissolved oxygen could be controlled by set-point controllers. The results of laboratory analyses of cell mass, optical density, glucose concentration and acids were entered into the computer when received. Entering of sample analyses, or other off-line data, automatically triggered calculations based on these variables.

The hardware of the LFCS consisted of a PDP 11/23 minicomputer, a graphic colour terminal, a digital 8-pen colour plotter, an alarm and report printer and the necessary interface.

Experimental Design

Response surface methodology (RSM) was used in the study of E. coli growth (Olkku et al., 1983 , Box and Wilson, 1951, Cochran and Cox, 1957). The experimental plan used was a 2^3 central composite orthogonal design with 18 experiments having the following independent variables: Initial glucose concentration of the medium (Z_1), 0...40 g/l, pH (Z_2), 6.0...8.5 and temperature (Z_3), 30...40 °C (Table 1).

During the fermentations stirring speed and aeration were kept constant at 500 RPM and 0.5 VVM, respectively.

For each single fermentation a logistic model for the bacterial growth was derived from the results using a nonlinear curve-fitting technique. Using multiple regression analysis, second order polynomial mathematical models were derived from the results. The coded values of glucose concentration (Z_1), pH (Z_2) and temperature (Z_3) were set as the independent variables and the parameters of the logistic growth model, generation time during the exponential growth phase, final concentrations of biomass and acids and total glucose utilization time, were the dependent variables.

The regression models obtained were used to investigate some characteristics of the process and the influence of glucose concentration, pH and temperature on the growth of E. coli, as well as to develop optimal conditions for the bacterial growth

TABLE 1 Experimental plan and the independent variables.

exp no.	Z_1	Z_2	Z_3	glucose (g/l)	pH ( )	T (°C)
1	-1	-1	-1	6	6.4	32
2	1	-1	-1	34	6.4	32
3	-1	1	-1	6	8.1	32
4	1	1	-1	34	8.1	32
5	-1	-1	1	6	6.4	38
6	1	-1	1	34	6.4	38
7	-1	1	1	6	8.1	38
8	1	1	1	34	8.1	38
9	-1.42	0	0	0	7.3	35
10	1.42	0	0	40	7.3	35
11	0	-1.42	0	20	6.0	35
12	0	1.42	0	20	8.5	35
13	0	0	-1.42	20	7.3	30
14	0	0	1.42	20	7.3	40
15	0	0	0	20	7.3	35
16	0	0	0	20	7.3	35
17	0	0	0	20	7.3	35
18	0	0	0	20	7.3	35

RESULTS AND DISCUSSION

Fermentations with static conditions

A Kinetic Growth Model

From the results of 18 static experiments a logistic glucose-, pH- and temperature-dependent model (1) for bacterial growth was derived.

The kinetic growth model (1) was used for studying the effects of initial glucose concentration, pH and temperature on the final biomass concentration and on the growth rate of E. coli.

According to model (1), increase in glucose concentration from 0 to 40 g/l increased the final biomass concentration from about 2 to about 4 g/l at pH 7.3 and at 35°C (Fig. 1). On the other hand, the amount of glucose had no influence on the growth rate in the beginning of the fermentation process.

pH had an influence both on the final biomass concentration and on the growth rate at 35 °C when the glucose concentration was 20 g/l. According to model (1) the highest biomass concentration (3.3 g/l) was reached at pH 7.3 and the lowest concentration (2.5 g/l) at pH6. The growth was slowest at pH 8.5 (Fig. 2).

$$X = A/\left[1 + \exp\,(B + Ct)\right] \qquad (1a)$$

$$\begin{bmatrix} A \\ B \\ C \end{bmatrix} = \begin{bmatrix} 0.582 & 0.170 & 0.104 & -0.047 & -0.269 & -0.032 & 0.328 & 0.110 & -0.038 & 3.276 \\ -0.029 & 0.039 & -0.355 & 0.171 & 0.253 & -0.112 & -0.062 & 0.048 & -0.123 & 3.117 \\ 0.065 & 0.083 & 0.022 & -0.029 & -0.072 & 0.027 & 0.070 & 0.014 & 0.061 & -0.779 \end{bmatrix} \begin{bmatrix} Z_1 \\ Z_2 \\ Z_3 \\ Z_1Z_1 \\ Z_2Z_2 \\ Z_3Z_3 \\ Z_1Z_2 \\ Z_1Z_3 \\ Z_2Z_3 \\ 1 \end{bmatrix} \qquad (1b),\ (1c),\ (1d)$$

where X is biomass concentration (g/l)
t is time (h)
A, B, C are model parameters
Z_1, Z_2 and Z_3 are coded values of independent variables

Equation (1a) can be rearranged to give the following form (Edwards and Wilke, 1968):

$$dX/dt = -C \cdot X\left[1-(X/A)\right] \qquad (2)$$
$$X(o) = X_o$$

Temperature had no influence on the final biomass concentration within the range tested, but the growth rate increased with increasing temperature (Fig. 3).

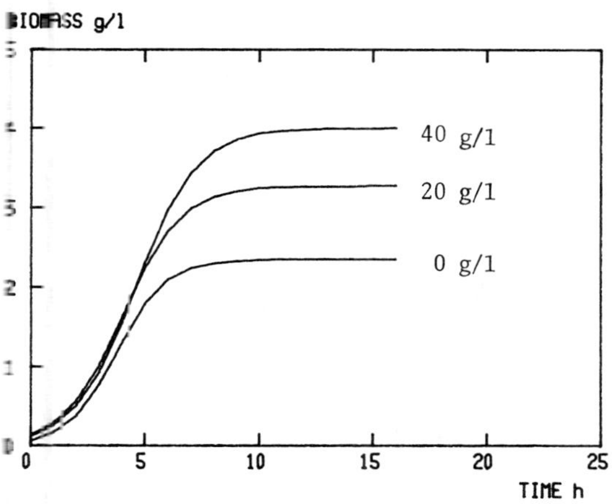

Fig. 1. The influence of initial glucose concentration on the growth of Escherichia coli at ph 7.3 and at 35 °C according to the model (1)

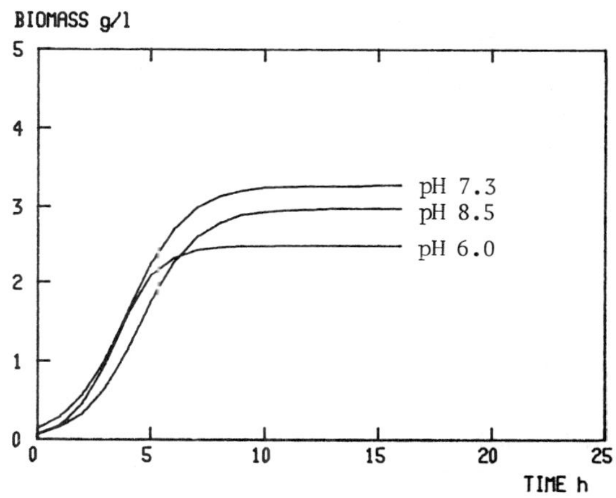

Fig. 2. The influence of pH on the growth of Escherichia coli at 35°C according to the model (1) (initial glucose concentration 20 g/l)

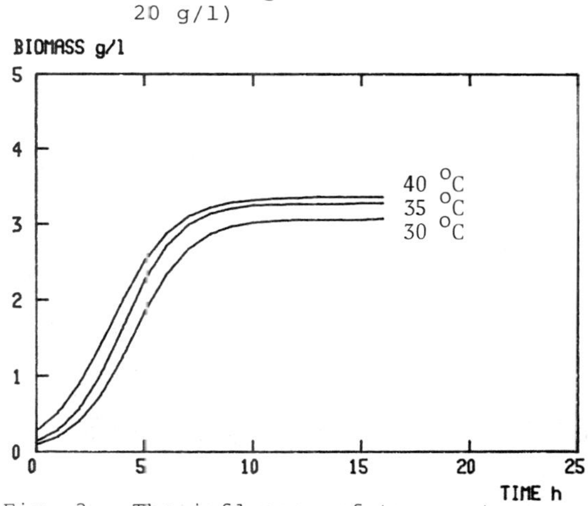

Fig. 3. The influence of temperature on the growth of Escherichia coli at pH 7.3 according to the model (1) (initial glucose concentration 20 g/l)

Static Regression Models

According to the regression analysis and equation (3), initial glucose concentration had no influence on the generation time of E. coli during the exponential phase of growth.

$$t_g(\text{min}) = 41.417 - 6.342Z_3 + 5.440Z_2 + 5.122Z_2^2 \qquad (3)$$

R = 0.94

where t_g is generation time during the exponential phase of growth, Z_1, Z_2 are coded values of independent variables, and R is the correlation coefficient of the regression.

Growth was fastest at pH 6.5...7.0 and at 40 °C and was considerably slower at higher pH and at lower temperatures. Under optimum conditions the generation time was less than 35 min (Fig. 4). Specific growth rate can be calculated from the values of generation time using equation (4).

$$\mu = (\ln 2)/(t_g) \qquad (4)$$

The regression model (5) suggested that temperature had no effect on the final biomass concentration:

$$X_f(g/l) = 3.224 + 0.582Z_1 + 0.328Z_1Z_2 - 0.269Z_2^2 + 0.170Z_2 \qquad (5)$$

R = 0.94

X_f is the final biomass concentration

At pH 8.0...8.5 the increase in initial glucose concentration increased the biomass yield almost linearly. At pH values near to 6, on the other hand, the biomass yield (~2.5 g/l) seemed to be independent of initial glucose concentration (Fig. 5). At low pH with low initial sugar concentration all the glucose was utilized, whereas with a high initial concentration some glucose remained when the final biomass concentration was reached. This suggests that under these conditions the medium had some undefined growth-limiting factor. Under optimum conditions the highest biomass concentration obtained was 4.3 g/l (Fig. 5).

Figure 6, based on the regression model (6), shows the effects of fermentation pH and initial glucose concentration on the time required for glucose utilization.

$$t(h) = 22.794 + 16.280Z_1 - 12.280Z_2 - 11.188Z_1Z_2 \qquad (6)$$

R = 0.84

where t is the time reguired for glucose utilization.

Glucose utilization was fastest at high pH (>8) and very slow below pH 7. Temperature also had some effect on the glucose utilization rate, but this was small compared with the effect of pH. Thus the glucose utilization model (6) based on regression shows a dependence only on pH and of course on the initial glucose concentration (Fig. 6.)

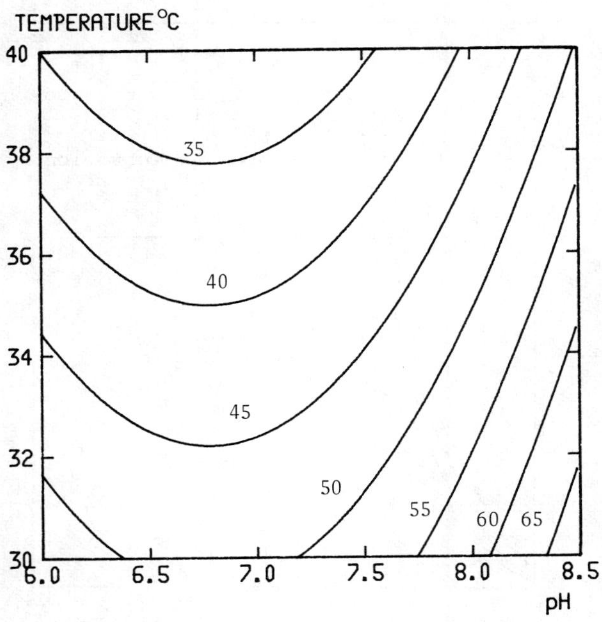

Fig. 4. Effects of fermentation temperature and pH on generation time (min) during exponential growth according to the regression model(3)

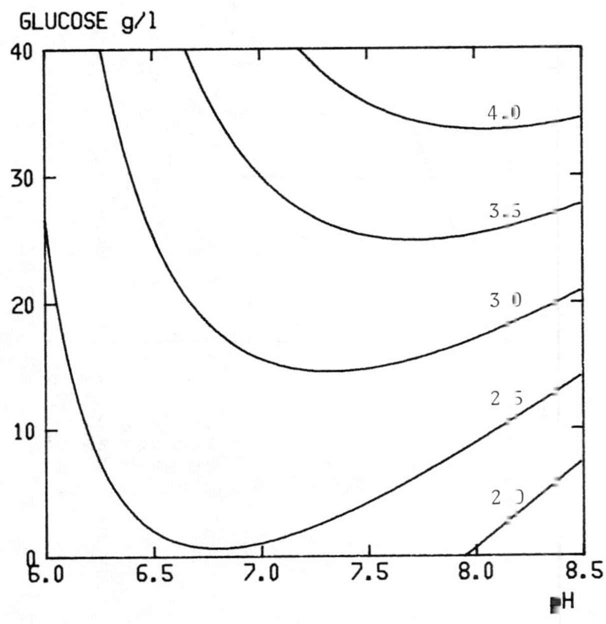

Fig. 5. Effects of pH and initial glucose concentration on the final biomass concentration (g/l) according to the regression model (5)

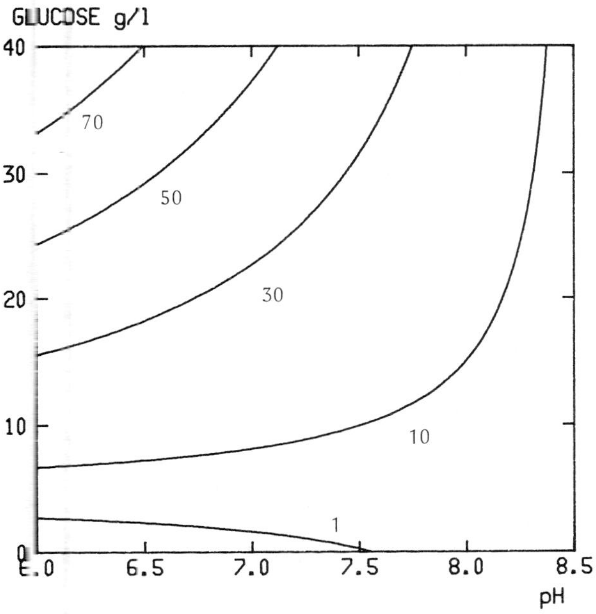

Fig. 6. Effects of pH and initial glucose concentration on the time (h) required for glucose utilization according to the regression model (6)

The regression model (7) of the final amount of acids shows strong dependence only on the initial glucose concentration (Fig. 7.).

$$\text{Acids} = 329.882 + 231.155Z_1 \quad (7)$$

$$R = 0.97$$

Figure 8 shows the effect of pH on both glucose utilization and the production of acids. The results can be explained by feedback inhibition caused by acids as pH decreases.

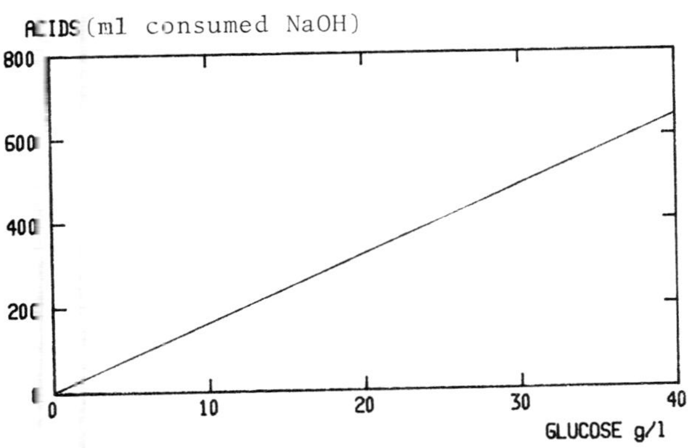

Fig. 7. Final amount of acids as a function of initial glucose concentration according to regression model (7)

Fermentations with Dynamic Control Profiles

The results from the static experiments were used to construct pH and temperature profiles for further fermentations. Figure 9 shows the results from an experiment in which the conditions were chosen to obtain rapid growth (T=40 °C, pH 6.0, Fig. 4) with minimum glucose consumption (Figs. 6,8). When the growth rate decreased (at 5 h), conditions were changed to obtain a high rate of glucose utilization and acids production (pH 8.5). According to Fig. 5 an initial glucose concentration of 20 g/l would have yielded a biomass concentration of 2.5...3.3 g/l in static experiments.The fermentation with the profile, however, yielded a biomass concentration exceeding 4 g/l. The final amount of acids was also considerably higher than in the static experiments. Subsequent profile fermentations supported these results.

Another approach to the construction of a profile was to use a "natural" pH-profile of *E. coli* growing on LB-broth. The profile was obtained by allowing the pH to change freely during an experiment in which there was no glucose in the medium. The pH changed from 7.0 to 8.5 during bacterial growth. This pH-profile was then imitated in an experiment in which the medium had an initial glucose concentration of 20 g/l (Fig. 10.). The results were similar to those obtained with a one-step pH-profile constructed on the basis of mathematical models (Fig. 9).

By comparing the measured and calculated biomass concentration values it was observed that the logistic growth model (1) predicted biomass concentration well in the different static fermentations. However, the model was inadequate in the case of fermentations with dynamic control profiles. Apparently the model does not contain enough information concerning process dynamics. However, it could be used together with static regression models for studying chosen characteristics of the process.

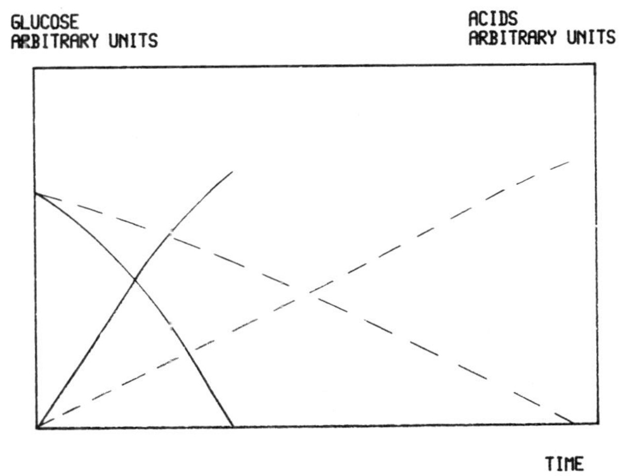

Fig. 8. General trends in glucose utilization and production of acids at low (---) and high (——) pH.

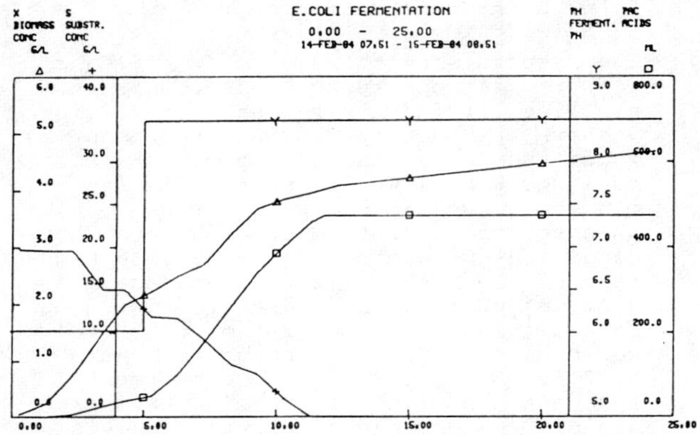

Fig. 9. Fermentation of Escherichia coli K-12 HB101 in a 10 l laboratory fermenter. Initial glucose concentration was 20 g/l, temperature 40 ^{o}C, pH-shift from 6 to 8.5 after 5 h growth. The generation time was 31 min during exponential growth.

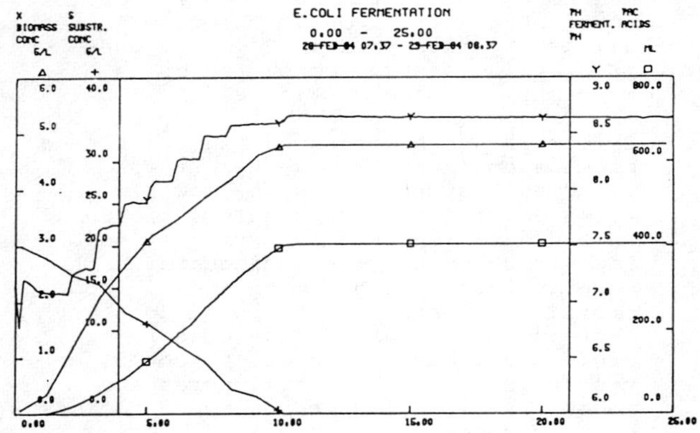

Fig. 10. Fermentation of Escherichia coli K-12 HB101 in a 10 l laboratory fermenter with dynamic pH-profile from 7.0 to 8.5. Initial glucose concentration was 20 g/l and temperature 37 ^{o}C. The generation time was 31 min during exponential growth.

CONCLUSIONS

The logistic glucose-, pH and temperature-dependant model for bacterial growth and static regression models obtained by employing response surface methodology, were found to be useful for studying an E. coli fermentation process. Good knowledge of the influence of initial glucose concentration, pH and temperature on the fermentation behaviour in static conditions was obtained. Regression models were used to determine the desired process conditions and they turned out to be useful for selecting control profiles. The results showed the importance of the investigation and employment of dynamic control profiles during fermentation processes. The use of suitable control profiles increased biomass and acids yields by 30...40 % compared with static fermentations. Mathematical modelling and dynamic optimization to determine optimal control strategy will present a major challenge for future research into fermentation processes.

REFERENCES

Bertani, G. (1951). Studies on lysogenesis J. of Bacteriol. 62, 293-300.

Box, B.E.P. and Wilson, K.B. (1951). On the experimental attainment of optimum conditions. J. Roy. Statist. Soc. 13, 1-45

Cheruy, A. and Durand, A. (1979). Optimization of erythromycin biosynthesis by controlling pH and temperature: Theoretical aspects and practical application. Biotechnol. Bioeng. Symp. No. 9, 303-320

Cochran, W.G. and Cox, G.M. (1957). Experimental designs 2nd ed., John Wiley & Sons Inc., New York

Edwards, V.H. and Wilke, C.R. (1968). Mathematical representation of batch culture data. Biotechnol. Bioeng. 10, 205-232.

Oinas, R. (1982). Application of computers to the monitoring, control and reporting of fermentation processes. Process Biochemistry, November/ December 1982, 10-11, 18, 36.

Olkku, J., Hagqvist, A. and Linko, P. (1983). Steady state modelling of HTST cooking extrusion employing response surface methodology. J. Food Eng. 2, 105-128.

Reinikainen, P., Markkanen, P., Holmberg A. (1984). Determination of optimal pH and temperature profiles for bacterial growth utilizing on-line computer control. In Third Congress on Biotechnology, Volume II, Verlag Chemie, Weinheim, 329-334.

OPTIMIZATION OF THE OPERATING VARIABLES OF AN INDUSTRIAL WASTE-WATER TREATMENT PROCESS

A. Johnson and M. Voetter

Department of Applied Physics, Delft University of Technology, Prins Bernhardlaan 6, 2628 BW Delft, The Netherlands

Abstract. The Complex Method of optimization has been applied to minimize the operating costs of a large (1200 m^3/hr effluent with an average of 1500 gCOD/m^3) chemical effluent processing plant. The process, an oxygen enriched activated sludge process, consists of 3 parallel trains of UNOX bioreactors, each of four compartments, followed by 3 settling tanks.

The five most important operating variables for a given recycle and waste sludge flowrate are the amount of gas vented to the atmosphere from the last compartment of each bioreactor train and the submersion depths of the electrically driven aerators in each bioreactor compartment. Using the cost prices of oxygen and electricity these five operating variables have been incorporated into a cost function which, subject to various constraints, must be minimized. The constraints are described by a steady-state, nonlinear model in which the concentrations of the dissolved oxygen in each compartment of each reactor are expressed in terms of the operating variables. The model parameters have been estimated from steady-state measurements.

The result of the optimization is a simple strategy, independent of any of the model parameters, which fixes the depths of submersion of the aerators. The amount of gas vented to the atmosphere can be calculated regularly offline by means of a numerical search procedure. Savings in the order of Hfl. 15,000 to Hfl. 30,000 per year can be predicted from this study.

Keywords. Waste-water treatment; activated sludge process; parameter estimation; modelling; static optimization; pure oxygen aeration.

INTRODUCTION

The waste-water treatment plant at Shell's Pernis refinery in Holland was commissioned in 1983 to process effluents from various parts of the refinery. It consists of three parallel trains of aerobic bioreactors, each of four enclosed compartments, followed by three settling tanks. To facilitate the aeration process, not air but pure oxygen (ca. 800 kg/hr) is fed to gas space of the first compartments (the so-called UNOX process). As the gas flows through each compartment, becoming more contaminated with carbon dioxide, four electrically driven surface aerators (ca. 350 KW total) bring the gas into contact with the waste-water and activated sludge in the compartments. The depth of submersion of each aerator is adjustable.

Since considerable costs (ca. Hfl. 10^6/year) are incurred in the aeration and supply of oxygen to the plant, it was decided to investigate how the operating variables should be chosen in order to minimize the costs. Per train, the five most important operating variables for a given recycle and waste sludge flowrate are the vent gas that is expelled from the last compartment to the atmosphere and the four submersion depths of the aerators. Using the cost prices of oxygen and electricity the five operating variables have been incorporated into a cost function which, subject to various constraints, must be minimized. This optimization problem has been solved by the Complex Method (Box, 1965; Beveridge and Schechter, 1970) as applied by Saguy (1982) to optimize other biological processes.

The most important constraint mentioned above, is that the dissolved oxygen (DO) in the reactor water of each compartment must be above a limiting value. To describe this restriction, a nonlinear steady-state model has been developed that relates the DO in each compartment to the operating variables. The activated sludge process (ASP) has been the subject of much study. In particular, for control purposes modelling the DO mechanism is important for two reasons; firstly since good control of the DO is important for the successful and economical operation of the bioreactor (Goodwin, 1982; Otho and co-workers, 1977) and secondly because using the DO an estimate can be made of some of the biological variables which are difficult to measure (Holmberg, 1981; Holmberg and Ranta, 1982; Marsili-Libelli, 1982; Ko, McInnis and Goodwin, 1982). The model presented here differs from those usually encountered in the literature

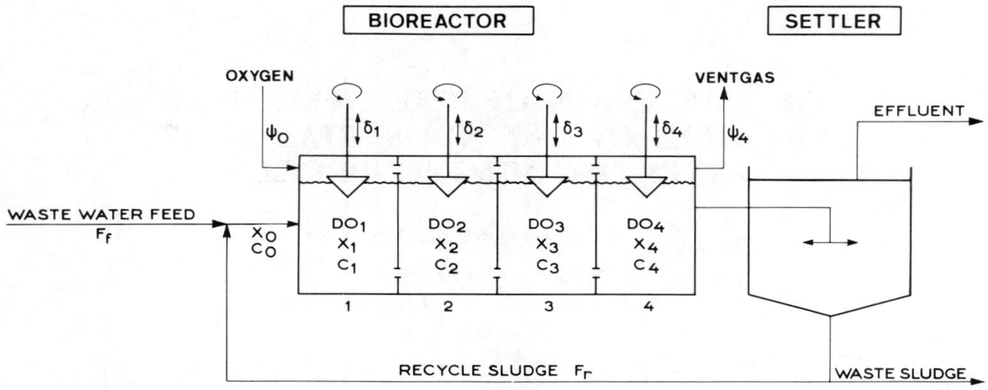

Fig. 1. Schematic diagram of the waste-water treatment process.

where air, and not oxygen, is supplied to the bioreactors.

The contents of this paper are as follows. Firstly, the cost criterion and constraints are formulated. Then a model is developed which allows the cost criterion and constraints to be described in terms of the operating variables. In the following section a series of steady-state measurements are used to estimate the parameters of the model. Finally, the numerical optimization technique and results are presented.

COST CRITERION AND CONSTRAINTS

After pretreatment the waste-water is pumped to one of three identical bioreactors (see Fig. 1). Note that contrary to normal ASP practice, the UNOX bioreactors are closed over on top. Pure oxygen is fed to the gas space of the first compartment; by means of a pressure controller in the oxygen feed line the gas pressure in the compartments is maintained a few centimeters watergauge above atmospheric pressure. The oxygen dissolves in the waste-water where aerobic decomposition of the waste takes place while carbon dioxide and nitrogen are released. This exchange of gases is accelerated by the electrically driven aerators. In the fourth compartment only about a quarter of the gas space is oxygen, this gas being vented through a control valve to atmosphere.

The process operating variables are

- the sludge recycle flowrate (controlled by a pump between 0 - 600 m^3/hr)
- the waste sludge flowrate (controlled by a valve between 0 - 30 m^3/hr)
- submersion depths of each aerator (20 cm range)
- the vent gas (controlled by a valve between 0 - 300 m^3/hr)

The numbers given are approximate values and refer to one of the three identical trains. For a given recycle and waste sludge flowrate the static optimization problem is expressed as

$$\underset{\psi_4,\delta_1,\delta_2,\delta_3,\delta_4}{\text{Min!}}\ J = \xi_{el}\cdot\{Q_1(\delta_1)+Q_2(\delta_2)+Q_3(\delta_3)+Q_4(\delta_4)\} + \xi_{O_2}\cdot\psi_O(\psi_4,\delta_1,\delta_2,\delta_3,\delta_4,\underline{v}) \quad (1)$$

$$\text{Subj.}\left\{\begin{matrix} DO_n(\psi_4,\delta_1,\delta_2,\delta_3,\delta_4,\underline{v}) \geqslant DO_{min} \\ \delta_{min} \leqslant \delta_n \leqslant \delta_{max} \end{matrix}\right\} n = 1,2,3,4$$

Here J is the positive cost function (Hfl/hr) that must be minimized in order to optimise the submersion depths δ_n(cm) and the vent gas flowrate ψ_4(m^3/hr). ξ_{el}(Hfl/kWh) and ξ_{O_2}(Hfl/m^3O_2) are the costs of electricity and oxygen, Q_n(kW) is the power consumed by the n-th aerator and ψ_O(m^3/hr) the oxygen supply rate. The constraints ensure that the DO(kgO_2/m^3) is above a certain limiting value DO_{min} (the process is aerobic), and that the aerators are not submerged beyond or extended above their allowable range. In the vector $\underline{v} \triangleq (F_f,F_r,C_O,C_1,C_2,C_3,C_4,X_O,X_1,X_2,X_3,X_4)$ are collected together those variables which we cannot (or will not) manipulate such as the feed flowrate of waste water F_f(m^3/hr), the sludge recycle flowrate F_r(m^3/hr) and the substrate and biomass concentrations C_n(kgCOD/l) and X_n(kgVSS/l) in each compartment n respectively. Index 0 refers to a stream entering the bioreactor.

The functions $Q_n(\delta_n)$, n = 1,2,3,4 are unknown; how they are estimated is described later. ψ_O and DO_n have been calculated using a nonlinear steady-state model developed in the next section. Since the DO concentration is usually kept above 4 g/m^3 in the UNOX process, DO_{min} is taken to be 4 g/m^3. Steady-state measurements of F_f, F_r, C_O, C_4 and X_O are available; the other variables of $\underline{v}$ have been estimated.

As can be seen from the cost criterion the biological activity is assumed to be independent of the DO concentration. This is justified since the DO is held far above the limiting value, 1-2 g/m^3 (Olsson and Hansson, 1976).

The optimization problem just outlined will be treated henceforth as a static optimization problem even though an ASP is a dynamic system frequently perturbed by variations in input flow and concentrations. This is possible because of the damping action of two large (time constant ca. 10 hr.) equalisation tanks present in the pretreatment section (not described here).

THE MODEL

In this section a model is developed to describe the gas exchange in the compartments of one bioreactor. Assuming perfect mixing in both the gas and liquid phase in every compartment the following balances hold for the gas phase of compartment n

$$\rho_{O_2}\cdot(\phi_{n-1}\cdot P_{O_2,n-1} - \phi_n\cdot P_{O_2,n}) - V\cdot OTR_n = 0 \quad (2)$$

$$\rho_{CO_2}\cdot(\phi_{n-1}\cdot P_{CO_2,n-1} - \phi_n\cdot P_{CO_2,n}) - V\cdot CTR_n = 0 \quad (3)$$

and for the liquid phase of the same compartment

$$(F_f+F_r)\cdot(DO_{n-1}-DO_n) + V\cdot OTR_n - V\cdot OUR_n = 0 \quad (4)$$

$$(F_f+F_r)\cdot(DC_{n-1}-DC_n) + V\cdot CTR_n + V\cdot CER_n = 0 \quad (5)$$

Here ρ_{O_2} and ρ_{CO_2}(kg/m^{3}atm.) are the specific gravity of O_2 and CO_2 at reactor temperature, ϕ_n(m^3/hr) is the gas flowrate from compartment n to n+1, $P_{gas,n}$ (atm.) the partial pressure of a component of the gas mixture, V(m^3) the liquid volume in each compartment, OTR_n(kgO_2/m^{3}hr) is the oxygen transfer rate and CTR_n(kgCO_2/m^{3}hr) the carbon transfer rate from gas to liquid in compartment n. Furthermore, DC_n(kgCO_2/m^3) is the total dissolved CO_2, comprising of aqueous CO_2, HCO_3^- and $CO_3^=$ ions, OUR_n(kgO_2/m^{3}hr) the oxygen uptake rate and CER_n(kgCO_2/m^{3}hr) the carbon evolution rate. The mass balances, eqns. (2)-(5) include a convective term related to the convective mass transport, a term accounting for gas dissolving or evolving from the liquid and (in equations (4) and (5)) a term to describe the reaction taking place.

Moreover, applying Dalton's Law to the gas phase of each compartment gives

$$P_{O_2,n} + P_{CO_2,n} + P_{N_2} + P_{H_2O} = P_{Reactor} \quad (6)$$

Since no denitrification takes place, the only source of nitrogen gas in the system is the dissolved nitrogen in the waste-water feed, which we assume to be in equilibrium with the nitrogen in the atmosphere. Therefore, P_{N_2} will be small (±0.05 atm.) with only minor variations. It is assumed that P_{N_2} has the same value in each compartment. The last two terms of equation (6) are also constant since the reactor temperature and pressure are constant. OTR is given by

$$OTR_n = kl_{O_2,n}a_n\cdot(KH_{O_2}\cdot P_{O_2,n} - DO_n) \quad (7)$$

where $kl_{O_2,n}$ is the mass transfer coefficient (m/hr), a_n is the specific surface area (m^{-1}) and KH_{O_2} is Henry's coefficient (kg/m^{3}atm.). Both $kl_{O_2,n}$ and a_n depend on the submersion depth of the n-th aerator. We assume here that the following relationship holds

$$kl_{O_2,n}a_n = \alpha_{O_2}\cdot K_{O_2}\cdot Q_n(\delta_n) \quad (8)$$

where K_{O_2}, the aerator constant for oxygen (1/kWh), is calculated from manufacturer's data. The dimensionless constant α_{O_2} is included because it has been observed that in practice the rate that gases dissolve depends strongly upon the composition of the waste-water. α_{O_2} takes values between 0.4 - 0.8 for industrial waste-water (Adams and co-workers, 1978); for pure water α_{O_2} = 1. Due to the uncertainty surrounding α_{O_2}, this parameter will be one of those we will have to estimate from measurements of the process variables.

The situation regarding the CTR is similar to that of the OTR. However, because carbon dioxide dissociates (depending upon the pH) in water, a complication arises. Only the non-dissociated CO_2 interacts with the gas phase so that

$$CTR_n = \alpha_{CO_2}\cdot K_{CO_2}\cdot Q_n(\delta_r)\cdot(KH_{CO_2}\cdot P_{CO_2} - \Omega\cdot DC_n) \quad (9)$$

Here, Ω is the fraction of non-dissociated dissolved CO_2, and we assume one value for all compartments (Ω = 0.8 for pH = 5.5). α_{CO_2} is an unknown parameter that will be estimated. We discuss KH_{CO_2} in the next chapter.

Component balances may be written for OUR_n and CER_n:

$$OUR_n\cdot V = (F_f+F_r)\{(C_{n-1}-C_n) + 1.56(X_{n-1}-X_n)\} \quad (10)$$

$$CER_n\cdot V = (F_f+F_r)\{1.03(C_{n-1}-C_n) + 1.80(X_{n-1}-X_n)\} \quad (11)$$

The numerical coefficients may be derived from the (assumed) biomass and substrate compositions, and are assumed to hold for all compartments.

The flow into the first compartment is a mixture of fresh waste-water containing no dissolved CO_2 or O_2, and recycled active sludge. Since there is no aeration in the settling tank, the dissolved oxygen in the water leaving the bioreactor will be consumed very quickly; furthermore we assume that its carbon dioxide concentration remains unaltered in the settling tank. With these assumptions we have that

$$DO_0 = 0 \quad (12)$$

$$DC_0 = \frac{F_r}{F_r+F_f}\cdot DC_4 \quad (13)$$

Finally, since only pure oxygen is fed to the first compartment

$$P_{O_2,0} = P_{Reactor} \tag{14}$$

$$P_{CO_2,0} = 0 \tag{15}$$

The model describing the dissolved oxygen and carbon dioxide concentrations and the composition of the gas phase is given by equations (2) - (15), with n = 1,2,3,4.

PARAMETER ESTIMATION

The following quantities must be estimated before the model can be used:

- the functions $Q_n(\delta_n)$, n = 1,2,3,4
- the variables collected together in $\underline{v}$
- the unknown parameters α_{O_2}, α_{CO_2}, KH_{CO_2}

Tables 1 and 2 show typical measurements taken from the process. $Q_n(\delta_n)$ are found by linear interpolation of the data of Table 1. Concerning $\underline{v}$, only F_f, F_r, X_0, C_0 and C_4 can be measured directly; the remaining variables are computed using the following model, based on Monod kinetics:

$$\frac{F_f+F_r}{V}\cdot(C_{n-1}-C_n) = \frac{X_n}{Y}\cdot G\cdot(C_n-C^-) \tag{16}$$

$$\frac{F_f+F_r}{V}\cdot(X_{n-1}-X_n) = X_n\cdot\{G\cdot(C_n-C^-)-\mu_s\} \tag{17}$$

$$\} \; n = 1,2,3,4$$

The non-decomposable part, C^-(kgCOD/m^3), of the substrate is assumed to be equal to C_4 (which implies that the bioreaction is almost completed), an assumption which has been verified experimentally.

TABLE 1. Measured power consumption of the aerators.

Submersion depth δ	Power consumption Q	
	Compartment 1	Compartment 2,3,4
0 cm (middle position)	41 kW	33 kW
-10 cm (extended position)	24 kW	22 kW

TABLE 2. Measurements taken from the process.

DO_1	5.2±1.5	g/m^3	δ_1	0	cm
DO_4	3.0±1.5	g/m^3	δ_2	-4	cm
F_f	400	m^3/hr	δ_3	-10	cm
F_r	400	m^3/hr	δ_4	-10	cm
X_0	6500	gVSS/m^3	ϕ_0	216	m^3/hr
C_0	805	gCOD/m^3	ϕ_4	60	m^3/hr
C_4	360	gCOD/m^3	$pH_{reactor}$	5.5	
$P_{O_2,4}$	0.25	atm.	$T_{reactor}$	30	°C

TABLE 3. Estimation of X and C.

Parameters: G = 0.20 m^3/kgCODhr
Y = 0.40 kgVSS/kgCOD
μ_s = 2.9·10^{-3} hr^{-1}

Compartment	0	1	2	3	4
X (gVSS/m^3)	6500	6619	6623	6605	6582
C (gCOD/m^3)	805	447	377	363	351

Values for the growth constant G (m^3/kgCODhr) yield Y (kgVSS/kgCOD) and the term for endogene respiration and cell decay μ_s (hr^{-1}) are given in Table 3. It should be explained that it is possible to derive two additional criteria for the biomass and substrate concentrations (and thus for G, Y and μ_s) from our model developed in the previous section.

Firstly, by adding equation (10) for each compartment gives

$$OUR_{tot}\cdot V = (F_f+F_r)\cdot\{(C_0-C_4) + 1.56(X_0- X_4)\} \tag{18}$$

where $OUR_{tot}\cdot V$, the total oxygen uptake, can be found from a mass balance over the whole of the reactor to be 253 kg/hr, based on data from Table 2. This implies X_4 = 6582 gVSS/m^3.

Secondly, we consider how the biological activity is distributed between the four compartments, which has a strong influence on the DO concentrations. As the simulated DO_1 and DO_4 have to match the measured data, this leads to another constraint on the permissible values for G, Y and μ_s; the data of Table 3, which are based on earlier investigations (Stougie, 1983), do not infringe this or the previous constraint.

Returning now to α_{O_2}, α_{CO_2} and KH_{CO_2}, these parameters are estimated using the model of the previous section. Table 5 shows the values obtained, together with a sensitivity analysis (see Table 4 for the data used).

TABLE 4. Constants used in the simulations.

Constant	Value	Remarks
KH_{O_2}	0.035 kg/m^{3}atm.	} pure water; literature
KH_{CO_2}	1.2 kg/m^{3}atm.	
ρ_{O_2}	1.28 kg/m^{3}atm.	at 30°C
ρ_{CO_2}	1.76 kg/m^{3}atm.	at 30°C
$P_{reactor}$	1.0 atm.	Atmospheric pressure
P_{H_2O}	0.05 atm.	at 30°C
P_{N_2}	0.05 atm.	calculated from N_2-balance
K_{O_2}	0.294 kWh^{-1}	} from manufacturer's data
K_{CO_2}	0.262 kWh^{-1}	
V	1000 m^3	

TABLE 5. Simulations of the process.

Operating variables: δ_1 = 0 cm; δ_2 = -4 cm; $\delta_3 = \delta_4$ = -10 cm; Φ_4 = 60 m^3/hr								
No.	α_{O_2}	α_{CO_2}	KH_{CO_2} kg/m^3/atm.	DO_1 g/m^3	DO_4 g/m^3	$P_{O_2,4}$ atm.	Φ_o m^3/hr
1	0.8	0.8	0.56	5.2	3.3	0.25	215
2	1.0	0.8	0.56	8.0	4.4	0.25	216
3	0.8	1.0	0.56	5.2	3.3	0.25	215
4	0.8	0.8	0.70	6.4	6.7	0.35	223
5	0.8	0.8	1.20	9.0	13.3	0.55	239

Some remarks concerning Table 5 follow. Firstly, we see that our estimated KH_{CO_2} deviates considerably from the values often quoted in the literature, probably because of the composition of the waste-water. Secondly, it is apparent that the model is very insensitive to variations in α_{CO_2}. This can be explained since the dissolved and gaseous CO_2 are almost in equilibrium with each other in every compartment. Because of the insensitivity, α_{CO_2} is assumed for our purposes to equal α_{O_2}. Finally we remark that it would be expected that KH_{O_2}, just like KH_{CO_2}, would deviate from its normal value for pure water. However, it is not possible to also estimate KH_{O_2} from the available measurements.

OPTIMIZATION

The 5-dimensional optimization problem, equation (1) is now completely defined. It is solved using the numerical technique known as the Complex Method (Box, 1965). The parameters used are those appearing in Table 5, no. 1; additional constants are as given below in Table 6.

TABLE 6. Constants used with optimization.

Constant	Value
ξ_{el}	0.137 Hfl/kWh
ξ_{O_2}	0.101 Hfl/m^3
DO_{min}	4 g/m^3
δ_{min}	-10 cm
δ_{max}	10 cm

Table 7 shows the results of the optimization, together with the effects of suboptimal operating variables. The DO concentration in each compartment is only influenced by the vent gas flowrate and the submersion depth of the aerator in that compartment. It is therefore possible to state that for the minimum cost either the aerator submersion depth or the DO must be minimal in every compartment, i.e.

$$(DO_n - DO_{min}) \cdot (\delta_n - \delta_{min}) = 0 , \quad n = 1,2,3,4 \qquad (19)$$

TABLE 7. Simulations with optimal (no. 6) and sub-optimal (no. 7-11) operating variables.

No.	Operating variables δ_1 (cm)	δ_2 (cm)	δ_3 (cm)	δ_4 (cm)	Φ_4 m^3/hr	COST J Hfl/hr
6	-2.6	-10	-10	-10	67	36.12
7	-1.6	-10	-10	-10	67	36.34
8	-2.6	-9	-10	-10	67	36.27
9	-2.6	-10	-9	-10	67	36.27
10	-2.6	-10	-10	-9	67	36.29
11	-2.6	-10	-10	-10	75	36.53

No.	DO_1 g/m^3	DO_2 g/m^3	DO_3 g/m^3	DO_4 g/m^3	Φ_0 m^3/hr	$P_{O_2,4}$ atm.
6	4.0	6.3	6.7	4.0	219	0.27
7	4.6	6.4	6.7	4.0	219	0.27
8	4.0	6.7	6.7	4.0	219	0.27
9	4.0	6.3	6.9	4.0	219	0.27
10	4.0	6.3	6.7	4.2	219	0.27
11	4.3	7.6	7.2	4.7	223	0.30

These four extra equations lead to two important simplifications when practical application of the optimization is envisaged. Firstly, the optimization problem is reduced from 5 to 1 dimension. This would permit a very efficient one dimensional optimization algorithm to be used (Wilde and Beightler, 1967). Then secondly, the selection of a setpoint for control of the aerator submersion depths at their optimal values is extremely simple. A problem is that the DO concentration is not measured in every compartment.

The optimal vent gas flowrate cannot be expressed in a criterium such as equation (19). Fig. 2 shows the total costs as a function of the vent gas flow while all the aerators are submerged to their optimal depths. The edged minimum of the figure can be explained by refering to Figs. 3 and 4, showing the corresponding submersion depths and DO concentrations.

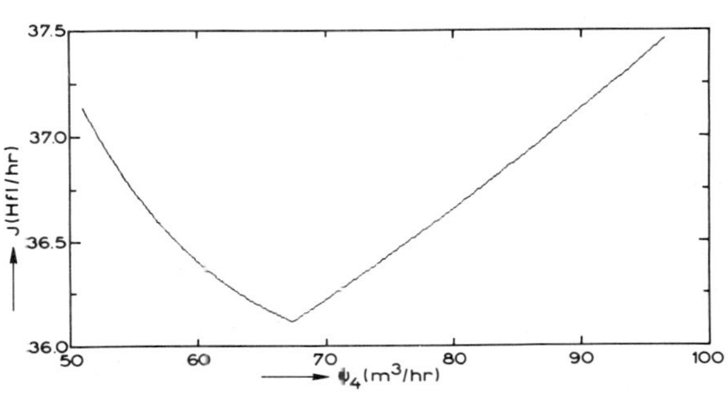

Fig. 2. Total costs vs. vent gas flow, with optimally submerged aerators.

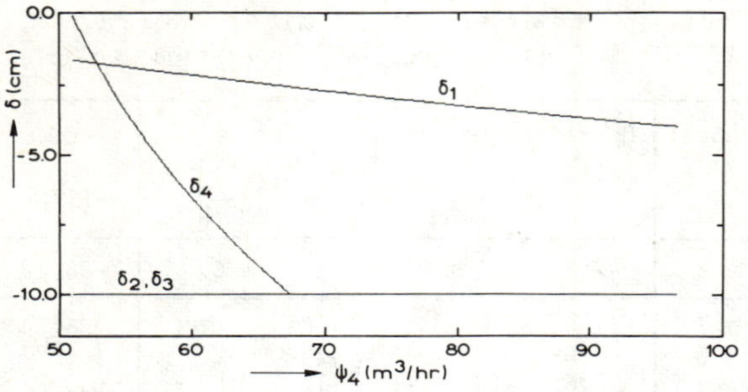

Fig. 3. Optimal submersion depths vs. vent gas flow.

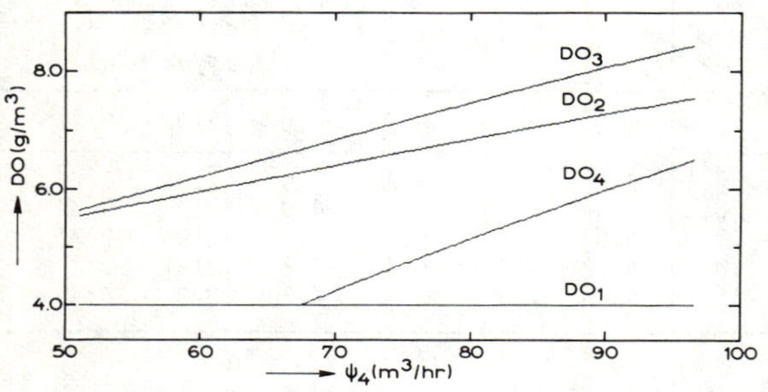

Fig. 4. DO concentrations vs. vent gas flow, with optimally submerged aerators.

It can be seen that at ventgas flowrates lower than the optimum the constraint on DO_4 is active, and at flowrates higher than the optimum that on δ_4. A consequence of this change is the edge; the optimum lies in this case against 5 constraints. Of course, a higher or lower bioreactor loading could alter the position of the optimum point in Fig. 2 to the smooth area, where only 4 constraints are active.

CONCLUSIONS

The Complex Method of numerical optimization has been used to minimize the aeration costs of an oxygen enriched industrial waste-water treatment process. Savings are estimated to be in the order of 15,000 - 30,000 Hfl/year.

Sensitivity analysis suggests that there is a simple strategy to ensure the optimal value for four of the five operating variables (the submersion depths of the aerators). This strategy is to hold the DO at its minimal allowable value in every compartment; an alternative to measuring the DO in every compartment is to estimate its value from other measurements.

The optimal value of the remaining operating variable, the gas vented to atmosphere from the last compartment, is closely related to the operating state and parameters of the bioreactor.

This research is being continued with the aim of implementing on-line the optimization described in this paper. In particular, more accurate and robust parameter estimation techniques are being investigated.

ACKNOWLEDGEMENT

The authors wish to acknowledge the cooperation of Shell Nederland B.V. in this research and their permission to publish the results contained in this paper.

REFERENCES

Adams, C.E, W.W. Eckenfelder, J.H. Koon, and S. E. Shelby (1978). Oxygen activated sludge considerations for industrial applications. AIChE Symposium series, 74, 178, 93-101.

Beveridge, G.S.G., and R.S. Schechter (1970). Optimization: theory and practice. McGraw-Hill, New York.

Box, M.J. (1965). A new method of constrained optimization and a comparison with other methods. Computer J., 8, 42-52.

Goodwin, G.C. (1982). Adaptive control and identification of the dissolved oxygen process. Automatica, 18, 181-193.

Holmberg, A. (1981). Microprocessor based estimation of oxygen utilization in the activated sludge process. International Journal of Systems Science, 12, 703-718.

Holmberg, A., and J. Ranta (1982). Procedures for parameter and state estimation of microbial growth process models. Automatica, 18, 181-193.

Ko, K.Y., B.C. McInnis, G.C. and Goodwin (1982). Adaptive control and identification of the dissolved oxygen process. Automatica, 18, 727-730.

Marsili-Libelli, S. (1982). On-line estimation of bioactivities in activated sludge processes. Proceedings 1st IFAC workshop on modelling and control of biotechnical processes, Helsinki, 121-126.

Olsson, G., and O. Hansson (1976). Modelling and identification of an activated sludge process. Preprints 4th IFAC workshop on identification and system parameter estimation, Tbilisi, USSR, 134-136.

Otho, T., S. Nogita, H. Iwaki, and S. Kato (1977). Experiences of dissolved oxygen control of a diffused air aeration plant. Progress in Water Technology, 9, 399-403.

Saguy, I. (1982). Utilization of the Complex Method to optimize a fermentation process. Biotech. and Bioeng., 24, 1519-1525.

Stougie, R. (1983). Modelling an industrial activated sludge process (in Dutch). Degree thesis, Department of Applied Physics, Delft University of Technology.

Wilde, D.J., C.S. Beightler (1967). Foundations of optimization. Prentice Hall, Englewood Cliffs, New York.

FUZZY CONTROL APPLICATION TO GLUTAMIC ACID FERMENTATION

T. Nakamura, T. Kuratani and Y. Morita

Central Research Laboratories, Ajinomoto Co., Inc., 1–1 Suzuki-cho, Kawasaki-ku, Kawasaki 210, Japan

ABSTRACT

Consumption rate of ammonia in glutamic acid fermentation was found to be correlated with consumption rate of sugar. This gives a good control strategy for sugar feed rate to keep sugar concentration in a suitable range in fed-batch culture of glutamic acid fermentation.

Slight changes of culture conditions, however, make some troubles in this control scheme. Fuzzy theory application could be found to be effective in such cases.

KEYWORDS

ammonia consumption, sugar feed control, fuzzy control, fuzzy theory, fuzzy reasoning, regression coefficient, emergent sugar feed

INTRODUCTION

In a fed-batch culture of glutamic acid fermentation, sugar concentration must be within a suitable range. At present, the on-line measurement of sugar concentration is considered to be impossible, so operators must analyze the sugar concentration of a sample taken at an appropriate interval and determine the sugar feed rate.

K. T. Hong, et al. (1984)[1)] automatically controlled sugar feed rate by using oxygen uptake rate. However this method is difficult to apply to commercial plants, because a gas analyzer troubles operators with its maintenance.

Therefore, the consumption of ammonia used for nitrogen source and pH control, which was easy to measure on-line, was applied to determination of the sugar feed rate. Most cultivations worked well by the above method. However, when sugar feedstocks were exchanged or activity of microorganism inoculated were lower or higher than usual, sugar concentration deviated from the desired range and operators had to cope with these troubles using their intuitions and experiences.

Generally speaking, the fuzzy control is suitable for processes operated with skilled operators' intuitions and experiences. Recently, the examples with fuzzy theory applied to processes like this have been increasing. It was applied to the rotary cement kiln control by F. L. Smith Co., Ltd. (1981)[2),3)]. It was also applied to controll the amount of chemicals fed to a cleaning bed in the water purification process by Fuji Electric Co., Ltd. and FUJIFACOM Corp. (1984)[4)]. Hitachi, Ltd. controlled train operations by applying the predictive fuzzy theory and succeeded to make trains operated better than PID control (1983)[5)]. However, fuzzy control has never been applied to fermentation processes which have been operated with operators' intuitions and experiences.

In the Central Research Laboratories of AJINOMOTO CO., INC., the first application of fuzzy theory was successfully made for the sugar feed control process.

In this paper, firstly, the determination method of sugar feed rate by using ammonia consumption will be explained. Secondly, fuzzy control strategy used here will be mentioned. Thirdly, the sugar feed control system combined with both methods and the results of field tests will be shown.

DECISION OF SUGAR FEED RATE BY USING OF THE AMOUNT OF AMMONIA CONSUMPTION

The theoretical equation of mass balance of glutamic acid fermentation is as follows:

$$2C_6H_{12}O_6 + 3O_2 + 2NH_3 \longrightarrow 2C_5H_9O_4N + 2CO_2 + 6H_2O \qquad (1)$$

Methods

The cultivations are carried out in a 1 kl bench plant. An industrial strain of Brevibacterium sp. is used, and cane molasses as carbon source. Other conditions for cultivations are the same as those of commercial plant. The method of cultivations is as follows: Firstly, a batch culture is started from certain concentrations of cell mass and sugar. Secondly, surfactant is added when cell grows to a suitable concentration. Thirdly, when the sugar concentration decreases to a value set in advance, fed-batch culture is begun. The pH is maintained at 7.8 by automatic feeding of gaseous ammonia. The temperature is maintained at 35°C. Sugar feed rate during the fed-batch culture is controlled by an operator, according to the result of sugar analysis sampled every two hours. Sugar concentration is analyzed by a HPLC (Hitachi, Ltd., JAPAN). The amount of ammonia consumption is measured with a vortex flowmeter (Oval Engineering Co., Ltd., JAPAN), and that of fed sugar is with electromagnetic flowmeter (Oval Engineering Co., Ltd.). Dissolved oxygen (D.O.) is measured with a D.O. meter(DKK,JAPAN). Furthermore, oxygen and carbon dioxide in the exit gas are analyzed by a paramagnetic oxygen analyzer (Fuji Electric Co., Ltd., JAPAN) and an infrared gas analyzer (Fuji Electric Co., Ltd.), reaspectively. The result obtained from the amount of ammonia consumption is compared with those from the amounts of oxygen uptake and carbon dioxide evolution.

Correlation between the Amounts of Sugar Consumption and Ammonia Consumption

The correlation between the amounts of sugar

consumption(ΣS_t) and ammonia consumption(ΣN_t) was examined. The result is shown in Fig. 1. Eq.(2) shows its regression equation, which fits well after surfactant addition. The correlation coefficient of eq.(2) after surfactant addition is 0.994, which is almost the same as that obtained between the amounts of sugar consumption and oxygen uptake or that between the amounts of sugar consumption and carbon dioxide evolution.

$$\Sigma S_t = 8.0 * \Sigma N_t + 20.0 \qquad (2)$$

Sugar Feed Strategy by Using of the Amount of Ammonia Consumption

The amount of sugar consumption is able to be estimated by measuring on-line the amount of ammonia consumption using eq.(2). Sugar feed rate(F_t) in a period, from t to t+Δt, can be shown in eq.(4), by solving eq.(3) assuming that the sugar concentration at t equals that at t+Δt.

$$S_t/V_t = (S_t + S_0 * F_t * \Delta t - 8.0*(\Sigma N_{t+\Delta t} - \Sigma N_t)) / (V_t + F_t * \Delta t) \qquad (3)$$

$$F_t = (8.0 *(\Sigma N_{t+\Delta t} - \Sigma N_t)) / ((S_0 - S_t/V_t) * \Delta t) \qquad (4)$$

where S_t is the amount of sugar in fermentor at t, ΣN_t that of ammonia consumed by t, V_t the culture volume, and S_0 the sugar concentration of feedstock. In eq.(3), the volume of anti-foamer added and water produced during the reaction(1) are assumed to be negligible. Therefore, by giving the sugar concentration and culture volume at the starting point of sugar feed control and by assuming $F_{t+\Delta t} = F_t$, sugar feed rate can be calculated every Δt on the basis of eq.(4). As a result, sugar concentration is in a suitable range.

APPLICATION OF THE FUZZY THEORY TO SUGAR FEED CONTROL

Usually, when sugar concentration almost exceeded the upper or lower limit of the suitable range, operators had to cope with the troubles using their intuitions and experiences on the basis of those informations, such as D.O., sugar feed rate, and the time until D.O. sharply varied. The operations at this situation are as follows:
(a) If sugar concentration exceeded the upper limit in the suitable range, sugar feed rate was set to a slightly lower level than before. If it rapidly approached the lower limit in the range, the rate was set to a slightly higher level than before. (These mean modifying the regression coefficient.)
(b) The sudden change in sugar feed rate was temporally carried out in order to put sugar concentration back to the middle of the range. (This is an emergent sugar feed.)

Some operations based on operators' intuitions and experiences work better than that based on mathematical equations. Therefore, fuzzy theory was applied to make the operations (a) and (b) practical, when the sugar concentration deviated from the range.

Reasoning based on the Fuzzy Theory

The details of the fuzzy reasoning used here are described as follows:
The empirical rules of the modification are assumed to be two, as shown below.

R1: if t = "NORMAL" and
d = "BIG" and
l = "HIGH"
then Δf = "PM"

R2: if t = "LONG" and
d = "BIG" and
l = "HIGH"
then Δf = "PS"

where t, d, l, and Δf mean the time elapsed from surfactant addition, increase rate of D.O., D.O. variation, and the proportion of modification of the regression coefficient in rules, respectively, and NORMAL, LONG, BIG, HIGH, PM(Positive Medium), PS(Positive Small) are fuzzy variables.

When t = t_0, d = d_0, and l = l_0 are observed, PM*(Δf) of R1 is calculated according to eq.(5):

$$PM^*(\Delta f) = (LONG(t_0) \wedge BIG(d_0) \wedge HIGH(l_0)) * PM(\Delta f) \qquad (5)$$

where PM(Δf), LONG(t), BIG(d), and HIGH(l), are membership functions of fuzzy sets, PM, LONG, BIG, and HIGH, respectively, and "$\wedge$" in A $\wedge$ B shows the smaller of A and B. Furthermore, (LONG(t_0) $\wedge$ BIG (d_0) $\wedge$ HIGH(l_0)) shows a satisfactory factor of R1 for observed variables. This calcuration is also carried out for R2, and DF(Δf) is calculated according to eq.(6):

$$DF(\Delta f) = PM^*(\Delta f) + PS^*(\Delta f) \qquad (6)$$

By the use of DF(Δf), the proportion of the modification of the regression coefficient(Δf_0), is determined as below (Cf. Fig. 2):

$$\Delta f_0 = \int \Delta f * DF(\Delta f)\, d(\Delta f) \Big/ \int DF(\Delta f)\, d(\Delta f) \qquad (7)$$

However, Δf_0 is set to zero on the condition that both PM*(Δf)/PM(Δf) and PS*(Δf)/PS(Δf) become below an appropriate threshold.

On actual operations, Δf_0 is calculated on the basis of rules concerning with the modification of the regression coefficient. However, if satisfactory factors of all empirical rules for observed variables become below an appropriate threshold, then Δf_0 is set to zero and the modification and an emergent sugar feed are not carried out. The reasoning like this prevents frequent modifications of the regression coeficient and the unrequired emergent sugar feed, which realize securer oprations. However, if Δf_0 is not zero, then the regression coefficient is modified and an emergent sugar feed is carried out after reasoning based on the emergent operation rules.

Fuzzy Sets and Operation Rules

Rules and membership functions were made up from the informations gathered from skilled operators. TABLE 1 shows fuzzy variables used in rules Some of membership functions are shown in Fig 2. In the present system, the numbers of rules for modifying the regression coefficient are four except R1 and R2, and those for emergent sugar feed are fourteen.

Furthermore, a convenient conversational typed editor, which was easy to input rules and membership functions, was prepared. As a result, operators could easily add rules and membership functions.

SUGAR FEED CONTROL SYSTEM

Hardware. Fig. 3 shows the hardware construction of the sugar feed control system. The amount of ammonia consumption is measured with a vortex flowmeter and taken into an industrial microcomputer YEWMAC 200 (YEW, JAPAN) through A/D converter. D.O. is measured with a D.O. meter, and also taken into YEWMAC 200. The setting value of sugar feed rate calculated with YEWMAC 200 is transferred to FIC at sugar feed line through D/A converter. The starting signal of culture, that of sugar feed, and the stop signal of culture are taken into YEWMAC 200 through D/I. YEWMAC 200 has a CRT for supervision, a printer for recording of operation conditions, and a floppy disk drive for copying rules and membership functions made with the editor of other microcomputer.

Software. Fig. 4 shows the algorism of sugar feed control. When cell mass concentration, X, reaches the setting value, X_{SET}, surfactant is added. After this operation, when sugar concentration, S, reaches the setting value, S_{SET}, sugar feed control is set on work. The amount of sugar consumption is calculated, from second to second, by using ammonia feed rate taken in through A/D converter. D.O. signals which are noisy, are smoothed out by means of moving average.

The amount of ammonia consumption and smoothed D.O. are memorized every Δt, and the sugar concentration is checked with the fuzzy reasoning. If the concentration is detected out of the range, the regression coefficient of eq.(2) is modified, and at the same time an emergent sugar feed is done. If not, the sugar feed rate is calculated from eq.(4) by the use of the amount of ammonia consumption within Δt, and the value is transferred to FIC.

FIELD TESTS

Field tests were carried out in a 1 kl bench plant. Whatever lot was used as sugar feedstocks, sugar concentration was in a suitable range. Most of the cultivations were controlled using only the amount of ammonia consumption. But, in a few cultivations fuzzy reasonings were carried out and the modifications of the regression coefficient and emergent sugar feeds were done. Fig. 5 shows the result of one of the latter cases. When the culture time became 16.5 hours, the sugar concentration decreased to the lower limit of the range and then fuzzy reasoning was carried out by using increase rate of D.O. etc. After the regression coefficient was modified and an emergent sugar feed was done by the reasoning, sugar concentration was controlled within the range.

CONCLUSIONS

Both the determination of sugar feed rate with the amount of ammonia consumption and the fuzzy theory can be applied to sugar feed control of glutamic acid fermentation. The characteristics of this method are as follows:

1 Sugar feeding rate can be controlled with ammonia, which can be measured on-line, and be free from the maintenance.

2 The operations by the use of skilled operators' intuitions and experiences can be implemented into a control system by the fuzzy theory.

3 The system can control sugar feed rate in spite of changes in both sugar lots or activity of the microorganism.

ACKNOWLEDGEMENTS

The authors would like to thank Prof. H. Yamasaki(University of Tokyo) for his valuable advice, and thank Dr. T. Yasuda, Dr. Y. Iida, Dr. H. Yoshii, and Dr. Y. Yamada of Central Research Lab., AJINOMOTO CO., INC., for their useful suggestions.

We are also grateful to Mr. H. Obana, Mr. T. Suzuki, and Mr. Y. Kunimoto of Kawasaki Plant, AJINOMOTO CO., INC., for their continuous supports.

NOMENCLATURES

CRT	: cathode ray tube
d	: increase rate of D.O.
D.O.	: dissolved oxygen
Δf	: the proportion of modification of the regression coefficient in rules
Δf_0	: the proportion of modification of the regression coefficient
FDD	: floppy disk drive
FIC	: flow rate indicating controller
FR	: flow rate recorder
F_t	: sugar feed rate from t to $t+\Delta t$
HPLC	: high performance liquid chromatography
l	: D.O. variation
ΣN_t	: the amount of ammonia consumed by t
PHRC	: pH recording controller
S	: sugar concentration
S_0	: sugar concentration of feedstock
S_{SET}	: setting value of sugar concentration at the start point of sugar feed control
S_t	: the amount of sugar in fermentor at t
ΣS_t	: the amount of sugar consumed by t
t	: time or the time elasped from surfactant addition
V_t	: culture volume
X	: cell mass concentration
X_{SET}	: setting value of cell mass concentration at the point of surfactant addition

REFERENCES

1) Hong, K.T., S.H. Park, J.H. Lee, C.Y. Choi, and J.C. Bae (1984). Control of Sugar Feeding for Glutamic Acid Fermentation. J. Ferment. Technol., 62, 49-54.

2) Larsen, P.M. (1980). Industrial Application of Fuzzy Logic Control. Int. J. Man-Machine Studies, 12, 3-10.

3) Umbers, I.G. and P.J. King (1980). An Analysis of Human Decision-Making in Cement Kiln Control and the Implications for Automation. Int. J. Man-Machine Studies, 12, 11-23.

4) Yagishita, O., O. Itoh, and M. Sugeno (1984). Application of Fuzzy Reasoning to the Water Purification Process. Systems and Control, 28, 597-604.

5) Yasunobu, S., S. Miyamoto, and H. Ihara (1983). Fuzzy Control for Automatic Train Operation System. 4th IPAC/IFIP/IFRS Int. Conf. on Transportation System, 39-44.

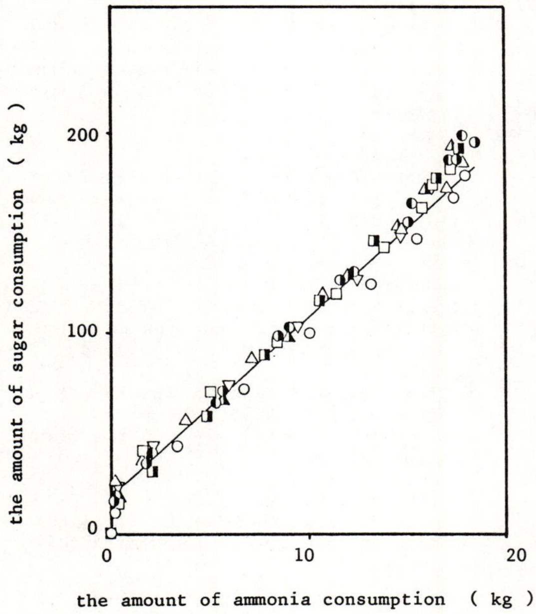

Fig. 1 Correlation between the amounts of sugar consumption and ammonia consumption

TABLE 1 Variables

INPUT OR CONTROL VARIABLES	FUZZY VARIABLES
t	LONG NORMAL SHORT
d	BIG SMALL
l	VHIGH (Very HIGH) HIGH LOW
r (flow rate)	HIGH NORMAL LOW
Δf	PB (Positive Big) PM (Positive Medium) PS (Positive Small) NS (Negative Small) NM (Negative Medium) NB (Negative Big)
Δr	PB PM PS NS NM NB

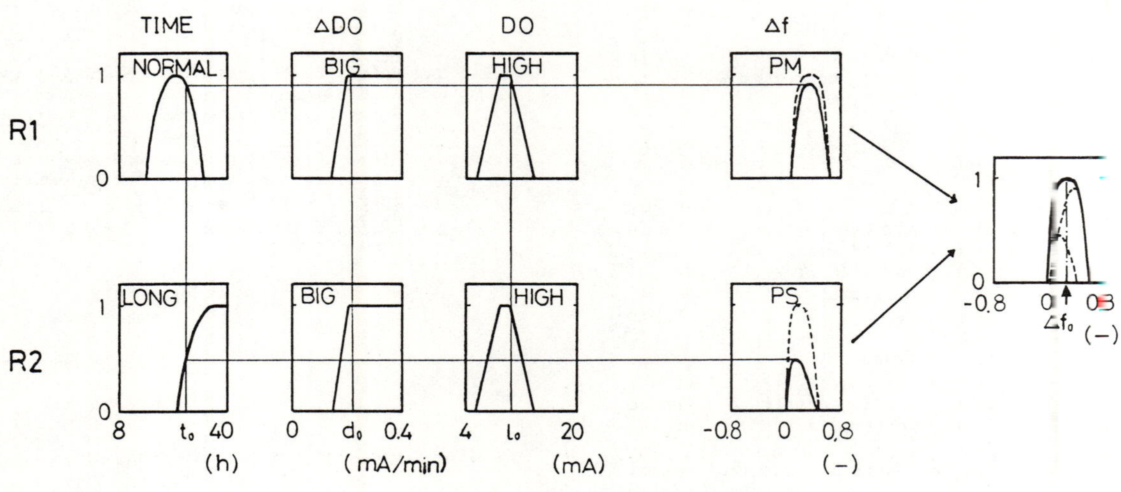

Fig. 2 A method of fuzzy reasoning

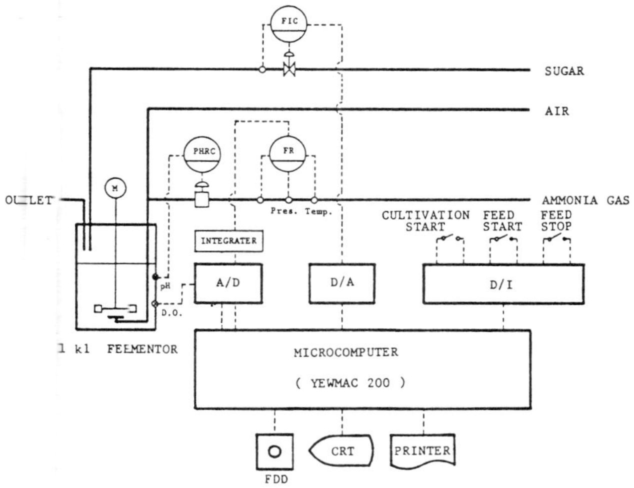

Fig. 3 A schematic diagram of sugar feed control system

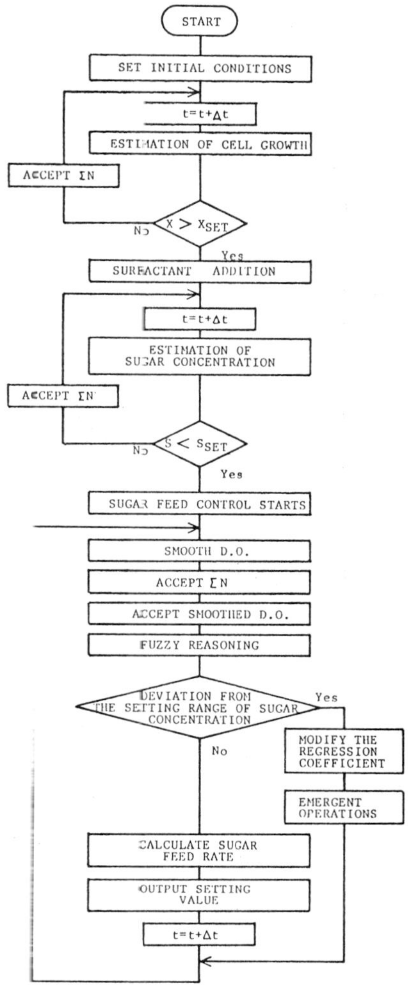

Fig. 4 A flow diagram of sugar feed control

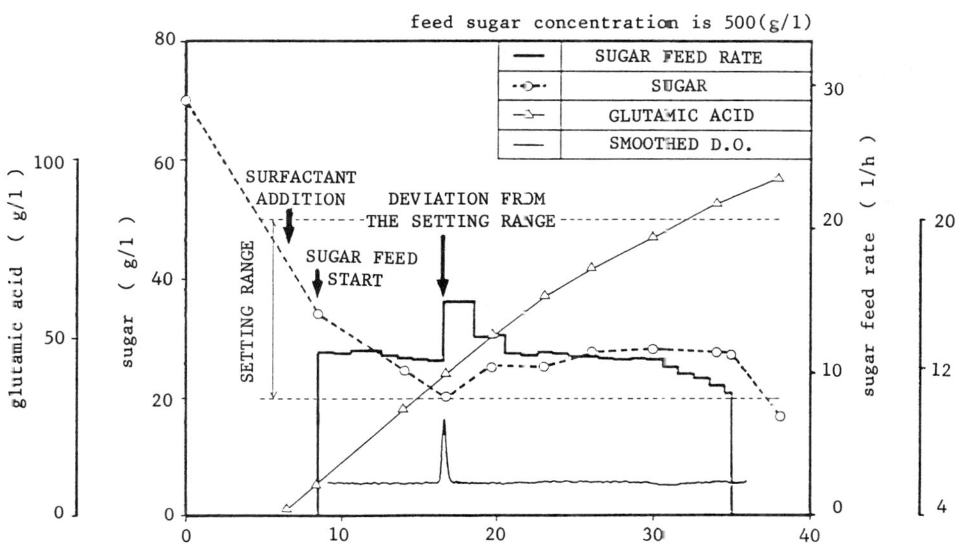

Fig. 5 The result of a field test

DIAGNOSTIC CHECKING USING A MODEL FOR A MULTI-ENZYME SYSTEM

D. Peferoen* and J. A. Spriet**

IWONL, University of Ghent, Coupure Links 653, 9000 Ghent, Belgium
***NFWO, University of Ghent, Coupure Links 653, 9000 Ghent, Belgium*

Abstract. Gramicidin-S is an antibiotic produced by the micro-organism Bacillus Brevis ATCC 9999. The biosynthesis is believed to occur according to the so-called 'multi-enzyme thiotemplate mechanism'; a multi-enzyme system composed of two proteins. The system is of great theoretical importance because a large number of antibiotics are synthesized according to that mechanism. Fermentation research has been mainly done through biochemical quantification of the enzymes and the product. Unfortunately, the measurement techniques are lengthy and involved processes which are difficult to validate. In the paper a mathematical model of the biosynthesis mechanism is built and used to check the adequateness of the quantification techniques. In a first step the thiotemplate mechanism which is essentially a biochemical pathway is modelled mathematically. The differential equations are derived according to proper chemical rules so that it may be accepted that they summarize adequately the a priori knowledge concerning the multi-enzyme system. Proper care has to be taken with simplification. Indeed a full implementation would yield a model that is untractable. Through proper assumptions on aggregation and time constants a set of five differential equations are retained. In a second step, the model is used in a simulation in order to see whether the measured enzyme concentrations can be used as inputs to yield the Gramicidin-S measurements as output. As quite a large number of parameters occur in the model and have to be fixed, a large amount of simulation power is required. An AD-10 simulator with extremely high speed simulation capabilities is utilized. The paper concludes with an assessment of the consistency of the results obtained with the biochemical measurement techniques.

Keywords. Diagnostic checking, antibiotics fermentation, multi-enzyme model, multi-enzyme thiotemplate mechanism, mathematical modelling, high speed simulation, consistency analysis.

INTRODUCTION

In biotechnology one is concerned with the production of useful substances through biological means. An important area in the field is the procurement of useful microbial products. One category of products that are provided through the activity of micro-organisms are antibiotics. The paper treats a problem that is related to the biosynthesis of a special class of antibiotics : those which are oligopeptides : e.g. bacitracin, tyrocidine, polymyxin, circulin, nisin, malformin, mycobacillin, gramicidin, etc... . Nowadays there is substantial evidence that the biosynthesis of those oligopeptides does not occur over the classic ribosomal system. At the contrary there seems to exist a special pathway that is utilized for the synthesis of the antibiotics. This pathway has been studied extensively mainly on the basis of the Gramicidin-S antibiotic. Nowadays the non-ribosomal synthesis apparatus is still studied intensively for a number of reasons : first, as it is simpler than the ribosomal system the synthesis pathway can perhaps be fruitfully exploited for the production of new antibiotics without having to modify the genetic part of the organism ; second, as some of the product antibiotics are industrially very important improving the understanding of the synthesis pathway may greatly benefit the overall production yields. Indeed in case of Gramicidin-S the oxygen supply is very critical for the production and optimizing its profile during fermentation is an important problem in the biotechnological area.

The problems mentioned have been attacked in the past in two manners. First an attempt has been made through systematic research efforts, to have a more fully understanding of the non-ribosomal synthesis pathway. No attempt has been made to keep the micro-organisms in a healthy environment. Indeed most of the time, research was done on cell extracts containing the molecules, more particularly enzymes that catalyze the oligopeptide synthesis. In that manner most elements of the pathway under consideration have been discovered and elucidated. At the contrary, in vivo studies and knowledge concerning the fermentation patterns of the micro-organisms using the special pathway have been done respectively obtained through mainly experimental means. In this paper it is tried to build a link between the two approaches. The key to the problem is believed to be a mathematical model that summarizes the theoretical findings on the non-ribosomal synthesis mechanism. As a first practical application it is looked whether theoretical evidence can be brought in accordance with experimental outcomes. The goal of the mathematical representation is thus diagnostic checking. Indeed a key to further improve fermentation practice is the measurement of the enzyme activity during growth and production. A lengthy procedure has been worked out to do just that. Unfortunately, it is difficult to validate the procedure and there is some evidence that it is prone to a lot of errors. A first and simple application of a link between theory and practice is thus a check whether theoretical knowledge can be reconciliated with the outcomes of the analysis technique for the estimation of the activity of the enzymes. Further modelling activity and more complicated goals could be strived at in a further stage.

The paper is organized as follows. First a short

account in verbal and pictorial terms is given of the non-ribosomal synthesis system. Second the mathematical model is built of the system, condensing the a priori theoretical knowledge concerning the synthesis mechanism. Then the measurement technique for the enzymes is given in full detail. The fermentation that has been done to obtain the data is discussed under the heading : experiment. In addition some details are given on materials and methods. The next section summarizes the details concerning the simulation effort. Then results are given and conclusions drawn.

THE MULTI-ENZYM THIOTEMPLATE MECHANISM

Generalities

Certain strains of micro-organisms produce small peptides as antibiotics. The classic system for peptide synthesis is the ribosomal protein synthesis system. It has been found by Winnick and co-workers (1961) that the oligopeptide antibiotics are synthesized by certain Bacilli-strains through a complete different non-ribosomal system. This system has been termed the multi-enzym thiotemplate mechanism. Through different research efforts, the mechanism has been elucidated for a large part. Most work has been done on the cyclic decapeptide Gramicidin-S. Still not all details are known and active research continues. The Gramicidin-S molecule, termed GS in the sequel, is given below

D-Phe ⌢ L-Pro ⌢ L-Val ⌢ L-Orn ⌢ L-Leu
L-Leu ⌣ L-Orn ⌣ L-Val ⌣ L-Pro ⌣ D-Phe

It is a secondary metabolite of Bacillus Brevis (var. Nagano, G.B., and ATCC 9999). It is a broad spectrum antibiotic that is active against G^+ but that is not in full use yet because of its high toxicity. Bacitracine however, which is also a product of a non-ribosomal synthesis system, is extensively used. The multi-enzym thiotemplate mechanism is similar to the classic ribosomal system in the sense that peptides are synthesized on a matrix based on building blocks thanks to the availability of energy-rich chemical bonds. The building blocks are the constitutive amino acids. The matrix are enzymes. In case of GS, there are two enzymes involved, GS-synthetase I and GS-synthetase II, which will be designated as GS-SI and GS-SII. Both enzymes have multiple active sides and they cooperate to bring about the synthesis of the oligopeptide GS. The energy is provided by ATP and energy-rich thioesterbonds on the enzymes. The stochiometric relationship of the synthesis is given by the following equation :

$$10 \text{ amino-acids} + 10 \text{ ATP} \xrightarrow[\text{GS-SI,GS-SII}]{Mg^{2+},Mn^{2+}} \text{GS} + 10 \text{ AMP} + 10 \text{ PP}_i \qquad (1)$$

One mole of ATP is required for the production of each peptide bond. ATP is not directly used to provide the energy for the peptide bonds. Instead the energy of the ATP-PP_i exchange reaction is used to generate thioesterbonds on the enzymes. Those bonds are energy-rich and provide then the required power for generating the peptide bonds. Survey information on peptide antibiotics can be found in Katz and Demain (1977) and in Shoji (1978). The precise mechanism of biosynthesis is discussed in the following section.

The Synthesis of Gramicidin-S

Figure 1 gives an approximate picture of the synthesis of Gramicidin-S. As it is the basis for the mathematical model that is described in one of the following sections, a verbal description of the various steps is given. The authors are well aware that some details of the synthesis are still hypothetical ; however scientific evidence exists for the main parts and proper hypotheses have been made for the remaining details. The picture is a state-of-the-art about the knowledge concerning the Gramicidin-S synthesis.

I -A schematic representation is given of GS-SI and GS-SII. The larger molecule has a broken line as a representation for the phosphopantetheïne-arm. On GSI there is one thiolgroup that needs special attention on GSII, there are six. The phosphopantetheïne-arm ends on a thiolgroup.

II -The amino-acid building blocks are activated by ATP in the presence of the enzymes. One by one, the substrate amino-acids are bound in a non-covalent fashion as amino-acyladenylates. L-Phe on GS-SI and the other ones L-Pro, L-Val, L-Orn and L-Leu on GS-SII. On the latter enzymes, it is believed that once the first amino-acid L-Pro is activated and bound, the enzymatic activity of GS-SII is increased and the second, third and fourth amino-acid are activated and bound more rapidly.

III -The amino-acyl parts are transferred to secondary acceptors - SH-thiolgroups, in order to be bound covalently with a thio-esterbound. It is believed that GS-SII changes in conformation so that subsequent steps become possible.

IV -GS-SI brings about the racimisation of L-Phe to D-Phe. It is not known how this occurs At this point the two enzymes are fully loaded and the formation of an oligo-peptide can begin.

V -GS-SI and GS-SII are believed to form a synthetase complex. This complex may well play a vital role in the further synthesis.

VI -First there is transfer from D-Phe from a thiolgroup on GS-SI to a free thiolgroup on GS-SII

VII -The formation of a first pentapeptide is possible now. Strange enough the first bond is formed without the intervention of the phosphopantetheïne acid-cofactor. D-Phe is transferred to the iminogroup of L-Pro with the formation of a dipeptide molecule.

VIII -For the further synthesis the cofactor is required. It has a free thiolgroup. The dipeptide is transferred to the thiolgroup of the phosphopantetheïne 'arm', which now will 'turn' along the molecule and pick up the other amino-acids. First the arm is brough to L-Val and there a tripeptide is formed, which is transferred on the arm. In the same manner a tetrapeptide and a pentapeptide are produced. This piece of peptide is probably attached to a special thiolgroup where it waits till a second peptide is provided.

It is not clear yet whether the cyclization is an intra-mollecular on intermollecular reaction. In this paper it is assumed that steps I to VIII occur a second time so that a second peptide becomes available for the final step in the Gramicidin-S synthesis.

IX -The final step is a head to tail cyclization of two pentapeptides to a final Gramicidin-S molecule.

This is an eventual scenario of the formation of Gramicidin-S. Major references are the articles by Froyshov and coworkers (1978) and by Kleinkauf and Koischwitz (1980).

MATHEMATICAL MODELLING

The modelling of the multi-enzym thiotemplate mechanism is by no means a trivial task. Modelling proceeds usually from three information sources (Spriet and Vansteenkiste, 1982) : a priori knowledge, measurements, data and modelling goals. The a priori basis for the model is the information contained in the previous section together with chemical and biochemical kinetics - Roels (1983). Unfortunately, knowledge concerning the non-ribosomal system is not yet advanced enough to write down immediately the model equations and parameters. Measurements and data are not available in sufficient amounts to support the modelling effort. We can however make some simplifying assumptions on the basis of the modelling goal. Indeed we are not looking for a mechanistic representation but only for an approximate model of the kinetics in order to use it for diagnostic checking. Based on the a priori knowledge mentioned and literature data such an approximate model will be built. We proceed in the following manner ; first the verbal qualitative model is translated in a chemical model. Then kinetics are used to obtain a full set of differential equations and finally simplifications are made in order to reduce the model set to a reasonable size.

Table I gives a list of all occurring intermediate compounds, their chemical representation and their concentration.

TABLE I Model variables

Symbol	Concentration µmol/gr DCW	Description
ATP	C_{ATP}	Energy carrier in the micro-organism in loaded form
AMP	C_{AMP}	Energy carrier in unloaded form
PP_i	C_{PP_i}	diphosphate residu
A	C_A	building block on GS-SI; phenylalanine
G	C_G	lumped building block on GS-SII representing a combination of four amino-acids
E_1	C_{E_1}	GS-synthetase I
E_2	C_{E_2}	GS-synthetase II
A-AMP	C_{A-AMP}	activated amino-acid, AMP complex for GS-SI
G-AMP	C_{G-AMP}	activated amino-acid, AMP complex for GS-SII
E_1-A	C_{E_1-A}	loaded GS-SI
E_2-G	C_{E_2-G}	loaded GS-SII
E_2-G-A	C_{E_2-G-A}	GS-synthetase II with polypeptide
E_2-K	C_{E_2-K}	GS-synthetase II with pentapeptide in waiting position
E_2-K-G	C_{E_2-K-G}	GS-SII loaded with a pentapeptide and a building block
E_2-K-G-A	$C_{E_2-K-G-A}$	GS-SII loaded with a pentapeptide and a polypeptide
E_2-K-K	C_{E_2-K-K}	GS-SII loaded with two pentapeptides
GS	C_{GS}	Gramicidine

The non-ribosomal peptide synthesis system is approximated by the following chemical reaction equations :

$$ATP+A \underset{k'_1}{\overset{k_1}{\rightleftharpoons}} A\text{-}AMP+PP_i \qquad (1)$$

$$ATP+G \underset{k'_2}{\overset{k_2}{\rightleftharpoons}} G\text{-}AMP+PP_i \qquad (2)$$

$$E_1+A\text{-}AMP \underset{k'_3}{\overset{k_3}{\rightleftharpoons}} E_1\text{-}A+AMP \qquad (3)$$

$$E_2+G\text{-}AMP \underset{k'_4}{\overset{k_4}{\rightleftharpoons}} E_2\text{-}G+AMP \qquad (4)$$

$$E_2\text{-}G+E_1\text{-}A \underset{k'_5}{\overset{k_5}{\rightleftharpoons}} E_2\text{-}G\text{-}A+E_1 \qquad (5)$$

$$E_2\text{-}G\text{-}A \underset{k'_6}{\overset{k_6}{\rightleftharpoons}} E_2\text{-}K \qquad (6)$$

$$E_2\text{-}K+G\text{-}AMP \underset{k'_7}{\overset{k_7}{\rightleftharpoons}} E_2\text{-}K\text{-}G+AMP \qquad (7)$$

$$E_2\text{-}K\text{-}G+E_1\text{-}A \underset{k'_8}{\overset{k_8}{\rightleftharpoons}} E_2\text{-}K\text{-}G\text{-}A+E_1 \qquad (8)$$

$$E_2\text{-}K\text{-}G\text{-}A \underset{k'_9}{\overset{k_9}{\rightleftharpoons}} E_2\text{-}K\text{-}K \qquad (9)$$

$$E_2\text{-}K\text{-}K \underset{k'_{10}}{\overset{k_{10}}{\rightleftharpoons}} GS + E_2 \qquad (10)$$

As can be seen the equations mimic the verbal description of the previous paragraph. A simplifying assumption has been made : the four amino-acids L-Pro, L-Val, L-Orn, L-Leu are lumped in a single building block G. There is evidence that this is acceptable as binding of one amino-acid increases the activity of the enzyme and the others are bound in a more rapid fashion. In order to reduce the complexity of the equation, it is assumed that chemical equations (5), (6), (8), (9), (10) are essentially irriversible.
Thus $k'_5 = k'_6 = k'_8 = k'_9 = k'_{10} = 0$. This is acceptable if one realizes what is happening there. Based on mass balances and chemical kinetics, it is now a straightforward exercise to write down the mathematical equations :

$$dC_{A-AMP}/dt = k_1 C_A \cdot C_{ATP} + k'_3 C_{E_1-A} \cdot C_{AMP} - k'_1 C_{A-AMP} \cdot C_{PP_i} - k_3 C_{E_1} \cdot C_{A-AMP} \qquad (11)$$

$$dC_{G-AMP}/dt = k_2 C_G \cdot C_{ATP} + k'_4 C_{E_2-G} \cdot C_{AMP} + k'_7 C_{E_2-K-G} \cdot C_{AMP} - k'_2 C_{G-AMP} \cdot C_{PP_i} - k_4 C_{E_2} \cdot C_{G-AMP} - k_7 C_{E_2-K} \cdot C_{G-AMP} \qquad (12)$$

$$dC_{E_1-A}/dt = k_3 C_{E_1} \cdot C_{A-AMP} - k'_3 C_{E_1-A} \cdot C_{AMP} - k_8 C_{E_1-A} \cdot C_{E_2-K-G} - k_5 C_{E_2-G} \cdot C_{E_1-A} \qquad (13)$$

$$dC_{E_2-G}/dt = k_4 C_{E_2} \cdot C_{G-AMP} - k'_4 C_{E_2-G} \cdot C_{AMP} - k_5 C_{E_2-G} \cdot C_{E_1-A} \qquad (14)$$

$$dC_{E_2\text{-}G\text{-}A}/dt = k_5 C_{E_2\text{-}G}\cdot C_{E_1\text{-}A} - k_6 C_{E_2\text{-}G\text{-}A} \tag{15}$$

$$dC_{E_2\text{-}K}/dt = k_6 C_{E_2\text{-}G\text{-}A} + k'_7 C_{E_2\text{-}K\text{-}G}\cdot C_{AMP} - k_7 C_{E_2\text{-}K}\cdot C_{G\text{-}AMP} \tag{16}$$

$$dC_{E_2\text{-}K\text{-}G}/dt = k_7 C_{E_2\text{-}K}\cdot C_{G\text{-}AMP} - k'_7 C_{E_2\text{-}K\text{-}G}\cdot C_{AMP} - k_8 C_{E_2\text{-}K\text{-}G}\cdot C_{E_1\text{-}A} \tag{17}$$

$$dC_{E_2\text{-}K\text{-}G\text{-}A}/dt = k_8 C_{E_2\text{-}K\text{-}G}\cdot C_{E_1\text{-}A} - k_9 C_{E_2\text{-}K\text{-}G\text{-}A} \tag{18}$$

$$dC_{E_2\text{-}K\text{-}K}/dt = k_9 C_{E_2\text{-}K\text{-}G\text{-}A} - k_{10} C_{E_2\text{-}K\text{-}K} \tag{19}$$

$$dC_{GS}/dt = k_{10} C_{E_2\text{-}K\text{-}K} \tag{20}$$

Notice that there are a large number of constants to be estimated or taken from the literature. Notice also that the model is still rather complex as 10 differential equations are obtained. After a first run with constants taken as realistically as possible it was found that the model was stiff due to the large differences in time constants. Consequently, it was decided to distinguish fast and slow reactions. It was decided that the slow reactions are the interenzyme reaction, the formation of a pentapeptide and the cyclization to Gramicidin-S. Consequently, Eqs. (1), (2), (3), (4) and (7) can be considered in equilibrium when compared with the other ones. In that manner, the equations simplify to the following set :

$$C_{A\text{-}AMP} = \frac{K_1\cdot C_A\cdot C_{ATP}}{C_{PP_i}} \tag{21}$$

$$C_{G\text{-}AMP} = \frac{K_2\cdot C_G\cdot C_{ATP}}{C_{PP_i}} \tag{22}$$

$$C_{E_1\text{-}A} = \frac{K_3\cdot C_{E1}\cdot C_{A\text{-}AMP}}{C_{AMP}} \tag{23}$$

$$C_{E_2\text{-}G} = \frac{K_4\cdot C_{E2}\cdot C_{G\text{-}AMP}}{C_{AMP}} \tag{24}$$

$$dC_{E_2\text{-}G\text{-}A}/dt = k_5 C_{E_2\text{-}G}\cdot C_{E_1\text{-}A} - k_6 C_{E_2\text{-}G\text{-}A} \tag{25}$$

$$dC_{E_2\text{-}K}/dt = k_6 C_{E_2\text{-}G\text{-}A} - k_8 C_{E_2\text{-}K\text{-}G}\cdot C_{E_1\text{-}A} \tag{26}$$

$$C_{E_2\text{-}K\text{-}G} = \frac{K_7\cdot C_{E_2\text{-}K}\cdot C_{G\text{-}AMP}}{C_{AMP}} \tag{27}$$

$$dC_{E_2\text{-}K\text{-}G\text{-}A}/dt = k_8 C_{E_2\text{-}K\text{-}G}\cdot C_{E_1\text{-}A} - k_9 C_{E_2\text{-}K\text{-}G\text{-}A} \tag{28}$$

$$dC_{E_2\text{-}K\text{-}K}/dt = k_9 C_{E_2\text{-}K\text{-}G\text{-}A} - k_{10} C_{E_2\text{-}K\text{-}K} \tag{29}$$

$$dC_{GS}/dt = k_{10} C_{E_2\text{-}K\text{-}K} \tag{30}$$

This is the final mathematical model that will be used in the simulation.

MEASUREMENT OF SYNTHETASES

Based on general principles of enzymology and on radioactive measurement techniques the following procedure was proposed - Vandamme (1976).

Preparation of Enzym Extracts

During fermentation cells are harvested, centrifugated at 10,000 g for 10 minutes, washed and centrifugated again. Wet cells are freezed at -20°C. Cell extracts are prepared at 4°C. Buffer A (pH 7,6) is prepared : triethanolamine HCl (10 mmol/l), $MgCl_2$ (10 mmol/l), dithiotreitol (1 mmol/l), EDTA (0.75 mmol/l) and chlooramfenicol (10 mg/l). Cells are suspended in 6 volumes of buffer A ; lysosyme (12 mg/gr wet weight) is added. After incubation during 20 min. at 30°C, the suspension is centrifugated at 20,000 g. The precipitate is again treated with buffer A and lysosyme in the same manner and centrifugated. Further work is done on the combined cell lysates. Then a purification step follows based on ammoniumsulfate-fractionation. To the previously obtained extract a cold saturated $(NH_4)_2\ SO_4$-solution (pH 7,6) is added to 55% of saturation. After incubation for 30 minutes on ice, the mixture is centrifugated for 15 min. at 20,000g. The supernatans is eliminated. Buffer B is prepared (pH 7,6) : triethanolamine HCl (20 mmol/l), $MgCl_2$ (10 mmol/l), dithiotreitol (1 mmol/l), EDTA (0,25 mmol/l) and sucrose (10%). The precipitate is dissolved in buffer B and water is added till a solution of 1 mg protein/ml is obtained. The protein analysis on the purified extract is done according to Lowry and coworkers (1951).

GS-Synthetase I and II Activity

The determination of the GS-synthetase I and II activity is based on the by the enzymes catalyzed ATP-pyrophosphate (PP_i) exchange reaction - Eq. (1) and (2) in the previous section :

$$\text{amino-acid} + \text{ATP} \rightleftharpoons \text{amino-acid-AMP} + PP_i$$

For GS-SI, D-phenylanaline is used as amino-acid; for GS-SII, L-ornithine is utilized. After incubation, ATP is taken from the mixture selectively after the reaction has been forced in the backward direction through addition of radioactive labeled $^{32}PP_i$. Incorporated radioactivity is measured in a Hewlett Packard Scintillation Counter as a number of cpm (counts per min.). This is used then to evaluate specific enzyme activity. If no enzyme is present, the reaction will not proceed. Addition of ^{32}P-pyrophosphate will have no effect and the total amount of ATP will show no radioactivity. At the contrary if active enzyme is present, the contrary will occur. The full procedure is as follows. To small testtubes one adds : 50 µmol triethanolamine HCl - buffer (pH 7,5), 2,5 µmol KF, 2,5 µmol $MgCl_2$, 2,5 µmol ATP, 0,5 µmol dithiotreitol, 2,5 µmol radioactive Na-pyrophosphate ^{32}P (specific activity 0,05 μC_i/2,5 µmol ^{32}P or 45 dpm/nmol ^{32}P). The volume taken by this solution is 0.3 ml. To two of the testtubes 0.1 ml H_2O is added ; to two others 2,5 µmol L-ornithine or D-phenylalanine (0.1 ml of 25 mmol/l solution) is added. Next the enzyme solution (0,1 ml) is added (final concentration 0,2 mg protein/ml) to all testtubes. Incubation is done for 15 minutes and then the enzymatic reaction is stopped through addition of 0,3 ml of ice-cold 2% thrichlooraceticacid solution. Then extraction of ATP is strived at. In each testtube 0,1 ml of a 15% suspension of active coal (Norit A) is added. After intense mixing the suspensions are filtered on cottonwoolfilters. Testtubes are washed with 1 ml of water. Next the cottonwoolfilters are eluated twice with 1 ml of H_2O and the radioactive water is eliminated. Washed active coal on the filter is then twice eluated with 0,5 ml ethanol-ammonium-hydroxide-water solution (5:1:4). The radioactive eluate is added to scintillation bottles which contain 10 ml H_2O. Next the radioactivity is measured in a Hewlett Packard Liquid Scintillation Counter. The result is given in specific activity : this is the number of units per mg protein. A unit of enzym activity, is the amount of enzym which incorporates 1 mmol ^{32}P per minute in ATP. Under the circumstances mentioned the incorporation is assumed to be a linear function of time. Consequently, the next equation can be put forth

$$SA = \frac{cpm}{E.A.\eta.t} \quad (31)$$

SA : specific activity in nmol $^{32}P\ min^{-1} mg^{-1}$

E : amount of protein in incubation mixture (mg)

cpm : number of counts per minute of the sample minus the number of counts per minute in blanco (cpm)

A : specific activity of Na-pyrophosphate ^{32}P (dpm/nmol^{32}P)

η : count-efficiency

t : duration of incubation (min)

This concludes the determination of the specific enzyme activity in the samples.

EXPERIMENT

An experiment was done in a computer-coupled 28-liter New Brunswick fermentor. 18 liter of a medium composed of 3% peptone and 3% yeastextract was sterilized and inoculated with 10 ml of a Bacillus Brevis ATCC 9999 spore-suspension (± 10^{10} spores/ml). Fermentation occurred at 37°C ; aeration was kept constant at 8 liter/mm and agitation at 500 rpm. Samples were taken at regular intervals for determination of dry weight (DCW), optical density (O.D), GS-titer and soluble GS synthetase activity.

Dry weight is obtained as follows : a sample is centrifugated at 10,000 g for 10 minutes, washed and centrifugated again. The pellet is transferred to weighted small plates filled with sand. After drying at 105°C for 20 hours the plates are weighted again and the sample dry weight (DCW) is obtained. GS is extracted from the cells by adding 2 ml culture medium to 2 ml extraction mixture (ethanol and 0.2 NHCl ; 9:1 v/v) and shaking it for three hours on a rotary mixer. Quantification is done with the agar-diffusion method with Bacillus subtilis ATCC 6C51 as test organism. The procedure is not explained in detail here because it is lengthy. It yields however reliable results as the quantification method has been statistically verified. For details we refer to Matteo and coworkers (1975). The results are shown on Fig. 2. Dry weight values, specific enzyme activity values and specific Gramicidin-S values are shown as obtained from the samples taken during the fermentation. It is noted that enzyme activity is observed first during the exponential growth phase while the Gramicidin-S is formed much later on.

SIMULATION

Through simulation it is looked whether it is possible to bring the measurements into line with the mathematical model developed above. First a number of assumptions are made. It is believed that the regulatory mechanisms of the catabolic activity is sufficiently powerful and rapid as to keep the concentrations of a number of substances at nominal level : C_G, C_A, C_{ATP}, C_{AMP}, C_{PP_i}. For the concentrations of the free enzymes Gaussian curves are fitted to the main peaks that have been measured. The levels of these peaks are considered unknowns - see Eq. (32), (33) and (34).

$$N(M,S)(t) = - \frac{1}{S\sqrt{2\pi}} \exp\left[\frac{1}{2}\left(\frac{t-M}{S}\right)^2\right] \quad (32)$$

$$C_{E_1}(t) = p\ N(M_1,S_1)(t) \quad (33)$$

$$C_{E_2}(t) = q\ N(M_2,S_2)(t) \quad (34)$$

M_1 and M_2 are the time values of maximum enzyme activity as measured with the procedure outlined in a previous paragraph. S_1 and S_2 are obtained from a determination of the experimental peak with Δ at half the height of the maximum peak value. It can easily be proved that $\Delta_i = 2.355\ S_i$. This set-up leaves the model with a number of unknown constants : C_G, C_A, C_{ATP}, C_{AMP}, C_{PP_i} ; k_5, k_6, k_8, k_9, k_{10} ; K_1, K_2, K_3, K_4, K_7, p, q. This is considered here as the freedom that is available to match the model to the data.

Matching the model to the data requires that the model be solved a great number of times with different parameter values. Extensive simulation is thus required. Initial runs with a continuous model simulation language like CSMP on a PDP 11-34 indicated that the run-time of the model was too high. It was found that a classic minicomputer was not fast enough to provide adequate simulation power for the problem at hand. Use was made here of a special processor, the AD-10 simulator. The PDP-11 serves as a host computer for the proper use of a digital simulator that is specially designed for fast continuous system simulation. The AD-10 is characterized by a large high-speed interleaved memory, very fast arithmetic processor speeds, many high speed input-output channels and by a special parallel architecture. Programming of the machine is facilitated through a high level modular programming system : MPS 10. The MPS-10 user needs no knowledge of the AD-10's multiprocessor architecture or assembly language to make full use of the AD-10's ultra fast speed. Extensive simulation interaction is provided with the simulation model coefficients, parameters, integration methods, real-time solution display, etc. ... in ways that are normally provided on other computers (analog, hybrid or general purpose digital) or simulation language environments. Basically after preparation of the necessary programs and files on the host computer and after down-loading the AD-10, the simulator can be used much like analogue machines and commands can be easily provided through the host in order to produce fast simulation and interactive changes to these simulations. The AD-10 can be seen as a very fast differential equation solver, connected to a general purpose digital machine, serving as host on which supervising algorithms can be programmed. The run-time of the model defined above is less than a second. Use of the AD-10 simulator thus solves the simulation problem in an adequate manner.

Besides solving the differential equations, a procedure must be provided to systematically adjust the unknown parameters in order to fit the model output of the Gramicidin-S production to the experimental data. A program is used that minimizes the cost given by Eq. (35) :

$$J(\underline{p}) = \sum_{i=1}^{N} (C_{GS}(k,\underline{p}) - C^m_{GS}(k))^2 \quad (35)$$

$C_{GS}(k,\underline{p})$; model output for C_{GS} at instant of measurement t_k

$C^m_{GS}(k)$; measured value for Gramicidin-S at t_k

The minimizing procedure is a mixture of algorithms termed Newton Raphson, steepest descent and Marquardt. For an approximation of the first derivative the ideas of Broyden are used. For more details on algorithmic optimization algorithms, we refer to the work of Birta (1934). The implementation was done as follows : the AD-10 did the integration of the model and provided $C_{GS}(k,\underline{p})$-values to the host. The host contained the optimization algorithm and called on the AD-10 to calculate $C_{GS}(k,\underline{p})$. In that manner the optimization of the many parameters could be done in a reasonable amount of time. The overall routine was run many different times starting from different initial conditions in order to find out whether local minima existed. If a minimum was found it was checked by manual means.

RESULTS

The major result is that the data cannot be brought into accordance with the model. Fig. 3 summarizes the findings. As can be seen, the fit between the experimental GS-data and the GS-curve resulting from the simulation is acceptable. The final level, the time shift and the slope are all properly modelled. Unfortunately, the enzyme levels are unacceptable. E_2-K-K reaches a peak value of 81.6 µmol/gr. DCW. This means that 22.8 gr. E_2-K-K are produced per gr. DCW and this is clearly unacceptable as it must be a fraction of the total dry weight. Through simulation it was found that the time shift is the major cause of trouble. In order to bring the data in accordance with the GS measurements it was assumed that the measurement of the synthetases only detects free enzyme. This is highly unlikely but the most optimistic hypothesis from the viewpoint of diagnostic checking. The hypothesis allows for the fit but then yields unacceptable levels of the intermediate products. The time lapse between the appearance of the enzymes and the appearance of the product is thus highly improbable and is the major reason why the model cannot be reconciliated with the data in a consistent manner.

CONCLUSION

The paper shows that mathematical models can be used for diagnostic checking. A priori knowledge concerning the multi-enzym thiotemplate mechanism cannot be brought in accordance with experimental data. One can question the model itself and it is agreed upon that further modelling efforts are required. However the experimental procedure for measuring synthetase activity should be carefully reconsidered in view of the present findings. It is planned to use electrophoresis to check the procedure outlined in the paper.

REFERENCES

Birta, L.G. (1984). Optimization in Simulation Studies. In T.I. Oren, B.P. Zeigler and M.S. Elzas (Eds.), Simulation and Model-Based Methodologies: An Integrative View, Springer Verlag, Berlin, pp. 451-474.

Frøyshov, Ø, T.L. Zimmerman and G.G. Laland (1978). Biosynthesis of Microbial Peptides by the Thiotemplate Mechanism. In H.R. Von Arnstein (Ed.), Amino Acid and Protein Biosynthesis II, Vol.18, University Park Press, Baltimore.

Katz, E. and A. Demain (1977). The peptide antibiotics of Bacillus : Chemistry, Biogenesis and Possible Functions. Bacteriological Reviews, American Society for Microbiology, pp. 449-474.

Kleinkauf, H. and H. Koischwitz (1980). Gramicidine S-Synthetase. In H. Bisswanger and E. Schmincke-Ott (Eds.), Multifunctional Proteins, J. Wiley & Sons Inc.

Lowry, O.H., N.J. Rosebrough, A.L. Farr and R.J. Randall (1951). Protein Measurement with the Folm-Phenol Reagent. J. Biol. Chem., 193, 265-275.

Matteo, C.C., M. Glade, A. Tanaka, J.M. Piret and A.L. Demain (1975). Microbiological studies on the formation of Gramicidin S synthetases. Biotechnol. and Bioeng., 17, 129-142.

Roels, J.A. (1983). Energetics and Kinetics in Biotechnology. Elsevier Biomedical Press, Amsterdam, 1983.

Shoji, J. (1978). Recent Chemical Studies on Peptide Antibiotics from the Genus Bacillus. Advances in Applied Microbiology, Vol.24, Academic Press, 187-214.

Spriet, J.A. and G.C. Vansteenkiste (1982). Computer-aided modelling and simulation. Academic Press, London.

Vandamme, E.J. (1976). Productie, Biosynthese en Werkingsmechanisme van Bacillus Oligopeptide Antibiotica. Aggregaatsthesis voor het Hoger Onderwijs, R.U.G.

Winnick, R.E., H. Lis and T. Winnick (1961) Biosynthesis of Gramicidin S. Biochimica et Biophysica Acta, 49, 451-461.

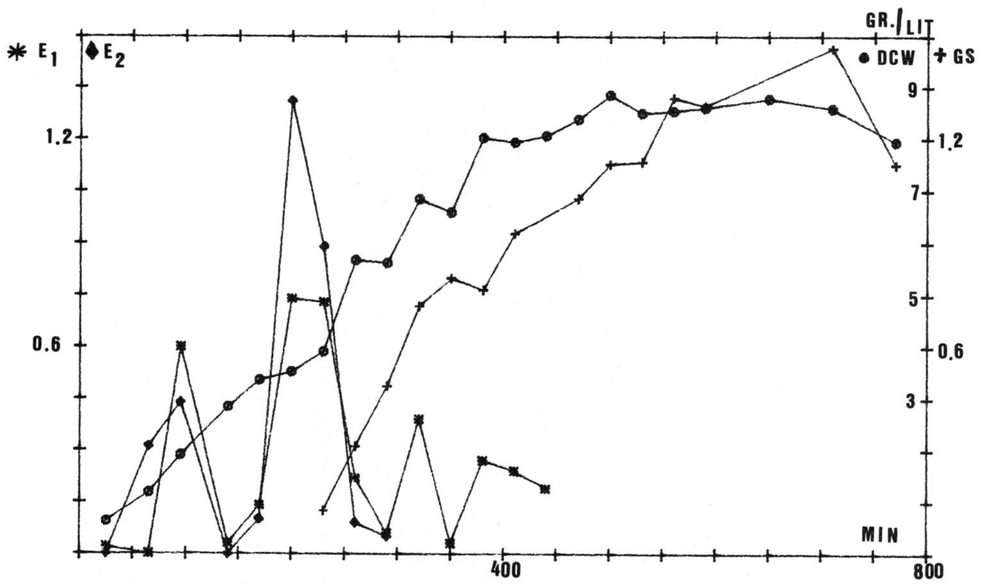

Fig. 2. Experimental fermentation data

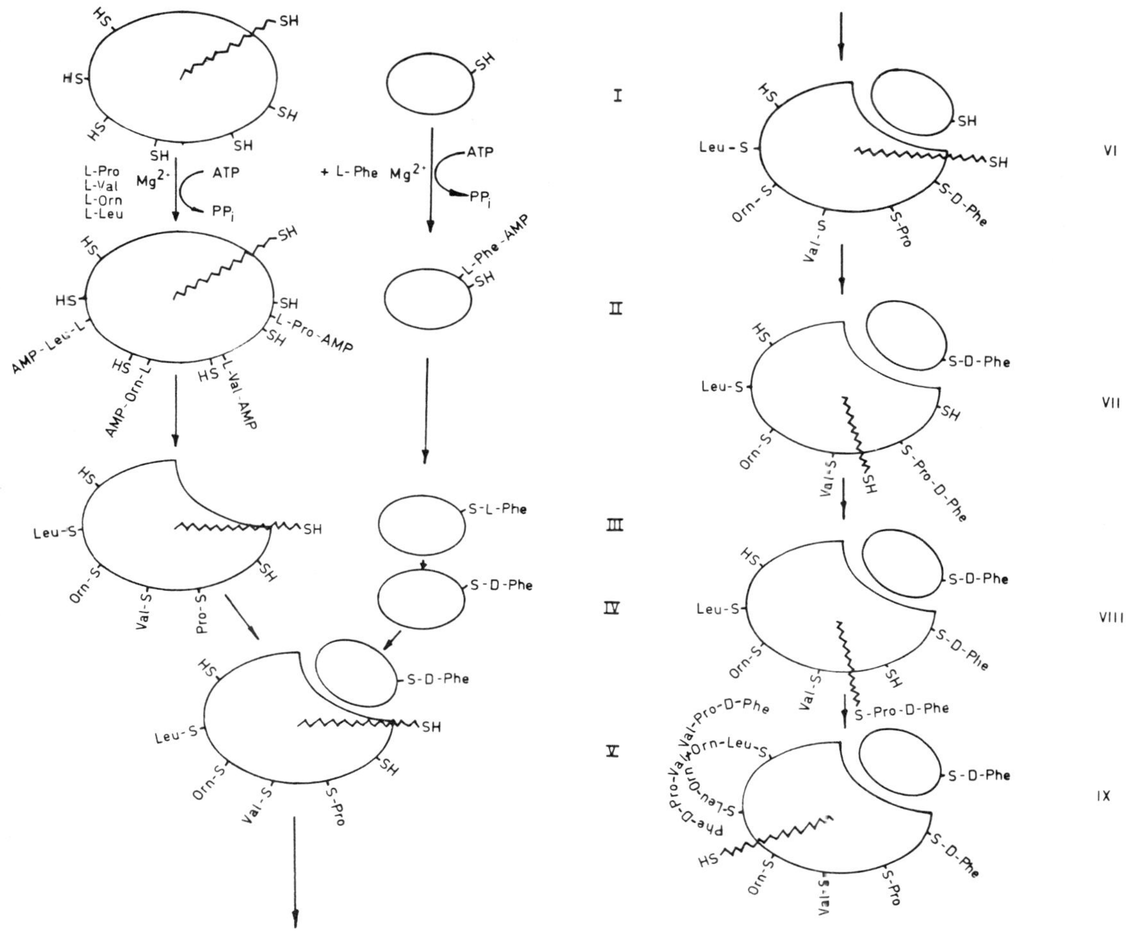

Fig. 1. The multi-enzym thiotemplate mechanism

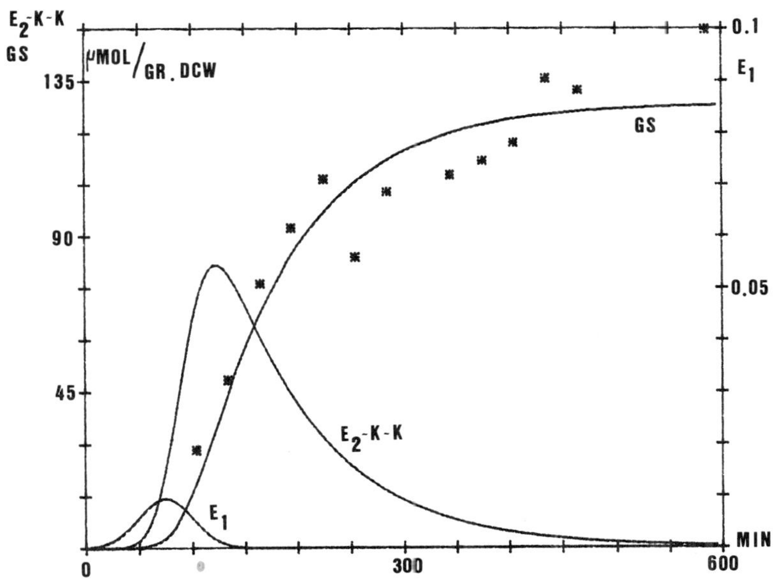

Fig. 3. Fit between model and GS-data

YEAST BIOTECHNOLOGICAL PROCESSES MONITORED BY ANALYSIS OF SEGREGATED DATA WITH STRUCTURED MODELS

E. Martegani*, L. Mariani and L. Alberghina***

**Laboratory of Comparative Biochemistry, Department of General Physiology and Biochemistry, University of Milan, Via G. Celoria, 26, 20133 Milano, Italy*
***Department of Electrical Engineering, University of Padua and LADSEB-CNR, Corso Stati Uniti, 4, 35020 Padova, Italy*

ABSTRACT.

Several biological parameters have been measured in populations of budding yeast growing in a glucose-limited chemostat, both in steady state and during perturbed growth conditions. The population structure has been determined with a segregated data analysis performed on protein distribution and on volume distributions. A new structured model for budding yeast population is presented, in which the entire system is described only in terms of critical cell mass parameters. This approach allows to integrate in a fairly simple way the population model with the bioreactor performance and to describe transient states of growth.

KEYWORDS.: Structured mathematical models; yeast biotechnological processes; protein and volume distributions.

INTRODUCTION.

Yeasts are eukaryotic microorganisms widely used for biotechnical processes both for the production of biomass and for the production of specific (bio)chemicals. Their application range increases continuosly with the development of recombinant DNA techniques that now allow to modify in a programmed way the genome of yeasts such as *Saccharomyces cerevisiae* and *Kluyveromyces lactis* (Struhl and collegues, 1979; Das and Hollenberg, 1982).

The knowledge of biomass properties is of fundamental importance for monitoring and controlling biothechnical processes (Bailey, Agar and Hjorsto, 1982) but at present the control of bioprocesses is restricted to the control of some physical or chemical parameters, so that the monitoring and control of biological activities is reached in an indirect way. Moreover informations about the functional heterogeneity of the microbial population are seldom obtained and they are not utilized for this purpose (Cooney, 1979).

Instead many recent papers have shown that often there is an age-dependent physiological and biochemical heterogeneity of the population and they have stressed the importance of the age structure of a microbial population for maximizing the yield of bioconversions (Ranta, 1983; Dairaku and collegues, 1982).

The heterogeneity of cell population can be readily determined with methods that measure the parameters in the single cell (segregated parameters) and fast methods to measure intracellular parameters on a single cell basis are now available. Their use could became increasingly important as soon as defined relations are established between segregated parameter distributions and biochemical and physiological properties of the biomass (Bailey, Agar and Hjortso, 1982).

The line of research that has been developed in our laboratories aims to analyze protein distributions, obtained by flow cytometry, and cell volume distributions, obtained by a Coulter Counter, in order to find out relations between these distributions and the dynamics of growth of the yeast cell population. This is a first necessary step to establish relations between the distributions and the biochemical and physiological activities of the cell population in a reactor.

The studies carried out so far have shown that protein and volume distributions have the following properties (Alberghina, Martegani and Mariani, 1983 b; Ranzi, Compagno and Martegani, 1984):

i) are closely related to the age and size distribution;
ii) give informations about the law of growth of the single cell in the population;
iii) give informations about the heterogeneity of the population;
iv) do not change during a given steady-state of growth, but are characteristically different for different steady states;
v) quickly change as soon as the growth state of the population changes.

Moreover it is possible to analyze in a quantitative way protein distribution of yeast population growing exponentially in batch cultures (Alberghina, Martegani and Mariani, 1983 b; Martegani, Vanoni and Delia, 1984), with the aid of a structured model of the cell cycle in yeast previously developed in our laboratories (Vanoni and collegues, 1983).

In this work we exploit the sensitivity and reliability of these distributions for detecting perturbations of growth in yeast population.

In particular we have tested these distributions during chemostat growth of *S. cerevisiae*, both in conditions of steady state and during nutritional perturbations and in oscillating continuous cultures.

In order to analyze in a quantitative way the transients states that follow such growth perturbations we develop an extension of the previously described model. It takes into account the existence of two threshold values of protein: a critical protein content required for the beginning of budded phase and a critical protein content required for mitosis. In fact a number of observations have suggested that the progression of the cell cycle in all eukaryotic cells is determined by the sequential attainment of these two critical protein levels and that three different regimes are possible according to the values assumed by the two thresholds and by the minimal duration of the G_2 phase (Alberghina, Mariani and Martegani, 1980). The correlation between the two thresholds and the lenght of budded phase allows us to describe the whole system only in terms of critical protein content. A developement aimed to the analysis of transient states is presented.

POPULATION DYNAMICS DURING GLUCOSE-LIMITED -CHEMOSTAT GROWTH.

Cultures of *Saccharomyces cerevisiae* S288C were grown in a chemostat Bioflo C30 (New Brunswick, New Jersey) with a constant volume of 750 ml. Temperature was set to 30° C and the colture was aerated with an air flow of 0.8 1/min. The medium reservoir was filled with 10 1 of YEP-glucose medium containing 5 gr/1 of glucose. Steady state of growth were obtained for dilution rates D (h^{-1}) ranging from 0.07 to 0.43 h^{-1}; at D = 0.48 a washout regime was observed. To obtain protein distribution, aliquots of culture were withdrawn, briefly sonicated and fixed in 70 °/o ethanol. The cell were washed in phosphate buffered saline and stained with fluoresceine isothiocyanate (Alberghina and collegues, 1983 a) and analyzed with a FACS IV cytofluorimeter equipped with a 5 W argon laser.

Cell volume distributions were obtained with a Coulter Counter Channelyzer as described (Alberghina and collegues, 1983 a). All the distributions were stored on a floppy disk for further analysis.

A bud scar analysis was performed for each steady state and used to calculate the percentage of budded cells, the length of budded phase (T_B), the parent cell cycle time (T_P) and the daughter cell cycle time (T_D) (Table I) (Vanoni and collegues, 1983).

TABLE I

Temporal parameters of yeast populations growing in glucose-limited chemostat

D (h^{-1})	T (min)	B °/o	T_B	T_P	T_D
0.07	594	25	191	295	1384
0.20	207	65	152	190	249
0.25	166	65	120	150	197
0.31	134	66	98	123	169
0.35	118	64	85	107	145
0.40	103	67	78	99	147
0.43	96	63	67	84	119
0.48 *	86	70	65	80	100
0.55 $	75	68	58	72	85

*) : wash-out regime
$) : batch growth

Volume and protein distributions were determined in steady state cultures at various times intervals for each different dilution rates. The distributions obtained for each growth rate are found to be completely superimposable. Thus each growth rate is characterized by a specific protein and volume distribution (Fig. 1) (Ranzi, Compagno and Martegani, 1984). An increase in the growth rate is accompanied by a slight increase in the average cell size and average protein content per cell, determined either as mean fluorescence channel of the protein distribution or by a chemical assay (Table II, second and third columns).

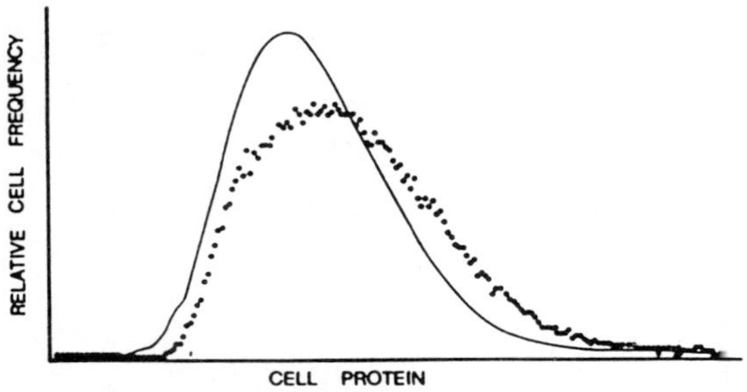

Fig. 1 - Protein distributions observed at dilution rate 0.2 (solid line) and 0.35 (dotted line) for yeast cultures growing exponentially in glucose-limited chemostat.

TABLE II

Cell-mass parameters for yeast populations growing in glucose limited chemostat

D(h^{-1})	$\bar{P}^a$	$\bar{P}_{ch}^b$	P_s^c	P_m^c	P_m/P_s
0.07	1.60	0.97	1.79	2.24	1.25
0.20	1.75	1.00	1.42	2.35	1.65
0.25	1.90	1.04	1.58	2.62	1.65
0.31	1.80	1.07	1.45	2.40	1.65
0.35	1.75	1.08	1.46	2.40	1.64
0.40	2.00	1.11	1.45	2.45	1.68
0.43	2.05	1.22	1.76	2.85	1.62
0.48 *	3.80	1.89	3.15	5.31	1.68
0.55 $	3.80	1.90	3.09	5.27	1.70

a): chemical assay
b): mean channel of protein distribution normalized to 1 for D = 0.2
c): values calculated from $\bar{P}$
*): wash-out regime
$): batch growth

By using the equations previously derived for unequally dividing cells (Vanoni and collegues, 1983) we have also estimated the average protein at the beginning of the budded phase (P_s) and at cell division (P_m). (Table II, fourth and fifth columns).

An analysis of these data show some interesting features of yeast populations at steady state. During chemostat growth, for a wide range of rates, the budded phase appear to be a constant fraction of the mass doubling time T, and approximatively T_B = 0.73 T. As a consequence the percentage of budded cell does not change much, at least for moderate to high rates of growth.

The relative constancy of the budded phase implies that the ratio between the two threshold protein levels is a constant in most of the considered steady states as shown in the experimental data reported in Table II (last column). These findings indicate that the *S. cerevisiae* cell cycle follows, in a wide range of D, the second pattern of cycle described by Alberghina, Mariani and Martegani (1980), in which both G_1 and G_2 phases expand at increasing mass doubling times.

The absolute value of the P_s threshold appears to be strongly dependent from nutrient composition in batch cultures (Vanoni and collegues, 1983) and upon the availability of the limiting substrate (glucose) in chemostat (Table II). A similar pattern has been reported for batch growth (Vanoni and collegues, 1983). The P_s value is low, about 1.5 pg/cell until glucose is limiting, and in washout regime rise to 3 pg/cell, a value close to that found in batch cultures growing exponentially in a similar medium (YEP-glucose).

This observation suggests a simple way to relate the dynamics of the population to the continuous reactor dynamics if the structured model can describe the population dynamics only in term of P_s and u and if these two parameters are related to substrate concentration according:

(1) $\mu = f(s)$ (Monod's law)

(2) $P_s = f'(s)$

where f'(s) is an empirically derived function.

A qualitative analysis of perturbed states of growth in chemostat strengthens this hypothesis.

During both a glucose pulse or a step-up of dilution rate there is an immediate increase of glucose concentration, due to a transient loss of balance between glucose input and the glucose metabolized by the cells. Since glucose is no more limiting, P_s should increase and as a consequence the average protein content of the cell population increases, while the perturbation of the cell cycle progression alters the normal parent to daughter ratio. (Ranzi, Compagno and Martegani; 1984).

An example of such a perturbation is shown in fig. 2. The average protein content per cell increases immediately and significant changes in both protein and volume distributions are detectable after 10-15 min. At increasing times the parent to daughter ratio is modified so to yieald bimodal protein and volume distributions.

The control of steady state continuous cultures of budding yeast is complicated by the insurgence of stable oscillatory regimens. (Meyer and Beyeler, 1984). The onset of oscillations can be triggered by a step change of D or of temperature (Von Meyenburg, 1973). Furthermore at low dilution rates ($D < 0.2\ h^{-1}$) and in appropriate range of p 0_2, spontaneous oscillations arise.

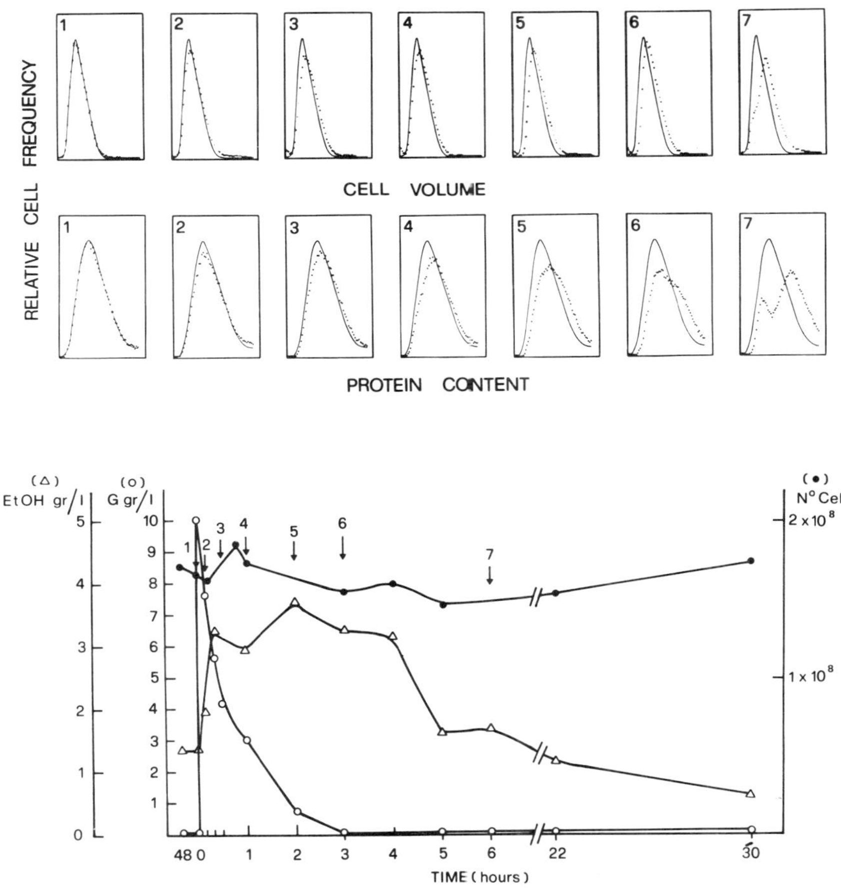

Fig. 2 - Effects of a pulse of glucose in steady-state culture. Protein and cell volume distributions which are completely superimposable at steady state (1) shift away from the characteristic distribution (solid line) just after 10 min. and are altered for several hours (from Ranzi, Compagno and Martegani, 1984).

These oscillations appear to be sustained by the rithmic change of many biological parameters, such as p 0_2, glucose and ethanol levels, etc., while the total cell number is much less influenced.

A segregated data analysis shows periodic alternations both of volume and protein distribution, indicating deep changes in the population structure, with an apparent partial synchronization of the subpopulations of the biomass. (Fig. 3).

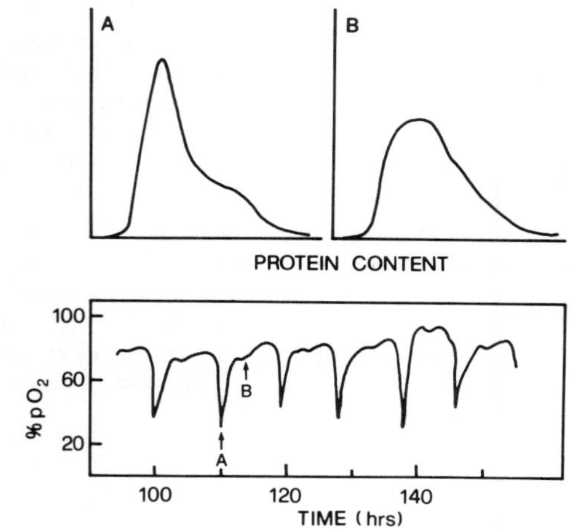

Fig. 3 - Oscillating culture in a glucose-limited chemostat. Dilution rate was 0.07 h^{-1}.
Lower panel: continuous registration of the p 0_2 in the culture. p 0_2 was measured amperometrically with an oxygen electrode (New Brunswich) and recorded with a MINC 11/23 computer.
Upper panel: Protein distributions measured at the time indicated by the arrows. (Ranzi and Compagno, unpublished results).

THE MODEL.

A structured model of the cell cycle, that takes into account the informations collected for the growth of the yeast *S. cerevisiae*, has been previously described (Alberghina, Martegani and Mariani, 1983 b). From the model it can be derived the theoretical age and cell size distributions wich have been used to develop a computer algorithm for the analysis of protein distributions during steady states of exponential growth in batch cultures.

The model takes into account the requirement of the attainement of a critical protein level (P_s) for budding, the unequal division and the increase in size with the genealogical age. Since this model has been developed to analyze populations in steady states of growth in batch, it considered that cell division occurred after a given time T_B following the activation of the P_s threshold, the T_B duration being a characteristic of each steady state of growth.

The findings previously reported on the growth in a chemostat indicate that in these growth conditions each steady state of growth is characterized by the values assumed by both threshold P_s and P_m wich cause a change of T_B in populations during transient states.

In order to develop a new computer algorithm able to analyze the age and protein distributions of yeast populations also in transient states, the matematical model of cell cycle needs to be modified to take explicitly into account the two thresholds.

The following statements are made:

Al-Protein synthesis depends on the environment by means of the specific growth rate function u (t) and it follows the equation:

$$(3) \qquad dp(t)/dt = \mu(t)\, p(t)$$

with $\mu = D$ in a steady state in a chemostat.

A2-A critical protein level P_s, whose value increases with the genealogical age i (Popolo, Vanoni and Alberghina 198 ; Johnston and collegues, 1979), is required for bud emergence.

A3-A critical protein level P_m is required for mitosis and cell division. Because the length of budded phase is assumed to be the same both for daughter and parent cells, independently from the genealogical age, also P_m should increase with genealogical age (Lord and Wheals, 1981).

A4- At cell division the protein synthetised during the budded phase are given to the daughter.

A5-There is no cell loss.

Starting from these assumptions the critical protein level required for budding is given by:

(4) P_s for all the daughters;
$P_s (1 + Q)$ for 1st generation parent;
$P_s (1 + 2Q)$ for 2nd generation parent etc,
where Q represent the relative increase in value of P_s for parent cells with one bud scar with respect to daughter cells.

In order to simplify the analysis in the following the model will take into account only 3 generations of daughters and 3 generations of parent cells, wich under normal growth conditions amount to more than 90 $^o/o$ of total cells.

Thus the critical protein content is set to $P_s (1 + 2Q)$ for all the generations of parents older than the second one.

Accordingly the critical protein level for cell division is:

(5) P_m for all the daughters; $P_m (1 + Q)$ for 1st generation parent and $P_m (1 + 2Q)$ for 2nd generation parents and all the others.

Let:

$$(6) \quad h = P_m/P_s \quad \text{with} \quad h < 2(1+Q)(1+2Q).$$

The length of budded phase is the same for all the subpopulations and equal to:

$$(7) \quad T_B = (\ln h)/\mu.$$

Under the hypothesis done in (4) we have only six subpopulations: three subpopulations of parents of increasing genealogical ages (P_1, P_2, P_3) and their daughters (D_1, D_2, D_3). The protein content at birth p (0) and the cycle time for the different subpopulation in a steady state having the growth rate μ are given by:

$$(8) \quad P_{D1}(0) = P_m - P_s = (h-1) P_s$$

$$(9) \quad P_{D2}(0) = (P_m - P_s)(1+Q) = (h-1)(1+Q) P_s$$

$$(10) \quad P_{D3}(0) = (P_m - P_s)(1+2Q) = (h-1)(1+2Q) P_s$$

$$(11) \quad P_{P1}(0) = P_s$$

$$(12) \quad P_{P2}(0) = P_s (1+Q)$$

(13) $$P_{P3}(0) = P_s(1+2Q)$$

(14) $$T_{D1} = \frac{1}{\mu}\ln\frac{h}{(h-1)}$$

(15) $$T_{D2} = \frac{1}{\mu}\ln\frac{h}{(h-1)(1+Q)}$$

(16) $$T_{D3} = \frac{1}{\mu}\ln\frac{h}{(h-1)(1+2Q)}$$

(17) $$T_{P1} = \frac{1}{\mu}\ln h(1+Q)$$

(18) $$T_{P2} = \frac{1}{\mu}\ln\left(h\frac{1+2Q}{1+Q}\right)$$

(19) $$T_{P3} = \frac{1}{\mu}\ln h$$

IDEAL AGE AND PROTEIN DENSITY FUNCTIONS.

The age is assumed as chronological age and it is measured ad the time (τ) to the next division, and the age density functions are:

(20) $$n_{Pi}(\tau) = n_{Pi}(o)e^{\mu\tau} \quad o < \tau < T_{Pi} \quad i = 1,2,3$$

(21) $$n_{Di}(\tau) = n_{Di}(o)e^{\mu\tau} \quad o < \tau < T_{Di} \quad i = 1,2,3.$$

Taking into account that at division each daughter originates a D_1 daughter and a P_1 parent, while a P_2 parent originates a D_2 daughter and a P_2 parent etc. the following age density functions are obtained, where $n_{D1}(o) = A$:

(22) $$n_{D1}(\tau) = Ae^{\mu\tau}; \quad o < \tau < T_{D1}$$

(23) $$n_{D2}(\tau) = \frac{A}{h}e^{\mu\tau}; \quad o < \tau < T_{D2}$$

(24) $$n_{D3}(\tau) = \frac{A}{h(h-1)}e^{\mu\tau}; \quad o < \tau < T_{D3}$$

(25) $$n_{P1}(\tau) = \frac{A}{(h-1)(1+Q)}e^{\mu\tau}; \quad o < \tau < T_{P1}$$

(26) $$n_{P2}(\tau) = \frac{A}{h(h-1)(1+2Q)}e^{\mu\tau}; \quad o < \tau < T_{P2}$$

(27) $$n_{P3}(\tau) = \frac{A}{h(h-1)^2(1+2Q)}e^{\mu\tau}; \quad o < \tau < T_{P3}.$$

A must satisfies:

(28) $$\sum_{i=1}^{6}\int_0^{T_i} n_i(0)e^{\mu\tau}d\tau = 1$$

and it is given by:

(29) $$A = \frac{\mu\{h^2(1+Q)(1+2Q) - Q[h(1+2Q)+1]\}}{(h-1)^2(1+Q)(1+2Q)}$$

From the age density functions and assuming a relative protein content $x = p/P_s$, the relative protein density functions can be obtained as previously indicated (Alberghina, Martegani and Mariani, 1983 b):

(30) $$n_{D1}(x) = \frac{Ah}{\mu P_s x^2}; \quad h-1 < x < h$$

(31) $$n_{D2}(x) = \frac{A}{\mu P_s x^2}; \quad (h-1)(1+Q) < x < h$$

(32) $$n_{D3}(x) = \frac{A}{(h-1)\mu P_s x^2}; \quad (h-1)(1+2Q) < x < h$$

(33) $$n_{P1}(x) = \frac{Ah}{(h-1)\mu P_s x^2}; \quad 1 < x < h(1+2Q)$$

(34) $$n_{P2}(x) = \frac{A}{(h-1)\mu P_s x^2}; \quad 1+Q < x < h(1+2Q)$$

(35) $$n_{P3}(x) = \frac{A}{(h-1)^2\mu P_s x^2}; \quad 1+2Q < x < h(1+2Q).$$

Examples of an ideal age density function and of an ideal protein density function derived in this way are shown in fig. 4.

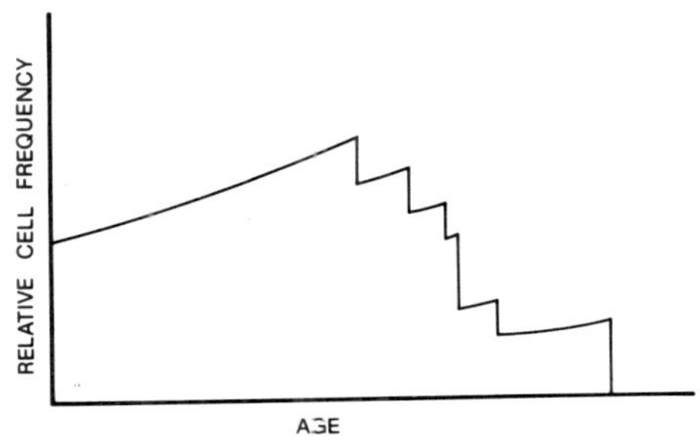

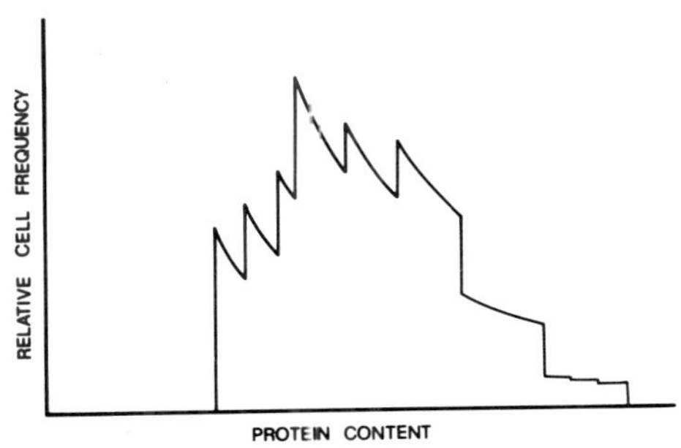

Fig. 4 - Ideal age and protein distributions derived from the model. D = 0.2; h = 1.65; Q = 0.2.

From (7), (29), (30) - (35) it follows that the percentage of budded cells in a balanced population is given by:

(36) $$B\,\% = (h-1)100.$$

This model is a simpler and more convenient version of the previous one, since it considers only parameters related to cell size and not related to both cell size and cycle times and further it describes the cell population as given by a limited number (6) of subpopulations. Nevertheless it allows equally an satisfactory analysis of experimental results.

Infact starting from the theoretical protein density function (eqs. 30 - 35), it is possible to obtain a good fitting of experimental protein distributions of yeast population growing in chemostat, providied that an adequate momentary size distributions function is considered (Fig. 5) as for the previous fittings (Alberghina and collegues, 1983 a; Alberghina, Martegani and Mariani, 1983 b).

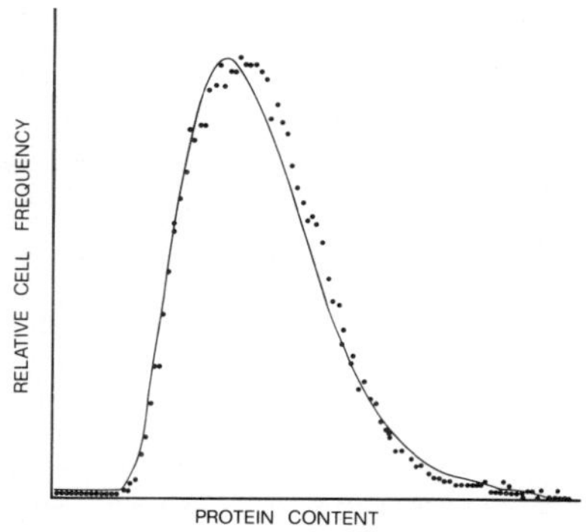

Fig. 5 - Fitting of experimental protein distribution with the model (D = 0.35; Coefficient of variation of the momentary size distribution = 0.2; Q = 0.2; h = 1.65).

Besides the relation (36), that links the percentage of budded cells to the mass parameters of the cell, is accurately verified for all the growth conditions tested. For instance, from the data reported in table II, when D = 0.35, h is equal to 1.64 and B °/o is calculated from (36) to be 64 °/o and when for D = 0.07, h = 1.251, B °/o = 25, in very good agreement with the experimental findings of Table I.

This model could easily be extended to the analysis of tra sient states by employing an analytical approach similar to that proposed by Hjorsto and Bailey (1983), that is utilizing a general population balance equations of the form:

$$\frac{\partial W}{\partial t} + \frac{\partial}{\partial p}[r(p)\,W] = -DW \qquad (37)$$

where W is the cell number distribution, and r (p) is the single cell growth rate and p is the protein content of the single cell.

The transient state could be imposed by a step change of the critical protein content P_s and/or by a change of μ. However the computations are very complex because we have to take in account all the different subpopulations and also different classes of cell between each subpopulation (Hjorstu and Bailey, 1983). Work to develop a suitable simulation program is in progress.

CONCLUSIONS.

Flow citometry allows to measure single cell parameter in significant population samples and thus gives detailed informations on the structure and the heterogeneity of the microbial population. Protein distribution has been found to be a very sensitive parameter to monitor growth kinetics in batch cultures of budding yeast. Recent findings and the resulted discussed here indicates that protein distribution as well as volume distribution, are useful parameters to monitor steady and transients states of growth also in continuous cultures. In particular a very fast response is observed during perturbed growth conditions, thus suggesting a very interesting possibility to develop control procedures based on these segregated parameters. A difficult task along this line is the identification of correlations between the cell enviroment and the bioreactor dynamics with the population structure, in such a way to be able to make previsions about the population structure changes in response to enviromental changes. In this paper we present a possible approach to solve this problem. From a detailed analysis of experimental data of cell parameters related to cell cycle progression we propose a new structured model able to describe the system only in terms of growth rate and of critical cell mass parameters. An interesting relation between these parameters and the enviroment is presented. Further work is now in progress to develop suitable programs to simulate transients states using this new model.

ACKNOWLEDGEMENT. This work was partially supported by the C.N.R. grant n. 83.018.47.

REFERENCES

- Alberghina L., L. Mariani, and E. Martegani (1980). Analysi of a model of a cell cycle in enkaryotes. J. Theor. Biol. 87, 171 - 188.
- Alberghina, L., L. Mariani, E. Martegani, and M. Vanon (1983 a). Analysis of protein distribution in budding yeast Biotechnol. Bioeng., 25, 1295 - 1301.
- Alberghina, L., E. Martegani, and L. Mariani (1983 b) Analysis of protein distributions in populations of budding yeast based on a structures model of cell growth. in A. Halme (Ed.). Modeling and control of Biotechnical Processes, Pergamon Press, pp. 83 - 91.
- Bailey, J.E., D.W. Agar, and M.A. Hjorsto (1982) Acquisition and interpretation of flow microfluorometry data on microbial populations in Computer Application in fermentation technology. Society of Chemical Industry, pp. 157-165.
- Cooney, C.L., (1979). Computer application in fermentation technology. A perspective. Biotechnol. Bioeng.Symp., 9, 1 - 11.

-Daimaku, K., E. Izumoto, H. Morikawa, S. Shioya, and T. Takamatsu (1982). Optimal quality control of baker's yeast fed-batch culture using population dynamics. Biotechnol Bioeng., 24, 1661 - 1674.

-Das S. and C.P. Hollenberg (1982). A high frequency transsformation system for the yeast kluyveromyres lactis Current Genetics 6, 123-128.

-Hjortso M.A., and J.A. Bailey (1983).Transients responses of budding yeast populations. Math.Biosc., 63, 121 -148.

-Johnston, G.C., C.W. Ehrhardt, A. Lorincz, and B.L.A. Carter (1979). Regulation of cell size in the yeast Saccharomyces cerevisiae. J. Bacteriol, 137, 1 - 8.

-Lord, P.G. and Wheals A.E. (1981). Variability and individual cell cycles of Saccharomyces cerevisiae. J. Cell Sci., 50, 361 - 376.

-Martegani, E., M. Vanoni, and D. Delia (1984). A computer algorithm for the analysis of protein distributions in budding yeast. Cytometry, 5, 916 - 925.

-Meyer, C., and W. Beyeler (1984). Control strategies for continuons bioprocesses based on biological activities. Biotechnol. Bioeng , 26, 916 - 925.

-Monod J., (1949). The growth of bacterial cultures. Ann. Rev. Microbiol., 3, 371 - 394.

-Popolo, L., M. Vanoni, and L. Alberghina (1982) Control of the yeast cell cycle by protein synthesis. Exp. Cell Res., 142, 69 - 78.

-Ranta, J., (1983) Some properties of microbial age dynamics models. in A. Halme (Ed.) Modeling and Control of Biotechnical Processes, Pergamon Press, pp. 93 - 100.

-Ranzi, B.M., C. Compagno, and E. Martegani (1984) Protein and volume distributions as growth rate parameters in continuous cultures of budding yeasts. Third European Congress on Biotechnology, Vol. I, Verlag Chemie, pp. 719 - 724.

-Struhl, K., D.T. Stinchomb, S. Scherer, and R.W. Davis (1979) High frequency transformation of yeast: autonomous replication of hibride DNA molecules Proc. Natl. Acad. Sci USA 76: 1035 - 1039.

-Vanoni, M., M. Vai, L. Popolo, and L. Alberghina (1983) Structural heterogeneity in populations of the budding yeast Saccharomyces cerevisiae. J. Bacteriol., 156, 1282 - 1291.

-Von Meyenburg K. (1973) in B. Chance (Ed.) Biological and Biochemical Oscillators, Academic Press.

STATIC OPTIMIZATION OF A MULTISTAGE FERMENTATION PROCESS INVOLVING DISTRIBUTED FEEDING

J. P. Bovee*, M. Orobengoa and Y. Sevely**

LAAS du CNRS, 7, avenue du Colonel Roche, 31077 Toulouse Cédex, France
**Presently INRA, BP 39, 59651 Villeneuve d'Ascq Cédex, France*
***Université Paul Sabatier, Toulouse, France*

Abstract : Starting from a model of the chemical kinetic type (excluding biomass considered as a catalyser) the aim is to optimize an alcoholic fermentation process in a cascade of N fermentors.

The optimization criterion is a follows : for a given total volume of fermentation V_T, obtain the maximum flowrate Q_N at the output corresponding to $P_N = P_F$ and $S_N = 0$ (P ethanol concentration, S saccharose concentration).

We assume $\left[r_p\right]_k = \left[\frac{dP}{dt}\right]_k = a\left(P_M - P_k\right)$, and $S_k \leq S_M$ avoiding the inhibitory effects of product and substrate. We assume also a constant yield conversion and that it is possible to have a distributed feeding.

We are seeking for the optimal partition of the volumes V_k (k=1 to N) and the optimal distribution of the concentrations S_k and P_k in the steady state.

N, P_F and S_M are fixed and we study their influences on the results.

The main conclusion is that the optimum is got when the first stage only is fed. The volumes are equal when S has no inhibitory effect and, if not, V_1 is greater than others (which are equal). If we take all the volumes equal, we need a distributed feeding and the production decreases when S_M decreases and P_F increases.

The gain from two to three fermentors is not very important in normal work conditions and less from three to four : it is the reason why we have limited our study to N = 3.

Keywords : Multivariable systems, Non linear systems, Optimisation, Process control

INTRODUCTION

This paper concerns results which complete, in an non exhaustive way, those presented in Helsinki [1] and we recall here briefly.

Starting with a chemical like model applied to an alcoholic fermentation performed in anaerobiosis, substrate (S) consumption and ethanol (E) production by various yeasts [2]

$$r_s = \frac{dS}{dt} = k\, S^{\alpha}\, P^{\beta} \qquad (1)$$

$$r_p = \frac{dP}{dt} = -\eta\, r_s \qquad (2)$$

our goal was the static optimization of a cascade of N reactors fed by the first stage and the start up, in a fed batch way, to reach the optimal steady state as quick as possible.

The cost function was defined as follows

- P_N, the ethanol concentration in the outlet, must not be less than a given value P_d in order to minimize the cost of distillation 3 - Figure 1.

- S_N, the residual substrate must be near to zero due to technical and economic accounts.

Then, for a total volume V_T of the cascade we intend

- to determine the optimal partitionning V_k (k=1 to N) of this volume $V_T = \sum_{k=1}^{N} V_k$ and the distribution of concentrations (P_k, S_k).

- to maximize the outlet Q (constant through the whole cascade fed by the first stage only).

In this study N has been taken as a parameter whose influence upon the cost function was considered a posteriori. Moreover we assumed the production rate of ethanol.

$$r_p = a\left(P_M - P\right) \qquad (3)$$

Which is available for continuous reactions only [3] - Figure 2.

The main results were :

- all volumes are equal
- Q given, the greater N is, the lower is the total volume V_T.
- Q and N given, V_T increases quickly with P_N.

Hereafter we intend to study how the results are modified by two new hypothesis :

1) to avoid inhibition by high values of S, we shall suppose that, in each reactor, S_k will be less than S_M.

2) a distributed feeding is allowed.

STATE OF THE PROBLEM

The study conditions are the following

1) N, the number of the reactors, is given. Of course, it would be possible to include N as an optimization parameter, but it would be tedious. A more realistic position is, as previously, the a posteriori study of the cost function variations when N changes.

2) The characteristics of the outlet are given $Q_N = Q_F = 1$, $P_N = P_F$ and $S_N = S_F = 0$. We shall study the influence of P_F.

3) For each reactor k, $S_k \leqslant S_M$ to avoid the inhibitory effect : we shall study the influence of S_M.

4) The feed concentration is also constrained because substrate concentration is a costly operation and it is not realistic to allow very high limit of feed concentration. Moreover to simplify the problem and to be near realistic conditions we have supposed that all the reactors, if they are, are fed by the same concentration S^a.

To the enumerated constraints before, we must add, for the non linear programming solution:

5) $Q_k \leq Q_m \quad \forall\, k \leqslant m$ since the flow rate are positive.

6) $V_k > 0 \quad \forall k \in [1,N]$

To sum up, the optimal problem may be stated as follows :
the final state Q_N, P_N, S_N given :

i) what are the minimum total volume and its optimal partition.

ii) what are the flowrates associated to the distributed feed of the reactors, the concentration of the feed substrate being assumed the same for each reactor.

We recall the expression

$$r_{P_k} = a\left(P_M - P_k\right)$$

a and P_m depend on the yeast (in the following we shall take $a = 10^{-2}\ H^{-1}$ and $P_M = 100$ g/ℓ)

EQUATIONS FOR THE STEADY STATE

The mass balance for substrate and product, assuming the yield conversion of substrate into ethanol constant, which is almost true in a large range of substrate concentrations, gives the following equations - Figures 3a - 3b.

1rst reactor

$$Q_1 S_1^a - Q_1 S_1 - \frac{a}{\eta}\left(P_M - P_1\right) V_1 = 0 \qquad (1\text{-a})$$

$$- Q_1 P_1 + a\left(P_M - P_1\right) V_1 = 0 \qquad (1\text{-b})$$

reactor k

$$Q_{k-1} S_{k-1} + Q_k^a S_k^a - Q_k S_k - \frac{a}{\eta}\left(P_M - P_k\right) V_k = 0 \qquad (k\text{-a})$$

$$Q_{k-1} P_{k-1} - Q_k P_k + a\left(P_M - P_k\right) V_k = 0 \qquad (k\text{-b})$$

$$Q_k = Q_{k-1} + Q_k^a \qquad (k\text{-c})$$

reactor N : $Q_N = Q_F$, $S_N = 0$, $P_N = P_F$

$$Q_{N-1} S_{N-1} + Q_N^a S_N^a - \frac{a}{\eta}\left(P_M - P_F\right) V_N = 0 \qquad (N\text{-a})$$

$$Q_{N-1} P_{N-1} - Q_F P_F + a\left(P_M - P_F\right) V_N = 0 \qquad (N\text{-b})$$

$$Q_F = Q_{N-1} + Q_N^a \qquad (N\text{-c})$$

Traducing the mass balance concerning the substrate we get

$$Q_k\left(S_k + \frac{P_k}{\eta}\right) = \sum_{i=1}^{k} Q_i^a\ S_i^a$$

When all S_i^a are supposed equal to S^a we have

$$S_k + \frac{P_k}{\eta} = S^a$$

and for the last stage when $S_N = 0$ and $P_N = P_F$.

$$P_F = \eta\, S^a$$

in this practical case which will be considered hereafter P_F given, S^a is no longer a parameter of the optimization problem.

METHODS

The problem belongs to non linear programming optimization with equality constraints.

Three methods have been successfully tested to minimize the cost functions.

- The first one is a random walk method with direction exploitation associated with a random jumping method, as for initialization of the starting points. All the points are verified to be feasible : n being the dimension of the random vector and taking account of the m equality constraints (m < n), only (n-m) coordinates are to be generated. This method allows to reach the minium of the cost function quickly and accurately.

- A second method has also been tested : it consists in performing a serial of nested loops controlled by the different (n-m) variables.

- At the end we have concerned and adjusted a program specially adapted to this kind of problem with linear constraints based on Rosen's method (called gradient projection since it uses the projection of the gradient cost function on on the active constraints to get the minimum search direction).

This third method was technically more difficult to adjust than the two others, to get good convergence conditions.

RESULTS

Although some results are valid for two fermenters or more, we shall expose for clarity, accepting repetitions, successively the case of two and three stages.

Two stages

The optimum is got when the first reactor only is fed.

The volume V_1 and V_2 are equal when S_1 is less than S_M.

When S_1 equals S_M, V_1 becomes larger than V_2 and the larger as S_M is little (i.e. the constraint is hard).

The total volume V_T, P_F given, decreases when S_M increases. This diminution exists only when $S_1 = S_M$ and is the greater as P_F is large - Figure 4.

S_M given, V_T increases approximatively as the square of P_F - Figure 5.

Three stages

The most important result, not at all obvious for us, is that, such as for two stages, the optimum is got when the first reactor only is fed.

The total volume V_T, P_F given, decreases when S_M increases. This diminution exists only when S_1 is constrained and is the more important as P_F is large.

We give – Figure 6 – an example of the variations of volumes V_T, V_1, V_2 and V_3 in terms of S_M for P_F = 60 g/ℓ.

When S_1 is constrained, the volume of the first stage V_1 is the same, P_F given, in the case of two or three fermentors.

The sum $V_2 + V_3$ is then lightly less than V_2 corresponding to two stages process.
We give hereafter – Table I – an example corresponding to P_F = 80 g/ℓ.

S_M		30	40	50	60	80	100
V_T	2 fermentors	260	250	247	247	247	247
	3 fermentors	251	233	222	216	213	213
V_1	2 fermentors	186	150	123	123	123	123
	3 fermentors	186	151	122	100	71	71
V_2	2 fermentors	75	100	123	123	123	123
(V_2+V_3)	3 fermentors	65	82	100	116	142	142

Table I

Moreover, when S_1 is constrained, P_1 has the same value for two or three fermentors. On the other hand, if $S_1 < S_M$, P_1 is less in the case of three fermentors. In all cases P_k increases regularly – Figure 7.

Then we have looked at the gain obtained in the optimal configuration with respect to the case where the three volumes would be equal. In this case, obviously, when S_1 is constrained, the second fermentor is to be fed. We have taken the volume of each reactor equal to the third of the total optimal volume and computed the outled flow.

The result, easily foreseeable, is that, P_F given, the production flow decreases the more P_F is large and S_M small – Figure 8.

About the distributed feeding with equal volumes we must notice two phenomena.

– the first one is there exist cases without solution : it happens when S_M is very small with respect to P_F.

– the second one is that, when the solution exists, S_M given and P_F increasing :

– for P_F small enough to let S_1 less than the constraint S_M, we have $Q_1 = 1$ and $Q_2 = 0$ as it has been said previously.

– when P_F increases beyond this value, Q_1 decreases, reaches a minimum and increases again while Q_2 increases, reaches a maximum before to decrease – Figure 9.

The explanation of this second phenomenon is in the expression of $r_p = a\,(P_M-P)$ which traduces the inhibition by ethanol : when $P_3 = P_F$ and then P_2 are close to P_M, the substrate is no longer consumed and then the feeding of the 2nd fermentor is no longer necessary.

As far as the first phenomenon is concerned we must notice that when S_M is very small not only S_1 but also S_2 are constrained then 2 two first reactors are equivalent to one with volume twice than the last one.
A solutiion will exist if S_M is not less than the value for which, in the case of two fermentors we have $V_1 = 2\,V_2$ (for instance, for P_F = 60 g/ℓ when S_M = 32 g/ℓ).

In conclusion, we do insist on the point that the results we have presented have been obtained from hypothesis we recall here :

$$r_p = a\,(P_M-P)$$

$$Q_F,\ P_F \text{ and } S_F\ (S_F=0) \text{ given}$$

$$S_k \leq S_M$$

In view of these results we have estimated it was unnecessary to pursue this study above three fermentors : the gain obtained which is not very important from two to three reactors is still less for four stages.

REFERENCES

[1] J.P. Bovee, Y. Sevely (1982). A new trend in alcoholic fermentation : modelling and optimal steady state for continuous cascade ethanol production. 1rst IFAC Workshop on Modelling and Control of Biotechnical Processes, Helsinki – Otaniemi, August.

[2] J.P. Bovee, P. Strehaiano, G. Goma, Y. Sevely (1984). Alcoholic fermentation : modelling base on sole substrate and product measurement. Biotech – Bioeng. vol. 26, n° 3, pp. 328-334, April 1984.

[3] J.P. Bovee (1981). Modélisation, identification et première approche de l'optimisation d'une unité pilote de fermentation alcoolique Thèse de Docteur-Ingénieur, Toulouse 25 Nov. 1981.

[4] O. Levenspiel (1967). Chemical Reaction Engineering Academic Press, London 1967.

[5] S.S. Rao (1978). Optimization : theory and Application. Wiley Eastern Ltd 1978.

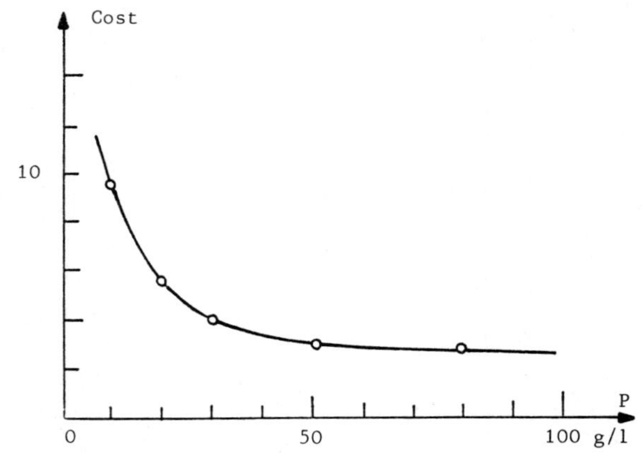

Figure 1. : Distillation cost as a function of P_F. From "Fermentation Kinetics and Process economics for the production of ethanol. G.R. Cysewski and Ch. Wilke (Berkeley)

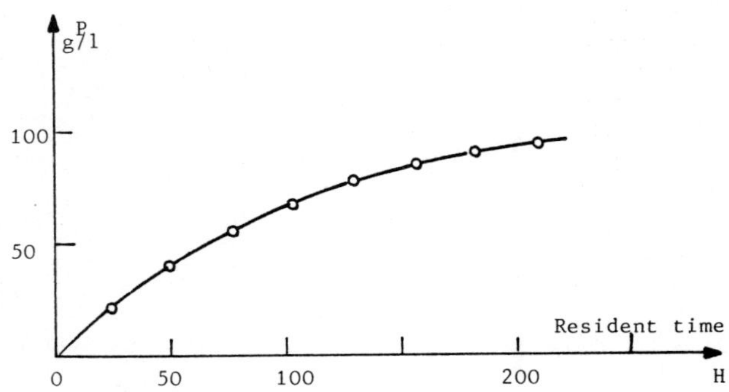

Figure 2. : Curve from which the model $r_p = a\left[P_M-P\right]$ has been derived. (experimental points are plotted).

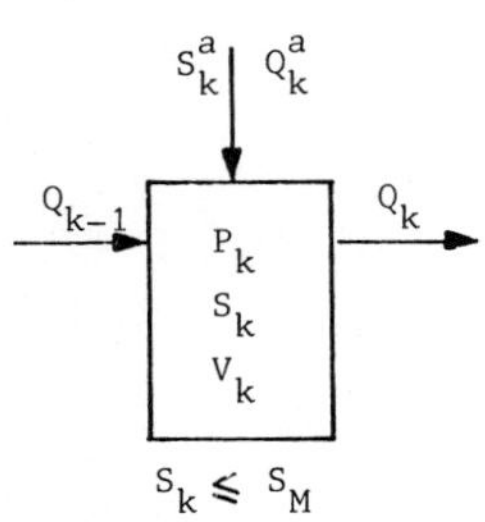

Figure 3.a : k^{th} fermentor

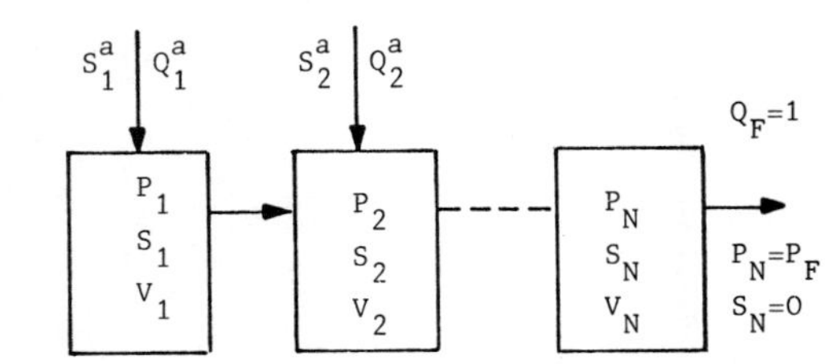

Figure 3.b : multistage fermentation process with distributed feeding

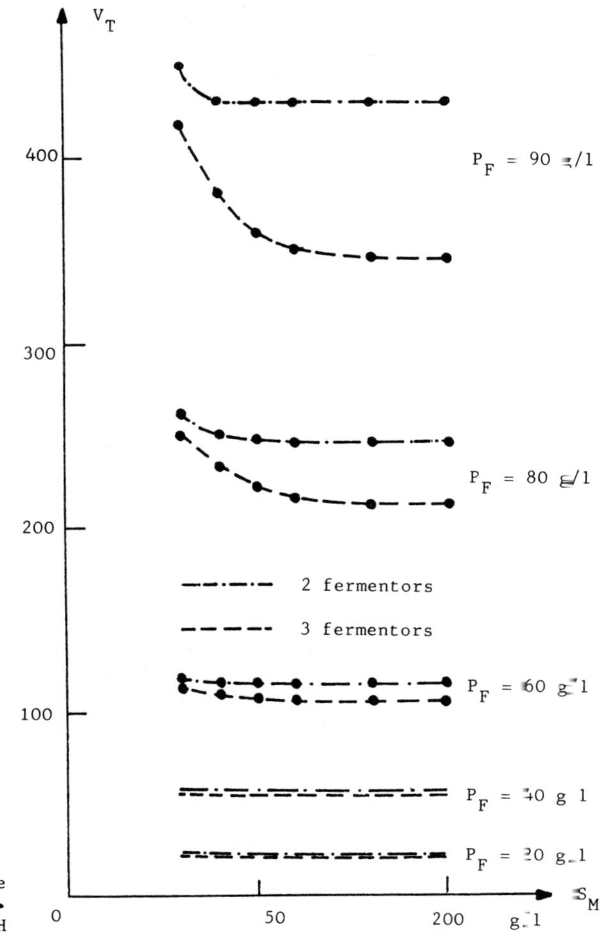

Figure 4 : V_T as a function of S_M for different P_F, $Q_F=1$, $S_F=0$, $r_P=10^{-2}\left[100-P\right]$ g/l.h., $\eta = 0,5$

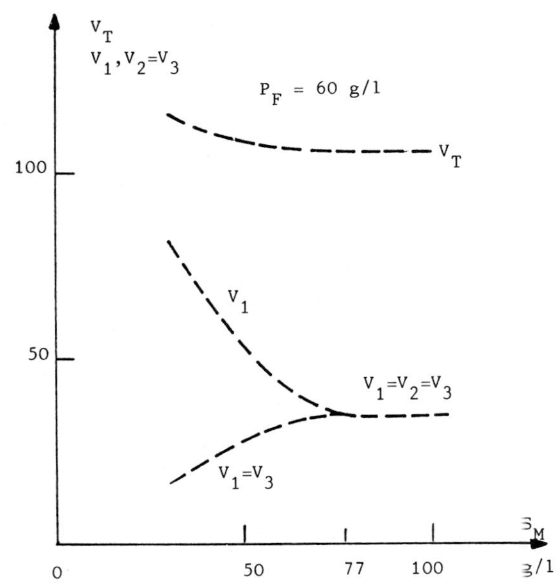

Figure 6. : V_T, V_1, V_2, V_3 as functions of S_M, $Q_F=1$, $P_F=60$, $S_F=0$, $r_P=10^{-2}\left[100-P\right]$ g/l.h., $\eta = 0,5$.

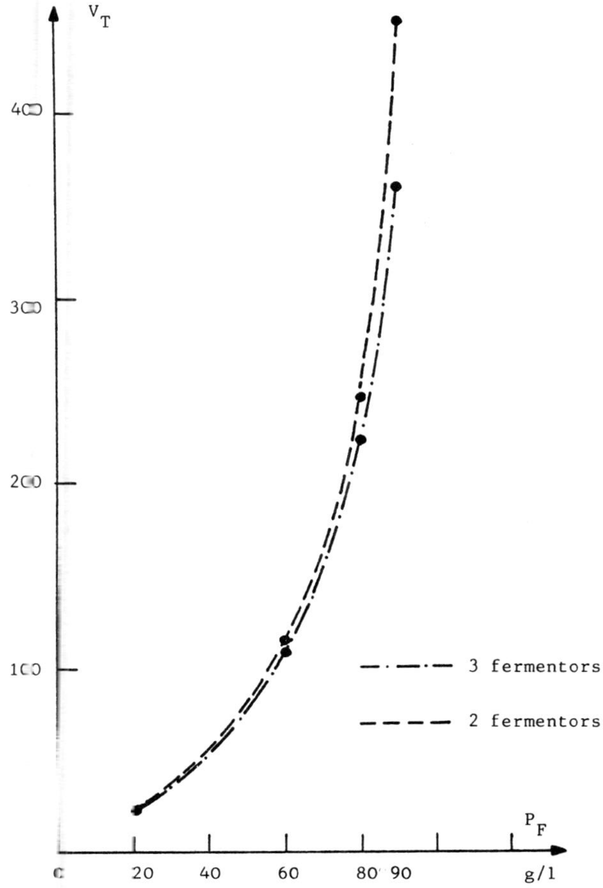

Figure 5. : V_T as a function of P_F for S_M=50g/l Q_F=1, S_F=0, $r_P=10^{-2}[100-P]$ g/l.h. η = 0.5.

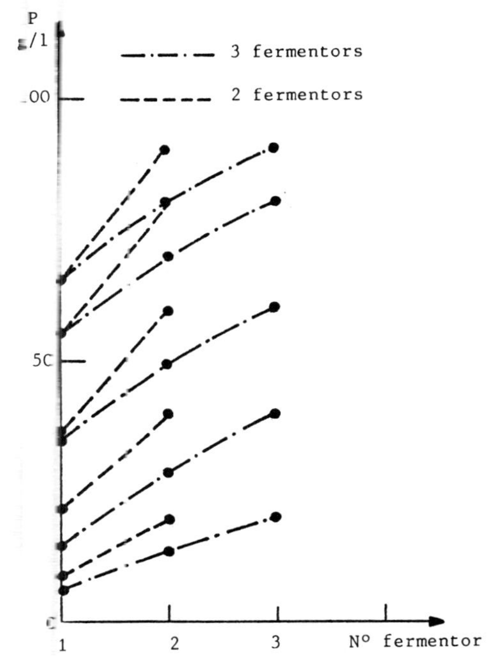

Figure 7. : Ethanol concentration along the cascade for two and three stages, for different P_F. S_M = 50 g/l.

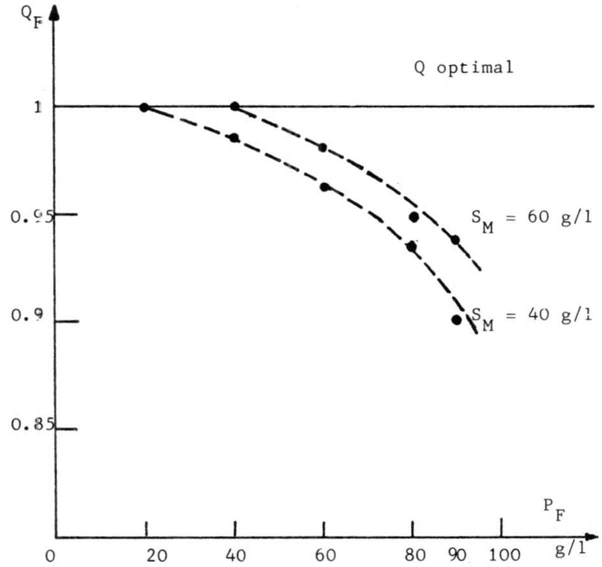

Figure 8. : Comparison of production flow in the optimal case first stage fed, $V_T = V_{opt}$ and in the case of equal volumes ($V = \frac{V_{opt}}{3}$) as functions of P_F for S_M = 40 and 60 g/l.

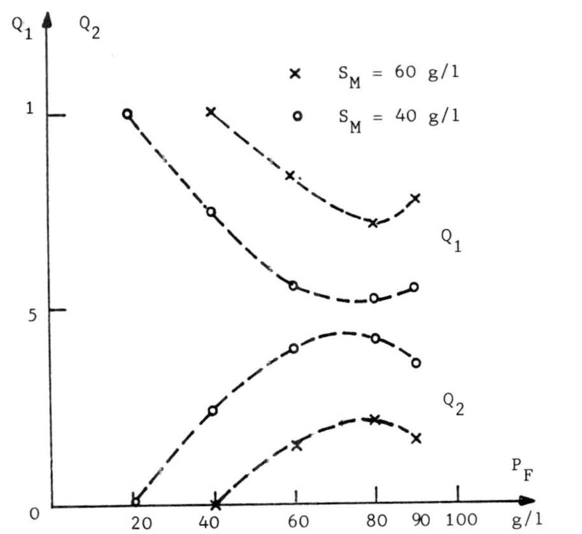

Figure 9. : Q_1 and Q_2 in the case of three fermentors fed by the two first ones as a function of P_F, S_M given S_M = 40 and 60 g/l.

AUTHOR INDEX

SUBJECT INDEX

IFAC Publications, Published and Forthcoming volumes

AKASHI: Control Science and Technology for the Progress of Society, 7 Volumes

ALONSO-CONCHEIRO: Real Time Digital Control Applications

ATHERTON: Multivariable Technological Systems

BABARY & LE LETTY: Control of Distributed Parameter Systems (1982)

BANKS & PRITCHARD: Control of Distributed Parameter Systems (1977)

BARKER & YOUNG: Identification and Systems Parameter Estimation (1985)

BASAR & PAU: Dynamic Modelling and Control of National Economies (1983)

BAYLIS: Safety of Computer Control Systems (1983)

BEKEY & SARIDIS: Identification and System Parameter Estimation (1982)

BINDER & PERRET: Components and Instruments for Distributed Computer Control Systems

BULL: Real Time Programming (1983)

BULL & WILLIAMS: Real Time Programming (1985)

CAMPBELL: Control Aspects of Prosthetics and Orthotics

Van CAUWENBERGHE: Instrumentation and Automation in the Paper, Rubber, Plastics and Polymerisation Industries (1980) (1983)

CHESTNUT, GENSER, KOPACEK & WIERZBICKI: Supplemental Ways for Improving International Stability

CHRETIEN: Automatic Control in Space (1985)

CICHOCKI & STRASZAK: Systems Analysis Applications to Complex Programs

CRONHJORT: Real Time Programming (1978)

CUENOD: Computer Aided Design of Control Systems

DA CUNHA: Planning and Operation of Electric Energy Systems

De GIORGIO & ROVEDA: Criteria for Selecting Appropriate Technologies under Different Cultural, Technical and Social Conditions

DI PILLO: Control Applications of Nonlinear Programming and Optimization

DUBUISSON: Information and Systems

ELLIS: Control Problems and Devices in Manufacturing Technology (1980)

FERRATE & FUENTE: Software for Computer Control (1982)

FLEISSNER: Systems Approach to Appropriate Technology Transfer

GELLIE, FERRATE & BASANEZ: Robot Control "SYROCO '85"

GELLIE & TAVAST: Distributed Computer Control Systems (1982)

GERTLER & KEVICZKY: A Bridge Between Control Science and Technology, 6 Volumes

GHONAIMY: Systems Approach for Development (1977)

HAASE: Real Time Programming (1980)

HAIMES & KINDLER: Water and Related Land Resource Systems

HALME: Modelling and Control of Biotechnical Processes

HARDT: Information Control Problems in Manufacturing Technology (1982)

HARRISON: Distributed Computer Control Systems (1979)

HASEGAWA: Real Time Programming (1981)

HASEGAWA & INOUE: Urban, Regional and National Planning—Environmental Aspects

HERBST: Automatic Control in Power Generation Distribution and Protection

ISERMANN: Identification and System Parameter Estimation (1979)

ISERMANN & KALTENECKER: Digital Computer Applications to Process Control

JANSSEN, PAU & STRASZAK: Dynamic Modelling and Control of National Economies (1980)

JOHANNSEN & RIJNSDORP: Analysis, Design, and Evaluation of Man-Machine Systems

JOHNSON: Modelling and Control of Biotechnological Processes

KLAMT & LAUBER: Control in Transportation Systems

KOTOB: Automatic Control in Petroleum, Petrochemical and Desalination Industries

LANDAU: Adaptive Systems in Control and Signal Processing

LARSEN & HANSEN: Computer Aided Design in Control and Engineering Systems

LAUBER: Safety of Computer Control Systems (1979)

LEININGER: Computer Aided Design of Multivariable Technological Systems

LEONHARD: Control in Power Electronics and Electrical Drives (1977)

LESKIEWICZ & ZAREMBA: Pneumatic and Hydraulic Components and Instruments in Automatic Control

MAFFEZZONI: Modelling and Control of Electric Power Plants

MAHALANABIS: Theory and Application of Digital Control

MANCINI, JOHANNSEN & MARTENSSON: Analysis, Design and Evaluation of Man–Machine Systems (1985)

MARTIN: Design of Work in Automated Manufacturing Systems

MILLER: Distributed Computer Control Systems (1981)

MUNDAY: Automatic Control in Space (1979)

NAJIM & ABDEL-FATTAH: Systems Approach for Development (1980)

NIEMI: A Link Between Science and Applications of Automatic Control, 4 Volumes

NORRIE & TURNER: Automation for Mineral Resource Development

NOVAK: Software for Computer Control (1979)

O'SHEA & POLIS: Automation in Mining, Mineral and Metal Processing (1980)

OSHIMA: Information Control Problems in Manufacturing Technology (1977)

PAUL: Digital Computer Applications to Process Control (1985)

PONOMARYOV: Artificial Intelligence

QUIRK: Safety of Computer Control Systems (1985)

RAUCH: Applications of Nonlinear Programming to Optimization and Control

RAUCH: Control Applications of Nonlinear Programming

REMBOLD: Information Control Problems in Manufacturing Technology (1979)

RIJNSDORP: Case Studies in Automation related to Humanization of Work

RIJNSDORP, PLOMP & MÖLLER: Training for Tomorrow — Educational Aspects of Computerized Automation

RODD: Distributed Computer Control Systems (1983)

SANCHEZ: Fuzzy Information, Knowledge Representation and Decision Analysis

SAWARAGI & AKASHI: Environmental Systems Planning, Design and Control

SINGH & TITLI: Control and Management of Integrated Industrial Complexes

SKELTON & OWENS: Model Error Concepts and Compensation

SMEDEMA: Real Time Programming (1977)

STRASZAK: Large Scale Systems: Theory and Applications (1983)

SUBRAMANYAM: Computer Applications in Large Scale Power Systems

SUSKI: Distributed Computer Control Systems (1985)

TITLI & SINGH: Large Scale Systems: Theory and Applications (1980)

UNBEHAUEN: Adaptive Control of Chemical Processes

VALADARES TAVARES & DA SILVA: Systems Analysis Applied to Water and Related Land Resources

WESTERLUND: Automation in Mining, Mineral and Metal Processing (1983)

van WOERKOM: Automatic Control in Space (1982)

YANG JIACHI: Control Science and Technology for Development

ZWICKY: Control in Power Electronics and Electrical Drives (1983)